Auditorium Acoustics and Architectural Design

Auditorium Acoustics and Architectural Design

Second Edition

Michael Barron

Routledge
Taylor & Francis Group

LONDON AND NEW YORK

First published in paperback 2024

First edition published 1993
by E & FN Spon

This edition first published 2010
by Spon Press

Published 2014 by Routledge
4 Park Square, Milton Park, Abingdon, Oxon OX14 4RN

and by Routledge
605 Third Avenue, New York, NY 10158

Routledge is an imprint of the Taylor & Francis Group, an informa business

British Library Cataloguing in Publication Data
A catalogue record for this book is available from the British Library

Library of Congress Cataloging in Publication Data
Barron, Michael.
    Auditorium acoustics and architectural design / Michael Barron. – 2nd ed.
        p. cm.
    Includes bibliographical references and index.
    1. Architectural acoustics. 2. Music-halls – Great Britain. 3. Music-halls. I. Title.
    NA2800.B36 2010
    725'.81-dc22   2009002327

ISBN: 978-0-419-24510-0 (hbk)
ISBN: 978-1-03-283669-0 (pbk)
ISBN: 978-0-203-87422-6 (ebk)

DOI: 10.4324/9780203874226

Typeset in Myriad by HWA Text and Data Management,
London

Contents

Preface

With hindsight, the timing of the first edition of this book was fortunate. The understanding of how sound behaves in auditoria and how people react to what they hear had reached a crucial watershed. For the first time, a consistent picture existed of the important factors for listening such as clarity and measurable quantities had been proposed relating to each of these factors. This was especially the case for music performance. Research, which had begun in the 1950s, had finally provided results that made it possible to say that the art and science of auditorium acoustics had reached a certain level of maturity.

A further development had taken place in the 1980s that was to have a major impact on design. Within design teams on auditorium projects, advice from acoustic consultants was being assigned greater importance. What was going to matter more for the client, a visually beautiful hall or a hall with good acoustics? This development however placed greater responsibility on the shoulders of the acoustic consultant, to which the response has often been to play safe and base designs more on successful precedents. New rectangular plan concert halls and horseshoe-plan opera houses have become popular again; both of these precedents come from the nineteenth century.

One can see the last 15 years as a period of consolidation, during which designers have been able to work with much greater confidence in their designs. Disappointing acoustics at the opening of a new auditorium is much less frequent now than was the case about 50 years ago. On the physical acoustic front during this period, the measurable quantities relevant to auditorium listening were included in a standard (ISO3382). Valuable books have also been published by others, in particular Beranek's distinguished surveys of the acoustics of world concert halls and opera houses (most recently in 2004). In addition a compendium of designs between 1982 and 2002 has been produced by Hoffman *et al.* – these references are to be found at the end of Chapter 1 (p.8).

In this second edition, greater reliance can also be placed on the advice offered. The text has been thoroughly reviewed and new developments included where appropriate on such topics as listener envelopment, scattering, auralization and greater use of computer simulation. Five new case studies of recent auditoria have been added: four concert halls and one opera house. The overall aim has remained as before, to illuminate the relationship between auditorium form and acoustic performance, all placed in a historical context and aimed at a wide readership.

For the production of this book, special thanks go to architectural firms that have supplied plans and sections, and auditorium managements who have allowed us to make measurements. I am particularly grateful to Dr. Leo Beranek for permitting me to use so many of the architectural drawings from his books. I am also grateful to the following individuals and organisations who have given permission for reproduction of and in some cases also kindly provided copies of copyright material (figure numbers are in parenthesis):

Dover Publications Inc., New York (1.3, 4.12); Elsevier (2.27, 2.31, 3.16); Elsevier (Journal of Sound and Vibration) (3.2, 4.33, 4.35); Prof. M. Morimoto (3.5); Neumann Berlin (3.6); J. Meyer (4.1); Hirzel Verlag (Acustica) (4.1, 6.6, 9.4, 9.10, D2); Dr. L.L. Beranek (4.3, 4.6, 4.7, 4.10, 4.18, 4.20, 4.25, 4.26, 4.42, 4.44, 9.18, 9.20); Concertgebouw, Amsterdam (4.8); Methuen Publishing, London (4.9, 6.1, 8.8, 8.9); Boston Symphony Hall, Mass. (4.11); Dr. H.A. Müller (4.13, 6.6); Fondation Le Corbusier, Paris (4.17); Beethovenhalle, Bonn (4.22); Estate of Prof. P. de Lange (4.23, 4.24, 4.32, 6.4, 6.5); Akademie der Künste, Berlin (4.29(b)); Pennsylvania State University, Special Collections Library (4.31, 8.5, 9.21, 9.22, 10.12, 10.13); Sir Harold Marshall (4.33, 4.35, 4.36, 9.4); Dr. H. Winkler (4.37); Shin Sugino, Toronto (photographer) (4.39); Artec Consultants Inc., New York (4.40, 4.41); Prof. H. Tachibana (4.43); Dr. T. Hidaka of Takenaka R&D Institute, Japan (4.45);Fairfield Halls, Croydon (5.34, 5.35); City of Nottingham (5.52, 5.53); Waterfront Hall, Belfast (5.69); T. Fütterer (6.9); Arup Associates, London (6.20, 6.21); Architectural Review, London (6.26, 8.26); Pentagram Design, London (8.6); Accademia Olimpica, Vicenza (8.7); National Theatre, London (8.27); Sheffield Theatres (8.42); Clive Totman, London (photographer) (8.49); Sir John Soane's Museum, London (9.1); Scientific American, New York (9.3); Dr. G. Naylor (9.16); Clive Boursnell, London (photographer) (9.19, 9.26); Tim Flach, London (photographer) (9.30, 9.31); Taylor & Francis Books (9.38); Marshall Day Acoustics, Auckland (10.2); Bickerdike Allen Partners, London (10.3); Rogers Stirk Harbour + Partners, London (10.5); IRCAM, Paris (10.6); Philips, Holland (10.11); Peutz, Holland (10.19); RHWL Architects, London (11.36); Dr. T. Houtgast (D.2).

In addition to those thanked in the preface to the first edition (reproduced below), warm thanks go to the following who helped with this new edition: Dr. Jin Jeon for assisting with the new auditorium measurements, Sir Harold Marshall for revisions to the foreword, Anders-Christian Gade for revisions to the text on acoustic conditions for performers and Rachel Mundell and Georg Herdt, who produced the new architectural plans and sections. Thank you to my co-workers during this period, Stephen Chiles and Jens Jørgen Dammerud, and especially to my consultancy partner, David Fleming, for being most supportive. Finally I owe many thanks to my university department, my project managers at HWA Text and Data Management, my patient editor, Georgina Johnson-Cook, and Penny for even more patience.

Michael Barron
Bath

Preface to the first edition

What form should one use to enhance to the utmost the brilliance, harmony and depth of sound from a symphony orchestra? Is there perhaps an ideal form on acoustics grounds for a drama theatre for Shakespeare's plays? Such questions must have inspired many of the more original solutions to the auditorium problem. A particularly appealing approach for instance, especially for those sensitive to the charms of classical architecture, is that certain ratios of proportion are propitious. The argument runs that since our ears appreciate harmony and harmony is based on simple proportions, so buildings to support harmony should likewise be built with favourable ratios of proportion. Sadly this charming argument is fallacious since the actual dimensions are also relevant. This means that a concert hall with preferred proportions would only be suitable for one musical key, a concert hall in C perhaps? However there is the further point that the dimensions of auditoria are in any case too large for such considerations to be significant; we can fortunately design halls for all keys.

There are logical sources for a solution, but we have three possible directions in which to look for inspiration. We could start from the characteristics of sources of sound: the properties of violins, wind instruments, the human voice and so on. Or we could respond to the wave nature of sound, by starting from an understanding of the way sound travels away from sound sources and establishing how it is complemented by reflections from wall and ceiling surfaces. And a further possibility is to use the ear as guiding principal: to determine what our ears find particularly desirable when listening to

music and what constitutes acceptable conditions for intelligible speech. One can cite examples in built auditoria of each of these guiding principles. Yet each concern on its own is inadequate. We now know that successful design is based on a whole series of independent considerations, that each of the elements between performer and listener has to be considered and understood. This book has its guiding concern to illuminate these questions and to search for the relationship between architectural form and acoustic behaviour. The presentation is directed at both the specialist and non-specialist.

The problem of achieving good concert hall acoustics has stimulated much valuable research since 1950. The results of this work have often been in the form of some preference in reflection characteristics or some new measurable acoustic quantity. In the case of speech, less thorough investigations have been made, but the majority of important issues have been resolved often for goals other than auditoria. However, translating acoustic requirements into three-dimensional built form has received rather less attention. It was this consideration which led to an acoustic survey of existing auditoria of all types: concert halls, recital halls, drama theatres, opera houses and multi-purpose spaces. The survey encompassed two forms of study: measurements of physical acoustic characteristics of auditoria and listening tests conducted at public performances with questionnaires. This book presents an account of the full survey within the framework of a thorough discussion of the concerns and principles of good acoustic design. No discussion of this sort is possible without extensive illustration; the reader will discover that all architectural drawings have been reproduced to the same scale of 1 : 500.

On a personal level, I have many people to thank for help of many kinds towards this book. Much of the original survey work was financed by the Science and Engineering Research Council while the author was at the Martin Centre for Architectural and Urban Studies at Cambridge University. Among many indulgent colleagues there, Lee-Jong Lee was an enthusiatic co-worker for two years and was responsible for the majority of the architectural drawings ot British auditoria. Also at Cambridge, Raf Orlowski and Tim Lewers helped out with field work among much else. Dean Hawkes oversaw operations with admirable detachment.

Particular thanks go to all those who attended concerts or plays armed with a questionnaire, often travelling great distances at their own expense for the privilege. Members of Bickerdike Allen Partners and Sandy Brown Associates were especially conscientious in their attendance, making the whole exercise feasible. The managements of all the auditoria studied in detail generously allowed access for acoustic measurements and gave permission for the author's photographs of their respective halls to be used here. They or the architects concerned must also be thanked for providing the architectural drawings used as the basis for reduced plans and sections.

In the preparation of this book, Professor Philip Doak was most supportive at moments of (minor!) crisis. I owe a special debt to my fellow consultancy partner, David Fleming, who has valiantly provided continual encouragement and much necessary constructive criticism. The quality of the final product also owes much to the support and faith of my editor at E & FN Spon, Martin Hyndman.

I am indebted to Anders-Christian Gade for contributing the section on musicians' acoustic requirements, to John Bradley for allowing me to use the results of his measurements in the Vienna Musikvereinssaal, and to Harold Marshall and Jerry Hyde among much else for permission to reproduce the results of measurements in the New Zealand concert halls in Christchurch and Wellington and the Segerstrom Hall, Orange County, California. My thanks go also to Prof. G.C. Izenour of Yale University for generous permission to reproduce figures from his book *Theater Design*. In a project which has lasted over ten years there are few of one's associates who have not in some way generously participated. For assorted assistance I must include mention of Derek Carruthers, Michael Forsyth, Thomas Fütterer, Dieter Gottlob, Jean-Dominique Polack, Maritz Vandenberg, H.J. Griffiths, Mark and Vicki Wooding and last, but far from least, Penny.

While every effort has been made to ensure the accuracy of the opinions and data presented in this book, the author can accept no responsibility for any inaccuracies it may be found to contain.

This reprint includes corrections to the known inaccuracies in the first edition. One consistent error needs to be brought to the attention of the more technical reader. The early decay time (EDT) is now considered a fundamental measure in spaces for music and in the first edition measured values were included in all figures containing measured results in concert spaces, such as Figures 3.25 and 5.6. However from comparison with EDT measurements by others in some of the same spaces, it became apparent that my measured values were slightly in error. This proved to be due to an error in the analysis program which derived EDT values. The measured data has therefore been reanalysed and the correct EDT's are included in this reprint in Figures 5.6 etc. Further details can be found in *Acustica*, **81,** 320–331 (1995). Fortunately the magnitude of the errors in EDT are generally very minor. The changes are significant at 125Hz in several halls and at midfrequencies in the two New Zealand halls discussed in section 4.10 and Appendix C. Since all the figures with objective results for concert spaces were being revised, the opportunity has been taken to improve the readability of the figures by indicating on them the preferred ranges for objective clarity and envelopment.

Michael Barron

Foreword

There was a time about 50 years ago when physics reigned unchallenged in the world of science. Its derivative profession, engineering, seemed to offer recipes for everything from earthquake engineering to genetics, and everything that could be thought of would sooner or later be realized. People contemplated with equanimity the possibility of creating harbours by subterranean thermonuclear blasts, moon walks, universal health – concert halls in which by following step-by-step instructions 'perfect acoustics' would be produced for all. Alas for the brave new world. Things have turned out to be much more complicated than they seemed in an age so full of hope and as naïve as the lyrics of the popular music of the day suggested.

One should have been warned by the failure of the acoustical predictions for the Royal Festival Hall that there was much more to understand than was clear from the state of knowledge then. But in the prevailing mood of optimism a miss of 40% in the predicted reverberation time was not sufficient to trigger much research. That had to wait for another monumental miss in the next decade. It was the failure of the design for the now historic New York Philharmonic Hall that sparked a new generation of effort to understand what makes a good concert hall. This book is largely the history of that effort.

Perhaps the most striking change in approach during the last 50 years has been a movement away from the physics of sound in rooms to a concentration on the effects of sound in the virtual space between the ears of the listener. It was extremely difficult to relinquish the linear thinking which had

proved so effective in other fields of engineering, and equally difficult to abandon the search for one main ingredient of goodness in concert hall sound which would provide the reductive answer to the designer's task. Even the use of the singular term 'listener' is a relic of this linear paradigm. We now know that audiences are far from homogeneous in their preferences – and the 'space between the ears' is only accessible at one remove.

Rather than a single path with a starting point and a single desired goal, the requirements of successful large performance spaces for music might better be thought of as a four or more dimensional maze with several ways in and at least as many desired outcomes. There are plenty of dead ends as well and it would be a brave person who would assert that they have all been identified at this time.

Michael Barron has been committed to the way through this maze for the past 35 or more years. Indeed, he was largely responsible for the recognition of a number of desired outcomes and for the refinement of the objective measures by which numbers can be set to the subjective impressions to be found there. Patiently, he has compiled both data and wisdom about auditorium design into what must be the most useful review of knowledge since Beranek's *Music Acoustics and Architecture* of 1962.

One discerns a consistent effort in this book to communicate with a broad audience, an effort equally concerned with music and with the rooms in which it takes place. In the late 20th century these rooms have become an essential and positive dimension of the acoustical form of music (as any musician will confirm this robust art also has form which is not acoustical). Of course the book is partisan to the approach which Michael Barron has found most productive of good halls and to those halls with which he is most familiar. But the range is sufficiently large to be considered comprehensive and those that find their theoretical approach omitted and/or their preferred objective measures unused will nevertheless be able to relate their work to Barron's results. Inevitably, there is a huge reduction of data required in a book such as this but information essential for the serious inquirer is not lost.

In the late 1960s, in the linear mode of thought, I believed that I had discovered the one essential source of premium concert hall sound, the lateral reflections. My hypothesis was subjected to critical scrutiny in Michael Barron's doctoral thesis. That work established the beneficial effects of lateral reflections and showed that the audibility of the lateral sound was dependent on the lateral to frontal energy fraction rather than on a lack of masking by frontal sound as I had previously thought. Later, in Goettingen, Gottlob and Siebrasse in their respective theses devised an experiment which further cemented the importance of the lateral reflection concept in place as one of the independent factors in determining concert hall preference. By then the multidimensional nature of preference had been established by Hawkes and Douglas, and the linear approach to subjective room acoustics was history.

Michael Barron went on, at Cambridge, to develop acoustic modelling to a level which it had never reached before. The idea of acoustic modelling using ultrasound was already quite venerable, dating from before the Second World War but was inordinately expensive. The Cambridge work showed that models at scales as small as 1:50 could produce useful data given suitable dry air conditions. The use of these models is now commonplace acoustics design practice and the development of digital data acquisition in conjunction with such models permits numerical compensation for the real air conditions, and so avoids the tedious problem of drying. Modelling is now a truly interactive design aid. It is certainly possible to know what will occur acoustically in the completed performance space but whether it will be liked or not is a question with several answers.

The intriguing uncertainties of audience preference for music do not exist in speech auditoria. There the outcome is uni-dimensional. You have speech clarity or you do not. The design problem is considerably easier than it is for music because speech rooms are usually smaller than music rooms and the signal is more consistent. It is when the speech communication is part of an art-form such as theatre, that the integration of architectural intentions and the acoustics, in a multidimensional way becomes essential. It is no surprise then to discover a substantial section of the book devoted to the history of theatre and the building forms associated with it. Technically, speech rooms are more straightforward but the cultural and symbolic functions of the space in which communication is to occur may well provide as much complexity as one finds in the design of rooms for music.

The maze is still there though not all its dimensions are yet named. The power dimension of the process of making a concert hall is yet to be addressed although buildings stand or are (literally in the case of the NY Philharmonic Hall) demolished because of it, acousticians succeed or fail, and architecture is, or is not created in the context of that power. In the numerous case histories included in this excellent book that is a question the reader is invited to bear in mind.

Emeritus Professor Sir Harold Marshall KNZM, FRSNZ, FASA, FNZIA
Auckland

1 Introduction

Our echoes roll from soul to soul,
And grow for ever and for ever.
Blow, bugle, blow, set the wild echoes flying,
And answer, echoes, answer, dying, dying, dying.
(The Princess, Tennyson)

'Theatre design is based on hard facts and is rarely a matter of inventing new forms of theatre. The architect works within a large number of constraints, some of which are self-imposed, and if inventiveness comes into it at all, it is in the way that theory is put into practice.' These words by the architect Peter Moro (1982) referred to theatre design, but they can equally well be applied to all forms of auditorium design. Many a designer must have greeted the prospect of a new auditorium with ideas of creating a totally new form or of returning to the purity of classical models, only to find his options becoming progressively more limited, with consultants from other disciplines criticizing one feature after another. As with all buildings, structural needs can limit freedom and architectural style is a balance between respect for precedents and current fashion. But for auditoria, there are the additional constraints in planning for good visual conditions and sightlines, of meeting certain social demands in regard to layout of seating and not least providing good acoustic conditions. The designer has to remain continually aware of the relationship he is creating between performer and audience, striving to make it as intimate as possible. Designing auditoria is a complex, elaborate but highly constrained exercise.

All auditoria rely on both visual and acoustic stimulation. Tennyson is referring above to the acoustic response in a mountainous landscape, in which echoes add to the visual imagery and create a sense of space. At the scale of concert halls and theatres, acoustic reflections are no less significant in creating a sense of space, but the ear does not perceive as discrete events or echoes the thousands of reflections it receives. It blends them into a total experience. Only certain aspects of the sound are used to establish the size and character of the enclosed space. The ear also establishes the direction of the sound source and extracts the information content where possible. A world in which the ear was unable to combine sound reflections was described by the Rev. Brewer (1854):

> If the ear ... were capable of appreciating every impulse, the confusion of sounds would be truly terrific. ... Every sound in our houses, every word in our churches, would be repeated ten thousand times. We should hear the direct sound of one syllable mingled with the reflections of another, and both recurring so frequently, that language would be a Babel of 'confusion worse confounded'. The voice of affection and of love, so tranquil, so soothing, and so gentle, would be a clatter more painful than the gibbering of a stammerer.

These are very perceptive comments for the time; the Rev. Brewer would have had to wait a century for a quantitative description of the workings of the ear, which enables us to avoid this 'clatter'. The discussion of auditorium design and development on the basis of precedent were well established in Brewer's day, even if not on a particularly scientific basis. Development from precedent dates of course from the classical Greek era, but sadly no contemporary discussion survives of the development of the earliest theatres. The earliest written record comes from the Roman Vitruvius in the first century BC (Vitruvius, 1960). He appears to base his geometric prescriptions for designing Greek and Roman theatres on an understanding of acoustics. But the validity of his arguments depends on favourable analogies with modern concepts. The most valuable information on ancient classical auditoria is to be gleaned from the surviving examples. The fan-shaped plan and the arena form both became highly developed in classical times and remain a constant point of reference for present design (Figure 1.1). The development of these dominant auditorium plans through the centuries is particularly fascinating (Barron, 1992).

Discussion of auditoria as a building type has a respectable history. The first of its genre, Dumont's *Parallèle de plans des plus belles salles de spectacles* of 1774 contains, as the title indicates, comparisons of plans plus a few sections of then existing theatres and opera houses. Dumont also proposes his own designs with vast concave domed ceilings, which would now give any acoustician nightmares. Louis (1782) included comparative plans in the account of his Bordeaux Grand Théâtre (Figure 1.2) Contant and Filippi's *Parallèle* of 1860 is a thorough review of the major west European theatres. Containing drawings and text, it discusses 33 theatres and became the standard work on proscenium theatre design. In the late nineteenth century, European theatres and opera houses experienced a building boom. By 1896 Sachs could fill his monumental three-volume *Modern opera houses and theatres* with extensive detail about more than 50 theatres completed since Contant's earlier survey. This great work remains a tribute to many of the great houses destroyed in two world wars. Of our own time the closest book in the same vein is Izenour's *Theater design* (1977). This is lavishly illustrated but is more partisan than its predecessors, with discussion

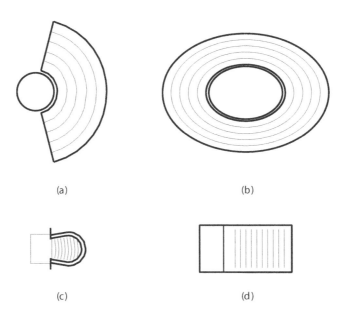

(a)

(b)

(c)

(d)

Figure 1.1 Historically dominant auditorium plan forms: (a) classical Greek theatre (fan-shape plan); (b) classical Roman arena; (c) baroque theatre (horseshoe plan); (d) nineteenth-century rectangular concert hall

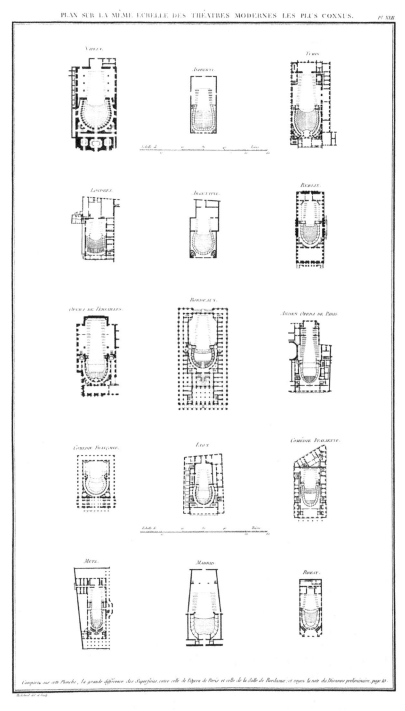

Figure 1.2 Plans of major eighteenth-century theatres to the same scale (V. Louis, 1782)

leading from ancient theatres to modern American multi-purpose performing arts centres. For an architectural analysis, Chapter 6 on 'Theatres' in Pevsner's *A history of building types* (1976) is typically erudite.

The literature for concert hall design is more recent than for theatres. Somehow the history of concert hall development had not been treated until Forsyth's valuable *Buildings for music* appeared in 1985. The first major review of the acoustics of concert hall buildings was Beranek's *Music, acoustics and architecture* (1962), which surveyed 47 world-famous concert halls and seven opera houses. Beranek pioneered analysis of auditorium acoustics on the basis of several independent subjective qualities, such as 'intimacy', 'liveness' and 'warmth'. Beranek has since revised his earlier book twice in 1996 and 2004; the most recent considers 82 concert halls and 18 opera houses. The strengths and weaknesses of his surveys are considered in section 3.2. Other reviews of concert hall designs include Talaske *et al*. (1982), Lord and Templeton (1986), Forsyth (1987) and Hoffman *et al*. (2003).

Beranek's 1962 book was the first to seriously attempt a complete explanation of auditorium acoustics and to answer the many misconceptions on the subject. Acoustics for concert halls, theatres and opera houses is frequently referred to as an inexact science, or worse, an area of myths. While modern technology and computers are considered to be reasonable design aids for a new auditorium, the success of old halls is often considered to be due to secret ingredients. The suggestion, for instance, that concert halls mature with age is still prevalent, as if they behaved like good claret. Wood is often quoted as the optimum material to line concert halls, just because violins are made of the same material. The presumed inaccessibility of acoustic knowledge is long standing. Vitruvius, commenting on the musical basis of acoustics, suggested that: 'Harmonics is an obscure and difficult branch of musical science, especially for those who do not know Greek'. Vitruvius considered that sound propagated curving upwards rather than just horizontally. Interestingly he justified the choice of

raked seating not primarily for visual reasons but to optimize the movement of the 'ascending voice'. The ancient Greeks had, of course, studied resonance of stretched strings and found that harmony depended on simple arithmetical ratios. Such mathematical purity led to the concept of 'the music of the spheres', which so appealed to the Renaissance mind. European scientists like Boyle and Newton in the seventeenth century furthered understanding of sound behaviour, and among other things demolished the notion of music in planetary space. In the 'Age of Reason' the first attempts were made at relating auditorium form to acoustic behaviour, with Patte (1782) proposing elliptical plans and Saunders (1790) circular ones. Both plan forms would now be considered dangerous due to focusing by concave surfaces.

Little further progress was made towards understanding the acoustics of rooms in the nineteenth century. A rare voice of inspired understanding is found in the comments of Dr R.B. Reid in 1835, when he said that 'any difficulty in the communication of sound in large rooms arises generally from the interruption of sound produced by a prolonged reverberation'. Reid successfully advised on acoustic treatment of the contemporary House of Commons, Westminster (see Bagenal and Wood, 1931). In Germany, Langhans published a remarkable book in 1810 in which he not only shows that he correctly appreciated how early reflections enhance intelligibility but also came close to understanding the independent role of reverberation (Izenour, 1977; Hartmann, 1990). This work appears not to have reached England or France. Another advance was made by Scott Russell (1838) who calculated the optimum floor profile in long section for good vision and hearing. The ideal profile is curved to provide equal-seeing and 'isoacoustic' conditions. (It is notable that Scott Russell equates vision and sound in this respect, as this is still assumed today.) But concerning useful advice for the architect, the situation remained woeful regarding suggestions for suitable room form and size for good acoustics. Charles Garnier, architect of the Paris Opéra, understandably complained in 1880:

It is not my fault that acoustics and I can never come to an understanding. I gave myself great pains to master this bizarre science, but after fifteen years' labour, I found myself hardly in advance of where I stood on the first day. ... I had read diligently in my books, and conferred industriously with philosophers – nowhere did I find a positive rule of action to guide me; on the contrary nothing but contradictory statements.

Somewhat ironically, in the face of such scientific ignorance, elaboration on the basis of precedent produced the rectangular 'shoebox' concert hall, some examples of which are still considered by many to have the world's best acoustics. In the case of theatres, many large playhouses built during the theatre-building boom around 1900 are still used and function surprisingly well acoustically. The state of the art in 1861, or rather its lack, was summarized by Roger Smith in his *Acoustics of public buildings*. Smith failed to reconcile the conflicting evidence from the men of science. Even in the major theoretical work of the period, Lord Rayleigh's *Theory of Sound*, which first appeared in 1877/8 but is still extensively referred to, the comments on room acoustics are limited to generalities (Vol.2, §287):

In connection with the acoustics of public buildings there are many points which remain obscure. ... In order to prevent reverberation it may often be necessary to introduce carpets or hangings to absorb sound. In some cases the presence of an audience is found sufficient to produce the desired effect.

The pleas of Garnier and Smith were answered on the other side of the Atlantic, when in 1895 Sabine began his pioneering work on room acoustics. At a time when physicists were making revolutionary early discoveries in atomic physics, the relative tardiness of concern for the commonplace experience of sound in rooms deserves comment. Two reasons can be advanced for this lack of progress. Firstly the fundamental relationship for sound in rooms requires a statistical rather than simply linear treatment. The linear approach, which can be applied to the single sound wave, has to be

applied a thousand times over to provide a complete picture of sound travelling from performer to listener. From this plethora of information, selection of the relevant attributes remains a problem to this day. The second problem until the 1920s was that of measurement. The ear is a highly sophisticated organ but its very adaptation to understanding sounds in enclosed spaces precludes its use as a fully analytic instrument in the scientific sense. The advent of microphones and electronics was obviously a great breakthrough.

In the circumstances, Sabine's achievement was all the more remarkable. W.C. Sabine was an Assistant Professor in the Physics Department of Harvard University. His previous work had been in optics and electricity, when the President of the University

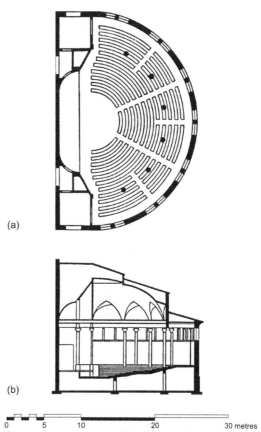

(a)

(b)

Figure 1.3 Lecture room of the Fogg Art Museum, Cambridge, MA. Plan and long section

asked him to find a solution to the appalling acoustics of the lecture room of the newly completed Fogg Art Museum (Figure 1.3). Sabine quickly realized that there was excessive reverberation in the lecture room (he measured the unoccupied reverberation time as 5.5 seconds; Sabine, 1922). Consultation of the literature would have revealed that this could be reduced by installing sound absorbing material, but Sabine chose to undertake a fundamental investigation into the behaviour of sound in enclosed spaces. He developed a technique for measuring the decay time of residual sound after an organ pipe was switched off, using nothing more than the observer's ears and a stop-watch. Measurements were made of this time, now known as the reverberation time, while different lengths of absorbent cushion were brought into the room from a neighbouring lecture theatre, up to the point where the time had been reduced to 0.75 seconds. Changes to improve the acoustics of the lecture room were made in 1898; the room was finally demolished in 1973 (Cavanaugh and Wilkes, 1999).

In the autumn of 1898, a new Boston Music Hall was being contemplated and Sabine was asked if he would be willing to advise on the acoustics (Beranek, 1979). Sabine hesitated but embarked on a hectic review of the experimental data he had collected in several Harvard rooms. On Saturday evening, 29 October 1898, he suddenly realized the explanation for his results, shouting out to the only other person in the house: 'Mother, it's a hyperbola!' He had discovered that the reverberation time was proportional to the reciprocal of the amount of absorption (Sabine, 1922: introduction in Dover edition). Thus was born the relationship now called the Sabine reverberation equation. He then accepted the commission to advise on what is now known as Boston Symphony Hall, a concert hall which has gained a reputation for having one of the best acoustics worldwide.

What is particularly striking about Sabine was his clarity of vision. In reading his writings, it is difficult to believe 'the meagreness and inconsistency of the current suggestions' (Sabine, 1922) when he began. Much of it reads as relevant to today's problems.

For example in his paper of 1898 (Sabine, 1922), the requirements for room acoustics were summarized much as they might be today:

> In order that hearing may be good in any auditorium, it is necessary that the sound should be sufficiently loud; that the simultaneous components of a complex sound should maintain their proper relative intensities; and that the successive sounds in rapidly moving articulation, either of speech or music, should be clear and distinct, free from each other and from extraneous noises. These three are the necessary, as they are the entirely sufficient, conditions for good hearing.

Sabine founded the quantitative approach, not only to internal room acoustics, but also to noise transmission through buildings. His premature death in 1919 meant he was unable to develop what we now appreciate as a necessary multi-dimensional approach to auditorium acoustics.

In the 1930s two influential books appeared which consolidated the progress of the 'new' science of room acoustics: *Planning for good acoustics* by Bagenal and Wood (1931) and *Architectural acoustics* by Knudsen (1932). Yet auditorium design still remained a somewhat hit-and-miss affair. The Royal Festival Hall, London, of 1951 was typical of several subsequent disappointments. The great successes of the previous century had been due to design by precedent as well as a few grains of fortune. Their excellence had still not been matched in spite of greater understanding. Reverberation time had been proved to be very important but by itself it did not guarantee good acoustics; further aspects required consideration for successful auditorium design. Auditorium form as well as size is now recognized as highly significant. This question of the interplay between architectural form and acoustic behaviour is one of the dominant themes of this book.

The successful search for other significant measurable quantities began around 1950. It had been realized that the unanswered questions concerned the manner in which the ear processes acoustic information. Subjective experiments were initiated with subjects listening and responding to controlled acoustic conditions. Auditorium acoustics had entered the realm of experimental psychology, a story which is elaborated in Chapter 3. Several new measurable quantities have now found wide acceptance as being significant for music listening. Much of this progress has been admirably recorded by Cremer and Müller in *Principles and applications of room acoustics* (1982). More recently, Meyer's *Acoustics and the performance of music* (2009) takes as its basis the characteristics of musical instruments (spectral, directional etc.) and how this is significant for performance.

A leading question however still remains regarding how the form and detailed acoustic characteristics of a design determine acoustic quality. Reverberation time can, with reasonable accuracy, be calculated and thus predicted from architectural drawings. But in the case of the newer measurable quantities such prediction is less straightforward. Acoustic scale models can be useful but a feel for the influence of design features is needed from the very start of development of a design. Computer models are now routinely used but they are not entirely reliable, particularly in inexperienced hands. A knowledge of the way in which form determines the various elements of subjective quality is required. In order to gain this knowledge, the most accessible resource is existing auditoria. In an area as complex as three-dimensional acoustic design, science is forced to rely on precedents.

Thinking along these lines led to an acoustic survey of British auditoria conducted by the author mainly in the years 1982–3. Acoustic measurements were made in over 40 auditoria. In most of these, subjective tests were also conducted with listeners completing questionnaires at public performances. A further five British auditoria completed since 1990 have since been tested and reported on. While many auditoria worldwide are discussed in the following chapters, a detailed understanding of their acoustics is usually only possible where

extensive tests have been made. This has been possible in the British halls tested and is recorded here. The book starts with an introduction to acoustics in rooms, before treating acoustic design for symphony concert halls, recital halls, drama theatres, opera houses and finally multi-purpose halls.

The new scientific basis for auditorium acoustics has made acoustic design a more confident exercise, in spaces for both music and speech. But acoustic design is also an art involving choices, as Bagenal (1950) recorded: 'To design a concert hall is to go down into the arena and risk death from the violence of your contending passions'. One hesitates to guarantee excellence, but many of the disasters of the past should be avoidable with present knowledge.

Sophisticated acoustic design however is but one element of an auditorium project. The demands on the architect are considerable. The strongly constrained nature of auditorium design means that the grand gesture is generally out of place. Several auditoria exist around the world where a major world-class architect has wanted to create a great public monument, but the architecture gets in the way of an effective functioning auditorium. Respect by architects for acoustics has fortunately improved over time. One of the grandest music spaces in Britain is St George's Hall, Liverpool of 1854; a colleague of the architect, H.L. Elmes, commented that 'The specific adaptation of this grand interior for the acoustic performance of music was a thing that he (Elmes) held to be beneath his consideration', and this was despite him being a competent violinist!

The ideal requires a good rapport between the architect and his major consultants, a willingness to develop a scheme within the design team, and for the architect to view the needs of other disciplines as a stimulus rather than an irritation. This state of affairs has luckily become more common and depends on suitable stature for all involved. But success reaps many rewards, by providing a venue to sustain the essential live performances in the arts and frequently by rejuvenating the areas in cities around public performance spaces.

References

Bagenal, H. (1950) Concert halls. *Royal Institute of British Architects Journal*, January, pp. 83–93.

Bagenal, H. and Wood, A. (1931) *Planning for good acoustics*, Methuen, London.

Barron, M. (1992) Precedents in concert hall form. *Proceedings of the Institute of Acoustics*, **14**, Part 2, 147–156.

Beranek, L.L. (1962) *Music, acoustics and architecture*, John Wiley, New York.

Beranek, L.L. (1979) The acoustical design of Boston Symphony Hall. *Journal of the Acoustical Society of America*, **66**, 1220–1221.

Beranek, L.L. (1996) *Concert and opera halls: how Rathey sound*, Acoustical Society of America, Woodbury, New York.

Beranek, L.L. (2004) *Concert and opera houses: Music, acoustics and architecture*, 2nd edn, Springer, New York.

Brewer, E.C. (1854) *Sound and its phenomena*, Jarrold, London.

Cavanaugh, W.J. and Wilkes, J.A. (1999) *Architectural acoustics: principles and practice*, John Wiley, New York.

Contant, C. and de Filippi, J. (1860) *Parallèle des principaux théâtres modernes de l'Europe et des machines théâtrales françaises, allemandes et anglaises*, Paris (reissued by Benjamin Blom, New York, 1968).

Cremer, L. and Müller, H.A. (translated by T.J. Schultz) (1982) *Principles and applications of room acoustics*, Vols. 1 and 2, Applied Science, London.

Dumont, G.P.M. (1774) *Parallèle de plans des plus belles salles de spectacle d'Italie et de France, avec des détails de machines théâtrales*, Paris (facsimile by Benjamin Blom, New York, 1968).

Forsyth, M. (1985) *Buildings for music*, Cambridge University Press, England and MIT Press, Cambridge, MA.

Forsyth, M. (1987) *Auditoria: designing for the performing arts*, Mitchell, London.

Garnier, C. (1871) *Le Théâtre*, Librairie Hachette, Paris.

Garnier, C. (1880) *Le nouvel Opéra de Paris*, Paris.

Hartmann, G. (1990) Aus der Frühgeschichte der Raumakustik. *Acustica*, **72**, 247–257.

Hoffman, I.B., Storch, C.A. and Foulkes, T.J. (2003) *Halls for music performance: another two decades of experience 1982–2002*. Acoustical Society of America, Woodbury, New York.

Izenour, G.C. (1977) *Theater design*, McGraw-Hill, New York.

Knudsen, V.O. (1932) *Architectural acoustics*, John Wiley, New York.

Langhans, C.F. (1810) *Über Theater oder Bemerkungen über Katakustik*, Gottfried Haydn, Berlin.

Lord, P. and Templeton, D. (1986) *The architecture of sound*, Architectural Press, London.

Louis, V. (1782) *Salle de spectacle de Bordeaux*, Paris.

Meyer, J. (2009) *Acoustics and the performance of music*, 5th edition, Springer, New York.

Moro, P. (1982) Architect's statement on the Theatre Royal, Plymouth. *Architects' Journal* 13 October, p. 66.

Patte, P. (1782) *Essai sur l'architecture théâtrale*, Paris.

Pevsner, N. (1976) *A history of building types*, Thames and Hudson, London.

Rayleigh, Lord (J.W. Strutt) (1877/8) *The theory of sound*, Vols. 1 and 2, Macmillan, London (reprinted by Dover, New York, 1945).

Reid, R.B. (1835) On the construction of public buildings in reference to the communication of sound. *Transactions of the British Association*.

Sabine, W.C. (1922) *Collected papers on acoustics*, Harvard University Press (reprinted 1964, Dover, New York).

Sachs, E.O. (1896–8) *Modern opera houses and theatres*, Vols. 1–3, Batsford, London.

Saunders, G. (1790) *A treatise on theatres*, London (facsimile by Benjamin Blom, New York, 1968).

Scott Russell, J. (1838) Elementary considerations of some principles in the construction of buildings designed to accommodate spectators and auditors. *Edinburgh New Philosophical Journal*, **27**.

Smith, T.R. (1861) *Acoustics of public buildings*, Crosby Lockwood, London.

Talaske, R.H., Wetherill, E.A. and Cavanaugh, W.J. (eds) (1982) *Halls for music performance: two decades of experience 1962–1982*, American Institute of Physics, New York.

Vitruvius (translated by M.H. Morgan) (1960) *The ten books on architecture*, Dover Publications, New York.

2 Sound and rooms

2.1 The essence of sound waves

Most people remember the school experiment with an electric bell suspended in a vacuum jar. The experiment demonstrates that sound depends on air for its propagation; as the jar is evacuated the bell can no longer be heard. Sound propagation depends on a source of sound and a receiver, but the propagating medium is all around us. This, coupled with the fact that sound bends round corners and that the ear is only barely directional, makes our hearing such an invaluable sense to notify us of danger, to inform us of our surroundings and to communicate. And if there is a mystery associated with this sense then it resides in the complexities of the ear. This remarkable organ is still far from fully understood. At its centre it contains a converter to transform the vibrations conducted from the eardrum into digital nerve pulses. Nerve pulses travel between the ear and brain in both directions, allowing the listener to suppress his/her sensitivity to a continuous noise for instance. The ears' response to music is no less intriguing.

The subject of this chapter is the nature of sound and the manner in which it propagates in rooms. Of necessity the treatment of the subject here is rather selective. Many standard texts cover the standard issues in more detail. For those that like a well-illustrated coverage, the book by Egan (1988) has much to recommend it.

Sound is generated in most cases by a vibrating object. Human vocal chords or the vibrating reed of a clarinet are obvious examples, but any wall which transmits sound from a neighbouring room is also vibrating, even though the amplitude is minuscule. A sound wave consists of a pressure fluctuation alternatively positive and negative relative to atmospheric pressure. As shown in Figure 2.1, passage of a sound wave causes air particles to move backwards and forwards parallel to the direction of motion of the wave. Sound waves are therefore **longitudinal**, rather than transverse like waves travelling along a string, in which string elements move at right angles to the direction of the wave. (Water waves look like transverse waves but are more complicated because the water particles move both vertically and horizontally in roughly circular paths.) For a sound wave, movement of air particles causes localized areas of compression or rarefaction; compression implies a higher pressure, rarefaction a lower one. As long as the amplitudes are not unreasonable, all regions of compression (and rarefaction) travel at a fixed speed, the **speed of sound**, which at 20°C is 343 m/s, or 1125 ft/sec. A

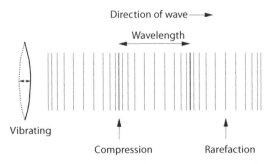

Figure 2.1 Basic sound wave showing alternate areas of compression and rarefaction

sound wave is characterized by amplitude, frequency and direction.

The **amplitude** of a sound wave is determined by the magnitude of the pressure fluctuation. But the range of pressures to which our ears can respond exceeds a ratio of one to a million and the response is not linear. A change from one to two units may be audible; a change from 100 to 101 will not be. Hence the ubiquitous **decibel**, named after Alexander Bell (1847–1922), inventor of the telephone. There are two key aspects of the decibel: it is a ratio measurement of energies or powers and it is logarithmic.

The decibel can be used in two ways: to measure differences or to measure sound (or other) amplitudes. In both cases, the choice of a logarithmic scale overcomes the problem of large numbers. The magnitude of the decibel for differences can be illustrated by some examples. A perceived doubling of loudness corresponds roughly to a change of 10 decibels, or 10 dB. Under controlled conditions, the smallest perceptible change of level is about 1 dB. Improvements in noise situations are generally only considered worthwhile if a 5 db reduction can be achieved.

In physical terms the effect of doubling the size of an orchestra is to produce a 3 db sound level increase. At the other extreme, the sound level difference across a masonry wall, such as a party wall, is typically around 50 dB.

The decibel is based on a ratio. For an absolute measurement of sound level, the ratio is taken between the measured sound pressure or intensity (sound intensity is a measure of energy) and a reference value. For sound pressure level, the reference value is 2.10^{-7} mbar (compared with atmospheric pressure of around 1000 mbar). In fact in acoustics the more usual unit for pressure is newtons per square metre, in which case 2.10^{-5} N/m^2 is the reference pressure. The reference value is chosen to be close to the quietest sound we can hear, so 0 db sound level is roughly the threshold of hearing. Conversational speech has a level of around 50 dB. A very loud sound of 120 db causes pain in the ears and would cause deafness if experienced regularly.

An extreme orchestral climax can reach 100 db for a member of the audience in a concert hall. The quietest sound in an auditorium is likely to be no lower than 20 dB, but it can be higher due to ventilation noise. A logarithmic scale for sound level approximately matches the loudness characteristic of the ear.

Frequency intrudes into all aspects of acoustics. A pure tone has a single frequency associated with it. All musical instruments however produce complex sounds made up of several frequencies, though the lowest of these normally determines the **pitch**, the name given to the perceived frequency. If a surface is vibrating, the frequency of vibration is the number of cycles per second, though this is now expressed in **hertz** (**Hz**). (Hertz and cycles per second are identical.) The surface generates a sound wave at the same frequency and in a room this will be judged by the ear as the same pitch as that frequency, in spite of the superposition of many acoustic reflections by the room. The ear can perceive frequencies between 20 Hz and 20 000 Hz, though the upper limit decreases with age. The ear also perceives frequency logarithmically; however in this case no new logarithmic measure is used. The fundamental musical interval is the **octave**, which corresponds to a doubling of frequency. Acoustic measurements are also conventionally made over octave intervals, with centre frequencies of 125, 250, 500, 1000 Hz etc.

The third fundamental quantity is **wavelength**. In Figure 2.1, the wavelength is the distance between adjacent pressure maxima (or minima). There is a simple relationship between frequency, wavelength and the speed of sound, which can be derived from Figure 2.1:

speed of sound = frequency × wavelength

The wavelength for a sound in the middle of the frequency range of 1000 Hz is thus 0.343 m (roughly 1 ft). The range of wavelength of audible sounds is between 17 m at 20 Hz and 17 mm at 20 kHz. This implies that these wavelengths are comparable to dimensions of room surfaces and common objects. At low frequencies with large wavelengths sound

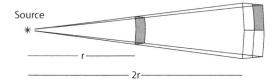

Figure 2.2 Sound spreading from a point occupies four times the area for each doubling of distance, leading to the inverse square law or 6 db decrease in level per doubling of distance

waves commonly bend round objects. But at high frequencies, objects are generally larger in dimension than wavelength and sound behaves much like light, travelling in straight lines and forming shadow zones.

Most sound sources radiate more energy in some directions than others; the human voice is typical in being directional. Violins and other stringed instruments tend to radiate most strongly at right angles to the sounding board. Yet once the sound has left the source, it behaves in the same way as it does from an omni-directional source. For every doubling of distance from the source, the area occupied by the energy increases by a factor of four (Figure 2.2). Intensity thus drops by a quarter and we get the 6 db decrease per doubling of distance characteristic, also known as spherical spreading. In rooms, not only the direct sound but also reflected sound rays behave in this way.

2.2 The nature of music and speech sounds

Both music and speech consist of brief sound events separated by silence. They both cover most of the audible frequency range, though anything above 10 000 Hz need not concern us, and for speech there is very little energy below 100 Hz. The individual sound events with speech are syllables and a typical speaking rate is five syllables per second. Music is obviously much more flexible.

Nearly all musical sounds with pitch depend on **resonance** for their generation. In a resonating musical instrument, a small amount of energy is supplied by the player which maintains oscillations at a particular frequency. The fundamental resonant frequency is determined by the length of string, tube etc.

A room can also be treated as a complex resonant system. To appreciate this, it is perhaps easiest to consider first the one-dimensional situation of a tube. If sound is produced within a tube, certain frequencies are enhanced at resonant frequencies determined by the length of the tube. If sound is produced in a room, certain frequencies are enhanced, which depend on the dimensions of the room. A room can thus be seen as a three-dimensional counterpart of a tube. However for rooms of the size of auditoria, this does not prove to be a particularly productive approach. The number of resonant frequencies in a room is so high (with many per hertz at frequencies above 100 Hz) that the phenomena which we normally associate with resonance are not observed. The word 'resonant' is also often used for rooms which are highly reverberant, in which sound decays only slowly. Though this use of the word is logical, it can be confusing and will therefore be avoided here. Resonance itself does not intrude much in room acoustics, except for certain frequency-specific absorbers (section 2.6.3).

The sound character or timbre of different musical instruments is related to their frequency spectrum. For a continuous sound the spectrum consists of a series of discrete frequencies (Figure 2.3). The lowest frequency, 415 Hz in the figure, is known as the **fundamental** or first **harmonic**. The higher frequencies are simple multiples of the fundamental frequency and are known as the second, third harmonic etc. Our ears interpret the mixture of frequencies in terms of its pitch, given by the fundamental frequency, while the relative strength of the harmonics characterizes the sound quality or timbre of the instrument. In the clarinet spectrum, the odd-numbered harmonics tend to be stronger than the even-numbered ones.

In the case of some instruments, such as the French horn, the relative strength of the harmonics varies in such a way that the frequency region of the **loudest** harmonic is roughly constant. This

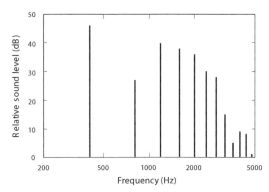

Figure 2.3 Spectrum of the clarinet in its middle range, fundamental frequency 415 Hz (after Meyer, 1978)

implies for instance that, when the horn is played in its lower register, the fundamental is weaker than higher harmonics (Figure 2.4). Maxima in the spectrum which do not change with fundamental frequency are known as **formants**, around 310 Hz in the case of Figure 2.4. The formant nature of an instrument further characterizes its sound. The human voice is unique in its scope for varying the formant frequency and uses this to produce the different vowel sounds.

Much more can be said about musical instrument sound, such as attack characteristics, the change of quality with dynamics etc. The interested reader is referred to Meyer (1978) and Olson (1967).

In fact as long as the appropriate frequency range is preserved in an auditorium, the detail of the sound character in frequency terms generally has little influence on room acoustic design. The harmonic structure of musical notes does explain however why the frequency range of interest extends much higher than the pitch range of musical instruments (Figure 2.5). The highest fundamental of the piccolo is 4186 Hz, yet a sound reproduction system which stopped at 5 kHz would be very low fidelity.

Sound sources are also characterized by their power. Sound generation is usually a very inefficient process; loudspeakers for instance are normally less than 2 per cent efficient. A symphony orchestra when playing *fortissimo* only generates about 2.5 watts of **acoustic power**, while the human voice is typically 25 μW. Both speech and music rely on contrasts of soft and loud, with the **dynamic range** in classical music of over 70 dB. The ability to hear a *pianissimo* can be as vital to the excitement of a live performance as a *fortissimo* climax.

The aspect of speech and music which substantially influences room acoustics is the temporal one. Speech sounds and musical notes can be represented diagrammatically as in Figure 2.6, which uses

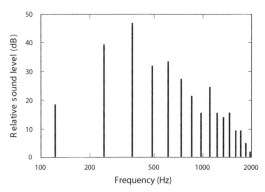

Figure 2.4 French horn spectrum in the lower register, fundamental frequency 123 Hz (after Meyer, 1978)

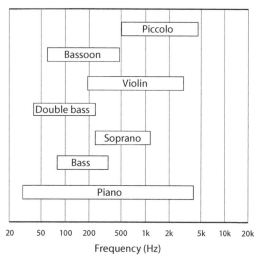

Figure 2.5 Frequency ranges of the fundamentals of musical instruments and voices

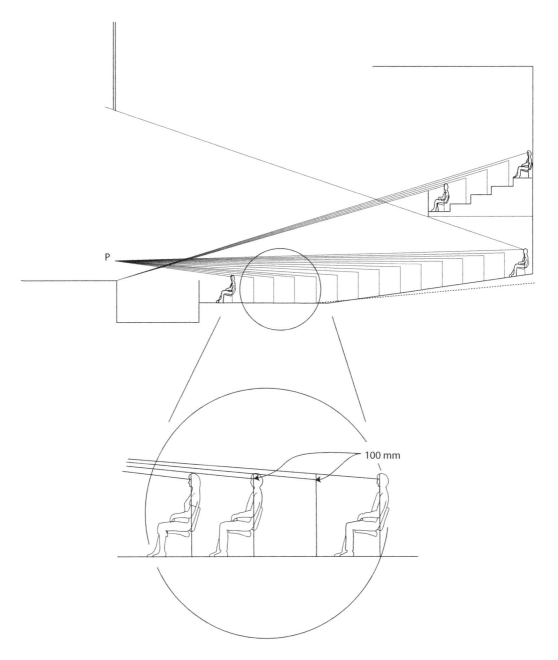

Figure 2.7 Sightline design. P is the setting-out point for the Stalls seating.

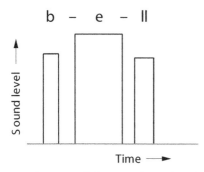

Figure 2.6 The temporal character of speech and music, illustrated with the word 'bell'

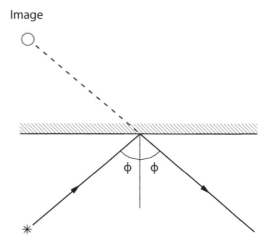

Figure 2.8 The geometry of reflection. The reflected sound behaves as if it had originated at the image point of the source. Angles, ϕ, between the normal to the wall and the incident or reflected sound are the same

the example of a word. Individual speech sounds (or musical notes) have particular amplitudes and durations. Vowels, for instance, tend to be longer and louder than consonants. The relative strengths and durations of concurrent speech sounds or notes are of crucial importance for room acoustics. They determine for instance the degree to which sound reflections can be tolerated without intelligibility or clarity being undermined. Such questions will be further considered below.

2.3 Sound propagation

For a source of sound outside, acoustic energy spreads into free space and since the surface area of a sphere is related to the square of the radius, we get inverse square law propagation. For every doubling of distance the sound level decreases 6 dB. When sound travels long distances outside, it is also influenced by wind and temperature effects. But even at the scale of the Greek classical theatres with typical distances of 50 m, there is no obvious evidence that environmental effects significantly influence the acoustics (except that wind increases the background noise).

In an enclosed space, the **direct sound** of course decreases in level in the same way as outside. The major concern for the direct sound in auditoria is that it is not affected by other audience members. This requirement is generally satisfied if there are adequate sightlines to the stage. The principles of

sightline design are simply stated. Since our eyes are below the top of our heads by on average 100 mm, floors should be raked to allow the sightline to pass over the heads of audience in the row in front (Figure 2.7). The setting-out point on the stage, P, influences the result; for stalls seating P is normally taken as 0.6–0.9 m above the stage front and for balconies the stage front itself is often used (Ham, 1987). The sightline criterion generates a curved floor rake, but this is normally approximated by a series of linear rakes. The maximum permissible rake in balconies is normally contained in legislation; 35° is a common limit. Sightline design becomes complicated in three dimensions; indeed optimal results are achieved with non-horizontal seat rows.

Most of the sound energy we receive in enclosed spaces has been reflected off wall and ceiling surfaces. The geometries of **reflection** for light and sound are identical (Figure 2.8). The reflected wave behaves as if it had originated from the image position. But while a small mirror will reflect light, for sound much larger surfaces are required owing to the much longer wavelengths involved. An acoustic mirror is a large, plane, massive surface of, for instance, concrete or plastered masonry. Sound reflected by one surface will continue to be reflected between the room surfaces, until its energy is removed by absorption. Figure 2.9 shows successive image positions for sound reflecting between two surfaces. Moving from the simple to a whole room: if we consider the situation in a large concert hall at the moment one second after a sound source has been switched off, our ears are listening to sound which has been reflected by roughly 20 successive surfaces.

Pure, or specular, reflection is one of four possible consequences when sound hits an obstacle.

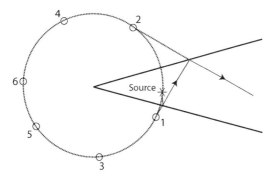

Figure 2.9 Reflection between two planes. Numbered image positions are for reflections that are initially off the lower plane. They all lie on a circle centred on the intersection point of the planes

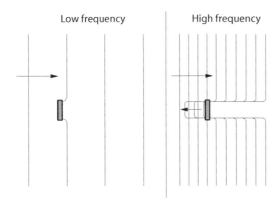

Figure 2.10 Wavefront diagrams of sound waves hitting an obstacle. Diffraction occurs at low frequencies where the obstacle is small relative to the wavelength

For a finite-size reflector, behaviour at low and high frequencies will be different. If the reflector is suspended, high-frequency sound is reflected like light, creating a shadow zone beyond the obstacle (Figure 2.10). But at low frequencies where the wavelength of sound is large compared with the size of the obstacle, bending or **diffraction** will take place and the wavefronts recombine as if the obstacle had not been there. At intermediate frequencies the behaviour becomes mathematically complicated. Bending, or diffraction, of sound waves is thus normally a low-frequency phenomenon.

A white matt surface, like a sheet of paper, is reflective to light but scatters it in all directions. For a surface to be acoustically **scattering** requires an irregularly profiled surface with projections of a depth between 0.3 and 0.6 m. Again the gross nature of projections required is due to the wavelengths of audible sound. A lightly profiled surface will only scatter sound at high frequencies. The word 'diffusing' is often used as an alternative to 'scattering', however a diffusing surface is easily confused with the concept of a diffuse sound field. Scattering surfaces can contribute to a diffuse sound field, but the interplay is complex. Scattering surfaces for sound are an important aspect of auditorium design.

Reflection, diffraction and scattering are all possible without energy loss. In practice, some **absorption** of sound energy occurs for reflection from all surfaces, while some materials are highly absorbent. The most common absorption mechanism is porous absorption; sound energy is dissipated in a porous material owing to the friction involved in movement of air particles in the pores. Typical porous absorbers are fabrics, curtains and carpets; the most efficient materials are mineral wool, fibreglass and acoustic open-cell foam. In auditoria the major absorbing surfaces are the clothed audience and performers, who at mid-frequencies absorb about 90 per cent of incident energy. It is normal practice in large auditoria to keep the amount of additional absorbent material to a minimum. Table 2.1 compares sound and light behaviour at surfaces.

Table 2.1 Surfaces for reflection and absorption with light and sound

	Reflection	*Scattered reflection*	*Absorption*
Light	Mirror	Matt white surface	Matt black surface
Sound	Hard massive smooth surface	Massive surface with projections of 0.3–0.6 m	Thick porous material

2.4 Sound in rooms

Much that is known to be important in auditorium acoustics can be discussed in terms of the acoustic response at a listener's position to a short sound produced on stage. The response to a longer musical note can be calculated from the response to a short impulse. As a listener, the first thing one hears is the **direct sound**, which travels in a straight line from the source (Figure 2.11). This is followed by a series of **early reflections** from the side walls, ceiling etc. Reflected sound has to travel further, so will arrive later; it will not be as loud as the direct component (unless some focusing of the reflection occurs). This response can be represented as a diagram of sound level against time (Figure 2.11). Such a figure is known as an **impulse response** or 'echogram', though this latter name is somewhat anomalous. The word **echo** is used for a reflection which is heard as a discrete event, such as can be heard outside when shouting some way from a large wall. An echo is an intelligible repetition and is not to be confused with reverberation, which is unintelligible. Most reflections in rooms are not heard as echoes.

The listener thus hears the direct sound followed by a series of early reflections whose paths can normally be precisely defined. Sound that is not absorbed continues to be reflected (Figure 2.12), and one sees on the impulse response that after the early reflections the number of reflections arriving

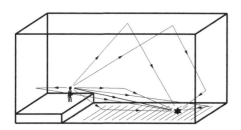

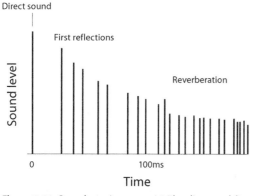

Figure 2.11 Sound rays in rooms. (a) The direct and first reflection paths; (b) sound level against time for a short sound as received by the listener

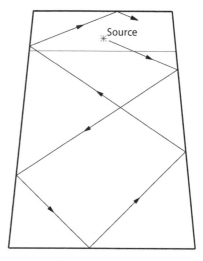

Figure 2.12 Multiple reflection within a room

within say each hundredth of a second (= 10 ms) progressively increases. In large halls, the number of reflections arriving at times later than one-tenth of a second after the direct sound (= 100 ms) is so high that individual reflections are no longer distinguishable. (The 100 ms time limit is shorter in smaller halls.) The late sound after about 100 ms is called the **reverberant sound**. It usually decays in a linear manner when plotted in decibels; its duration is described by the **reverberation time** (section 2.8). Reverberation is particularly obvious in large church or cathedral spaces.

The received sound can thus be divided into three components: the direct sound, early reflections and late reverberant sound. Design features of an auditorium often influence only one rather than all these components. There are several aspects of the impulse response which the ear perceives. The reverberation time is one of them. The reason for its importance is illustrated in Figure 2.13. The temporal nature of speech and music has been described in terms of elements with different amplitudes and durations (Figure 2.6). Figure 2.13 shows the instantaneous sound level in a room for such types of sound. When presented as a sound level in decibels, the build-up time is short, while the decay time is much longer. Reverberation thus provides the background of the decay of previous syllables or notes against which the new speech sounds or musical notes are heard. If the reverberation time is too long (Figure 2.13(b)), then one sound can be rendered inaudible (i.e. masked) by an earlier louder sound. But with too short a reverberation time, the sound quality becomes too stark, like listening in the open air. There is thus an optimum reverberation time mainly determined by programme, such as roughly 2 seconds for symphonic music and 1 second for speech.

But the subjective judgement of acoustics depends on much more than just reverberation time. For speech, the important concern is the proportion of energy in the impulse response which arrives early as a fraction of the whole. For speech to be intelligible requires this proportion to be high; in design terms strong early reflections may be

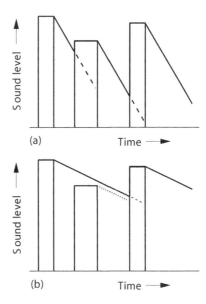

Figure 2.13 Instantaneous sound level with speech or music in a reverberant space. (a) Short reverberation time giving good clarity; (b) long reverberation time producing masking

needed. The comparable concern for music is clarity or definition. Also important for music is the sound level for the listener. Clearly this level depends on the orchestra and the music, but the received sound level for the listener is also influenced by the auditorium design and size. These various quantities, relating to intelligibility, clarity and sound level, can all be derived from the impulse response.

Unfortunately the impulse response **as such** cannot easily be interpreted as it stands. It can indicate the presence of an audible echo, but a visual inspection is not reliable for assessment of the amount of early reflections, for instance. Figure 2.14 shows two measured impulse responses in a hall at different distances from the source. At a position close to the source the direct sound dominates, the early reflections are well delayed and are weak relative to the direct sound. At remote locations, the first reflections arrive early and are almost as loud as the direct sound. Figure 2.15 shows the reason for this behaviour; it arises because the

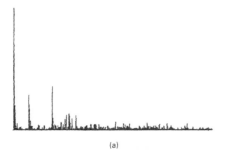

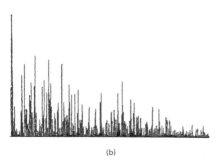

Figure 2.14 Measured (pressure squared) impulse responses in the Royal Festival Hall, London, measured at (a) 18 m and (b) 35 m from the source. Total duration is 200 ms

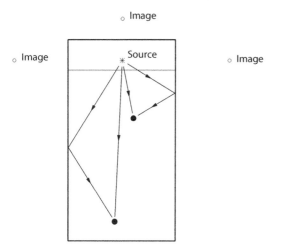

Figure 2.15 Reflections in an auditorium. Images of the source are fixed. The level of reflections relative to the direct sound is lower close to the source because of the large ratio of path distances. Delays, given by path-length differences, are also longer when closer to the source

image positions of reflections are fixed. Since the difference between the reflection path lengths and the direct path length is smaller for a more distant receiver, the delays and level differences of reflections are smaller for a distant receiver than for a receiver close to the source. In general, our ears are not as sensitive to distance as the impulse response plots would suggest.

2.5 Reflections and the ear

In a large concert hall a listener receives about 8000 reflections of a short sound within one second of the direct sound. Each reflection has a delay, a sound level and a direction associated with it. Clearly the ear is highly selective in interpreting this wealth of information. After many years of research and experience, the nature of the selection performed by the ear is becoming clearer.

As a warning sensor the ear needs to establish the direction of the sound source. To do this it is able to isolate the direct sound and localize on the first wavefront received. The localization ability of the ear is such that it can function when the direct sound is substantially weaker than the combined energy of later reflections. Because the wavelengths of sound encompass the dimensions of the head, different localization mechanisms operate at different frequencies. For localization from side to side the lateral position of the ears is used; at low frequencies time differences are predominantly used while at high frequencies sound level differences are exploited. But these mechanisms are useless for vertical localization, since in that case the sound signals at the two ears are virtually identical. For vertical localization, it appears that the ears use the frequency distortion provided by the outer ear, the pinna (Blauert, 1983).

For communication of speech (though the same mechanism also applies to clarity in music), the direct sound on its own has the disadvantage that it decreases substantially in level as one moves away from the source. However for the purposes of intelligibility and clarity, the ear is able to combine the **energy** of the early reflections with that of the direct sound. (Direct sound plus an equally loud early reflection is equivalent to a 3 db increase in the direct sound.) There is a certain time limit involved for a reflection to be 'early' which depends both on the processing performed by the ear and on the programme. A common value used for the limit within which reflections contribute to speech intelligibility is 50 ms, while 80 ms is usually applied for clarity in music. From the point of view of intelligibility alone, later reflections are detrimental but they are valuable and indeed necessary to create for the listener a sense of room sound.

The significance of the ability of the ear to sum energy of early reflections can be illustrated. In a large concert hall the direct sound level decreases by 12 db between 10 and 40 m from the source, while the early energy within 80 ms typically falls only 5 db (Figure 3.29). Hence the importance, particularly in large auditoria, of considering the early reflection distribution at all seating areas. The manifold ways in which the ear processes sound is a fascinating subject (see for instance Tobias, 1972). Much remains to be discovered.

It is worth comparing the aural and visual response to reflections. If we consider a visual reflection such as when a house is seen from across a lake on a still day, the house's reflection is seen as separate from the house itself and in a different vertical direction. The acoustic analogy for this is probably an echo that is heard as a discrete repetition of the original. However an echo is a special case of a sound reflection. The vast majority of sound reflections merge into the total sound experience. The original sound direction remains clear, but as discussed above, early reflections can add to clarity and intelligibility for speech. Later reflections create an audible background. Reflections also contribute towards the spatial character of the sound experience.

2.6 Sound propagation in detail

2.6.1 Direct sound

While sound propagation from the stage to the front row of a balcony is uninterrupted, for the majority of the audience in an auditorium the direct sound has had to travel at **grazing incidence** over a highly absorbent surface: the audience seated in front of the listener. Comments are frequently found in the literature that direct sound is absorbed by this grazing incidence. The situation at mid-frequencies and low frequencies has to be treated separately.

At mid-frequencies, there is no doubt that grazing incidence absorption can occur. But the problem of making elaborate experiments with seated audiences has so far prevented the question from being fully resolved. Measurements in unoccupied seating indicate little attenuation effect at mid-frequencies, Figure 2.16 (Schultz and Watters, 1964). With seated audiences the general rule has already been mentioned: good sightlines should provide good direct sound propagation. It does appear though that steeper seating rakes are acoustically preferable beyond the point of providing adequate vision (Cremer and Müller, 1982, Ch. 1.6). Sceptics on this point may comment that some

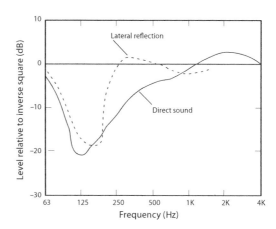

Figure 2.16 Measured propagation over theatre seats in excess of inverse square law behaviour (after Schultz and Watters, 1964)

of the world's most renowned concert halls have flat floors and poor sightlines! Yet within these halls, few would disagree that the best seats do have a good view of the whole orchestra. Unfortunately designing with steep seating rakes tends to restrict the number of seats which can be accommodated.

The situation at low frequencies is better documented. Severe low frequency attenuation was only discovered in 1962, when complaints of poor bass sound were investigated in the New York Philharmonic Hall (section 4.8). The discovery of grazing incidence attenuation, or **seat-dip effect** as is it frequently called, occurred because by chance the reflection from the suspended ceiling in Philharmonic Hall also lacked low frequencies. The grazing incidence effect occurs in all auditoria, but it had gone unnoticed. This is particularly surprising since one would expect the sound quality at a seat in the front row of a balcony (where the attenuation effect does not occur) to be markedly different from that at a seat directly below. That this is not distinctly obvious appears to be due to the fact that many reflections arrive from paths remote from the seating plane, which are thus not affected. Since the ear appears to be relatively insensitive to the arrival times of low frequency sound, later bass sound can compensate for a deficiency in the direct sound.

Figure 2.16 illustrates the extra attenuation due to grazing incidence propagation over audience seats. There is a sharp dip around 150–250 Hz of about 15 dB, as well as a broader-band attenuation up to about 800 Hz. These attenuations are established after sound has passed over only a few rows of seating, but thereafter they are virtually independent of the number of rows covered. The acoustic mechanisms involved seem to be a vertical resonance between the seat rows for the 15 dB dip, and effects due to propagation over an absorbent surface for the broader band effect. Two strategies appear to reduce the attenuation effect: using a steep seating rake such that the angle of arrival of the direct sound is greater than 15° to the seating plane (Bradley, 1991), or by placing the seating over a pit or well of about 1 m depth, with a wire mesh floor covered with light carpet (Davies and Cox,

2000); it appears that this last scheme has yet to be tried in an actual hall.

But the fact that low-frequency grazing attenuation was discovered so late indicates that in most halls it is of little subjective significance. This depends on the presence of enough reflections on paths remote from audience seating, a condition certainly fulfilled in the classical rectangular halls. The relative loudness of the bass can with advantage be enhanced by an extended low-frequency reverberation time (Barron, 1995, and section 2.8.1).

There is a further high-frequency effect on sound intensity which is only significant when sound has travelled many times round an auditorium. Absorption of energy as sound travels through the air is a high-frequency effect which causes the reverberation time at 2 kHz and above to be reduced (section 2.8.5). The major absorption effect is caused by the simultaneous presence of oxygen and water vapour in the air. Air absorption is thus unavoidable. The amount of absorption does in fact increase at low humidity levels, so that in dry, desert-type climates it is appropriate to humidify the air for acoustic, as well as comfort reasons.

2.6.2 Reflection from finite surfaces

Reflection from a finite surface depends on the relationship between the size of the reflector and the wavelength of sound. Perfect reflection occurs at high frequencies, whereas when the frequency is lowered, less and less energy is reflected along the geometrical reflection path. The distance of the reflector from the source and receiver also proves to be significant. The mathematical relationship governing this behaviour is given in Appendix A section A.1.1, and is particularly relevant to the design of suspended reflectors in auditoria. For reflection from elements of the hall envelope the situation is more complicated, particularly at low frequencies. A series of surfaces adjacent but modestly inclined relative to each other will often behave at low frequencies as a single curved surface.

A convex surface will disperse sound (Figure 2.17); it is thus a thoroughly safe feature in an

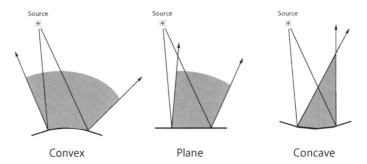

Convex   Plane   Concave

Figure 2.17 Reflection off convex, plane and concave surfaces

auditorium and is valuable in situations where reflection from a plane surface might produce undesirable effects. By contrast, concave surfaces are particularly dangerous because they focus sound. They were a common feature of nineteenth-century concert halls and theatres and indeed the concave domes so common in theatres and opera houses of the period often constitute their major acoustic fault. If the focal point of a concave surface is anywhere near audience or performers, the reflection is likely to be heard as an echo (Cremer and Müller, 1982, Ch. 1.3). Highly curved concave surfaces remote from audience do however act as dispersers of sound (Figure 2.18). The rule for acceptable reflection from a concave surface is simple: if neither the source nor receiver are within the extended circle of the concave surface, then the reflection will be weaker than from a plane surface.

Concave surfaces which focus on or near the audience are now taboo for auditoria. Remedial treatment is difficult. Sound absorbing or scattering treatments are common solutions but the level of sound reflected off a concave surface can still be too high even when absorbent material is applied to the surface. Providing a cure for focusing by a dome is demanding; an interesting solution with suspended panels was used by Reichardt *et al.* (1968) for the 'Haus des Lehrers' in former East Berlin. A quantitative method for calculating reflection levels from curved surfaces is included in Appendix A, section A.1.2.

2.6.3 Acoustic absorption

Acoustic absorption removes acoustic energy. There are three possible mechanisms: porous absorption, panel absorption and Helmholtz resonance. Porous absorption, as already mentioned, occurs with any porous material. In auditoria the major absorbent surface is the audience, whose clothes act as efficient porous absorbers. Absorption is measured by the **absorption coefficient** (α), which is simply the

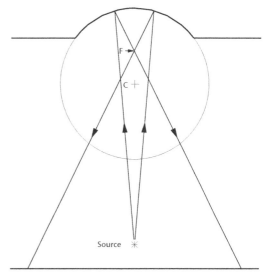

Figure 2.18 Focusing off a remote concave surface. F is the focus point and C is the centre of the circular dome. The situation is safe if the source and receiver are outside the extended circle of the concave surface (after Cremer and Müller, 1982)

fraction of incident energy absorbed. The frequency characteristic for porous absorbers is shown in Figure 2.19. The 'S-shape' characteristic is also found for audience absorption. Porous absorbers are thus efficient at high but not low frequencies. The low-frequency range can be extended by making the material thicker, but this becomes impracticable when absorption over the whole of the important frequency range is required. A much more economical solution is to position the porous material over an air-space; this is almost as effective as continuous porous material of the same total depth. Porous material over an air-space has more application in suspended ceilings for offices than for auditoria.

Any thin panel will absorb some energy in the low frequencies; vibration of the panel converts acoustic energy into heat. A simple tap with the knuckle reveals whether one is dealing with a panel or a solid wall, say, of masonry. Figure 2.20 shows the construction of a purpose-built panel absorber;

there is a frequency of maximum absorption which can be varied by selection of the panel mass and the depth of the airspace. Traditionally it was thought that 'wood was good' (section 4.3), that a timber finish resonated to enhance musical tone (as it does with a violin); the reality is the opposite, panelling absorbs acoustic energy. The absorption characteristic is included in Figure 2.19. Panel absorbers can complement porous absorbers to give absorption over the whole frequency range, though the maximum absorption coefficient of panel absorbers is not great. Panel absorption strongly influences the low-frequency reverberation time, as discussed further below.

In spite of its grand name, a Helmholtz resonator is familiar to all in the form of an empty wine bottle. When one blows across the top of the bottle, a clear note is produced. If a small amount of porous material is placed in the neck of the bottle, it no longer acts as a resonator aiding sound generation but as a tuned acoustic absorber, highly efficient at absorbing at the resonant frequency but absorbing little at other frequencies. A Helmholtz resonator thus requires a neck region coupled to an enclosed volume. These resonators have been little used in auditoria because of the complications of design, but the Queen Elizabeth Hall, London, is an exception (section 6.5). Further details on acoustic absorption are found in many acoustic texts.

2.6.4 Scattering

A textured surface will provide some scattering of sound, but the degree of texturing must be high for efficient scattering. The general principle is that the deeper the scattering treatment, the lower the frequency down to which the surface will scatter sound. Depths for projections of between 0.3 and 0.6 m are necessary to cover the majority of the audible frequency range. Substantial surface modelling is to be found in concert halls from the nineteenth century in line with the architectural style of the time. Most of the halls from this period are famous for their acoustics. The high degree of profiling of the walls and ceilings render these

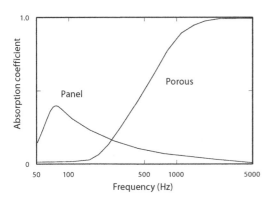

Figure 2.19 Absorption coefficients of porous and panel absorbers as a function of frequency. Making the porous material thicker shifts its characteristic towards low frequencies

Figure 2.20 Construction of a panel absorber, with a plywood panel mounted on battens. Porous absorbent material is sometimes placed in the cavity

Figure 2.21 A pair of suspended quadratic residue reflectors in the Glasgow Royal Concert Hall, Scotland

profile is based on a prime number. The use of QRD's in concert halls was pioneered by Marshall in Wellington, New Zealand (section 4.10). Both in that hall and in Figure 2.21 the sequence was based on the prime number 7. Quadratic residue diffusers and other types of Schroeder diffusers have also been used in stage enclosures and opera house pits (D'Antonio and Cox, 2000). Some further details about these diffusers are contained in Appendix A.2.

Computer models for prediction of the acoustic behaviour of enclosures require that the reflection behaviour of wall and other surfaces be defined. Early computer models assumed specular (i.e. mirror-like) reflection at all surfaces but this ignored the fact that many room surfaces scatter sound. Most current computer models now include a parameter to account for scattering: the scattering coefficient, d. Surfaces which are moderately scattering (Figure 2.22) tend to reflect with a proportion of reflected energy along the direction of the specular reflection with the remaining energy scattered, often fairly randomly. The scattering coefficient is simply the proportion of reflected sound that is scattered as a fraction of the total reflected energy. Wall and other surfaces are specified for the computer model in terms of both their absorption and scattering coefficients.

Progress has been made to enable scattering coefficients to be measured in the laboratory (Vorländer and Mommertz, 2000; Hargreaves *et al.*, 2000). Audience seating is highly scattering, while most finite-size surfaces are only slightly scattering.

Another recent development is the ability to predict reflection behaviour by surfaces (using boundary element methods). Out of this technique have come 'wavy' surfaces optimized for

surfaces acoustically scattering. The debate about how important this surface treatment is for acoustic quality has been running for at least half a century and still remains far from resolved. It is only recently that progress has been made towards quantifying the scattering offered by particular profiles. Regular profiles must however be avoided as they are frequency selective in their scattering characteristics.

A major development was made by Schroeder (1979) who proposed a group of diffusers whose reflection properties can be calculated in advance. These Schroeder diffusers consist of equal-width slots or wells with depths according to mathematical number theory sequences. The best known is the Quadratic Residue Diffuser (QRD), Figure 2.21, whose

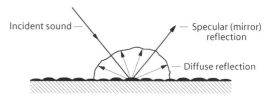

Figure 2.22 Partially scattered reflection from a textured surface showing specular and scattered components

Figure 2.23 A scattering wall in a rehearsal hall with circular geometry in plan. The shape of the wall is based on a wave motif optimized for diffusion using boundary element techniques. (Architects: Patel Taylor; acousticians: Arup Acoustics; diffuser design and installation: RPG.)

their directional reflection characteristics (Cox and D'Antonio, 2004). An impressive use of this technology is for surfaces designed to suppress focusing caused by concave walls, Figure 2.23 (Orlowski, 2000).

2.7 Acoustic defects

2.7.1 Echoes

While many acoustic defects are a matter of degree, such as too much reverberation or inadequate loudness, certain phenomena are always disagreeable. Echoes can certainly be obvious and pernicious. An echo is a reflection which is heard as a discrete event. The most famous example in Great Britain was the echo in the Royal Albert Hall, London. The glass dome was, as the longest free span yet built, the pride of British engineering when it opened in 1871. But acoustically it was a classic case of a surface which acted as a huge concave acoustic mirror. Sound originating from the stage was focused close to audience seating. With a delay of around 1/6 second and with the intensity of the echo in some seating locations actually exceeding that of the direct sound, this constituted an echo on the grand scale, sufficient to make a mockery of all performers' efforts. An account of the efforts to suppress this echo is given in section 5.1.

For a reflection to be heard as an echo it must arrive at least 50 ms later than the direct sound. The reflection also has to be more prominent than its neighbours (Figure 2.24). In all rooms there are numerous reflections arriving at more than 50 ms after the direct sound. To be perceived as an echo requires either reflection from a large surface by

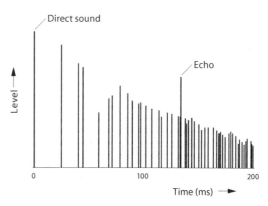

Figure 2.24 Impulse response with an echo

a path simpler than other reflections of the same delay, or reflection involving focusing. In the first category, reflection off the back wall or back wall and adjacent soffit is a frequent cause of echoes on stage, though obviously these are only heard as echoes in large rooms. Focused echoes are a common problem in halls with domes or barrel-vaulted ceilings, or in fan-shaped plans with curved rear walls. Focusing of reflections by curved surfaces is discussed in section 2.6.2 above.

The matter of predicting whether reflections will be audible as echoes is not trivial (Cremer and Müller, 1982, Ch. II.7; Dietsch and Kraak, 1986). The refined localization abilities of the ear can be tuned into detecting echoes. Whereas at first one might not hear an echo, once the echo is detected one may be unable to ignore it. Subjective assessment of impulse responses produced by models can be a useful aid to detect possible echo paths for new auditoria.

Curing echoes in completed buildings can be difficult. Either absorptive or scattering treatment of a reflecting surface can be used or the surface can be remodelled to direct the reflection away from performers and audience. In most auditoria, large areas of acoustic absorbent should be avoided as they lead to a quieter sound and 'holes' in the directional sound distribution. Substantial focusing would require substantial scattering treatment, which may be visually intrusive. To quell an echo from the back wall, inclining the wall upwards or downwards to redirect the reflection is worth considering (Figure 2.25). Scattering treatment may be equally suitable, if sufficiently effective. Nowadays echoes should always be avoided rather than cured at a later date.

2.7.2 Flutter echoes

While echoes involve large delays and long path lengths, flutter echoes normally involve short path lengths but iterated many times. A common flutter echo situation occurs between two parallel walls (Figure 2.26). The repeated reflections create a characteristic 'twang', which is obvious to anyone who

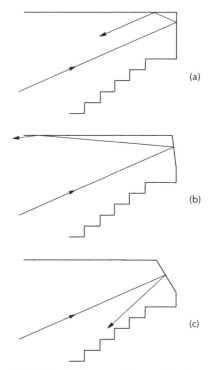

Figure 2.25 Echo off the rear wall and ceiling in an auditorium. (a) A vertical wall and horizontal ceiling return a reflection back towards the source; (b) slight inclination of the rear wall often renders the reflection harmless; (c) major inclination of the wall reflects onto nearby absorbent audience

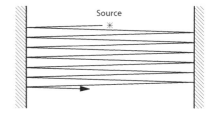

Figure 2.26 Flutter echo between parallel walls

has heard it by clapping their hands in corridors or rooms with flat parallel walls. More complicated repetitive paths are possible but rare.

In auditoria, the obvious concern is that musicians should not be between two parallel surfaces. But even with parallel stage walls, flutter echoes are only likely to be heard when just a few musicians are present. A flutter echo is one of the few acoustic problems which is relatively easy to correct. Reorientation of one of the parallel surfaces by only 5° is adequate to cure it, as is modest application of absorbent or scattering treatment.

2.7.3 Background noise

Auditoria present the engineer with some of the quietest spaces to service. On urban sites, noise intrusion from the exterior also needs to be considered. Traffic noise, railway noise or aircraft noise can necessitate substantial expenditure to

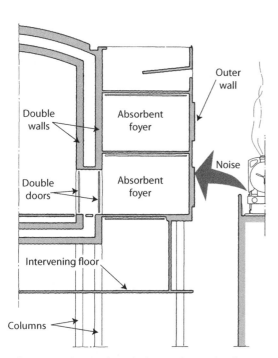

Figure 2.27 Section through the Royal Festival Hall, London, showing sound insulation measures

achieve inaudibility inside the auditorium. A typical approach is to place the 'box' of the auditorium within the box provided by the shell of the building; the intervening spaces are readily used for foyers (Figure 2.27). Methods for limiting noise ingress are no different in approach than for other building types. There are many books which cover the standard techniques. Vibration from underground trains in particular can also produce problems; the solution in severe cases can be to mount the whole auditorium on resilient pads. This vibration isolation has been used for the recent concert halls in Glasgow, Birmingham and Manchester.

The main contributor to background noise is usually the ventilation system. Ventilation systems generate noise due to fan noise and noise generated by the air-flow itself. Controlling the noise involves substantial attenuation between air-handling units and the auditorium, as well as low air outlet velocities.

Several criteria are available for background noise, which are based on individual octave band sound levels. The most common European criterion is the NR (Noise Rating) but there is also the American NC (Noise Criterion), a revised version the PNC (Preferred Noise Criterion) and more recently the RC (Room Criterion). A sample of NR curves is given in Figure 2.28; octave band levels are compared with the target curve and should not exceed it at any octave. Criteria for different auditoria are given in Table 2.2.

The audience itself also generates background noise from breathing. Though for one individual this will be insignificant, for a whole audience it can exceed the criteria mentioned. According to measurements by Kleiner (1980), the noise generated by the audience in a typical concert hall could reach NR25, that is 10 db greater than the recommended criterion (Figure 2.28). This is not however cause for complacency regarding ventilation noise. Part of the magic of a live event is the moments when concentrated anticipation by the audience creates a 'dramatic hush'. At these instants one does not wish to be introduced to the deficiencies of the air conditioning system.

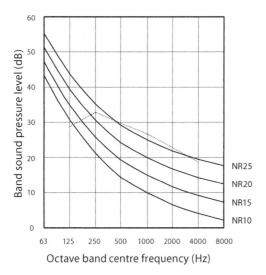

Figure 2.28 Noise rating curves relevant to auditoria and (dotted line) audience noise spectrum (after Kleiner, 1980; auditorium volume = 20 000 m³ with 1600 seats, reverberation time = 2 seconds)

Partly stimulated by the realization that audience noise is not a fixed quantity, that audiences can be very quiet, the idea has been proposed that ventilation noise should be lowered beyond traditional criteria. With a lower background noise the dynamic range of the performance can perhaps be increased, creating a more exciting sound experience. The first British hall responding to these ideas was Birmingham Symphony Hall of 1991 (section 5.14), where it appears they have achieved a ventilation noise level of NR10 (Newton and James, 1992). Critics of this design approach point out that such low levels just make coughing and other noises that much more irritating. But an intriguing discovery is that audiences in Birmingham have responded by becoming quieter than elsewhere, by perhaps as much as 5 dB.

Table 2.2 Background noise rating criteria for auditoria

Large concert halls	NR15
Opera houses and drama theatres	NR20
Small auditoria (less than 500 seats)	NR25

2.8 Reverberation time design

2.8.1 The criteria

The difference in acoustic character between a cathedral and a domestic living room is immediately obvious. The time for sound to decay to inaudibility in a cathedral can be timed with a stop-watch, much as Sabine did in his pioneering experiments. The audible decay time approximates to the reverberation time – it is 11 seconds in St Paul's Cathedral, London, for instance. Appropriate times in auditoria for orchestral music and speech are much shorter. The reverberation time of a domestic living room is about 0.5 seconds.

The room response, which can be heard after a loud, short-duration sound or after a continuous sound is turned off, is known as **terminal reverberation**. This is used for measurement and the reverberation time is defined as the time for the sound level to decay to one millionth of its energy (Figure 2.29). Yet in music performance or speaking, this terminal response is rarely heard. What is more significant is the level of the reverberant sound which constitutes the background for the **next** musical note or speech syllable (Figure 2.13). The term **running reverberation** has been used for this, for which the early part of the decay is more significant. In many

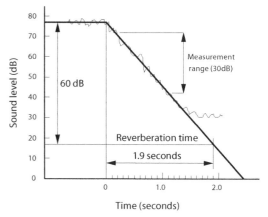

Figure 2.29 Reverberation time definition with a sample decay. The slope of the decay is, in practice, measured between –5 db and –35 db of the initial level

rooms however the reverberation time as traditionally measured is appropriate for both. Despite the many advances of recent years, reverberation time remains the single most valuable measurable quantity for the acoustics of an enclosed space. It has the advantage that it is generally constant throughout the room. An immediate preliminary assessment can be made of the suitability of a space for music or speech from knowledge of the reverberation time.

Though one refers to a 'linear sound decay', as shown in Figure 2.29, the decay of sound **energy** and **pressure** are exponential (Figure 2.30(a)). The pressure reduces by the same **fraction** for equal time intervals; so, if it has reduced to 1/10th after 1/3rd of a second, it will be reduced to 1/100th of its initial value after 2/3rd of a second. When converted into decibels, the decay approaches a straight line (Figure 2.30(b)). In reality the measured decay contains random modulations (Figure 2.31(a)); the reverberation time is measured from the slope of the best-fit straight line.

In some spaces, the decay is non-linear and a reverberation time can no longer be quoted (Figure 2.31(b)). The commonest cause of non-linear decays is coupled spaces with different reverberant conditions. If the decay is measured in the less reverberant space, the decay rate is initially that of the measured space but followed by the decay of the adjacent coupled space. An example of a coupled space is a proscenium theatre auditorium with a reverberant flytower, coupled through the proscenium opening. The orchestra pit and opera house auditorium are another example of coupled spaces.

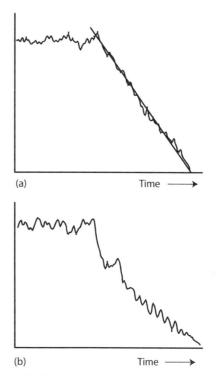

Figure 2.31 Measured reverberant decays. (a) Linear decay; (b) double slope decay

The appropriate reverberation time for an auditorium should be determined principally on the basis of programme (Table 2.3). Many references are found in the literature to optimum reverberation time also being a function of hall volume. In terms of major auditorium spaces, this approach only

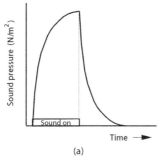
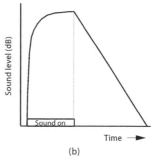

Figure 2.30 Sound build-up and decay displayed by (a) sound pressure and (b) sound level

Table 2.3 Recommended occupied reverberation times (seconds)

Organ music	>2.5
Romantic classical music	1.8–2.2
Early classical music	1.6–1.8
Opera	1.3–1.8
Chamber music	1.4–1.7
Drama theatre	0.7–1.0

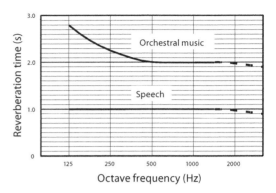

Figure 2.32 General recommended reverberation frequency characteristics for speech and orchestral music use (the latter with a maximum increase of 40% at 125 Hz)

appears relevant in small recital halls (section 6.2). A major dilemma exists for reverberation time design for multi-purpose spaces (Chapter 10). With orchestral music requiring a 2 second reverberation time and speech only 1 second, use of a single space for both speech **and** music is usually not possible without electronic assistance. Selecting a compromise time for such a situation can result in acoustics which are neither able to support intelligible speech nor sufficiently live by musical standards.

If a single figure is quoted for the reverberation time, it generally refers to the mid-frequency value, averaged between 500 and 1000 Hz. Reverberation time will vary with frequency (Figure 2.32). At high frequencies above 1 kHz, the reverberation time inevitably decreases due to air absorption (section 2.6.1). At low frequencies, the situation can be controlled by the designer. For speech there is good reason to keep the reverberation characteristic constant with frequency; a rise in the bass undermines intelligibility. But for music a bass rise in reverberation time is considered by most people as desirable. In this respect there used to be something of an Atlantic divide, with Americans considering that one cannot have too much bass, while some consultants in Europe strove for a flat characteristic with music. For orchestral music, the American argument seems to have won the day. A rational argument for this approach is that a bass reverberation time rise is needed to achieve a constant **perceived** decay time across frequency. This argument is based on the frequency characteristics of the ear: though at high sound levels the ear is roughly equally sensitive to different frequencies,

at low sound levels it is much less sensitive to bass frequencies (otherwise we would hear our own heart beats). A longer bass reverberation time can compensate for this. Up to a 40 per cent rise at 125 Hz compared with mid-frequencies is considered appropriate for orchestral music (Figure 2.32).

2.8.2 The Sabine reverberation time equation

While greater theoretical minds in the nineteenth century probably imagined that the problem was not amenable to simple mathematical analysis, Sabine discovered around the year 1900 (Sabine, 1922) that only two quantities determined the reverberation time: the room volume (V) and the **total acoustic absorption** (A). The total absorption is simply the sum of all the sound absorbing areas, determined by adding the product of area (S) and absorption coefficient (α) of **all** surfaces in the room ($A = \Sigma S\alpha = S_1\alpha_1 + S_2\alpha_2 + ...$). The Sabine equation, when volume and acoustic absorption are measured in m³ and m², is:

$$\text{Reverberation time (seconds)} = \frac{0.16V}{A}$$

This equation is the basis of virtually all reverberation time prediction in auditoria. The calculation is performed at each octave frequency of interest, at least from 125 to 2000 Hz. Sabine's equation does in fact involve a fallacy in that for a room with fully absorbent walls, floor and ceiling ($\alpha = 1$), a finite time is predicted when it should be zero. Alternative equations are available (particularly that by Eyring) to deal with highly absorbent rooms like recording studios. In auditoria, the coefficients assume use of the Sabine formula.

The major absorbing surface in an auditorium is the audience. Historically due to Sabine, the absorption of audience was calculated on the basis of the number of seats using absorption per person. This, combined with the fact that coefficients were too small, led to reverberation times in many concert halls which were shorter than predicted. One of the last cases of this occurred in the Royal Festival Hall, London, of 1951 (section 5.5).

2.8.3 Volume per seat as a design aid?

If one takes the Sabine equation and ignores the small proportion of incidental absorption by room surfaces other than audience, then, if audience absorption is calculated on the basis of absorption per person ($A = n\alpha_p$, where n is the audience size and α_p is the absorption per person), the consequence is that the reverberation time **should** be proportional to the auditorium volume per seat. For a long reverberation time a large volume per seat is

needed. This concept of volume/seat is often used for acoustic design; particularly at the early stages of design, it can provide a valuable rule-of-thumb. For concert halls a value of at least 10 m³/seat is often recommended. The equivalent values for opera are 7–9 m³/seat and 4 m³/seat for drama theatre.

But how reliable is this approach? Figure 2.33(a) shows the relationship between measured occupied reverberation time and volume/seat in British concert halls. Figure 2.33(b) shows a more convincing relationship between measured reverberation time and auditorium volume divided by acoustic seating area. The meaning of this last quantity is explained in the next section.

The alternative view was due to Beranek (1969) who was investigating the reasons for inaccurate reverberation time prediction in concert halls. He found that more accurate predictions were made when audience was treated like other materials, on the basis of absorption per unit area, or absorption coefficient. The absorption coefficient of seating and audience is relatively insensitive to seating density. For precise reverberation time calculations, absorption per person or per seat is little used nowadays and no coefficients for absorption per person will be quoted here.

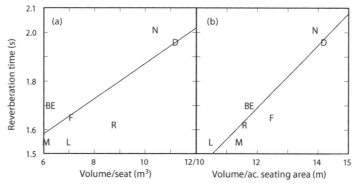

Figure 2.33 Occupied mid-frequency reverberation time vs. (a) volume/seat and (b) volume/acoustic seating area for British concert halls where data is available for halls without added absorption. Hall labels as in Figure 5.1, plus E, Usher Hall, Edinburgh and N, Bridgewater Hall, Manchester

2.8.4 Absorption by seating and audience

A major problem of reverberation time prediction is the choice of appropriate absorption coefficients for unoccupied seating or seated audience. A complication arises in the measurement of actual seats. Acoustic test chambers can only accommodate a maximum of about 24 seats, much smaller than the numbers found in real auditoria. The problem occurs not because of the numbers of seats but because the edge of a seating block absorbs sound in addition to the seating area when viewed from above. The perimeter-to-area ratio of the seating block proves to be important and this ratio is much higher in a test chamber with a small number of seats than it is in a typical auditorium.

One proposal for measuring seating absorption in test chambers is to place barriers around the seating block to screen the edges (also mentioned in the standard ISO354). However Bradley (1992) and Barron and Coleman (2001) have shown that barriers introduce problems of their own, leading to inaccurate measurements. Bradley (1992) proposes a measurement method for test chambers using a sequence of different seating arrangements. The Bradley method appears to be the most reliable one at present, though it has not yet been widely accepted. Many absorption measurements made in test chambers in the past are unlikely to be dependable.

Beranek (1969) derived absorption coefficients for unoccupied seating and seated audience from measured reverberation times in auditoria. A revised set of coefficients were published with Hidaka (1998); these figures appear currently to be the most reliable. Mention was made above that the edges of seating blocks also absorb sound. To deal with this, Beranek proposes that the seating area is enlarged by adding a strip around the seating block perimeter in plan. A perimeter strip 0.5 m wide is added to the true seating area, except where it abuts walls and along balcony fronts (Figure 2.34). The perimeter strip is only added where there is floor area in aisles etc. to which the additional strip can be applied. This enlarged seating area, which is the effective absorbing area of audience, will be called the **acoustic seating area**. The absorption coefficients quoted below are applied to this acoustic seating area.

2.8.5 Prediction of reverberation time

In the recently revised coefficients for seating (Beranek and Hidaka, 1998), the degree of upholstery is included as a parameter. Light upholstery is categorized as 25 mm of seat upholstery, 15 mm of upholstery on the front side of the seat back, a hard rear side of the seat back and solid arm rests. For medium upholstery the thicknesses of upholstery increase to 50 and 25 mm, with again a hard rear side of the seat back and solid arm rests. Heavy upholstery is defined as 100 mm of seat upholstery, 75 mm upholstery on the front side of the seat back, the rear side of the seat back possibly upholstered and 20 mm upholstery on the arm rests.

The recommended coefficients for seating and audience are given in Figure 2.35 and Table 2.4. The heavy upholstery audience coefficients also appear to be suitable for the proscenium opening in a theatre, when the stage area contains many drapes etc. to make it acoustically absorbent. The table also includes coefficients for other materials commonly found in auditoria. There is no entry in Table 2.4 for carpet. This is because carpet under audience

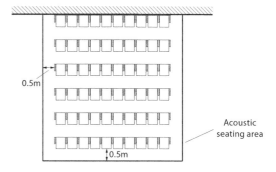

Figure 2.34 'Acoustic seating area' based on adding a 0.5 m perimeter strip round the true seating area

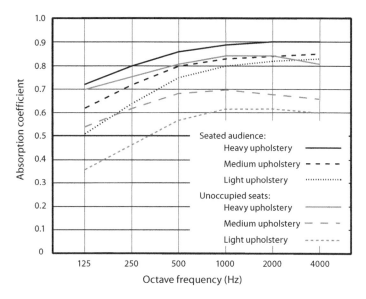

Figure 2.35 Absorption coefficients of seated audience and unoccupied seating after Beranek and Hidaka (1998)

seating is generally considered not to influence the total absorption. However the question of the significance of carpet cannot be considered to be fully resolved.

For accurate calculation of high frequency reverberation time, it is necessary to take account of the small absorption of sound that occurs when it passes through air. The Sabine equation including this component is:

$$\text{Reverberation time (seconds)} = \frac{0.16V}{\sum S\alpha + 4mV}$$

The air absorption coefficient, m, depends on temperature and humidity. Sample values of m are included in Table 2.4, taken from ISO 9613-1 (1993). Air absorption usually only influences reverberation time at 2 kHz and above. At 4 kHz the reverberation time becomes sensitive to temperature and humidity.

A sample reverberation time calculation is given in Appendix A.3. Calculation should be made to the nearest 0.1 seconds, which is probably the smallest detectable difference.

To achieve the criterion, mid-frequency reverberation time requires correct selection of the auditorium volume, which in turn usually influences the ceiling height. The need for a flat reverberation time characteristic for speech or a bass rise for music, Figure 2.32, influences the required area of low-frequency absorbent. By itself, audience acts as a porous absorber, which from Figure 2.19 is efficient at absorbing high frequencies but not low frequencies. Substantial additional low-frequency absorption is required to produce a flat reverberation time with frequency. Panel absorption is the most likely mechanism to be exploited, though resonator absorbers have been used (section 6.5). Although in a space for speech use one might well include purpose-designed panel absorbers (Figure 2.20), incidental light-weight panel materials, such as thin plaster for a suspended ceiling, also exhibit low-frequency absorption. In a concert hall where we want a long bass reverberation time, it is necessary that all walls, ceilings and suspended elements are sufficiently massive to minimize low-frequency absorption. This approach obviously has cost implications.

The accuracy of reverberation time prediction is seldom perfect. The theory assumes a diffuse sound field, in which energy is travelling with equal

Table 2.4 Absorption coefficients of surfaces in auditoria

	Frequency (Hz)					
	125	250	500	1000	2000	4000
Seated audience*:						
Heavy upholstery	0.72	0.80	0.86	0.89	0.90	0.90
Medium upholstery	0.62	0.72	0.80	0.83	0.84	0.85
Light upholstery	0.51	0.64	0.75	0.80	0.82	0.83
Unoccupied seating*:						
Heavy upholstery	0.70	0.76	0.81	0.84	0.84	0.81
Medium upholstery	0.54	0.62	0.68	0.70	0.68	0.66
Light upholstery	0.36	0.47	0.57	0.62	0.62	0.60
Plaster or thick wood*	0.10	0.06	0.05	0.05	0.05	0.04
Plaster on concrete block*	0.06	0.05	0.05	0.04	0.04	0.04
Concrete	0.02	0.02	0.03	0.04	0.05	0.05
Thin wood (approximate)	0.30	0.19	0.09	0.06	0.06	0.06
Curtain (velour, draped)	0.06	0.31	0.44	0.80	0.75	0.65
Absorbing power of orchestra (m²), 92 players on a 170 m² stage with vertical walls*	22	37	44	64	102	132
Air absorption coefficient, 4m (m⁻¹)	.000	.001	.003	.004	.009	.027

* After Beranek (1969) and Beranek and Hidaka (1998)

probability in all directions. Where this is not fulfilled, reverberation times different from those predicted may result. Acoustic scale models or good computer models can help monitor the accuracy of prediction. If one is to err on one side as a precautionary measure, it should be towards a longer reverberation time for music use and a shorter one for speech. Extra absorbent can if necessary reduce an excessively live space, while for speech a dead space will have better intelligibility.

2.8.6 Simplified reverberation time prediction

While full calculation is necessary for serious predictions, a simplified method was proposed by Kosten (1966), based on published data in concert halls. It is a tenet of good auditorium design that as little additional mid-frequency absorption should be introduced beyond the unavoidable absorption associated with performers and audience. Kosten found that total acoustic absorption was closely related to audience area. He introduced an **equivalent absorption coefficient** (α_{eq}), which is used to calculate the total absorption by multiplying it by the acoustic seating area, S_T; S_T equals the acoustic seating area (section 2.8.4) plus the orchestral stage area. The equivalent absorption coefficient is thus the absorption coefficient for audience plus an addition to take account of incidental absorption by the remaining hard surfaces in the room. Kosten's simplified equation for the occupied reverberation time (RT) is:

$$\text{Reverberation time (seconds)} = \frac{0.16V}{S_T\alpha_{eq}}$$

The air absorption component has been omitted here. The Kosten approach allows rapid calculation valuable for preliminary design. Kosten's original value for α_{eq} at 500/1000 Hz of 1.07 for occupied seating appears to be too low. For the data used in Figure 2.33(b), the calculated mean value of α_{eq} is slightly higher at 1.14, with a standard deviation of 0.05 (note S_T includes the stage area). This value agrees with the one effectively proposed by Beranek (2004, p. 633).

Simplification of the formula can be taken yet one step further for the case of an auditorium without overhanging balconies. If the floor is fully occupied by audience, access aisles and performers, then S_A equals the floor area. Placing this in the equation above with Kosten's α_{eq} value leads directly to a specification for the mean height (h) of the auditorium, namely $h = 7.1 \times RT$ (m). Thus in a space for speech with a 1 second reverberation time, the average height should be about 7.1 m. Acoustic design does not come much simpler!

2.9 Sound level distribution

Loudness is an obvious acoustic concern, yet for many years it was not felt to be significant in the auditorium context. For speech, the requirement is that it should be loud enough relative to background noise. In the case of symphonic music, many listeners respond enthusiastically to louder sound; it appears to render the experience more intimate (Barron, 1988). The subjective sense of **loudness** is determined by the objective **sound level**, measured in decibels. Traditional theory of reverberation provides a formula for sound level distribution.

Obviously the sound level in a room depends on the sound energy being generated. But the acoustic characteristics of the room also influence the resulting sound level. Both the distance from the source and the total acoustic absorption (A) affect the result. The total sound received is divided into two components: the direct sound and the reflected component. (The total level is derived by decibel addition; where the direct and reflected level are the same the total level is 3 db greater, for positions close to the source the direct sound dominates, for distant positions the reflected component dominates.) The direct sound decreases 6 db for every doubling of distance, whereas the reflected component has traditionally been assumed to be constant throughout the space (Figure 2.36). The reflected component is a function of total absorption; an auditorium with a larger audience will have

a quieter sound. The distance at which the direct and reflected components are the same is known as the **reverberation radius**. It is around 5 m in a full-size concert hall. Thus nearly all listeners receive more energy which is reflected rather than direct.

This theoretical behaviour for the total sound level makes various assumptions of uniformity of sound travel etc. In reality individual design details cause local variations. Perhaps the most obvious of these is that sound levels decrease under balcony overhangs. Several examples can be found in Chapter 5. A more universal deviation from the traditional theory has been discovered during the process of making measurements in the halls reported in detail here. It became clear that the assumption of constant reflected sound level throughout the space was not valid in auditoria. Typical average measured behaviour is shown in Figure 2.36, with a decrease with distance from the source (Barron and Lee, 1988). The revised theory predicts sound levels at the rear of large concert halls 3 db less than traditional theory. Since halving the size of the orchestra produces only a 3 db reduction, this discrepancy is surely significant.

The theoretical expression for the revised reflected sound level in Figure 2.36 is:

$$\text{Reflected sound pressure level} = SWL + 10.\log\frac{4}{A} - \frac{0.174r}{T}$$

SWL is the sound power of the source (in watts) expressed in decibels, A is again the total acoustic absorption, r is the source–receiver distance and T is the reverberation time (see Appendix B.3). The traditional expression omits the last term. As already mentioned the dominant quantity determining the sound level is the total acoustic absorption. A situation for which the revised theory might be significant runs as follows: in a drama theatre, which has a lot of extra absorbing material introduced to reduce the reverberation time, the sound levels at the furthest seats might be too low relative to the background noise level.

The revised theory as shown in Figure 2.36 does not by itself help to design better auditoria. Indeed there is some evidence that listeners compensate

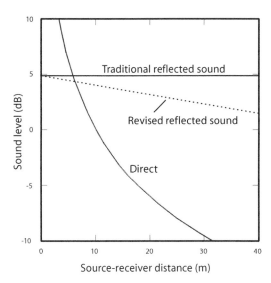

Figure 2.36 Sound level components as a function of source–receiver distance in a hall. Typical measured values of the reflected sound level lie closer to the revised line than the constant value according to traditional theory

for distance when they assess loudness (section 3.2). But the revised values provide a yardstick against which to compare measured results. Some design features cause unusually quiet or loud sound; the theory provides a value for what is usual. For example, insufficient early reflections in large halls lead to a quiet sound, which may be perceived as a lack of intimacy. These aspects are discussed further in section 3.10.5.

References

Barron, M. (1988) Subjective study of British symphony concert halls. *Acustica*, **66**, 1–14.

Barron, M. (1995) Bass sound in concert auditoria. *Journal of the Acoustical Society of America*, **97**, 1088–1098.

Barron, M. and Coleman, S. (2001) Measurements of the absorption by auditorium seating – a model study. *Journal of Sound and Vibration*, **239**, 573–587.

Barron, M. and Lee, L.-J. (1988) Energy relations in concert auditoria, I. *Journal of the Acoustical Society of America*, **84**, 618–628.

Beranek, L.L. (1969) Audience and chair absorption in large halls, II. *Journal of the Acoustical Society of America*, **45**, 13–19.

Beranek, L.L. (2004) *Concert and opera houses: Music, acoustics and architecture*, 2nd edn, Springer, New York.

Beranek, L.L. and Hidaka, T. (1998) Sound absorption in concert halls by seats, occupied and unoccupied, and by the hall's interior surfaces. *Journal of the Acoustical Society of America*, **104**, 3169–3177.

Blauert, J. (1983) *Spatial hearing*, MIT Press, Cambridge, MA.

Bradley, J.S. (1991) Some further investigations of the seat dip effect. *Journal of the Acoustical Society of America*, **90**, 324–333.

Bradley, J.S. (1992) Predicting theater chair absorption from reverberation chamber measurements. *Journal of the Acoustical Society of America*, **91**, 1514–1524.

Cox, T.J. and d'Antonio, P. (2004) *Acoustic absorbers and diffusers: theory, design and application*, Spon Press, London and New York.

Cremer, L. and Müller, H.A. (translated by T.J. Schultz) (1982) *Principles and applications of room acoustics*, Vol 1, Applied Science, London.

D'Antonio, P. and Cox, T.J. (2000) Diffuser application in rooms. *Applied Acoustics*, **60**, 113–142.

Davies, W.J. and Cox, T.J. (2000) Reducing seat dip attenuation. *Journal of the Acoustical Society of America*, **108,** 2211–2218.

Dietsch, L. and Kraak, W. (1986) Ein objectives Kriterium zur Erfassung von Echostörungen bei Musik- und Sprachdarbietungen. *Acustica*, **60**, 205–216.

Egan, M.D. (1988) *Architectural acoustics*, McGraw-Hill, New York.

Ham, R. (1987) *Theatres*, Architectural Press, London.

Hargreaves, T.J., Cox, T.J., Lam, Y.W. and D'Antonio, P. (2000) Surface diffusion coefficients for room acoustics: free-field measures. *Journal of the Acoustical Society of America*, **108**, 1710–1720.

Kleiner, M. (1980) On the audience induced background noise level in auditoria. *Acustica*, **46**, 82–88.

Kosten, C.W. (1966) New method for the calculation of the reverberation time of halls for public assembly. *Acustica*, **16**, 325–330.

Meyer, J. (1978) *Acoustics and the performance of music*, Verlag das Musikinstrument, Frankfurt am Main.

Newton, J.P. and James, A.W. (1992) Audience noise – how low can you get? *Proceedings of the Institute of Acoustics*, **14**, Part 2, 65–72.

Olson, H.F. (1967) *Music, physics and engineering*, 2nd edn, Dover, New York.

Orlowski, R. (2000) New sound diffusers in practice. *Acoustics Bulletin*, **25**, No. 5, 21–22.

Reichardt, W., Budach, P. and Winkler, H. (1968) Raumakustische Modelluntersuchungen mit dem Impuls-Schalltest beim Neubau des Kongress- und Konzertsaales im 'Haus des Lehrers' am Alexanderplatz, Berlin. *Acustica*, **20**, 149–158.

Sabine, W.C. (1922) *Collected papers on acoustics*, Harvard University Press (reprinted 1964, Dover, New York).

Schroeder, M.R. (1979) Binaural dissimilarity and optimum ceilings for concert halls: more lateral diffusion. *Journal of the Acoustical Society of America*, **65**, 958–963.

Schultz, T.J. and Watters, B.G. (1964) Propagation of sound across audience seating. *Journal of the Acoustical Society of America*, **36**, 885–896.

Tobias, J.V. (ed) (1972) *Foundations of modern auditory theory*, Vol II. Academic Press, New York.

Vorländer, M. and Mommertz, E. (2000) Definition and measurement of random-incidence scattering coefficients. *Applied Acoustics*, **60**, 187–199.

3 Acoustics for the symphony concert hall

3.1 Introduction

To achieve good acoustics in large concert halls is a major challenge for both acousticians and architects. In spite of many advances, areas of uncertainty still remain. Whenever a new hall is completed, there is anxiety before music is played for the first time. The acoustic reputation of an auditorium can be volatile, especially when a hall opens. A concerted campaign by a hostile critic can create enormous problems, preventing a rational approach to resolving acoustic shortcomings. In practice, audiences are generally not a group with sufficient cohesion to undermine the popularity of mediocre halls where good concert programmes are presented. Performing musicians on the other hand can and do make their views known, and it seems that above all it behoves a designer to ensure that the players will feel that their performing efforts are justified. One should add here that the acoustic requirements for the musicians and audience are different.

Music and speech share several characteristics. They both consist of bursts of sound separated by quiet; they both occupy similar frequency ranges. These similarities mean that the physical behaviour of sound in a room is virtually the same for both. The differences occur in the way our ears interpret the physical sound waves. In the case of music, tonality and harmony are fundamental aspects (both investigated in depth by Helmholtz in *Sensations of tone* of 1863). With speech, tone is only used for recognition purposes and phrasing. But tonality has little influence on design since frequency is unaffected by reflection etc. It is the manner in which our ears interpret the complex temporal pattern of reflections which should concentrate our attention. In the case of speech, intelligibility is paramount and this is known to be associated with the proportion of energy which arrives early, both in the direct sound and early reflections. The corresponding quality for music is clarity or definition, which can be related to a similar energy proportion. But many other qualities are also important for concert hall listening.

It has taken several years to evolve a methodology to resolve the concert hall problem. Much of this progress derives from techniques developed by experimental psychologists. Beranek (1962) was one of the first to list a series of 18 important subjective attributes: such qualities as warmth, liveness, intimacy, brilliance and ensemble. But how are these to be rated to provide an overall acoustic quality judgement? Beranek derived a linear additive system, which is essentially one-dimensional but thus too limiting. Figure 3.1(a) shows a diagrammatic representation of three conditions, in which C is twice as different from A as is B. But if we need to represent A, B and C which are equally different from each other but in different respects, then a two-dimensional diagram is needed, Figure 3.1(b). The two-dimensional data (b) can be forced into one dimension (a), but with some loss of information.

A fruitful analogy is with the flavour of wine, which is clearly multi-dimensional. The scales in Figure 3.1(b) have been labelled for sweetness and body. A display of this sort allows individual preferences to be accommodated, in a way which

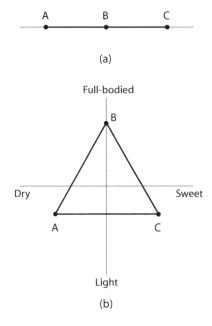

A   B   C

(a)

Full-bodied

B

Dry   Sweet

A   C

Light

(b)

Figure 3.1 Three perceived conditions displayed diagrammatically. (a) Along a single dimension; (b) in two dimensions

the ears then interpret in a highly selective fashion. The intermediate step that one needs to introduce is the **objective measure**. The first objective measure to be proposed was reverberation time and it still remains the most important. A large room volume causes a long reverberation time, which is interpreted as a high sense of reverberance.

There are several cases in which design criteria can be specified on the basis of simple subjective effects. An obvious example is concave domes causing strong echoes. But the slow development of concert hall acoustics bears witness to the difficulties of this strategy. In several cases it is not possible to go directly from design to subjective effect and a more scientific approach is required, which requires objective measures. A group of new objective measures going beyond reverberation time are now considered to be significant and these have informed many of the comments made in following sections and chapters. The logic of the argument depends on this approach, though many of the conclusions can be treated in simple design terms. The non-technical reader may prefer just to glance over the following two sections and section 3.10 concerning subjective and objective descriptions and direct his/her attention to the central sections of the chapter. Those wishing for a more detailed analysis will find most aspects treated in some depth by Cremer and Müller (1982) and Kuttruff (2000).

3.2 The subjective dimension

The subjective requirement for speech is that it should be intelligible. The success of a design can be assessed quantitatively by conducting intelligibility tests using words embedded in simple phrases. Physically measurable quantities exist which predict with some accuracy what the result of such an intelligibility test would be. With music no such direct assessment is possible. One can of course say that 'the acoustics are excellent', but as soon as two acoustic conditions are compared the answers will refer to a whole host of different characteristics. As already suggested, the appreciation of music acoustics is multi-dimensional.

is not possible in a single dimension. On a single dimension there is only one judgement criterion for all. Similarly with concert hall acoustics, experiments have demonstrated that listeners are probably hearing the same subjective effects, but that they place different individual weightings on these effects for their judgement of preference. There is thus no ideal concert hall, just as there is no ideal wine for all.

In the search for design strategies for good acoustics, it is convenient to be able to relate a physical characteristic of the hall to a subjective quality. For instance, large room volume produces a reverberant sound. A traditional association in the 1930s was that 'wood is good' and that wood was responsible for producing 'singing tone', but this has now been fully discounted. The connection between hall design and subjective impression does however contain two separate links: hall design determines the acoustic situation at the listeners' ears, which

In concert hall acoustics there are at least five independent dimensions. This was first established by Hawkes and Douglas (1971) and in the last three decades the nature of these different dimensions has been refined. The major concerns are that the **clarity** should be adequate to enable musical detail to be appreciated, that the **reverberant response** of the room should be suitable, that the sound should provide the listener with an **impression of space**, that the listener should sense the acoustic experience as **intimate** and that he/she should judge it as having adequate **loudness**. This list is by no means complete or definitive. It omits any reference to tone colour or timbre, which is certainly also important. Yet the five qualities: clarity, reverberance, spatial impression, intimacy and loudness provide a useful starting point for discussion.

These principal subjective dimensions for concert hall listening have been derived by questionnaire techniques, the history of which will be summarized below. To illustrate the nature of subjective experience, an alternative starting point is to look at the subjective effects associated with a single reflection. This might be considered as the first step in a description based on the full impulse response which contains many reflections. A single reflection is characterized by sound level (loudness), delay relative to the direct sound and direction. For a reflection arriving from the side, the subjective effects can be plotted on a single figure, Figure 3.2 (Barron, 1971). The direct sound is used as a reference for both level and delay. Very quiet reflections are inaudible, and therefore fall below threshold. The subjective response to loud reflections depends on the delay. Late, loud reflections are perceived as disturbing echoes. Early loud reflections cause what is called 'image shift': the sound appears to be coming from a direction intermediate between the direct and reflected sound directions. A lesser image shift also occurs for short delays less than about 7 ms.

For delays around 20 ms, the tone of the music appears to sharpen and become more harsh. **Tone colouration** is a sharpening of the timbre; it imparts a shrill, slightly metallic character to sound. String

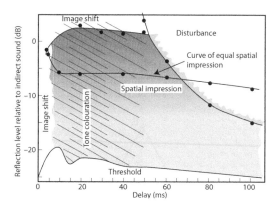

Figure 3.2 The subjective effects with music of a single side reflection as a function of reflection level and delay relative to the direct sound (azimuth angle 40°)

tone is particularly affected. In halls it is associated with reflections from large plane surfaces, whereas the effect is unlikely to be audible with reflections from scattering surfaces. The subjective effect is stronger for overhead reflections than for reflections from the side. Tone colouration constitutes one of the major reasons for avoiding suspended horizontal reflecting surfaces in concert halls.

But the principal subjective effect of an early reflection of typical level arriving from the side is what is called **spatial impression** in Figure 3.2. With a lateral reflection, one has the sensation that the source broadens, that one is involved in a three-dimensional sound experience. This effect was first suggested as significant by Marshall (1967) and is now considered to be a property of the best concert halls, including the classical rectangular designs. The physiological process involved in spatial impression is of interest. The hearing system responds favourably to dissimilarity at the two ears and appears to perform a cross-correlation type analysis (Blauert, 1983).

The results in Figure 3.2 were derived from a simulation experiment. The acoustic sound field in a room is very complex, yet our ears are highly selective in the way they interpret the acoustic pressure on the ear drums. It was realized around 1950 that, in order to discover what makes good concert hall acoustics, an understanding was required of the hearing system. This led to the first major progress since Sabine's work on reverberation time. Much of this work used simulation systems in which the acoustic sound field of a room was recreated in the laboratory. The standard technique is to use an anechoic (i.e. no echo) chamber in which the walls are covered with highly absorbent material (Figure 3.3). An array of loudspeakers is fed with signals processed from a music recording, itself preferably without any room reflections. In the simulation, a loudspeaker directly in front of the listener is used for the direct sound, reflections are created by introducing time delays for signals fed through additional loudspeakers; reverberation can be simulated in various ways, such as from a reverberation plate or electronic reverberator.

A seminal result of early simulation experiments was reported by Haas (1951) for speech. It was found that the ear uses the first arriving sound to locate the source, but that the early reflections contribute to intelligibility. Indeed the directional illusion is maintained even when the reflection is louder than the direct sound. This subjective work is the basis of electronic sound reinforcement for speech. The results for concert halls of subjective experiments into the determinants of clarity, reverberance, spatial impression etc. will be discussed below under objective measures.

Though simulation experiments have been crucial to our understanding of subjective response, they are not able to treat the whole problem. The simulation is not perfect. Beranek (1962) initiated a survey into 54 of the world's concert halls and opera houses. Views on the acoustic merits of these halls were solicited from conductors, performers and music critics. These formed the basis of a subjective rating for each hall. Weightings of various objective acoustic characteristics were then derived to establish the determinants of good concert acoustics. The major result was that while reverberation time was important, the highest weighting (40 per cent of the total) was associated with the initial-time-delay-gap. This time-delay-gap is simply the delay of the first reflection in the main body of the Stalls (Figure 3.4). The best-liked concert halls, which included the classical rectangular halls, were found to have short initial-time-delay-gaps of 21 ms or

Figure 3.3 Subjective experiment being conducted in an anechoic chamber with absorbent foam wedges on all room surfaces

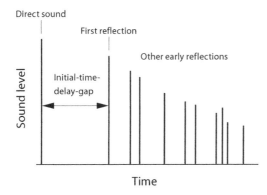

Figure 3.4 Impulse response (as in Figure 2.11(b)) illustrating the initial-time-delay-gap

less. The delay-gap was considered to be a measure of perceived acoustic intimacy. Other aspects considered to be important were warmth (provided by a long bass reverberation time), loudness of direct and reverberant sound, balance and blend, diffusion and ensemble. The results of Beranek's survey were used for the design of Philharmonic Hall, New York (section 4.8).

Philharmonic Hall only survived for 14 years, while some of the techniques and conclusions in Beranek's book have been criticized. It can be questioned whether initial-time-delay-gap really is in fact significant. Subjective/objective surveys conducted since do not confirm its importance. A further concern is whether the criterion can be applied within an individual concert hall. The delay of the first reflection in rooms is a function of, among many things, the distance from the source. Simple geometry shows that the delay will be long for seat positions close to the source and short for distant seats (Figure 2.14). According to Beranek, this should mean perceived intimacy is greatest at remote seats, whereas subjective surveys show the opposite to be the case (Hawkes and Douglas, 1971; Barron, 1988). Not surprisingly within a hall, intimacy is related to proximity.

In 1967 Marshall proposed that early reflection direction might also be significant with lateral reflections being preferred. Our ears are, after all,

located on the sides of our head rather than the top and bottom. Marshall's suggestions also offered an alternative interpretation for the subjective preference for classical halls; perhaps excellence is associated with strong early reflections arriving from the side? Subsequent experiments have confirmed Marshall's basic thesis. In objective terms, the proportion of early sound arriving from the side as well as sound level appear to be decisive (Barron, 1971; Barron and Marshall, 1981). The higher the proportion of lateral sound and the louder the sound, the greater is the perceived subjective **source broadening**. The effect can actually be assessed in experiments by asking listeners to judge the apparent source width (ASW). Marshall (1968) also suggested that the sensation of spatial impression contributed to the sense of acoustical warmth, which is normally associated with bass sound. Experiments suggest that the low frequencies are particularly important for the subjective spatial effect. Owing to the attenuation effect at grazing incidence over seating (section 2.6.1), this implies the need for lateral reflections on paths remote from audience seating.

The idea that early reflections from the side might be important managed to diverted attention from the long-held assumption that the later reverberant sound contributed a spatial effect for listeners. Morimoto and Maekawa (1989) were the first to provide evidence that there are two spatial effects occurring: that the early reflections provide a sense of the source becoming broader while the later sound gives a sense of being enveloped by the sound. Bradley and Soulodre (1995a) have confirmed this division; they propose the label **listener envelopment** or LEV for the spatial effect of the later reverberant sound. Spatial impression thus includes two spatial effects, as illustrated in Figure 3.5.

Returning to the total concert hall experience, the major conceptual criticism of Beranek's analysis is that subjective judgements are not linearly additive. If they were, a disturbing echo might be compensated by improved intimacy, for instance. The multi-dimensional nature of concert experience is now generally accepted. In particular, it allows for the observed differences in preference among

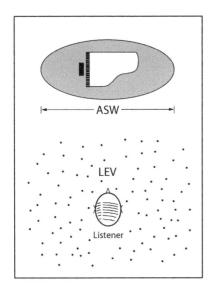

Figure 3.5 Diagrammatic representation comparing the spatial effects of source broadening (apparent source width – ASW) and listener envelopment (LEV) (after Morimoto and Maekawa, 1989)

Figure 3.6 View of a 'dummy head' with reproduced outer ears and microphones in the ear canals (manufactured by Neumann)

listeners. For instance, musicians tend to place greater emphasis on the ability to hear musical detail (i.e. clarity) than others.

The major experimental aid for subjective surveys has been the 'dummy head' (Figure 3.6). By arranging microphones in the canals of model dummy ears, recordings can be made which reproduce with great accuracy the sensation of listening at the point where the dummy head is placed. By replaying recordings made in actual concert halls, the acoustics of halls can be studied in the laboratory free of non-acoustic influences. Two major German studies were made at Göttingen and Berlin. The detailed results of each need not be elaborated here; they are well covered in Cremer and Müller (1982), discussed by Schroeder, Gottlob and Siebrasse (1974) and Plenge *et al.* (1975) and reviewed in Barron (1988). The results led to the consensus view expounded here.

No further dummy head surveys have been made since 1980. For subjective data, some authors have relied on memory by musicians and critics (Beranek, 1962; Haan and Fricke, 1997). More

recently some pioneering studies have been made with questionnaire surveys of concert audiences (Sémidor, 1997; Cox and Shield, 1999). This seemingly obvious approach has its difficulties such as absence of a shared vocabulary and widely differing ranges of listening experience which can be biased in some cases to the hall in question.

The approach used by this author was to have questionnaires completed by experienced listeners attending concerts (mainly acoustic consultants). This might appear to be a biased group though it was clear that they belonged to two groups: those that preferred intimacy and those that preferred reverberance. The implication is that for the best acoustics, each group should be satisfied.

The questionnaire used is shown in Figure 3.7. **Clarity** is described as the ability to hear musical detail, **reverberance** as the degree of perceived reverberation in a temporal sense and **envelopment** as the spatial aspect of the perceived sound. **Intimacy** refers to one's degree of identification with the performance, whether one felt acoustically involved or detached from it, while **loudness**

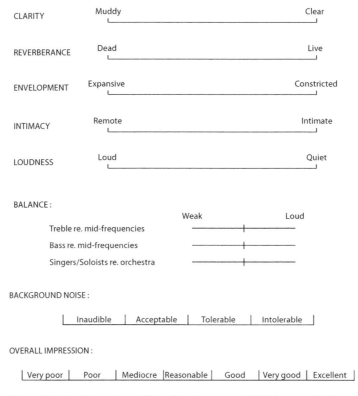

Figure 3.7 Questionnaire used for subjective survey of British concert halls

is required to be assessed relative to the orchestral forces involved. **Overall Impression** was reassuringly found to be related to responses on the other scales.

Eleven concert halls from before 1990, which are discussed in Chapter 5, formed the basis of an initial study (Barron, 1988). One intriguing result emerging from that study was that the judgement of loudness is virtually independent of position in a typical hall. This was observed in spite of the fact that the sound level decreases significantly (section 2.9 and Figure 5.6 etc.) towards the rear of a hall. Work by experimental psychologists independent of auditorium research, such as Zahorik and Wightman (2001), provides further evidence for loudness constancy within rooms. One implication of loudness constancy (Barron, 2007) would be that the criterion for sound level in concert halls needs to be

dependent on distance from the stage, rather than simply greater than 0 db (as proposed in section 3.10).

The subjective survey has also been extended to more recent British concert halls as well as British opera houses, theatres and multi-purpose spaces. In each case the results have informed comments on subjective characteristics to be found in the appropriate chapters below.

As mentioned above, recent work shows that two spatial effects occur in concert conditions: source broadening and listener envelopment. The latest version of the questionnaire replaces the single 'envelopment' scale in Figure 3.7 by two scales as follows: the first labelled 'source breadth' with extremes of 'broad' and 'narrow' and the second 'listener envelopment' with extremes of 'expansive' and 'constricted'. Distinguishing between these two effects is obviously more demanding, though it is likely that the majority of experienced listeners can separate the two. No large study has yet been made using this revised questionnaire.

3.3 Objective descriptions

Design details in auditoria ultimately determine subjective response. In some cases we can link cause and effect directly, such as the effect of the scattering wall treatment in nineteenth-century rectangular halls in creating a subjective sense of hearing sound from all directions. The need for an objective acoustic description arises because the link is generally not simple. Even in the case of reverberation time we find that it depends on two independent factors: room volume and total surface absorption. Before Sabine's work there were many comments about drapery etc. reducing reverberation, but only by scientific experiment could the full relationship be established. Objective measures offer an intermediate description between design and subjective effect. The design creates a sound field at the listener's position according to the behaviour of sound in rooms. This can be described in objective acoustic terms. Our interest is in the objective attributes significant for our ears. With the help of subjectively relevant objective measures, we can hope to establish how design determines the sound field and how the ear will then interpret it.

The concept of reverberation time dates from the year 1900 (Sabine, 1922). Reverberation time has been readily measured since the 1950s using a loudspeaker, microphone and level recorder, which plots sound level in decibels against time. It is normally measured over a series of octave frequencies. Reverberation time is a convenient measure since it

is usually constant throughout the space and so can be quoted as a characteristic of the hall as a whole. The significance of actual values of reverberation time is also well understood (Table 2.3) and the reverberation time can normally be predicted with reasonable accuracy from architectural drawings by the Sabine equation (section 2.8.2).

There is though an important criticism of reverberation time for music use: that subjectively we generally respond to 'running reverberation' which applies during continuous speech or music, whereas the reverberation time itself relates to 'terminal reverberation', the decay we hear when a sound is turned off (see section 2.8.1). The early decay time described below deals with running reverberation.

The search for objective measures in addition to reverberation time has been a long one. These new measures lack many of the convenient aspects of reverberation time. They are all dependent on position in the auditorium and a mean value is often not of great use. The significance of results of the various new measures is only now becoming clear and prediction from drawings is at best very approximate.

However with more sophisticated prediction techniques, such as acoustic scale modelling or computer modelling (section 3.9), the measures can be forecast in advance. Commercial testing systems for the newer measures are now also available. The new measures are included in the standard ISO3382 from 1997, which has and will do much to further their acceptance. The number of new measures proposed has been large but, for simplicity, only the most promising will be discussed here. These particular measures are the most broadly accepted among acousticians, but there are several alternative views. The major new measures are:

Equation 1 Objective clarity

$$Objective\ clarity = 10.\log \frac{\text{Energy arriving within 80 ms of direct sound}}{\text{Energy arriving later than 80 ms after direct sound}} \ dB$$

Equation 2 Objective source broadening

$$Objective\ source\ broadening = \frac{\text{Energy arriving laterally within 80 ms of direct sound}}{\text{Total energy arriving within 80 ms of direct sound}}$$

1   **Early decay time** (**EDT**) is a measure of the rate of sound decay, expressed in the same way as a reverberation time, based on the first 10 db portion of the decay. (Reverberation time, RT, is based on 30 db of decay.) In a highly diffuse space where the decay is completely linear, the two quantities, RT and EDT, would be identical. The early decay time has been shown to be better related to the subjective judgement of reverberation, also called 'reverberance', than the traditional reverberation time (Atal, Schroeder and Sessler, 1965).

2. **Objective clarity** or **early-to-late sound index** (**C$_{80}$**) (Equation 1) relates to the balance between perceived clarity and reverberance, which can be particularly delicate for music listening. The energy within 80 ms of the direct sound includes that of the direct sound and early reflections. This measure has a direct equivalent in speech work.

3. **Objective source broadening** or **early lateral energy fraction** (Equation 2) relates to perceived source broadening as described above. The fraction is sometimes labelled as LF. The spatial effect is also influenced by the sound level of the music. Objective measures for listener envelopment are discussed in section 3.10.3.

4   **Total sound level** in db for a standard sound source (also known as 'strength', *G*) is related to the judgement of loudness. The same orchestra will produce different sound levels in different halls and there is evidence that listeners are quite sensitive to small differences.

The objective measures are mainly considered at middle frequencies (in Chapter 5 averaged over the octaves 500–2000 Hz). In addition to these quantities are questions of tonal balance. A hall is described as possessing acoustic **warmth** if there is adequate sense of the bass sound; as an objective measure the relative level of bass compared with mid-frequency sound may be most significant here (section 5.17). A sense of acoustic brilliance is correspondingly associated with strong high-frequency sound.

The relationship between these newer measures and subjective response is in each case clear, with the exception of subjective intimacy. In the author's survey (Barron, 1988), intimacy was found to be best related to sound level, rather than the initial-time-delay-gap (discussed in the previous section). The sense of loudness is also predictably related to level and there is an overlap between judgements of intimacy and loudness. Only in the case of loudness though do listeners appear to compensate for distance from the source, as mentioned in the last section.

Of the four new measures listed above, numbers 1, 2 and 4 can all be derived from the response of the hall to an impulsive sound (section 2.4). Each of these measures is based on energy, which is derived from the (squared) impulse response (such as Figure 3.8) by a process of what is known as 'integration'. Conceptually this involves nothing more than measuring the black area in Figure 3.8. To measure the early decay time (EDT), integration of the impulse response has to be made in reverse time, starting at late time. Figure 3.9 illustrates a typical response; the EDT is derived by measuring the slope of the curve from zero to –10 dB. The same technique can in principle be used for measuring the reverberation time with the advantage that the integrated impulse response, as in Figure 3.9, contains no superimposed random fluctuations.

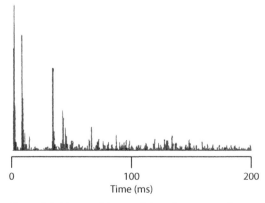

Figure 3.8 Measured (squared) impulse response in a concert hall

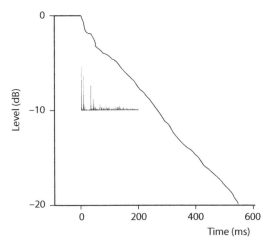

Figure 3.9 Integrated impulse response for the same position as that for Figure 3.8. Integration is made in reverse time. Superimposed is the impulse response to the same time scale; note in the early decay that there are 'steps' when strong reflections arrive

The total sound level comes from the total area of the impulse response. The total sound can also be measured by feeding a calibrated level noise signal to a loudspeaker and measuring the sound level distribution in the audience seating areas. Measurement of total sound level enables the sound level produced by the same orchestra to be predicted both within a hall and between different halls. In quantitative terms, the reference quantity for sound level is most conveniently chosen as the direct sound level at 10 m from the source. Thus:

Total sound level for a standard source (dB) =

(total sound level at the measurement position)
– (sound level of direct sound at 10 m from source)

In concert halls it is considered that sound should be judged loud enough if the measured level is greater than this direct sound level at 10 m.

3.4 Size, volume and form

W.C. Sabine is rightly considered as the father of architectural acoustics. Reading his papers (1922) one is struck by the breadth of his approach to auditorium design, but time tends to simplify. It was probably inevitable that his successors should limit their attention to his most concrete achievement: the concept of reverberation time, which for many years remained the only measurable quantity. Designing for an appropriate reverberation time produces a specification for the auditorium volume only; the form of the auditorium is barely relevant. It is the exception rather than the rule in which the form of an auditorium is sufficiently extreme for the prediction formula to be inaccurate. Many designers took this apparent *carte blanche* to sanction whatever auditorium shapes were proposed for non-acoustic reasons. The results were often very disappointing, though slow progress was gradually made. In the immediate post-war period, the most active firm in auditorium consultancy was Bolt, Beranek & Newman in the USA. With growing skill at remedying acoustic faults, Beranek summed up his firm's view as follows: 'In the 1950s and 1960s, one of our hopes was that we would be able to take halls of different shapes and sizes and make them all reasonably good acoustically' (Games, 1985, p. 162). That hope was never realized. Though auditorium form is not of great consequence at small scale, in large halls with over 1000 seats, plan and section shape become progressively more important as the seat count rises.

The implications of the Sabine formula for reverberation time are clear-cut in the case of concert halls. The optimum mid-frequency reverberation time for symphony concert halls is generally considered to be 1.8–2.2 seconds (though longer values in directed reflection designs may prove necessary, as outlined below). To achieve times as long as this requires a large room volume, generally achieved with a high ceiling, higher than needed for visual purposes. For instance, in a hall with no balcony and the floor area covered wholly by the players and audience, a ceiling height of 13 m is necessary. Audience seating is highly absorbent so that, in order to limit expensive volume, minimum additional absorbent material would normally be included in a concert hall. (Carpet on the aisles falls outside this dictum, since due to the effective absorption around the edge of seating it appears to have little influence on reverberation time.)

The recommended frequency characteristic for reverberation time is shown in Figure 3.10. At high frequencies there is an unavoidable decrease caused by air absorption, but again minimum additional surface absorbent should be installed in order to maintain brilliance. A rise in low-frequency reverberation time contributes to a sense of acoustic warmth. A value at 125 Hz up to 40 per cent above that at mid-frequencies is normally recommended, though it is somewhat a matter of taste. As the absorption of audience is less at low frequencies, a low-frequency rise will occur in a massively built hall. To maximize bass reverberation time, bounding walls and ceiling have to be made massive enough to exhibit minimal panel absorption.

Two maximum limiting values are conventionally quoted for concert hall size: one based on audience capacity, the other on linear dimensions. The difficulty of successful acoustic design increases considerably as seating numbers rise. An absolute maximum of 3000 seats is normally given for concert halls. In fact it is not the audience number so much as the area occupied by audience which is important; the area determines the total absorption that affects both reverberation time and sound level. The figure of 3000 refers to a typical seating density (around 0.5 m²/seat); if a more generous seating standard is used, only a smaller audience can be accommodated.

The second limiting value applied to concert halls concerns the distance of audience from the stage. A maximum of 40 m is usually specified for concert conditions. This limit is as much for visual as acoustic reasons.

According to the revised theory proposed below (section 3.10.5), these two limiting values lead to the sound level criterion given in the previous section, namely that the total sound level should exceed the direct sound level at 10 m. Using the Kosten method (section 2.8.2), a hall with a 2 second reverberation time with 3000 seats of average size (acoustic area per seat of 0.58 m²) requires a volume of 24 600 m³. Revised theory predicts for this hall a total sound level of 0.8 db at 40 m from the source, just above the criterion of 0 db mentioned above. It is gratifying that these various criteria are compatible.

With the reverberation criterion satisfied, one finds that satisfying the other requirements such as appropriate clarity and intimacy becomes frustrated at large scale. This is called here the **large concert hall problem** which has dominated design since the 1950s. The large concert hall problem is to provide sufficient early reflections for all seat locations in spite of large room volumes, which tend to remove useful reflecting surfaces away from seating areas. Solutions to the problem can be found by design of the seating layout and design of reflecting surfaces around seating areas.

One reason for the conflict between early reflections and gross size is illustrated in the impulse responses in Figure 3.11, which show the effect of simply scaling up an auditorium form: reflection delays become longer, so that fewer reflections arrive sufficiently early to contribute towards definition or clarity (80 ms is normally used as the limit). In addition, owing to more acoustic absorption in the space, levels of all sound components become lower, leading to a quieter sound overall (this second effect is not shown in Figure 3.11).

Concert hall seat capacity has become a major issue in the last twenty years. Some consultants lobby hard to keep numbers down, to ensure optimum acoustic conditions. Others have accepted the challenge of large capacities and achieved

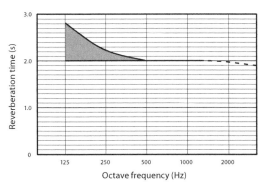

Figure 3.10 Recommended reverberation time characteristic for large concert halls, based on a mid-frequency value of 2 seconds. The preferred degree of bass rise is a matter of taste

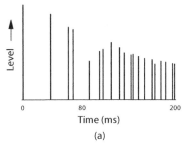

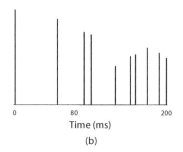

Figure 3.11 Notional impulse responses: (a) in a small hall and (b) in the same form scaled up by 50% linearly. The time limit for early reflections for music is generally taken as 80 ms

impressive acoustics (see for example section 10.6). There are viable solutions to the large concert hall problem, as is clear from the later parts of both Chapters 4 and 5.

3.5 Early reflection design

Early reflections in concert halls contribute to the sense of clarity, intimacy and loudness and, if they arrive from the side, to the sense of source broadening. All these characteristics are required for the best acoustics. For an acoustic reflection to occur is principally a matter of simple geometry. There will always exist one, but only one, path for reflection from an infinite surface. For reflection from a finite surface, the surface has to be tangential to an ellipse, which has the source and receiver positions at its foci (Figure 3.12). Since an ellipse has the property that the sum of the distances from the foci to any point on the ellipse is constant (this allows an ellipse to be drawn with two pins and a piece of string), a particular ellipse defines a certain reflection delay. In three dimensions, one is dealing with a rotated ellipse, known as an ellipsoid. Reflector size influences the range of frequencies over which the reflector is efficient. As discussed in section 2.6.2, small reflectors are effective only at high frequencies.

To achieve a sense of source broadening, one or more strong reflections are required from the side. 'Lateral' reflections from above and behind are acceptable, so that reflections off side wall/ceiling

cornices and side balcony cornices can be of value. In detail, a reflection is lateral when there is a time difference between its arrival at the near and far ear: in other words when there is an inter-aural time delay. There should be a sufficient angle between the reflection path and the plane bisecting the head when the listener faces the sound source (the plane is known as the median plane). The larger the angle, the greater is the spatial effect. This has a bearing on the behaviour of simple plans (Figure 3.13). With the fan-shape plan the angle of the side wall reflection to the direct sound is small and source broadening tends to be small, whereas in the reverse-splay plan the angle is high and source broadening is correspondingly greater. Though the reverse splay presents unrealistic problems for stage design to be used as an overall plan, it can be used with great effect as the envelope for remote seating.

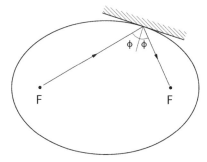

Figure 3.12 A surface has to be tangential to an ellipse for reflection to occur. The source and receiver are at the foci, F, of the ellipse

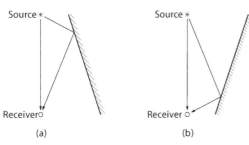

(a)   (b)

Figure 3.13 Angle of a lateral reflection in (a) a fan-shape plan and (b) a reverse splay plan (after Marshall, 1968)

Low frequencies should be present in lateral reflections for optimum spatial effect (section 3.2). Since low-frequency sound is attenuated as it passes over seating planes, this requires lateral reflections on paths remote from audience seating.

While the actual delay of the first reflection (the initial-time-delay-gap discussed in section 3.2) may not be of great importance, this is only within certain limits. A certain maximum value applies to allow for sufficient early energy within 80 ms; a delay gap in excess of 50 ms would be borderline for concert use. Application of this criterion leads to some interesting design plans. In a simple symmetrical plan form, the seats most distant from reflecting surfaces lie along the centre line of the hall. In Figure 3.14, a reflecting surface needs to be within or tangential to ellipse k to reflect sound from the source to position K. However for the remote seat N but with the same reflection delay, the relevant ellipse n is

much larger. From this diagram, the most effective location for a reflector able to serve all receivers is behind the source, such as reflector A in Figure 3.14. Such a reflector placed above the stage was used in the Salle Pleyel, Paris, of 1927, the London Royal Festival Hall of 1951 and others. Though it succeeds in maintaining loudness, the approach has two serious drawbacks which have deterred many subsequent consultants: an overhead reflector near the stage creates tone colouration (section 3.2) and a narrow source image. Reflections arriving more laterally are preferable in both respects.

The question of location for reflecting surfaces deserves consideration, Figure 3.15. Placing surfaces above the stage mentioned above was common around 1950, but has now fallen out of favour, at least as a means of providing additional reflections to distant seats. With the building in 1963 of the first terraced concert hall, the Berlin Philharmonie (section 4.9), the alternative of reflecting surfaces near to listeners was introduced. There are two advantages of surfaces near listeners: there is much greater control of reflection conditions throughout the audience seating areas and, with the situation in plan as shown in Figure 3.15(b), the side reflections arrive more laterally, creating greater source broadening.

For small halls a simple plan form can be used (Figure 3.16). The rectangular plan has a consistent reputation for good acoustics; the fan-shape on the other hand is not generally recommended. Figure 3.17 shows a disadvantage of a wide angle of fan

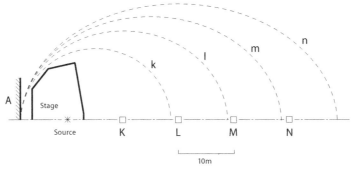

Figure 3.14 Ellipses superimposed on a half-plan. Ellipse k is relevant for a 50 ms reflection to receiver K, etc. Reflecting surfaces need to be within the relevant ellipse for the delay to be within the 50 ms limit

in that it provides image sources only in a concentrated region; for the listener this will produce only small source broadening. The narrow angle fan or parallel-sided hall generates reflections coming from lateral positions and this is thought to be one of its virtues. The relative merits of the rectangular and fan-shaped plan are further discussed in sections 4.3 and 4.5.

The horse-shoe plan, so popular in theatres, has little to recommend it for music; when enlarged

enough to achieve a satisfactory reverberation time, the reflection pattern becomes poor, and there are focusing problems associated with the concave rear wall. The elongated hexagon offers a compromise with the visual advantages of the fan shape and the acoustic advantages of the reverse splay. A further compromise scheme is the gross fan-shape plan but with stepped parallel walls (Figure 3.18). This offers an improvement relative to the simple fan shape but probably still carries some of its faults.

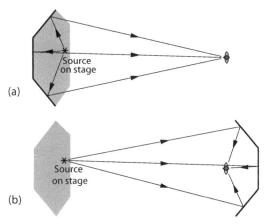

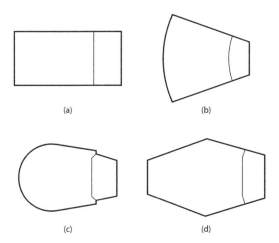

Figure 3.15 Comparison of the early reflection situation with (a) surfaces near the stage and (b) surfaces near the listener

Figure 3.16 Simple plan forms for concert halls

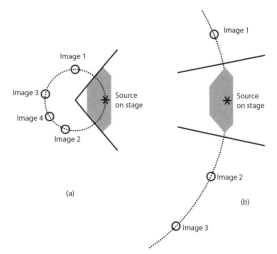

Figure 3.17 Image positions for two surfaces. For a wide angle fan in (a), the images are on a small circle; for near parallel surfaces (b), the image positions cover a large area

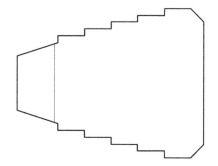

Figure 3.18 Stepped parallel walls with a gross fan-shape plan

Beyond a certain size of hall, some subdivision of the audience space becomes desirable. A design which can be rationalized on the basis of the expanding ellipse model is the large De Doelen Concert Hall of 1966 (2230 seats, section 4.7), Figure 3.19. The elongated hexagonal envelope provides reflections to remote seats, but these reflections arrive too late at seats in front of the stage. These seat locations are serviced by a subsidiary smaller hexagonal form surrounding the stage and front stalls seating.

The expanding ellipse model also leads directly to another stepped plan scheme, formalized by Cremer (1986) as a trapezium terraces room, Figure 3.20. At each seating level, there is a reverse splay surface to provide a lateral reflection. It is necessary to tilt balcony fronts downwards so that these reflections reach seating areas.

While lateral reflections are particularly desirable, some reflected energy from the ceiling is appropriate and often necessary. The ceiling is usually the largest room surface. The main concern is to avoid tone colouration effects. If the ceiling reflection is either first or prominent in the reflection sequence, it should preferably be diffused. Concert hall cross-sections with inclined ceiling sections to render their reflections lateral have been used in some recent concert halls, though it is debatable whether the net reflection pattern including cornice reflections from a horizontal ceiling and vertical wall is in fact inferior (Figure 3.21). A similar reflection path to that in Figure 3.21(b) can be provided by side walls and balcony soffits (Figure 4.5) offering additional early reflections to audience in the stalls.

Manipulating early reflections is not however without its risks, particularly if large plane surfaces

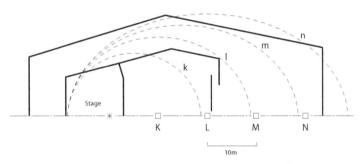

Figure 3.19 Half-plan of De Doelen Concert Hall, Rotterdam, with superimposed ellipses corresponding to a 50 ms delay reflection. Ellipse k is relevant for receiver K, etc

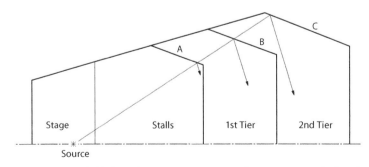

Figure 3.20 The trapezium terraces half-plan (after Cremer, 1986). To achieve geometrical side reflections, surfaces A, B and C have to be inclined slightly down from the vertical

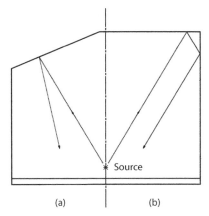

Figure 3.21 Two hall half cross-sections showing reflections off (a) an inclined ceiling and (b) a horizontal ceiling

(a)　(b)

are used. There is always a risk of tone colouration. Although the potential for colouration is greater for overhead reflections, it can still be perceived by the sensitive listener receiving a strong lateral reflection from a plane surface. Much more obvious subjectively is the case of false localization, where the sound of a particular instrument suddenly appears to come from the reflector instead of from the stage (a case of extreme image shift in Figure 3.2). This can occur, for instance, when trumpets playing in their upper register (where they are highly directional) point at such a reflector. A remedy for both these faults is to add scattering treatment to the reflector, though this increases the spread of the reflection at the expense of reflection level on the original path.

One design strategy for concert halls is to have surfaces oriented so that they direct reflections down onto the audience; these are often called directed reflection designs. One by-product of directing early reflections onto audience is that it also influences the late sound. Because the audience is highly absorbent, less energy is available in this case for the later reverberant sound. It has been noted that the subjective sense of reverberation is best related to the early decay time (EDT), and one finds in these directed designs that the EDT is often shorter than the reverberation time. If the reverberation time itself happens to be somewhat short, the sense of reverberation can then become

subjectively inadequate. The only solution appears to be to extend the reverberation time (probably by increasing the volume), in order to leave an EDT which is long enough, even if the reverberation time becomes longer than the recommended maximum of 2.2 seconds (see section 4.10).

Manipulating early reflections is a necessary part of the acoustic design of concert halls beyond a certain size. Seats which are furthest from useful reflecting surfaces generally require most attention. But there remains the question of how many reflections are necessary, to which there is no qualitative answer. This is the point at which acoustic design can become an art, though acoustic models provide a valuable guide. The revised theory approach offers the guideline of average behaviour, against which model measurements can be compared (section 3.10.5).

3.6 The diffusion question

The quest to provide suitable early reflections has dominated concert hall design since the 1960s, yet certain aspects of the later reverberant sound also deserve attention. The reverberation time, and more pertinently the early decay time, should take care of the temporal aspects of reverberant sound, providing adequate liveness and sufficient room response. But the directional distribution of the reverberant sound is also important. Our ears can respond to spatial effects generated by early reflections and those produced by later reverberant sound. The differences between the sensations involved are quite subtle (section 3.2). Early reflections from the side create spatial sensations related to the source giving a sense that the sound source is broadening. The degree of this spatial effect increases with sound level such that it is generally inaudible in *pianissimo* playing but marked in a *fortissimo* (Marshall and Barron, 2001). The spatial sensations associated with reverberant sound are detached from the source and are best considered by reference to highly reverberant spaces like cathedrals. In extreme reverberant conditions the source can no longer be localized; one feels surrounded by, if not suspended in sound. Both the spatial effects caused by early and late reflections, known as source broadening and listener envelopment, are significant for concert hall listening and there is evidence that individual listeners place more importance on one rather than the other when asked to judge the spatial effects (Barron, 1988).

For the reverberant sound, it is important to perceive some sound both from the side and behind. When sound energy is travelling with equal probability in all directions, it is referred to as a **diffuse sound field**. An auditorium at first sight looks an unlikely situation for a diffuse sound field since acoustic absorption is concentrated on the floor in the form of seating and audience. However the limited evidence we have suggests that most concert halls are reasonably diffuse. Profiled or scattering wall and ceiling surfaces, complicated geometries and balconies can all contribute towards making a sound field more diffuse. A diffuse sound field is likely to be judged as satisfactory in spatial terms. To promote diffusion in an auditorium it is recommended to have generous space above all sections of audience and to have the possibility of sound reflections arriving at listeners from most directions. The use of steep seating rakes, which is otherwise desirable for direct sound, can limit the amount of energy arriving from behind, because listeners effectively have an absorbent surface behind them.

In nineteenth-century classical rectangular halls, the room surfaces are strongly profiled providing scattered reflections and subjectively the sound in these halls is perceived as highly diffuse. These halls have a high acoustic reputation and an ongoing topic of discussion is the importance of scattering surfaces for the best concert hall acoustics. Recent research by Haan and Fricke (1997) provides some evidence that scattering surfaces are associated with good acoustics, though their subjective data was limited to views of conductors and performers.

It had long been assumed that scattering surfaces and the sense of diffuse sound were inextricably linked, that extensive surface decoration was required for the subjective spatial effect. In 1967,

laboratory subjective experiments by Damaske indicated that the directional sensitivity of the hearing system is limited (see section 4.7). Sound from only four distributed directions can produce the preferred subjective sense of complete diffusion. The case of St David's Hall, Cardiff (section 5.11) provides an illustration of this with a high perceived sense of diffusion but no surface decoration except at ceiling level above an acoustically transparent ceiling, where the details are highly scattering. In this case the subdivided seating surrounded by reflecting surfaces is probably also significant.

Scattering (diffusing) treatment of some boundary surfaces does however have its place for both early and later reflections. The advantages of scattering surfaces for early reflections for avoiding tone colouration and false localization have already been mentioned. Scattering treatment can also optimize directional distribution of late reflections. Yet indiscriminate application of scattering surfaces is probably not appropriate and many designs achieve subjectively diffuse sound without conscious use of acoustical scattering treatment. This question is one on which current measured quantities have little to offer.

3.7 Balcony design

Balconies are a common feature of large concert halls. They allow the vertical dimension to be exploited, which in concert halls tends to be large to satisfy the reverberation time requirement. Balconies allow more people to be accommodated within a certain distance from the performers. They can also offer acoustic advantages to listeners not underneath them; balcony soffits and balcony fronts represent additional hard surfaces that can provide useful reflections.

Seats under a balcony overhang are however disadvantaged both visually and acoustically. In both senses the audience under the overhang can feel cut off from the main volume of the auditorium, with a loss of intimacy and sense of detachment from the acoustics of the main space. Seat prices frequently reflect this. In some cities there are examples of auditoria where the decision has been taken to have no overhangs at all. But the penalty of this strategy can be that many listeners are so far from the performers that the intimacy of the whole performance space is seriously compromised. One of the major design problems for a large auditorium is to balance the advantages and disadvantages of balcony overhangs.

To understand the acoustic effect of an overhang, it is useful to consider the early and late sound separately. Contrary to what one might first expect, the early sound is often little influenced by the presence of an overhang. Side-wall reflections are usually unaffected and the presence of a nearby rear wall is likely to compensate for the absence of other early reflections which are obscured. The late sound on the other hand is significantly modified by an overhang. In an exposed seat away from balconies, reverberant sound arrives at a listener from most directions. Under an overhang however the vertical angle from which sound can arrive at a seat is greatly reduced (Figure 3.22).

In terms of the subjective qualities considered important for music listening, there appear to be three noticeable effects under overhangs that are all linked to reduced late sound arriving at listeners (Barron, 1995a). With reduced late sound level the balance between clarity and reverberance shifts to less sense of reverberation and greater clarity. Objective clarity (the early-to-late sound index) increases and early decay time (a measure of reverberance) decreases under overhangs. There is also a reduction in sound level under an overhang but in general the reduced sense of reverberation is likely to be more noticeable than the reduced loudness.

The third effect of a balcony overhang is clear in qualitative terms but difficult to quantify: listeners are aware of the reduced solid angle from which they are receiving sound. They are likely to detect the absence of sound from above as well as sound from behind; both probably contribute to the sense of detachment felt under overhangs. Listener envelopment (section 3.2) can also reduce under balcony overhangs.

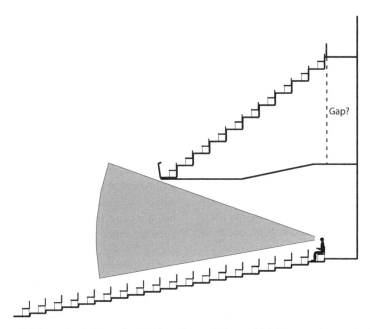

Figure 3.22 Long section through a balcony showing how the angle from which late sound can arrive is reduced below an overhang. A gap behind the balcony can provide some compensation

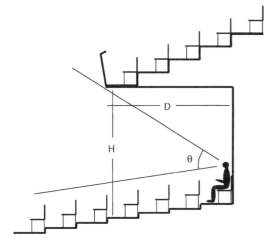

Figure 3.23 Basic balcony overhang proportions for a concert hall. Criteria are based on the ratio of depth to height, D/H, or θ, the vertical angle of view.

For balcony design, Beranek (1962) suggested a rule-of-thumb for concert halls that the depth beyond the overhang should not exceed the height of the opening (that $D/H \leq 1$ in Figure 3.23). More recent study (Barron, 1995a) suggests that the vertical angle of view (θ in Figure 3.23) may be a better parameter, with a recommended minimum value of 40°. The vertical angle of view approach takes account of the rake of the floor under the overhang.

Beyond restricting the degree of overhang, there is little that can be done to mitigate the effects of an overhang on the reverberant sound. It is possible to leave a gap behind the balcony to allow sound to filter round, providing a sense of some sound from behind for those below the overhang (Figure 3.22). A rare example of such a gap is to be found in the Michael Fowler Centre, Wellington, New Zealand (section 4.10). The sound level under an overhang can be increased by profiling the balcony soffit to provide extra reflections, but using geometry to boost late as opposed to early reflections is problematic. Maintaining a high opening at the overhang helps avoid reducing the reverberant sound, but that can of course remove one of the main design gains of the balcony. Balcony design in

Segerstrom Hall, California (section 10.6 and Barron, 1995a) is particularly interesting with heights at the openings to the overhangs of 6.1 and 7.2 m.

While Figure 3.23 considers geometry in just two dimensions, it is clear that the third dimension also influences sound behaviour under balcony overhangs. Such details as hall width, location of reflecting surfaces and the degree of diffusion in the main body of the hall must also be significant. These aspects remain to be investigated.

3.8 Design for the performers (written by A.-C. Gade)

It is only since around 1975 that the acoustic needs of the musicians have been subject to systematic investigation. Knowledge of performers' needs and how to fulfil them is not as well developed as knowledge about audience preferences. Yet certain guidelines have been developed, together with objective measurement procedures relating to stage conditions. The sound field inside the orchestra is extremely complex and probably impossible to describe in detail because of dependence on a wide interplay of factors: the orchestration of each piece (which varies from one bar to the next!), the directivity and sound powers of each instrument, the arrangement of the orchestra on the platform as well as the acoustics of the room. It is therefore wise to view the still immature scientific results against the background of many years of practical experience of stage conditions. It is in this spirit that the following practical design guidelines are given.

Meeting the needs of the musicians is primarily a question of proper design of the platform itself and of surfaces in its vicinity. These elements of the hall are of utmost importance for the two major acoustic concerns of performers: ease of hearing each other (necessary for good ensemble playing) and the feeling of support for sound from their own instrument. Other aspects of the musicians' acoustic experience, such as timbre and reverberance, are probably related to the design of the hall as a whole, so that if the audience is happy with these aspects the musicians are likely to be satisfied too.

Of course the surfaces around the platform may also be used to provide sound reflections to the audience. But such reflections have two serious drawbacks for listeners: potential tone colouration and a narrow source image (section 3.2). Adequate reflections for the audience should be available from the main body of the hall, leaving the platform area of the hall designed primarily to satisfy the musicians' needs. If the players are unable to perform at their best, the listeners' experience will definitely suffer.

3.8.1   Floor space, layout and risers

Selecting the platform floor area is a compromise between acoustic needs and comfort. As with the seating standard for audience, the area demanded per musician on the platform has also increased in recent years. While the platform in the Vienna Musikvereinssaal has an area of 130 m² and is still used successfully for modern full-sized orchestras, a newer hall like the Philharmonie Gasteig in Munich has a platform area of as much as 250 m². A variety of Parkinson's Law seems to afflict orchestral musicians, who will spread out to occupy the available area as well as tend to move closer and closer to the front edge of the stage (unless told otherwise by the conductor). But spreading out on a large floor space means greater distances and reduced acoustic communication between musicians. It has been found (Gade, 1989a) that beyond a distance of 8 metres, the delay of the direct sound becomes large enough to reduce the ease of ensemble playing. While aiming at a close spacing, one should not forget to consider the rise in sound level as one moves close to the more powerful instruments. Excessive levels near tympani or in front of brass sections may mask any other sound in the orchestra and even be dangerous for musicians placed in these areas. Local problems such as these should be avoided by adjusting the layout, by the use of risers or if all else fails by setting up small screens within the orchestra.

The tendency to move forwards is also dangerous for several reasons. The orchestra itself will get less benefit from the back wall and/or ceiling reflections. A late reflection off the stage back wall may also upset clarity for the audience. An orchestra on the edge of a stage also loses a reflection from the empty stage apron, which is especially useful to project sound from a soloist or a weak string section into the hall volume.

Beranek (1962, p. 498) says 'musicians like an area of about 20 sq. ft each', or 1.9 m². This is certainly too generous for violinists, for instance, but it may be used as an average figure to estimate the total platform area required: for a 100-piece orchestra almost 200 m² is necessary. More recent studies indicate the following net areas per player for different groups of instruments:

> 1.25 m² for upper string and wind instruments;
> 1.5 m² for cello and larger wind instruments;
> 1.8 m² for double bass;
> 10 m² for tympani, and up to 20 m² more for other percussion instruments.

For a full 100-piece orchestra (with a usual percussion section) this means a net covered area of about 150 m². If the platform is built at 200 m², this leaves ample extra space for soloists and extra percussion, space to compensate for losses due to risers, access routes, including a 1 m wide empty strip along the front of the platform. Also a narrow empty zone along the side and rear walls can be advantageous, as for certain instruments, for example the French horn, it is not comfortable to sit right against a wall.

For seated choirs, 0.5 m² per person is needed, so that a 100-person choir requires a further 50 m² of space. Choir seating may be placed at the rear of the stage with the stage front extended further outwards into the audience, as is done in the Barbican Concert Hall, London (section 5.10). Alternatively it is preferable to have a separate elevated choir balcony such as in the Royal Festival Hall, London (section 5.5) and St David's Hall, Cardiff (section 5.11). When not required as choir seating, these seats are sold to concertgoers, who appreciate the close visual contact with the orchestra.

In order to minimize distances within the orchestra, the platform should be neither too wide nor too deep. With a 200 m² stage, the average width should not exceed 18 m, resulting in a depth of about 12 m (for full orchestra but without choir). It is important for smaller groups that the effective floor area can be reduced. Ideally movable bounding walls, such as those on lifts at the Gulbenkian Great Hall, Lisbon, should be used (Barron, 1978). Where these are absent, the musicians should be persuaded to move backwards in order to limit distances between musicians and maximize the effect of reflections from the surrounding walls.

The provision of risers on the platform is important for good ensemble playing in large orchestras, because direct sound will propagate more freely between distant players when they are elevated rather than being hidden behind players sitting in between. For large distances between players (a value exceeding 8 m was mentioned above as causing time delay problems), it is clear that a weak direct sound cannot be fully compensated for by reflections, which will always be further delayed. Risers towards the sides of the platform for the back

rows of strings are therefore also recommended, a feature which is often implemented as semicircular risers, e.g. in the Berlin Philharmonie (Figure 3.24) or the Kitara Concert Hall in Sapporo, Japan.

If a separate choir balcony is not used, the riser system should allow for the full orchestra to be moved back and forth depending on whether a choir is present or not. This flexibility may be difficult and expensive to obtain if a hydraulic system is used, while a system of small, loose box elements (as in Derngate, Northampton, section 11.8) is cheaper and more flexible, but can be time- and cost-consuming to operate. In any case, it is important that the riser system allows for the ensemble, whatever its size, to be placed in a position on the platform where the benefit of reflections from surrounding surfaces is optimal. If the system lacks flexibility, it will result in wasted space, with steps in the wrong places for many of the required orchestral layouts.

To avoid loss of space, the horizontal depth of each riser is important: 1.25 m for upper strings and woodwind, 1.4 m for brass and cellos. Double

Figure 3.24 Stage of the Philharmonie Concert Hall, Berlin

bass players require slightly more depth, unless the music stand is placed on the step in front of them. If the rear of the platform is to double as choir seating, the 1.4 m can be split into two risers of 0.7 m. However 0.8 m is needed if the choir is to be seated. Percussion and tympani need a depth of about twice 1.4 m at uniform height. Riser height may be kept modest, at about 100 mm per step for woodwinds and slightly higher further back. But it is always advantageous if the heights are adjustable to suit the particular orchestra arrangement and the desired balance for the audience. It is often feared that risers will favour the audibility in the audience area of the wind and brass. But unless one is dealing with small halls and regional orchestras with severely undersized string sections, this is a factor which the conductor should be able to control.

Finally there is the question of the stage height above the stalls floor. The height should be more than 0.5 m, otherwise soloists' 'command' over the audience tends to be lost. Values in excess of 1.0 m lead to screening from view of the centre section of the orchestra from the front rows of the stalls. The stage height determines the setting-out point for sightlines and is thus influential for much of the seating layout (section 2.3).

3.8.2 Floor material

Musicians always speak unambiguously in favour of wood over an airspace as the appropriate construction for the platform floor, because it provides a 'warm' sound. The physics behind this notion has been the subject of few investigations (Askenfeld, 1986). Basically two processes with opposite effects are involved. The floor may be capable of acting as a sounding board for low-register instruments in direct contact with the floor, but it will also absorb sound reaching it through the air. Another possible effect may be the sensation of structural vibration set up in the floor adding positively to communication within the orchestra. The physical studies do not as yet suggest conclusive advantages, yet it would be unwise to contradict the request for this type of floor. In order to allow the floor to be as

acoustically 'alive', it should be chosen as thin and with the joist spacing supporting it being as large as possible, within limitations provided by loading requirements (especially grand pianos), rigidity (e.g. for television cameras) and local fire regulations. A platform thickness of 25 mm and a joist spacing of 600 mm may be regarded as a desirable minimal construction.

3.8.3 Side and back walls of the platform

As already mentioned, the wall surfaces in the vicinity of the platform should be mainly oriented so as to send reflected energy back to the musicians themselves. Where possible in order to avoid attenuation of reflections, non-horizontal reflection paths should be provided. This may be done either by tilting the upper part of the walls downwards, by tilting balcony fronts down or by providing balcony soffits around the stage to give reflections off the soffit and adjacent vertical surface.

In halls which contain a stagehouse, an orchestral shell is necessary. Yet the shell must be designed in such a way that the shell volume forms part of the same acoustic space as the hall. If it acts like a separate space the musicians will experience a lack of contact with the main hall acoustics and, with a high degree of enclosure, they may have so much support that they are tempted to play too softly. In other words, a rather critical balance exists between recessing the platform into an orchestra shell and exposing it in the main hall. (To confirm the situation objectively, the early decay time on the stage should not be less than 70 per cent of the value in the hall.) The shell itself should be made of fairly solid materials to avoid low-frequency absorption. A surface density of 20 kg/m^2 is recommended.

At this point it should be mentioned that orchestras, with members who have unfortunately experienced hearing loss and tinnitus due to excessive sound levels, are often reluctant to accept any advice for their new or existing hall. We are used to thinking in terms of sound absorption as a way to reduce sound levels, whereas more efficient sound reflection on stage may be a better solution. It is important to remember that musicians produce sound according to what they hear and if you only hear your closest neighbours, you are tempted to play louder – like when we shout in a telephone with poor transmission. In other words, the orchestra is more likely to cultivate a playing style with modest, balanced sound levels if the acoustic conditions allow them to communicate via efficient propagation of direct sound and early reflections. In fact measurements – even in orchestra pits – have shown that absorption added to wall surfaces near the orchestra only reduces support, but not the sound levels, which are always dominated by direct sound from the nearby instruments.

3.8.4   Ceiling reflectors

The most effective surface for providing early reflections on the platform is the ceiling. In halls with exposed platforms and high ceilings, it will often be necessary to suspend a single large reflector or an array of smaller reflectors over the platform. In the latter case, depending on the fractional area of the reflector array compared with the floor area beneath (the degree of perforation), the average height of the reflectors should if possible not exceed 8–10 m above the stage for them to be fully effective. The degree of perforation should be optimized. If it is too high, inadequate sound is reflected down and if it is too small, sound will be prevented from reaching the upper hall volume. Recent investigations seem to indicate that more small elements are preferable to fewer larger ones (Rindel, 1991). Here it was found that, with a degree of perforation of about 50 per cent, low frequencies were still reflected from an array of elements only 1.5 m² each. This arrangement provided more uniform coverage and less tone colouration, which is always a risk with overhead reflections. Slight convexity for the panels is also desirable for the same reasons.

Some acoustic designers are in favour of massive canopies – with built-in lighting and sound system equipment – rather than the more flimsy reflector arrays. Of course such a canopy will provide a higher reflection level if it is placed at the same height as the array; but this is seldom attractive for architectural as well as acoustical reasons. It may visually cut the hall in two, and the façade of the organ if present, and, if the canopy area is large, it may separate the volume above from the main space and so reduce reverberation. If placed higher than around 15 metres above the stage, the canopy reflection may arrive too late to assist ensemble for a large part of the orchestra.

For both types of overhead stage reflectors, it must also be considered whether or not the coverage area should be extended to the seating areas close to the stage; overhead reflections may contribute to clarity (and colouration!), but they are unlikely to improve spaciousness.

To avoid focusing and colouration, it is probably wise to provide the musicians with reflections from a number of surfaces surrounding the platform, such that it is not necessary to rely on or exaggerate the contribution from one particular surface or direction.

3.8.5   Scattering and absorption

A certain degree of scattering for reflections from surfaces surrounding the platform is required to avoid sound concentration, which would be detrimental to ensemble and orchestral balance. For the audience, a lack of diffusion may be heard as poor balance and blend between the various sections. It is in fact possible in principle to design reflectors specifically oriented so as to compensate for the different sound powers and directivities of the various instruments (Meyer and Biassoni de Serra, 1980). In practice, however, such a system would probably be highly complicated to use, since it would have to be readjusted for each orchestral layout. Because of the high sound levels emitted by brass and percussion instruments, it is often recommended to make small wall areas absorbent near these instruments. Ideally such areas should be adjustable between reflective and absorbent, to accommodate different arrangements. Another way of 'relieving the pressure' from loud brass and percussion instruments is to omit ceiling reflectors over the rear part of the platform.

3.8.6 Objective quantification

Most recommendations above are by necessity qualitative. But we need to know how, for instance, to trade off ceiling reflections against reflections from surrounding walls. An objective measure is required for stage acoustic conditions. Various measures have been proposed, of which only **objective support** (Gade, 1989a) seems to have gained wider recognition. Objective support is a measure relating to the musician's ability to hear himself and others. It has been found to correlate well with musicians' judgements on both these aspects. 'Early' objective support is defined as the energy of early reflections within 100 ms relative to the direct sound energy, for a microphone placed 1 m from an omni-directional source; the ratio is expressed in decibels. The equation describing early support is given in Appendix B.1.

Measurements in existing halls (Figure 3.25) indicate that early support is related to platform 'volume' (i.e. platform area × height to the ceiling or reflectors). (Five halls included in Figure 3.25 are discussed in detail in this book: the Usher Hall, Edinburgh, the Royal Festival Hall and Barbican Concert Hall, London, and St David's Hall, Cardiff are

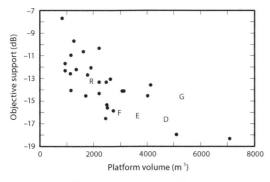

Figure 3.25 Objective support plotted against platform 'volume' in 32 halls (Gade). British halls are lettered: D, St David's Hall, Cardiff; E, Edinburgh Usher Hall; F, Royal Festival Hall, London; G, Derngate, Northampton; R, Barbican Concert Hall, London

considered in Chapter 5 and the Derngate Centre, Northampton in Chapter 11.) It can be seen that early support, the amount of early reflected energy, decreases as the volume of platform space increases. Subjective results relating to ease of ensemble indicate that objective support values between −13 and −11 db are optimal (Gade, 1989b). More recent research seems to confirm this result (Chiang *et al.*, 2003), but also indicates that a certain level of reverberant sound can be favourable for achieving ensemble (Ueno and Tachibana, 2003), given that the direct sound propagates freely between the players. This might be the reason why recent arena-shaped concert halls (e.g. Disney Hall, Los Angeles and the Kitara Concert Hall in Sapporo, Japan), which have semicircular riser systems but either no or only high canopies and only marginal wall surface area surrounding the stage, have been reported to be successful.

3.9 Acoustic modelling

While design on the basis of rules-of-thumb and of precedent provide valuable guides, only reverberation time can be predicted from architectural drawings, and even there cases arise when the prediction is inaccurate. It has now become commonplace to use some form of modelling to assist design. Traditionally physical models were used but computer models are gaining popularity as they become more sophisticated.

In their historical review, Cremer and Müller (1982) cite water-wave model experiments by Scott Russell as early as 1843. Similar two-dimensional studies in air were made by W.C. Sabine in 1912 using the Schlieren technique (Sabine, 1922). The value of these studies was limited though to those two dimensions. A major reason for wanting a model study is to aid with comprehension of three-dimensional behaviour. To do this, either acoustic testing in physical scale models of auditoria or computer analysis is necessary.

3.9.1   Acoustic scale models

Testing acoustics in scale models dates back to work by Spandöck in the 1930s (Barron, 1983). In the intervening years the principles of modelling have remained the same but major advances have been made in equipment and latterly by the introduction of digital signal processing. The virtue of acoustic scale modelling is that the complex behaviour of sound when it meets finite-size surfaces or obstacles depends on the relationship between the size of the surface and the wavelength of sound. Scale modelling depends on scaling the wavelength in line with the physical scaling.

The scaling procedure is based on two fundamental equations:

$$\text{Speed of sound} = \frac{\text{Distance}}{\text{Time}}$$
$$= \text{Frequency} \times \text{Wave length}$$

Air (or nitrogen which constitutes 80 per cent of air) is nearly always used as the propagating medium in the model so the speed of sound is the same both in the model and at full-size. Both distance and time are scaled together in the model but more significantly since wavelength has to be scaled with distance, frequency must be increased. In a 1:10 scale model, frequency is increased by a factor of 10. The particular attraction of acoustic scale modelling is that in most respects the modelling is perfect; the same acoustic behaviour occurs in the model in miniature.

The one complication is that, due to the frequency transposition, absorption of sound as it travels through air becomes an issue. This phenomenon of air absorption has already been mentioned as it influences high-frequency reverberation time (section 2.8.5). It is an effect which becomes progressively more marked the higher the frequency. The major component of the absorption, known as molecular absorption, can however be controlled since it depends on the simultaneous presence of oxygen and water vapour. The problem of air absorption in models can be almost eliminated by the use of low-humidity air (at around 2 per cent

r.h.) or by using a nitrogen atmosphere. For impulse responses, an alternative is to make corrections to recorded responses as one element of digital signal processing; unfortunately this has the disadvantage that it reduces dynamic range.

For the model itself, materials have to be used which match the absorption characteristics of full-size surfaces. To measure absorption by model materials, the same procedure is used as at full size, with a sample of material tested in a model-size reverberation chamber. It turns out that to reproduce an auditorium is simpler than most other spaces since most surfaces are acoustically hard and can be modelled with hard smooth surfaces. The most common model materials are varnished timber or plastic; plastic has the advantage that it contains no moisture.

The major absorbing element in an auditorium is audience seating or audience themselves. For seating upholstery, fabrics chosen by trial-and-error are generally used. It is also important to reproduce the seating physically since rows of seating scatter as well as absorb sound.

Model scales of between 1:8 and 1:50 have proved most appropriate. The traditional scales were 1:8 and 1:10; the first of these being a whole number of octaves and being suitable for tape recorders which have speeds based on factors of two (Figure 3.26). 1:10 is obviously convenient for architectural drawings. The disadvantage of the large model is the space it requires, the construction cost and the time it takes to build the model. One would not wish to base the model on the design too early in the design programme, but only having results from the model testing when the design is nearly fixed is not suitable either.

At the other extreme, modelling at 1:50 scale overcomes these problems and allows modelling to become fully integrated in the design process (Barron and Chinoy, 1979). A concert hall model at this scale will be about 1 m long (Figure 3.27). The range of measurements that can be made in small models is less than in large ones, predominantly objective rather than subjective measurements. This comes about since the maximum measurement

Figure 3.26 A 1:8 scale acoustic model of the Barbican Concert Hall, London

Figure 3.27 A 1:50 scale acoustic model of the Waterfront Hall, Belfast

frequency is around 100 kHz. Models with scale factors of 1:20 or 1:25 offer a good compromise.

The limiting measurement frequency of 100 kHz is of course well above any frequency we can hear with our ears. But it is the transducers, microphones and loudspeakers, which introduce the limit. Microphones with a diameter of less than 1/8 inch (3 mm) are rare, while loudspeakers for ultrasonic frequencies become highly directional. A very valuable sound source for models is an electrical spark discharge. This has the advantage that the source of sound can be very compact; the electrical energy in the spark determines the frequency of maximum acoustic energy.

Objective testing in models gives numerical values such as reverberation time, which can be compared with criteria developed at full size. A series of listener and source positions are tested. In reality a perfect match of reverberation time with prediction is rarely achieved, absorption often does not match at all measurement frequencies. Modelling is especially valuable for the other objective quantities, such as objective clarity and total sound level, which cannot be predicted from drawings; these can be corrected for the small reverberation-time errors that arise (Barron, 1997).

Subjective testing refers to listening to music that has been processed by the model. The traditional technique was to use a multiple-speed tape recorder: speeded-up speech or music was played through the model and slowed down again to be listened to over headphones. Other more sophisticated techniques involving computers have also been used. For recording in a model, an accurate 1/10th scale model head has even been developed to allow realistic recordings (Xiang and Blauert, 1991). Subjective testing is a highly seductive proposition, offering the chance to hear the sound of a space before construction. But this opportunity should not blind one to the fact that no system like this is perfect. Using subjective testing for comparison purposes is likely to be acceptable, but expecting to use the procedure to make an absolute judgement on acoustic quality of the design is probably over-ambitious.

Scale-model testing is now a mature technique for auditoria. Scale models automatically deal with the sound behaviour which is difficult to quantify, in particular scattering and diffraction at room surfaces. Testing a model represents a small cost as a proportion of total building cost and provides much greater confidence in the acoustic behaviour. As a design aid for major auditoria, the position of acoustic scale modelling looks secure for a while yet. Scale models have been tested to guide refurbishment since the year 2000 of two major London concert halls: the Royal Albert Hall and Royal Festival Hall.

3.9.2   Computer modelling

Computer models date from the 1960s when Krokstad and colleagues (1968) experimented with a model based on mirror (specular) reflections using a mainframe computer. The arrival of the PC has stimulated the development of commercially available computer programs, many directed at auditoria. The geometry of the hall is input and the acoustic characteristics (absorption and scattering coefficients) of each surface are specified; the program calculates reverberation time and the other objective measures.

The starting point for computer models is a geometrical model with rays or beams radiating out from a source, which are then reflected specularly at surfaces. In reality, surfaces rarely reflect specularly due to (i) surfaces being partly or highly scattering, (ii) surfaces being curved and (iii) diffraction occurring at reflection from finite surfaces. A fourth difficulty arises because it is necessary to treat the late sound statistically (that is as a mixture of reflections) and the transition from discrete early reflections to reverberation is difficult to handle.

The magnitude of the calculation problem can be appreciated from the following. A listener in a large concert hall receives about 8000 reflections in the first second after the original sound and each reflection which arrives after a second will have undergone about 20 reflections. The time of one second should of course be seen in the context of a typical concert hall reverberation time of 2 seconds. The number of sound ray impacts on surfaces that have to be considered is obviously very large, over 130 000 for the one-second period after the sound is emitted. If diffusion (scattering) is included in the model, each impact of a single incident ray generates several reflected rays. It is obvious that total computation time risks becoming excessive. The art of programming is to develop techniques to simplify the problem in ways that do not affect anything that is significant for the ear.

At the time of writing, many computer programs include scattering and some deal efficiently with the transition from early reflections to reverberation. Curved surfaces can also be handled by some programs. But the issue of diffraction remains to be tackled. The technique currently recommended for dealing with this problem is to make the geometrical model less detailed than one would for visual purposes. Those with experience of using computer models claim that they get reliable results using this technique. But there will always be some situations, a complex ceiling profile for instance, for which the choice of the appropriate degree of detail for the geometrical model is not obvious.

Computer models have come a long way from the early exercises on mainframe computers. Many are now using scattering coefficients, though measurements of these coefficients on actual surfaces are only now beginning to be published. A recent development has been comparisons (known as round-robin exercises) in which software writers use their programs to predict acoustic behaviour of the same space (Vorländer, 1995; Bork, 2000). Only a minority of programs have been found to predict accurately.

The equivalent of subjective testing with computers is generally known as **auralization**. The procedure uses measured results on human ears, what are known as external ear transfer functions, the effects of the head on sound arriving at the two ears from different directions (Blauert, 1983). These results are used to simulate for headphone listening the directions of individual reflections as calculated by the computer. With all reflections treated in this way, the listener should perceive the full sound picture for a particular seat in the modelled space. Auralization techniques are being progressively refined (Kleiner, Dalenbäck and Svensson, 1993).

Computer models are popular because they can be implemented quickly and enable acoustic consultants to engage with the geometrical design. The programs are cheap to use and produce elegant output that can be shown to clients; Figure 3.28 illustrates a typical rendering of a concert hall stage. It is though the hidden part of the program that one needs to be sure of. Reliability of programs will certainly increase in the coming years.

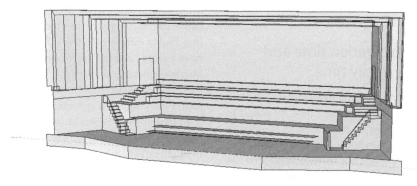

Figure 3.28 The stage area of a concert hall (Lighthouse Concert Hall, Poole) as rendered in the commercial computer simulation program CATT (courtesy of J.J. Dammerud)

3.10   Objective characteristics

In section 3.3, four 'new' measures for concert hall acoustics were introduced in addition to the traditional reverberation time. Further discussion of each of these proposed measures is given here. In Chapter 5 and beyond, the results of objective measurements are displayed. With five quantities measured at perhaps twelve positions in a hall often at five different frequencies, there is a potential cornucopia of data to display. Some data reduction is required. For measurements other than reverberation time and early decay time, a first step is to average values over a range of frequencies. There is then the question of whether the mean for the hall is relevant. The general approach taken here for the 'new' measures is that variations within a hall should not be concealed. The rationale for this is that for the listener it is the experience at an individual seat which is important, not listening conditions averaged over the auditorium.

The criteria appropriate to the new measures are not as widely agreed as those for reverberation time. The following tentative values appear acceptable from subjective and objective surveys:

Early decay time	between 1.8 and 2.2 seconds
Objective clarity	between −2 and +2 dB
Objective source broadening	between 0.1 and 0.35
Total sound level (for symphony concerts)	greater than 0 dB

3.10.1 Reverberation time and early decay time

Reverberation time, the traditional measure in room acoustics, roughly measures the time for sound to become inaudible when it is followed by silence. This terminal reverberation is not heard many times during a typical concert but reverberation time has retained its position for at least three reasons: it can be calculated from drawings with reasonable accuracy in most cases, it is a characteristic of the space

and measured values are well understood. The reverberation time is always shorter when auditoria are occupied by audience. Both unoccupied and occupied times are normally measured. In some halls it is necessary to predict the occupied value from the unoccupied because no measurement is available; the method used for results here is given in Appendix B.2.

The normal recommendation for the mid-frequency (500/1000 Hz) occupied reverberation time is in the range 1.8 – 2.2 seconds. The preferred frequency characteristic has a rise in reverberation time in the bass as shown in Figure 3.10.

For listeners, running reverberation is generally more significant, that is the reverberation which forms the background against which the latest musical notes are heard. This is measured by the early decay time (EDT), which is based on the first 10 db of the decay. The EDT is defined so that for a linear decay it takes the same value as the reverberation time. Because of this, the same criterion can be applied to both measures of between 1.8 and 2.2 seconds. Whereas the reverberation time (RT) tends to be constant throughout a hall, the EDT varies with position. Particularly low EDT values are normally found under deep balcony overhangs.

For listeners the actual value of the EDT is important. But looking at the ratio of the EDT to the RT can be instructive about the nature of the sound field in a concert hall (Barron, 1995b). At mid-frequencies, mean hall EDT/RT ratios in 17 British concert halls range between 0.79 and 1.06, with an overall average ratio of 0.96. The ratio proves to be an indication of diffuseness or directedness of the design. In a highly diffuse hall, the EDT/RT ratio tends to have a value close to 1 (as for example in the Vienna Musikvereinssaal, section 4.3, and the Cardiff St David's Hall, section 5.11). In a directed design in which early sound is reflected onto the audience, the ratio tends to be lower (section 4.10); to compensate, it may be necessary to have a longer reverberation time to give an acceptable EDT.

3.10.2 Objective clarity

Being able to hear musical detail is an important component of concert hall listening, yet we also prefer to hear a background sound, a response from the space. It is obvious from listening experiments that the direct sound as well as early reflections contribute to clarity, while late reverberant sound provides the background sound though it is in itself detrimental to clarity. A balance between the two is needed; this balance is also the basis for a preferred reverberation time. The time division between early and late is generally taken as 80 ms for music; the ear does not exhibit a precise cut-off at 80 ms but in most cases the sharp division for the objective measure is not a problem. Objective clarity is simply the ratio expressed in decibels of the early to the late sound energy received by listeners. Mathematical expressions for the objective measures are given in Appendix B.1. Also in Appendix B.1 there is discussion of an alternative measure for clarity: centre time. Centre time measurements are not presented here as they are closely related to early decay time.

With all measures, the appropriate frequency range has to be established. For clarity, there is evidence that the ear's temporal response to bass frequencies (125 and 250 Hz) is slight, and therefore of minor interest. To obtain a single number for objective clarity, the average of values at 500, 1000 and 2000 Hz is taken. In subjective terms, objective clarity relates both to how clear the sound is (which gives a minimum required value) as well as the balance between clarity and reverberance. There is thus some overlap in the subjective significance of objective clarity, reverberation time and EDT. The optimum range for symphony concerts for objective clarity appears to be between −2 and +2 dB, though the upper limit may be extended slightly higher.

Objective clarity is influenced by the reverberation time, as one would expect. Thus values in empty halls are higher than those in occupied conditions. Measured results are thus presented after correction for the change in reverberation time with occupancy (Appendix B.3). As mentioned in section 3.7, higher values of objective clarity tend to be found under balcony overhangs. The late reverberant sound level is reduced under a balcony, which has the effect of increasing objective clarity.

3.10.3 Spatial characteristics

In one respect, views on spatial hearing in concert halls have traced a circular path. Before 1960 audible spatial effects were associated with the late reverberant sound (Kuttruff, 2000); the experience of sound in a cathedral space clearly supports this connection. A long reverberation time and room surfaces that scatter sound were thought to enhance the spatial effect. Then in 1967 Marshall suggested that strong early reflections from the side were a component of sound in halls with the best acoustics. Whereas in the past there had been no guidelines available regarding the appropriate shape for symphony concert halls, here was a criterion with consequences for auditorium form. Marshall's ideas also provided an explanation for the high reputation of traditional rectangular plan halls.

Some designers (see Christchurch Town Hall, New Zealand, section 4.10, and Nottingham Royal Concert Hall, section 5.12) have developed radical designs providing high spatial impression, as the subjective effect was called. Spatial impression was found to involve a sense of the source becoming broader for loud sounds, as well as a sense for the listener of being surrounded by sound, a sense of envelopment. The two components of spatial impression are called 'source broadening' and 'listener envelopment'. The magnitude of the source broadening effect was found to be related to the proportion of sound which arrives from the side within the first 80 ms after the direct sound (Barron and Marshall, 1981). This ratio, the **early lateral energy fraction**, is called here objective source broadening. The ratio is taken between sound received by a figure-of-eight microphone and that received by an omni-directional microphone; the null of the figure-of-eight microphone points at the source. The importance of the bass for the spatial effect was mentioned in section 3.2; the mean of measurements at octave frequencies 125, 250, 500 and 1000 Hz is used.

Measured values of objective source broadening range from about 0.05 to 0.50, with an average value of around 0.18 (Barron, 2000). A minimum value of 0.10 is recommended. Too high a proportion of lateral sound can be unsettling, since it compromises identification with the performers. The proposed preferred range for objective source broadening is from 0.10 to 0.35.

Sound level is also important for source broadening; a measure combining both the lateral fraction and sound level is proposed in Appendix B.1. One could use this combined measure for assessing auditoria, but dealing with the spatial and level components separately has its benefits for design purposes. The spatial component is determined by hall geometry, the level is mainly determined by the area of absorbing audience as well as some geometrical features, such as balcony overhangs. In most cases it makes sense to discuss the two aspects individually.

As described in section 3.2, the supremacy of early lateral reflections was challenged in 1989. There is now general agreement that the spatial effects of later sound are also important. What was before 1960 considered to be the sole spatial effect is now rated as one of two spatial effects. In addition, the nature of the effect associated with early lateral reflections has been clarified: that early reflections create a sense of source broadening whereas the later sound is responsible for creating a sense of envelopment, now known as listener envelopment.

Bradley and Soulodre (1995a, 1995b) proposed the **late lateral level** as an objective measure of listener envelopment (Appendix B.1). This measure combines the spatial and level aspects; it is the sum of the late lateral energy fraction (converted into decibels) and the level of the late sound. In this way it is very similar in character to the early lateral energy fraction plus a level component except that the time period is after 80 ms rather than the first 80 ms following the direct sound. What new design consequences for concert halls does this new measure introduce?

Analysis of objective behaviour in halls shows that the proportion of late sound from the side does not vary much and takes on average the value one would expect from a diffuse sound field (Barron, 2001). This leads to the dominant component for the late lateral level being the sound level of the late sound. The most important quantity affecting the late sound level is the total acoustic absorption (section 2.9), which in practice generally means the area of the audience seating. Thus one expects to have a larger sense of envelopment in a small capacity hall compared with a large hall. Size matters in auditorium acoustics, listener envelopment appears to be yet one more aspect which is affected by the audience size.

Acoustic consultants seldom have much control over audience capacity and should be concerned about maintaining high sound levels for other reasons. Thus the new quantity proposed for listener envelopment has few design consequences that would not be already considered. One curiosity of listener envelopment is that to feel enveloped does not in fact depend on the listener actually receiving sound from behind. Interaction between the two spatial effects of source broadening and listener envelopment also occurs (Bradley, Reich and Norcross, 2000).

3.10.4 Total sound level

One of the surprising discoveries from subjective experiments has been that though sound levels do not vary much within and between concert auditoria, listeners are sensitive to changes of only a few decibels, even though they are effectively comparing experiences of live music separated by long periods of time (Barron, 1996). It is now standard to measure sound levels throughout the audience seating for a source of known power at a typical performing position.

The total sound level refers to the level of all sound (early plus late) received at a listening position. Some reference level is required; the chosen reference is the direct sound level at 10 m from the source. This choice has the advantage that measured total sound levels in concert halls vary between about 0 and +10 dB. Since the direct sound decreases 6 db per doubling of distance, this approach also allows for direct calculation of the 'amplification' of sound provided by the room. In practice the direct sound energy at the back of a large concert hall may be only 5 per cent of the total, hence the importance of good reflected energy design.

Sound level is normally measured over the full frequency range. For comparison with subjective loudness and intimacy, looking at the mid-frequency values seems appropriate (500–2000 Hz). The frequency spectrum of the total sound level may well determine perception of warmth (relative strength of bass) and brilliance (relative strength of high frequencies). The first of these is further discussed in section 5.17.

As a criterion for total sound level there was no broad agreement. Lehmann and Wilkens (1980) suggested a minimum value of +3 dB, but this is considered too severe as it eliminates 60 per cent of positions in British halls. A criterion for a minimum mid-frequency sound level of 0 db seems reasonable from experience. There is some casual evidence that low sound levels are more acceptable in spaces with diffuse sound fields. In small auditoria, sound levels can become too high for orchestral listening but this will only occur in halls of less than 1000 seats (section 6.2).

As with objective clarity, sound level is influenced by reverberation time (or more precisely by total acoustic absorption). Sound level values will be lower in halls occupied by audience. Measured results have been corrected for the change in reverberation time with occupancy (Appendix B.3). In addition, sound level values are often lower under balcony overhangs and so results for these positions are highlighted.

It has been already mentioned in section 2.9 that sound level decreases progressively with distance and that average behaviour conforms to a revised theory. Revised theory simply assumes that sound decays linearly following the arrival of the direct sound. A significant value of the theory is that it allows identification of the influence of

specific design features on sound level (Barron and Lee, 1988). Revised theory can also be applied to components of the total sound when it is split into its early and late parts.

3.10.5 Revised theory for the total, early and late sound

A subdivision of sound into an early and late part originates from the characteristics of our hearing system. The early part contributes to definition and clarity, while the late reverberant part provides an acoustic context against which the early sound is heard. The relevant time interval for early sound with music is 80 ms, while that for speech is usually taken as 50 ms. But in physical acoustic terms, there is also a distinction between the characters of early and late sound. The early component principally consists of discrete elements: the direct sound and a few early reflections. With the later sound,

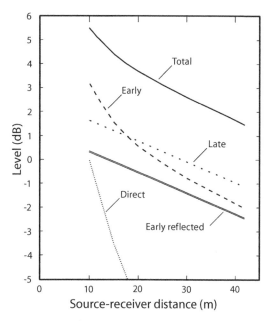

Figure 3.29 Behaviour according to revised theory of the various components of the sound, as a function of source–receiver distance. (Hall volume = 20 000 m³, reverberation time = 2 seconds.) The reference sound level of 0 db is the direct sound level at 10 m distance

the density of reflections (that is the number arriving in time intervals such as 150–160 ms after the direct sound) becomes so high that detailed sound paths are of little significance (except in the case of echoes). For the late sound it is the average behaviour with time (i.e. reverberation time), its level and gross directional distribution which are important. Revised theory also predicts the early and late sound energy values as a function of source–receiver distance (Appendix B.3).

This section considers the predicted behaviour according to the theory and discusses comparisons of measured values with the theory for early and late sound in some individual halls. Figure 3.29 shows revised theory results for a typical size hall. This and all remaining figures in this section have distance between source and receiver along the horizontal axis. The sound levels of all components decrease with distance, but the early sound level decreases faster than the late because it includes the direct sound, decreasing 6 db each doubling of distance. Hence expected objective clarity also decreases as one moves away from the stage. The behaviour in Figure 3.29 is of course average within a hall. In real halls, one measures slightly higher sound levels at positions near either side walls or rear walls. Early and late energy levels can readily be derived from measured results, since objective clarity is the ratio of early-to-late energy (converted into decibels) and total sound level is the sum of the two components.

Figures 3.30 and 3.31 show the results for early and late sound in the Lighthouse Concert Hall, Poole (section 5.9). In this hall, agreement with revised theory is good except for late sound at distant balcony seats. The lower late sound levels at these seats probably stem from the low ceiling height above them, which restricts possible directions for the arrival of late reflected sound.

Two clear examples of halls in which the early sound level is less at distant seats than predicted by revised theory are of interest. The measured early values in the Barbican Concert Hall, London (section 5.10), are shown in Figure 3.32. The highly scattering nature of the ceilings in this hall and in the Fairfield Hall, Croydon (section 5.8) prevent enough early

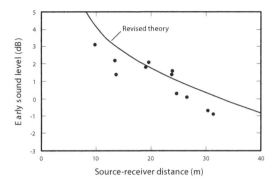

Figure 3.30 Early sound behaviour in the Lighthouse Concert Hall, Poole. The continuous line gives the revised theory behaviour

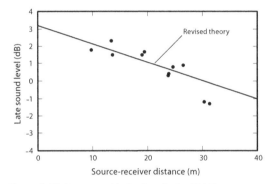

Figure 3.31 Late sound behaviour in the Lighthouse Concert Hall, Poole. The continuous line gives revised theory behaviour

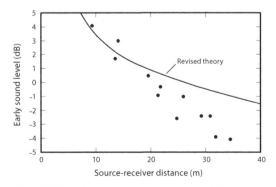

Figure 3.32 Early sound behaviour in the Barbican Concert Hall, London. The continuous line gives the revised theory behaviour

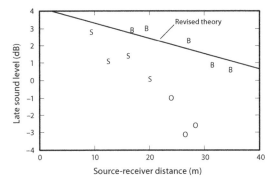

Figure 3.33 Late sound behaviour in the Colston Hall, Bristol. Points are labelled: S, Stalls, O, overhung Stalls and B, balcony. The continuous line gives revised theory behaviour

sound reaching balcony seats. Subjectively these seats are characterized by a lack of intimacy.

Turning to cases of marked late sound behaviour, the most obvious divergence between measurement and theory occurs under balcony overhangs. At these locations one finds deficient late sound, which is perceived as a diminished sense of reverberation. This effect is particularly marked in the Colston Hall, Bristol (section 5.6), where the overhang is deep and the balconies extend along the side walls to the sides of the stage. The objective effect in the stalls is to produce late sound which decreases rapidly with distance from the source. The points in Figure 3.33 can be considered to lie on two lines rather than one; the balcony is contributing to a subdivided acoustic space.

Further comparisons between subjective responses and objective observations are given in later chapters, particularly Chapter 5. Bradley (1986) has further interesting examples from Canadian auditoria.

3.10.6 Measurement procedure and display of results

Objective measurement results in British concert halls are presented in Chapter 5; subsequent chapters contain similar results for recital halls, drama theatres, opera houses and multi-purpose halls. Measurements in concert halls were made with a single loudspeaker source position and at 10 or more receiver positions (Barron and Lee, 1988). The source position was generally chosen at 3 m from the front of the stage; in this position there is usually no reflection off the stage. Though the directional characteristics of an orchestra are very complex, the normal solution is to use an omni-directional loudspeaker source.

Omni-directional microphones were used except for objective source broadening which requires in addition a figure-of-eight microphone. Most results were taken from recorded impulse responses with an impulsive signal fed to the loudspeaker (Barron, 1984). For sound level measurements, filtered noise is fed to the loudspeaker with sound level recorded by a standard sound level meter. Subsequent analysis of impulse responses is by computer. Objective clarity and sound level have been corrected from the unoccupied to the occupied reverberation time.

In Figure 3.34 an example of measured results is given from the Vienna Musikvereinssaal (measurements by courtesy of J.S. Bradley). The first graph shows the reverberation time as a function of octave frequency measured in the unoccupied and occupied hall. The value for the occupied hall is wherever possible taken from measurements by others; in some halls it has been necessary to predict the occupied value from the unoccupied measurement (Appendix B.2). Mean early decay time (EDT) values for the unoccupied condition are also plotted as circles. Since EDTs measured under balcony overhangs tend to be shorter than in exposed seats, the former are omitted from the mean. (In the hall in question in Figure 3.34, no balcony overhangs in fact exist.)

Objective clarity (the early-to-late sound index, C_{80}, average over the 500, 1000 and 2000 Hz octaves) is shown as a triangular 'pediment'; this gives the range of measured values and with the mean value at the apex. Here again, values at overhung seats tend to be different, higher than average for objective clarity, so these results are shown individually. The preferred range for objective clarity

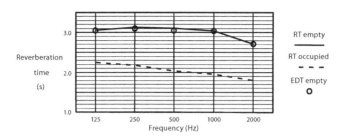

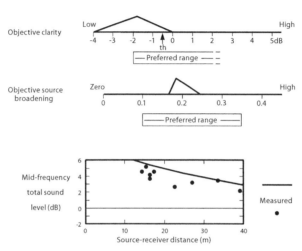

Figure 3.34 Measured results in the Grosser Musikvereinssaal, Vienna (courtesy of J.S. Bradley). Occupied reverberation time from Beranek (2004)

according to the suggestion made above is also shown. Further the mean value of theoretical objective clarity (according to revised theory at the actual measurement distances) is plotted as 'th'; significant differences between the mean measured and theory can often be related to design features.

Objective source broadening is displayed with a triangular pediment in the same way as objective clarity, with the range of measured values and the apex at the mean value. The mean value over four octaves 125–1000 Hz is used. In this case, special treatment for seats under balcony overhangs is unnecessary.

The final graph shows values by seat position for the total sound level, an average of values at 500,

1000 and 2000 Hz is used here. The minimum criterion value of 0 db is included as a horizontal line. The total sound level prediction according to revised theory is also shown as a line to allow comparison with the behaviour in average concert halls.

There are several unusual features about the objective behaviour in this famous Vienna concert hall. There is a large difference between the unoccupied and occupied reverberation time (some seating is not upholstered); also values for both objective clarity and sound level are lower than expected from revised theory. Possible reasons for these peculiarities are given with the detailed discussion of this hall in section 4.3.

References

Askenfeld, A. (1986) Stage floors and risers – supporting resonant bodies or sound traps? *Acoustics for choir and orchestra*. Royal Academy of Music, Stockholm, pp. 43–61.

Atal, B.S., Schroeder, M.R. and Sessler, G.M. (1965) Subjective reverberation time and its relation to sound decay. *Proceedings of the 5th International Congress on Acoustics*, Liège, Paper G32.

Barron, M. (1971) The subjective effects of first reflections in concert halls – the need for lateral reflections. *Journal of Sound and Vibration, 15*, 475–494.

Barron, M. (1978) The Gulbenkian Great Hall, Lisbon, II: an acoustic study of a concert hall with variable stage. *Journal of Sound and Vibration, 59*, 481–502.

Barron, M. (1983) Auditorium acoustic modelling now. *Applied Acoustics, 16*, 279–290.

Barron, M. (1984) Impulse testing techniques for auditoria. *Applied Acoustics, 17*, 165–181.

Barron, M. (1988) Subjective study of British symphony concert halls. *Acustica, 66*, 1–14.

Barron, M. (1995a) Balcony overhangs in concert auditoria. *Journal of the Acoustical Society of America, 98*, 2580–2589.

Barron, M. (1995b) Interpretation of early decay times in concert auditoria. *Acustica, 81*, 320–331.

Barron, M. (1996) Loudness in concert halls. *Acustica/Acta Acustica, 82*, S21–29.

Barron, M. (1997) Acoustic scale model testing over 21 years. *Acoustics Bulletin, 22*, No. 3, 5–12.

Barron, M. (2000) Measured early lateral energy fractions in concert halls and opera houses. *Journal of Sound and Vibration, 232*, 79–100.

Barron, M. (2001) Late lateral energy fractions and the envelopment question in concert halls. *Applied Acoustics, 62*, 185–202.

Barron, M. (2007) When is a concert hall too quiet? *Proceedings of the 19th International Congress on Acoustics, Madrid*, Paper RBA-06-006.

Barron, M. and Chinoy, C.B. (1979) 1:50 scale acoustic models for objective testing of auditoria. *Applied Acoustics, 12*, 361–375.

Barron, M. and Lee, L.-J. (1988) Energy relations in concert auditoriums, I. *Journal of the Acoustical Society of America, 84*, 618–628.

Barron, M. and Marshall, A.H. (1981) Spatial impression due to early lateral reflections in concert halls: the derivation of a physical measure. *Journal of Sound and Vibration, 77*, 211–232.

Beranek, L.L. (1962) *Music, acoustics and architecture*, John Wiley, New York.

Beranek, L.L. (2004) *Concert and opera houses: Music, acoustics and architecture*, 2nd edn, Springer, New York.

Blauert, J. (1983) *Spatial hearing*, MIT Press, Cambridge, MA.

Bork, I. (2000) A comparison of room simulation software – the 2nd round robin on acoustical computer simulation. *Acustica/Acta Acustica, 86*, 943–956.

Bradley, J.S. (1986) Progress in auditorium acoustics measurements. *Proceedings of the Vancouver Symposium on Acoustics and Theatre Planning for the Performing Arts*, August, pp. 89–94.

Bradley, J.S. and Soulodre, G.A. (1995a) The influence of late arriving energy on spatial impression. *Journal of the Acoustical Society of America, 97*, 2263–2271.

Bradley, J.S. and Soulodre, G.A. (1995b) Objective measures of listener envelopment. *Journal of the Acoustical Society of America, 98*, 2590–2597.

Bradley, J.S., Reich, R.D. and Norcross, S.G. (2000) On the combined effects of early- and late-arriving sound on spatial impression in concert halls. *Journal of the Acoustical Society of America, 108*, 651–661.

Chiang, W., Chen, S. and Huang, C. (2003) Subjective assessment of stage acoustics for solo and chamber music performances. *Acta Acustica, 89*, 848–856.

Cox, T. and Shield, B.M. (1999) Audience questionnaire survey of the acoustics of the Royal Festival Hall, London, England. *Acustica/Acta Acustica, 85*, 547–559.

Cremer, L. (1986) Der Trapezterrassenraum. *Acustica, 61*, 144–148.

Cremer, L. and Müller, H.A. (translated by T.J. Schultz) (1982) *Principles and applications of room acoustics, Vol 1*, Applied Science, London.

Damaske, P. (1967) Subjektive Untersuchungen von Schallfeldern. *Acustica*, **19**, 199–213.

Gade, A.C. (1989a) Investigations on musicians' room acoustic conditions in concert halls, I: Methods and laboratory experiments. *Acustica*, **69**, 193–203.

Gade, A.C. (1989b) Investigations on musicians' room acoustic conditions in concert halls, II: Field experiments and synthesis of results. *Acustica*, **69**, 249–262.

Games, S. (1985) *Behind the facade*, Ariel Books, London.

Haan, C.H. and Fricke, F.R. (1997) An evaluation of the importance of surface diffusivity in concert halls. *Applied Acoustics* **51**, 53-69.

Haas, H. (1951) Über den Einfluss eines Einfachechos auf die Hörsamkeit von Sprache. *Acustica*, **1**, 49–58.

Hawkes, R.J. and Douglas, H. (1971) Subjective acoustic experience in concert auditoria. *Acustica*, **24**, 235–250.

Helmholtz, H.L.F. (1863) *On the sensations of tone* (reprinted by Dover, New York, 1954).

Kleiner, M., Dalenbäck, B.-I. and Svensson, P. (1993) Auralization – an overview. *Journal of the Audio Engineering Society*, **41**, 861–875.

Krokstad, A., Strøm, S. and Sørsdal, S. (1968) Calculating the acoustical room response by use of a ray tracing technique. *Journal of Sound and Vibration*, **8**, 118–125.

Kuttruff, H. (2000) *Room acoustics*, 4th edn, Spon Press, London.

Lehmann, P. and Wilkens, H. (1980) Zusammenhang subjektiver Beurteilungen von Konzertsälen mit raumakustischen Kriterien. *Acustica*, **45**, 256–268.

Marshall, A.H. (1967) A note on the importance of room cross-section in concert halls. *Journal of Sound and Vibration*, **5**, 100–112.

Marshall, A.H. (1968) Levels of reflection masking in concert halls. *Journal of Sound and Vibration*, **7**, 116–118.

Marshall, A.H. and Barron, M. (2001). Spatial responsiveness in concert halls and the origins of spatial impression. *Applied Acoustics*, **62**, 91–108.

Meyer, J. and Biassoni de Serra, E.C. (1980) Zum Verdeckungseffekt bei Instrumentalmusikern. *Acustica*, **46**, 130–140.

Morimoto, M. and Maekawa, Z. (1989) Auditory spaciousness and envelopment. *Proceedings of the 13th International Congress on Acoustics*, Belgrade 1989, Vol. 2, 215–218.

Plenge, G., Lehmann, P., Wettschureck R. and Wilkens, H. (1975) New methods in architectural investigations to evaluate the acoustic qualities of concert halls. *Journal of the Acoustical Society of America*, **57**, 1292–1299.

Rindel, J.H. (1991) Design of new ceiling reflectors for improved ensemble in a concert hall. *Applied Acoustics*, **34**, 7–17.

Sabine, W.C. (1922) *Collected papers on acoustics*, Harvard University Press (reprinted 1964, Dover, New York).

Schroeder, M.R., Gottlob, D. and Siebrasse, K.F. (1974) Comparative study of European concert halls: correlation of subjective preference with geometric and acoustic parameters. *Journal of the Acoustical Society of America*, **56**, 1195–1201.

Sémidor, C. (1997) Comparison between the results of objective and subjective surveys in a dual-purpose hall. *Proceedings of the Institute of Acoustics*, **19**, Part 3, 29–38.

Ueno, K. and Tachibana, H. (2003) Experimental study on the evaluation of stage acoustics by musicians using a 6-channel sound simulation system. *Acoustical Science & Technology*, **24**, 130–138.

Vorländer, M. (1995) International round robin on room acoustical computer simulations. *Proceedings of the 15th International Congress on Acoustics*, Trondheim, Vol. II, 689–692.

Xiang, N. and Blauert, J. (1991) A miniature dummy head for binaural evaluation of tenth-scale acoustic models. *Applied Acoustics*, **33**, 123–140.

Zahorik, P. and Wightman, F.L. (2001) Loudness constancy with varying sound source distance. *Nature Neuroscience*, **4**, 78–83.

4 The development of the concert hall

4.1 Introduction

To the Renaissance mind, the harmony of music would beget buildings of sublime geometry. Such was the supposed universality of proportion that the planets were imagined to generate 'the music of the spheres'. In the concert hall, the least constrained of the auditorium types, is not architectural form linked most directly to the physics of music making? Regrettably the reality has proved more complex. No logical argument supports the use of particular precise proportions in music rooms; ratios such as 3:2 or 5:4 which as frequency relationships so delight our ears, have no magic qualities for auditorium spaces. Nor does resort to simple geometrical forms like ellipses provide an answer; the ellipse was tried and already found wanting in the eighteenth century. If proportion has a place in concert hall design it is for a host of different reasons which combine to leave particular values appropriate. The width of a hall might be related to the maximum acceptable width for a stage, the length to visual distance and the height to reverberation time considerations for instance. By chance the simple proportions of the ballroom proved to have acoustical virtues, which launched a golden era of concert halls before a science of room acoustics existed. Prior to the twentieth century, good acoustics owed more to precedents than theory.

The acoustic influences on design encompass the nature of the sound production, the way in which it is propagated through the room and probably most importantly the manner in which our ears process what they hear. Perception of harmony is a particularity of the ear, whereas an auditorium reacts identically to musical sounds and raucous noise. Of the many conflicts of interest involved in concert hall design, that between the demands of creating a space for a shared experience and the needs of a few thousand pairs of individual ears is possibly the most severe. It is a major challenge to a design team to reconcile these and other requirements.

Two main influences have shaped the concert auditorium: precedents and science. Design according to precedent was of necessity the norm prior to 1900, the date from which a science of room acoustics can be said to have originated. During the twentieth century, reliance more on precedents or more on science has been a theme for discussion with variations. Conservatives have relied on tried and tested precedents. More ambitious designers have responded to new scientific ideas, though in some cases these ideas have proved to be misguided. In terms of auditorium plans, nearly all concert halls can be related to one of four traditional forms, as illustrated in Figure 1.1: the fan-shape, the arena, the Baroque theatre and the rectangular plan (Barron, 1992). Examples of concert halls conforming to each plan are to be found in this chapter and the next one. Only two of these plans have emerged as suitable for concert performance.

Whereas today we design buildings for a particular style of music, indeed often mainly for music of earlier eras than our own, there are examples of buildings influencing musical form. An obvious musical example is that of plainsong which has survived as an art form for over 14 centuries. Its development must have evolved with church architecture. As building techniques improved after the Dark Ages, so aspirations grew to construct high churches with large enclosed volumes. The resulting long reverberation time not only makes speech transmission impossible, but also blurs most musical detail. In the slow-moving Gregorian chant, one tone is heard against the background of reverberation of the previous note. Harmony can be created from a single musical line. The chant was an admirable response to what must have been an unplanned acoustic environment. Regarding the relationship between the built form and the music, one can also observe that the form of chant was probably influenced by the individual church: the Ambrosian chant from the less reverberant Milan Cathedral is more decorated than the Roman Gregorian.

4.2 Early developments

Much of the early history of secular music was associated with court surroundings. Ballrooms were often used for concerts and the so-called classical concert hall can be seen as a development from this building type. Predictably ballrooms followed the rectangular plan form. The first commercial opera house was opened in Venice in 1637 (Chapter 9), but public performance of pure instrumental music arrived later. It is first found in England where music had been enthusiastically cultivated since the fifteenth century. Following the trauma of the English Civil War and eventual restoration of the monarchy, the earliest recorded public concert in Europe took place in London in 1672. During the next hundred years, London became the most vigorous capital for music, with the first purpose-built concert room in 1680 followed by many more. In the 1730s the fashion for music gardens sprang up, providing good music for all. The London examples at Vauxhall and Ranelagh Gardens were copied in several other European cities. None of these old London concert venues has survived, though they are well documented (Elkin, 1955; Forsyth, 1985).

The opportunities for historical detective work in acoustics are rare. A fascinating investigation has

been made by Meyer (1978) into the concert halls used for the first performances of Haydn's symphonies and the extent to which the compositions reflect the various acoustic characteristics of these spaces. The data collected by Meyer offer a chance to review a range of eighteenth-century auditoria. During Haydn's employment with the Esterházy family, the principal halls for which his symphonies were composed are in the Schloss Eisenstadt (Austria, 1760–65) and the Schloss Esterháza, Fertöd (Hungary, 1766–84). Both halls still exist much as they were 200 years ago. Haydn's London symphonies were performed in the Hanover Square Rooms (1791–94) and the King's Theatre, Haymarket (1794–95). Neither was still in use by 1900. The plans of all four halls are given in Figure 4.1, together with the mean ceiling height. The ceilings were nominally flat except in the case of the Hanover Square Rooms which was vaulted. All the halls had flat floors and no significant balconies. Owing to varying seat densities, the seating capacities in Table 4.1 are unreliable. The Hanover Square Rooms, built in 1775 by a partnership which included the (London) J.C. Bach, was to become the most famous concert hall for the next century. Its audience size was intended as 800, which in an area of 200 m^2 gives a density of 4 per m^2 (compared with a typical 2.1 per m^2 today). Somehow on exceptional occasions they fitted 1500 people into the hall! Contemporary pictures show a well-raked stage suitable for good sight and sound lines. The vaulted ceiling of the Hanover Square Rooms had a centre of curvature about 1 m above the floor. This would produce strong reflections and a non-uniform response, but an acoustic character liked by some (see section 6.4 for the response to a current example, where a contemporary has described the sound as 'excellent').

The room volumes of Haydn's halls varied from 1530 to 6800 m^3, and the calculated reverberation times between 1.0 and 1.7 seconds. With audience at a single level, we find the reverberation time closely related to the ceiling height (section 2.8.6). The size of orchestra specified and used by Haydn increased progressively during his career (Table 4.1). Of particular interest is the sound level

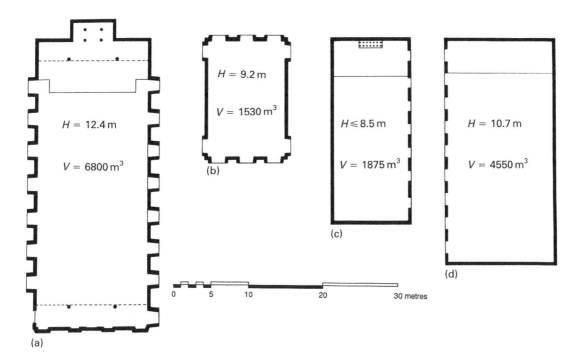

Figure 4.1 Plans of Haydn's concert halls (courtesy of J. Meyer). (a) Haydn-Saal, Schloss Eisenstadt; (b) Musiksaal, Schloss Esterhàza; (c) Hanover Square Rooms, London; (d) King's Theatre, London. H = ceiling height, V = room volume

Table 4.1 Details of the halls used for the performance of Haydn's symphonies (after Meyer, 1978)

	Concert hall			
	Schloss Eisenstadt	*Schloss Esterháza*	*Hanover Square Rooms, London*	*King's Theatre, London*
Dates used by Haydn	1760–65	1766–84	1791–94	1794–95
Volume (m³)	6800	1530	1875	4550
Approximate seat capacity	400	200	800	1050
Estimated reverberation time (s)	1.7	1.2	1.0	1.5
Orchestra size (without percussion)	16	22	36	57
Number of violins in orchestra	6	11	14	24
Forte level produced by above orchestra (dB)	84	90	91	92

produced in the various halls for these symphonies. It is intriguing to note that the levels calculated by Meyer for a *forte tutti* increased monotonically from 84 db at Eisenstadt to 92 db in the King's Theatre. So the orchestra size was increasing more than enough to compensate for increased audience sizes. A modern orchestra of 85 plus percussion produces about 90 db in the Vienna Musikvereinssaal during a *forte tutti*. Haydn, the father of the modern symphony, was thus for the majority of his life writing

Figure 4.2 Altes Gewandhaus, Leipzig

symphonies involving sound levels very close to present-day standards. However, in the name of authenticity many of our contemporary performances of classical symphonies use forces comparable to those of the classical period, but in concert halls which are substantially larger than those of the eighteenth century. In these circumstances the sound level for our contemporary audience will be significantly less than it was in Haydn's day, reducing the impact for the listener.

Meyer lists several examples which indicate Haydn's sensitivity to the acoustics of these various halls. In the small Esterháza hall with its short reverberation time, orchestral works would take on a chamber music character; scoring of this nature is recognizable in many of the symphonies composed there. In Symphony No. 61, first movement, there is the case of a *forte* chord followed immediately by a fast single *piano* line for the first violins. This *piano* line would only be audible in a dry acoustic. Whereas for the King's Theatre with its longer reverberation time, in the first movement of

Symphony No. 102 Haydn leaves a bar's rest after a *fortissimo* chord before the music continues *piano*, allowing the chord to reverberate. Meyer also suggests that Haydn was responsive to the nature of spatial impression (section 3.2) which is sensitive to sound level. He cites the example of a long unison orchestral passage at the beginning of Symphony No. 102, which starts *piano* followed by a long *crescendo* and *diminuendo*. To be fully effective, this would require the gradual onset and later collapse of spatial broadening associated with lateral reflections.

Public facilities comparable to those in London were found in continental Europe only in 1761 with a concert hall in Hamburg, though nothing is known of its size. The Leipzig Thomaskirche is of course famous for its association with J.S. Bach (1685–1750), for which his B-minor Mass and St Matthew Passion among other works were written. The town of Leipzig had no royal court associated with it, but enjoyed a vigorous musical life throughout the eighteenth century. After several false starts, a concert hall was constructed in an upper section of the Gewandhaus, or Drapers' Hall, in 1781 (Figure 4.2). This subsequently became known as the **Altes Gewandhaus** to distinguish it from the Neues Gewandhaus of a century later. It became particularly famous during the period 1835–47 when Mendelssohn was director there. The hall contained 400 seats, with rows running parallel to the long sides so that members of the audience faced each other. With a volume of 1800 m³, the occupied reverberation time was about 1.2 seconds.

By our standards the Altes Gewandhaus would rate as a recital hall, but its acoustic fame was such that its reputation established what became known as the Leipzig concert hall tradition. Bagenal and Wood (1931) discussed the characteristics which were considered to contribute to its excellent acoustics: a plan form with curved ends, wooden surfaces acting as resonators, a flat ceiling with cove margins and boxes at each end of the hall. These details were at the time much copied. Nowadays one places a different interpretation on many of them. The curved ends and boxes in the end walls do indeed prevent a slap echo from the back wall, but there are alternatives to concave surfaces which perform the same function without the risks of focusing. The idea that wooden panelling enhances tone by resonating was a major misconception (see section 2.6.3). In reality the panelling acts as a low-frequency absorber, limiting the bass reverberation time. Some panel absorption is desirable but today one generally allows for a modest rise in low-frequency reverberation time. Finally the ceiling coves probably had little effect. The principal comment one should make about the Altes Gewandhaus acoustics is that acoustic design for 400 seats is relatively uncritical. Nowadays one would usually aim for a slightly longer reverberation time. The plan form did however prove suitable to scale up for its successor.

Few of the halls of this period were large enough to present acoustic problems. In the nineteenth century, aspirations as well as social conditions led to altogether grander spaces for music performance. Berlioz dreamt of auditoria for 10 000

performers and 20 000 listeners. Closer to reality were the schemes of the German architect Schinkel (1781–1841) who espoused the Romantic movement of the time. His various grand projects involved huge volumes and large plan forms but were rejected on economic grounds. It is in England that auditoria on the grand scale were realized. Several northern town halls, such as St George's Hall in Liverpool, are of a size too large to be used much for music today, as well as having reverberation times too long for anything other than organ music. The all-time record for a grand performance space must go to Paxton's Crystal Palace at Sydenham in south London. The Crystal Palace, originally built for the 1851 Great Exhibition, was moved to Sydenham three years later, where it remained until destroyed by fire in 1936. Regular concerts were a popular feature, with Handel Festivals drawing the largest crowds. The 1882 Festival is reputed to have had 500 instrumentalists, 4000 choir and an audience of 87 769 (Forsyth, 1985). This combination incidentally would produce (without the choir) a sound level 8 dB less than that of a typical 100-piece orchestra in a hall seating 3000. The comparison assumes however good concert hall design, which the Crystal Palace patently lacked. The experience must have been rather subdued.

The major auditorium to survive from this period is the Royal Albert Hall in London of 1871, which demonstrated all too clearly the pitfalls of grand design: focusing by concave surfaces, excessive reverberation time and quiet sound (section 5.1). The experience of the nineteenth century showed that the rectangular plan form was the most successful acoustically, but that there were limits to the audience numbers which could be accommodated.

4.3 The classical rectangular concert hall

In small halls the room form matters relatively little for the acoustics. This probably applies to smaller rectangular plan auditoria built in Britain as civic halls during the nineteenth century. **Birmingham**

Town Hall of 1834, closely based on the Temple of Castor and Pollux in Rome, is an important example. It has a rich history which includes witnessing the premières of Mendelssohn's Elijah and Elgar's Dream of Gerontius. (The hall has recently been thoroughly refurbished, reopening in 2007 with a balcony on three sides and seat capacity of 1086.)

In the latter part of the nineteenth century the demand arose for large public concert halls, and those that derived from the rectangular plans and dimensions of ballrooms proved to have particularly favourable acoustics. The proportions were roughly of a double cube, i.e. of 1:1:2, often referred to as shoebox-shape. The balcony overhangs were modest and the style of the period led to highly decorated room surfaces. The following will concentrate on the four most significant of these classical halls plus the London hall contemporary with them. A technical analysis of the three of these halls that survive has been made by Bradley (1991). Many other rectangular halls were built which also gained good reputations, among them the Liverpool Philharmonic Hall (1849–1933), the Stadt-Casino, Basel (1876), the St Andrew's Hall, Glasgow (1877–1962) and the Grosser Tonhallesaal, Zurich (1895). Their acoustic character was similar to their contemporaries.

A major performing space in eighteenth- and nineteenth-century Vienna was the Redoutensaal in the Hapsburg royal palace, Schloss Hofburg. In 1752 this had been converted into a ballroom by Antonio Galli-Bibiena. It witnessed several famous premières, including that of Beethoven's Eighth Symphony in 1814. In 1870 the Gesellschaft der Musikfreunde (Society of the Friends of Music) opened a new building close to the Ringstrasse, containing a **Grosser** and Kleiner **Musikvereinssaal** by architect Theophil Ritter von Hansen. The latter hall is now known as the Brahmssaal, while the former has established the reputation as having one of the best acoustics in the world. Its name as the Goldener Saal derives both from its acoustics and the extensive gilding of its interior. It is the orchestral home of the Vienna Philharmonic Orchestra, and has included among its conductors Brahms, Bruckner and Mahler. It is likely that these and other Viennese composers were influenced for their compositions by the sound of the hall.

For its proportions (Table 4.2), the Grosser Musikvereinssaal followed those of the Redoutensaal extremely closely, Figures 4.3 and 4.4. The dimensions of the Redoutensaal of length 46 m, width 17 m and height 16 m were all scaled up about 12 per cent. The dimensions of the older hall were certainly not designed with concert music in mind, but the architect of the Musikvereinssaal must have appreciated its acoustic virtues. At a time before any theory of reverberation, the choice of ceiling height was certainly fortunate, given its strong influence on reverberation time. The distance to the furthest seat from the stage front is 40 m, again an accepted modern standard. In fact, the hall as we know it

Table 4.2 Details of the four most renowned classical concert halls (after Beranek, 1962 and 2004). Reverberation times are for the occupied halls at 500/1000 Hz.

	Concert hall			
	Grosser Musikvereins-saal, Vienna	Neues Gewandhaus, Leipzig	Concertgebouw, Amsterdam	Symphony Hall, Boston
Date	1870	1884–1944	1888	1900
Volume (m³)	15 000	10 600	18 770	18 750
Seat capacity	1680	1560	2037	2625
Length (m)	52.9	44.9	43.0	50.7
Width (m)	19.8	19.2	28.4	22.9
Height (m)	17.8	15.1	17.4	18.8
Reverberation time (s)	2.0	1.55	2.0	1.85

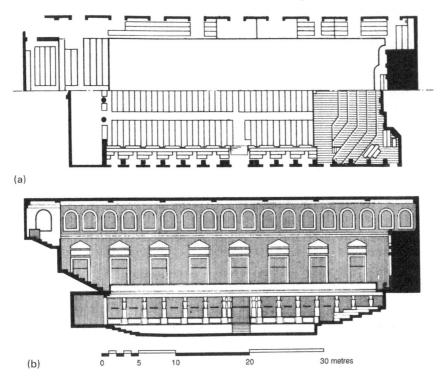

(a)

(b)

0   5   10   20   30 metres

Figure 4.3 Plan and long section of the Grosser Musikvereinssaal, Vienna

Figure 4.4 Grosser Musikvereinssaal, Vienna

today differs in several respects from the original design. A major renovation was undertaken in 1911 (Clements, 1999) primarily for reasons of fire safety. The opportunity was also taken then to cantilever the side balconies and move the caryatids from their position supporting the front of these balconies to locations flush with the side walls. This also allowed for an increase in usable platform area and, together with other modifications, for a modest increase in audience capacity.

Seating in the Grosser Musikvereinssaal is at three levels, though the gallery is basically a rear extension. The Stalls seating is mainly on a flat floor, surrounded by under-balcony seating at the level of the stage. The balcony runs round all four sides, enclosing the organ behind the stage. All surfaces are highly profiled, with the gilded caryatids along the lower side walls, numerous door surrounds, recessed windows round three sides at the upper level and a coffered ceiling. Such surface decoration produces a highly diffuse and blended sound experience, which is a hallmark of the Musikvereinssaal's acoustics. Indeed when seated in the side balcony with a view of only half the orchestra, the degree of reflected sound is so high that the visual loss appears much greater than the acoustic one. But

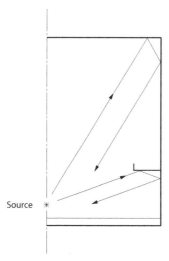

Figure 4.5 Lateral reflections from a balcony soffit and ceiling cornice on the half cross-section of a rectangular concert hall

this highly blended sound can be somewhat at the expense of clarity.

One feature which characterizes nearly all of the classical halls, but was not appreciated until recently, is the strength of the early lateral reflections they produce. Subjectively the sound source broadens considerably, with a sound image in *forte* passages extending well beyond the physical size of the orchestra. Two design features enhance the effect: the narrow width of the hall and in the stalls the possibility of balcony soffit reflections (Figure 4.5).

Yet in the hands of musicians unused to the hall, Clements (1999) records some disappointing experiences; Beranek (2004) refers to difficulties when visiting orchestras fail to restrain themselves in what is actually a fairly modest-size hall. The resident Vienna Philharmonic Orchestra have perfected their response to this subtle instrument and produce a sound that is frequently sublime, highly enveloping with delicate even textures and remarkable climaxes.

The objective behaviour of the Musikvereinssaal is given in Figure 3.34. The occupied mid-frequency reverberation time is right in the middle of the preferred range at 2.0 seconds, with a welcome gentle rise in the bass. It might be assumed that many of the internal surfaces of the hall were wood, whereas in reality there is little timber apart from the floor. Virtually all the wall and ceiling are of plaster, which hardly absorbs in the bass (Beranek, 2004). In rehearsal without an audience the acoustics are less ideal; the reverberation time rises to over 3 seconds. This occurs because many of the seats are not upholstered; all seats in the gallery and in the balcony facing the stage are wooden.

Turning to the other objective quantities in the hall, the early decay time is identical with the reverberation time, which is a symptom of a highly diffuse space. Objective source broadening is high as one expects from the comments above. However objective clarity and sound level are in many places lower than expected. This combination points to early sound which is quieter than anticipated, a surprising discovery but one which appears to do no harm to subjective quality for orchestras familiar with the hall.

A curious aspect in the history of concert halls is that the Vienna Musikvereinssaal, which has such a high reputation today, was little mentioned at the time. Neither Sabine (1922) nor Bagenal and Wood (1931) refer to it. Nor did the architects, Gropius and Schmieden, of the **Neues Gewandhaus** in **Leipzig**. This new Leipzig hall served as the world-wide model of excellence before the Second World War. From the point of view of reverberation time, this reputation proves to be somewhat surprising. The new hall in Leipzig was built because the existing Altes Gewandhaus had long outgrown its modest capacity, even with the addition of long side galleries with 170 extra seats. The design competition for a different site was won by Gropius and Schmieden. The building opened in 1884 with two concert halls, both at first floor level. The small hall was a very close replica of the old hall, which was itself later demolished. The large hall (Figure 4.6) held an audience of 1560 and immediately established its good reputation (Clements, 1998).

The architects record that 'with reference to the acoustics, ... the client intended that the form should duplicate the old hall with its highly regarded reputation' (Gropius and Schmieden, 1887). The first priority mentioned was to reproduce the box-like construction in wood which was considered (fallaciously) to allow the vibrations generated by

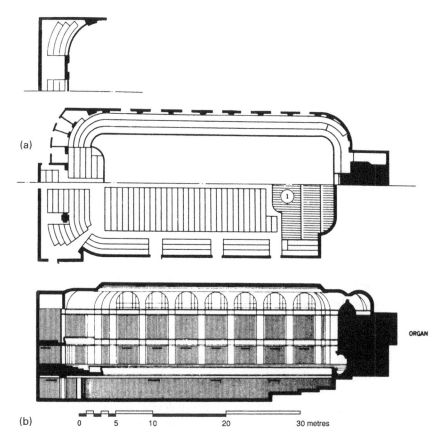

Figure 4.6 (a) Plan and (b) long section of the Neues Gewandhaus, Leipzig (1884–1944)

the orchestra to set the enclosure into sympathetic vibrations in the manner of a violin. Reflections from the rear wall were (correctly) considered acceptable only if they arrived within 1/12th second. As the rear wall reflection in the large hall would have a substantially longer delay, the rear wall was fragmented as much as possible and treated with some acoustical absorbent. Concern for reflection between parallel plane side walls led to (sound absorbing?) paintings being placed on these walls. The only auditorium referred to by the architects apart from the old hall was the Trocadero in Paris, which had the form of an amphitheatre.

In plan the length of the new hall was nominally twice its width, just as the Altes Gewandhaus had been. The curved ends were less pronounced but, as mentioned, attention had been paid to preventing echoes, with boxes at balcony level and a draped recess. Again there were ceiling coves, but here they were intersected by large clerestory windows. There was a high degree of moulding on the wall and ceiling surfaces. Bagenal (Bagenal and Wood, 1931) describes the experience of a concert in these terms:

> The Gewandhaus is a true instrument to music produced within it. ... There is no exaggeration in its reputed excellence for orchestral music. ... Tone is both 'full' and 'bright' and at the same time notes are distinct. ... To hear indeed the highly trained Leipzig orchestra in the Ninth Symphony, each phrase exactly presenting itself to the ear for the fraction of a second before it is resolved in the great onrush of the *scherzo*, to feel the control of sheer loudness maintained by the conductor, is a musical experience of considerable interest to the student of acoustics.

And by modern standards? Many of the details, including the precise proportions of the plan, are of no especial significance. Regarding vibrations being radiated by the box-like construction, Meyer and Cremer already demonstrated in 1933 that this was a myth. Nowadays we would take note of the diffusing nature of the hall envelope and the modest 18.9 m width. One would criticize the design for

poor sightlines from the gallery, and the then normal flat stalls floor. But the major surprise concerns the reverberation time which has been calculated as 1.55 seconds at mid-frequencies, probably falling slightly in the bass (Beranek, 1962; this result can in fact be derived by the Kosten method, section 2.8.6, but the measured unoccupied results by Meyer and Cremer of 1933, suitably corrected for occupancy, and analysis of recordings after the war by Kuhl are consistent with this figure).

While a reverberation time of 1.55 seconds would be suitable for late Classical or early Romantic music, it would certainly today be judged as too short for the main Romantic symphony repertoire. It is possible with the highly diffuse sound in the Gewandhaus that a shorter reverberation time was acceptable. Yet diffuse sound fields were not unique to the nineteenth century. One cannot help wondering whether musical taste has not developed since the demise of the Gewandhaus in 1944; our sensibilities have surely been influenced by daily exposure to recorded music. Although the building shell of the Neues Gewandhaus had survived the 1944 air raid, the East German authorities decided unfortunately to demolish it and build a new Gewandhaus in 1981 to a contemporary design (section 4.11).

Amsterdam's response to the Leipzig tradition has fortunately survived (Figures 4.7 and 4.8). Designed by A.L. van Gendt and completed in 1888, the **Concertgebouw** also has a high reputation for its acoustics, though it differs from its forebears in several respects. Principal among these is a width which is 45 per cent and 50 per cent larger than that of the Vienna and Leipzig halls, respectively. But providence again was kind in the selection of a ceiling height which provides a 2 second reverberation time when occupied with audience. The stalls seating is on a flat floor, though sightlines are somewhat compensated by a very high stage level of 1.5 m. A single high balcony skirts the hall in front of the stage. Behind the orchestral platform is extensive choir seating, which is now often used for audience. The acoustics for the audience in the stalls are generally considered slightly inferior to those of the Musikvereinssaal, with a sound which is live

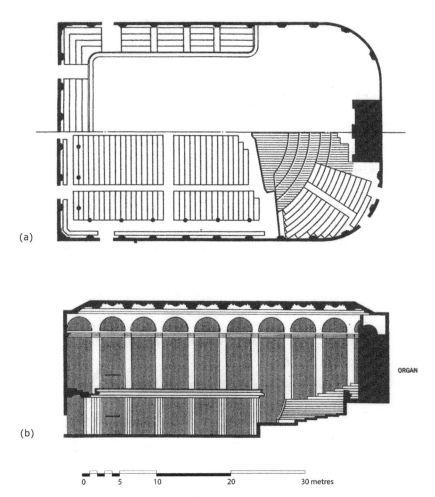

(a)

(b)

ORGAN

0	5	10	20	30 metres

Figure 4.7 (a) Plan and (b) long section of the Concertgebouw, Amsterdam

but lacking a little clarity. Conditions in the balcony though are good, both clear and well balanced. Beranek (1962) quotes musicians with differing views about the Concertgebouw. A likely reason for this is the lack of support they receive from above, with an unusually large distance of 16 m from the stage to the ceiling.

London's hall from this period was less distinguished and followed the rectangular plan less closely, but was viewed no less affectionately by its users. From 1895 it became the first home of London's annual Promenade concerts, which now take place in the Royal Albert Hall. The **Queen's Hall** (1893–1941) was designed by T.E. Knightley who described his scheme as follows (Elkin, 1944):

> The feature of the internal design ... is on the reversal of the lines usual in such cases – frequently a horseshoe; in this one a parallelogram with a curved end, wind instruments being the inspiration. For example, the end of the horn is normally convex, and that form has been adopted for the orchestra. The junction between the ceiling and wall is usually a hollow curve; in this case it will be the reverse.

Figure 4.8 Concertgebouw, Amsterdam

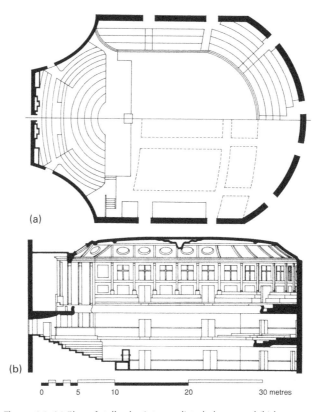

(a)

(b)

| 0 | 5 | 10 | 20 | 30 metres |

Figure 4.9 (a) Plan of stalls plus intermediate balcony and (b) long section of the Queen's Hall, London

In fact the plan for the hall (Figure 4.9) derived considerably from a design by the theatre architect C.J. Phipps for the same site but a different client; the dispute about the plan's origin had to be referred for arbitration to the Royal Institute of British Architects. The use of convex surfaces is frequently good acoustic practice, but the rear auditorium wall was concave and this must have performed much like the rear wall of the Usher Hall, Edinburgh, does to this day (section 5.2). The use of extensive wood panelling was a further detail which would not be used today, as it diminishes the sense of warmth. The audience capacity was initially 3000, becoming latterly a more relaxed 2050. With a volume of only 12 000 m³, the reverberation time was short, probably 1.4 seconds, which meant a clear but rather dry acoustic (Parkin, Scholes and Derbyshire, 1952). Allen (1969) records that the sound was best in the gallery. Given that the design owed much to Victorian theatre forms, this observation is in line with one made frequently of traditional opera houses.

The first application of a science of acoustics to concert hall design for **Boston Symphony Hall** is a much-quoted episode (Figures 4.10 and 4.11). The town of Boston, Massachusetts, USA, wanted to demolish their existing Music Hall of 1863 to make way for a road (Beranek, 1977, 1979 and 1988); in the event the old Music Hall has survived and is now called the Orpheum Theatre. The architects, McKim, Mead and White, had visited Europe and chose the Leipzig Neues Gewandhaus as a standard of excellence. Various conductors had advised them against pursuing their scheme of a concert hall in the form of a classical Greek theatre, since it was untried. Wallace Sabine was engaged somewhat by chance in 1898. He hesitated to accept, but spurred on by the challenge, he analysed over a couple of weeks all his reverberation time results from the Fogg Lecture Hall and other rooms in Harvard University (Chapter 1) and derived the first form of the now-famous Sabine equation for reverberation time. He then agreed to offer his assistance. Sabine

Figure 4.10 Symphony Hall, Boston, Massachusetts

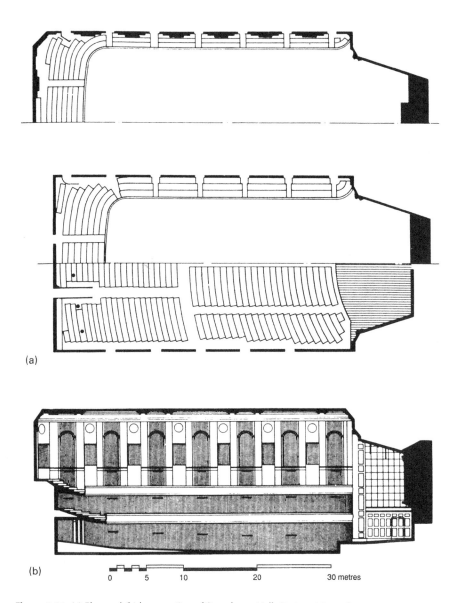

(a)

(b)

0   5   10   20   30 metres

Figure 4.11 (a) Plan and (b) long section of Symphony Hall, Boston, Massachusetts

next embarked on a frantic exercise collecting more absorption data to enable him to calculate reverberation times of auditoria.

The seat capacity for the new hall was to be 2600 compared with 1560 in the Leipzig Gewandhaus and this had led the architects to suggest scaling up all dimensions of the Gewandhaus. Sabine realized that this would imply a doubling of the auditorium volume, which in the absence of substantial (undesirable) extra absorbing material would lead to an

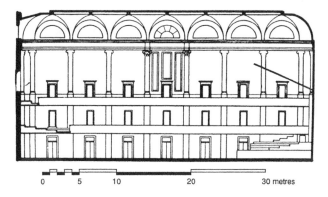

0   5   10   20   30 metres

Figure 4.12 Long section of the old Music Hall, Boston, Massachusetts

excessively reverberant hall. The only other hall referred to was the old Boston Music Hall of 1863 (Figure 4.12), which with 2361 seats was much closer in size to the new design. Beranek (1977) says that the old hall was well liked in Boston, while Sabine (1922) wrote that 'the old Music Hall was not a desirable model in every respect, even acoustically'. In fact the new hall bore a much closer resemblance to the old than to the Gewandhaus. It has two balconies rather than the one in the Gewandhaus. The principal modifications to the Music Hall design were to make it lower and longer and to remove the angled reflector above the orchestra. The stage area in Symphony Hall is mainly in a recess behind a notional proscenium frame. To compensate for the acoustic effects of increased length, the side walls of the stage enclosure were angled to project sound out to the audience. Symphony Hall has an early example in a concert hall of an optional raked stalls floor.

Sabine (1922) calculated the reverberation times of the Gewandhaus and old Music Hall as 2.30 and 2.44 seconds, with the new hall as 2.31 seconds. In the first and last cases these are now known to be substantial over-estimates; the occupied value in the Boston Symphony Hall is 1.8 seconds. Beranek (1977) suggests that the large gaps in Sabine's diary of the period may be due to the anguish this discrepancy caused him. In reality a reverberation time value of 2.3 seconds would have been excessive and

1.8 seconds is now generally considered the bottom of the optimum range. The main reason for Sabine's prediction error was that he was calculating seat absorption on a per-seat basis rather than on an absorption-per-square-metre basis. It took over 60 years for this confusion to be resolved (section 2.8.3).

Of particular interest is the question of what other acoustic advice Sabine gave for the design of Symphony Hall. As his general philosophy, he stated (1922) that 'in order that hearing may be good in any auditorium, it is necessary that the sound be sufficiently loud; that the simultaneous components of the complex sound should maintain their proper relative intensities; and that the successive sounds in rapidly moving articulation, either of speech or music, should be clear and distinct, free from each other and from extraneous noises'. Records on his specific advice are sadly few, but he did record that 'in the new hall the orchestra is not out in the main body of the room, and for this reason is slightly farther from the rear of the room [compared with the Gewandhaus and the old Music Hall]; but this is more than compensated for in respect to loudness by the orchestra being in a somewhat contracted stage recess, from the side walls of which the reflection is better because they are nearer and not occupied by an audience'.

Subjectively Boston Symphony Hall is rated among the best. Beranek (2004), who knows the hall intimately, describes the sound as clear, live, warm, brilliant and loud. He considers that only the other classical halls 'have the same growth of *crescendo* and quality of reverberation as Symphony Hall'. Acoustic measurements by Bradley (1991) point to a distribution of sound level which favours positions in the front half of the hall; he ascribes this to the stage enclosure in this hall. The balcony overhangs are also rather deeper than would now be recommended. Notwithstanding such minor blemishes, Boston Symphony Hall is considered by many to have the best acoustics in the Americas.

As an example of acoustic design, Symphony Hall clearly relied on precedent, but in its details it is hard to fault. It was a remarkable achievement for a man who had only worked in acoustics for three years before being engaged as the consultant. Sabine was not involved with any other large concert halls, and after his death in 1919 most 'students' of his work failed to appreciate his breadth of vision. His multi-dimensional view was many years ahead of his time.

Much has been written about the acoustics of classical shoebox halls. Some of it is misguided or irrelevant but to establish the reason for acoustic success of these halls is problematic. It is much easier to determine the reason for a failure than a success. One feature now considered irrelevant is the detailed proportions, though the width, length and height are all significant in their various ways. What is certain is that their good acoustics are not due to a single feature (Müller, 1992). In objective design terms, the favourable aspects appear to be that all seats are close to reflecting surfaces, that the hall width is small (which together with the effect of reflection from balcony cornices enhances spatial impression), that parallel side walls produce a high reflection density (Figure 4.13), that the hall surfaces are highly scattering and that balcony overhangs are shallow. With the highly diffused nature of sound in classical halls, it may be that a long reverberation time is less crucial than elsewhere, but values of more than 1.8 seconds are clearly preferred for the Romantic repertoire. In subjective terms, clarity is very much a function of reverberation time; poor clarity can be a subjective weakness of these designs. The sense of reverberation and envelopment by the sound are both very good. With many early reflections, the degree of intimacy and loudness is usually also high. Acoustical warmth is probably a function of relative bass level, which is determined by the materials of the hall and is generally independent of gross room form. There are a few audible differences between the halls, some of which can be linked to measurements (Bradley, 1991).

In terms of optimum acoustics, the classical hall offers one among what may be several options. Their acoustics are characterized by a basically diffuse sound, with a sense of later reverberation arriving from all directions. This contributes to a well-rounded sound, with acoustics which respond effortlessly

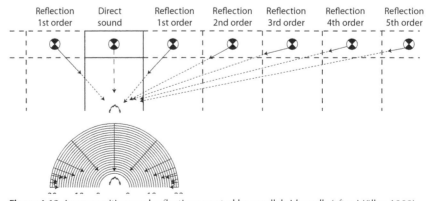

Figure 4.13 Image positions and reflections created by parallel side walls (after Müller, 1992)

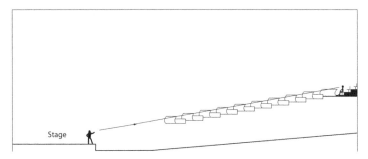

Figure 4.14 Stepped side-wall balconies obscure side-wall reflections. On this long section, the reflection off a plane wall without a balcony would be directly behind the direct sound path shown

to climaxes. Classical halls are kind, even flattering, to performers, concealing minor blemishes of performance. On the other hand several modern halls have enhanced early reflections, which offer a higher degree of clarity in reverberant spaces. To many people the modern halls offer a greater sense of identification with the performance: in other words more acoustic intimacy. Whether one prefers the diffuse or the somewhat focused sound is a matter of personal preference. Interestingly, some early halls moved towards the more focused design with reflectors above the stage (as in the old Boston Music Hall, Figure 4.12), but today we recognize that overhead reflections tend to create unwelcome colouration of tone, particularly for the strings.

Eulogy for the virtues of this design form, developed substantially before a science of auditorium acoustics existed, should not blind us to the true status of the achievement. In many of these halls the seating numbers are modest by present-day concert hall standards (up to 3000 seats). In addition the seat density of older halls is usually higher than is now acceptable. Achieving good acoustics is easier in smaller halls. Many of these old halls suffer from deficiencies at certain locations, particularly from poor sightlines in the side balconies. For instance in the Vienna Musikvereinssaal, most side balcony seats behind the front row have restricted view of the stage and the seats towards the back of the flat floor suffer from a lack of brilliance.

Any attempt to simply scale up the dimensions of the traditional classical hall runs into worse

sightline problems. One tempting solution to improve sightlines is to use side balconies that step down towards the stage. But while these can create better conditions in the side balconies, they tend to obscure wall reflections to the rear of the circle (Figure 4.14), resulting in disappointing sound in the balcony opposite the stage. The Royal Festival Hall in London is an interesting development within the constraints imposed by rectangular hall design (sections 4.6 and 5.5). The rectangular plan concert hall has had its champions throughout the twentieth century and has become more popular in recent years. This will be reviewed in section 4.12.

4.4 The directed sound hall

The arrival of the Modern Movement in architecture after the First World War meant the end of all the decorative mouldings, the statues in niches and the coffered ceilings of the classical halls. These decorated surfaces tend to create highly diffuse acoustic conditions, which are regarded by some to be a hallmark of the best acoustics. While acoustically scattering surfaces can normally be considered a safe expedient, there is enough evidence of successful halls with major unprofiled surfaces to indicate that fragmentation of **all** large surfaces is not essential for good acoustic design. Yet the opposite extreme of a hall with only large bare surfaces is unlikely to be satisfactory for concert use. The change of architectural style towards pure lines and smooth planes makes consultancy more difficult, as a demand for acoustic scattering surfaces may conflict with visual preferences.

In the heady early days of the Modern Movement, science had to provide the logical basis for much of design. Gustave Lyon rose to the challenge and conducted some flamboyant experiments to establish the value of acoustic reflections. Two observers were suspended below small balloons which could be moved independently. By the time they were 11 m apart, the speaking voice was found to be quite inaudible. On the other hand, over perfectly smooth water at night a normal voice could apparently be heard a mile away (Andrade, 1932). Lyon concluded that sound reflections are indispensable for sound transmission. In his design of the **Salle Pleyel, Paris** (1927; architects Auburtin, Granet and Mathon), he profiled the ceiling to optimize the reflection of sound onto the audience. He had also conducted experiments to establish the acceptable delay of a reflection, and applied the criterion of a maximum of 1/15th of a second, or 67 ms. For the stage of the Salle Pleyel he perceptively applied this knowledge to limit the maximum separation of performers on the diagonal dimension of the stage to 23 m (the distance travelled by sound in 67 ms).

In a contemporary record of his achievement, the journal *Building* (1928) noted that 'at once engineer and musician, M. Lyon has applied mathematics to the solution of the problem of constructing an auditorium capable of holding 3000 persons, each of whom shall be able to hear every note played or sung on the stage, and also be able to view the whole of it'. Andrade (1932) recorded his personal experience of clearly hearing a lecturer throughout the hall, including at the back of the gallery more than 45 metres away. These are impressive claims which the classical hall could not make. But does such a design, which allows speech transmission to so many, constitute a concert hall?

The long section in Figure 4.15 clearly shows the logic of the design, whereas the plan is an unexceptional modest fan shape (Figure 4.16). Each segment of the ceiling (labelled AB, BC and CD) directs a reflection onto a separate section of audience. The gross form approximates a cylindrical parabola, with the property that sound from the focus is

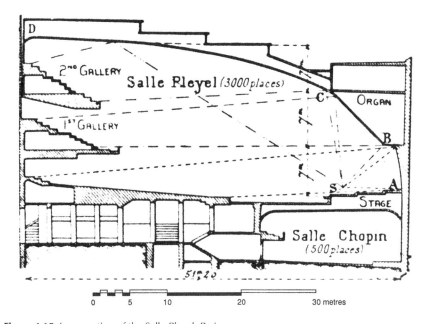

Figure 4.15 Long section of the Salle Pleyel, Paris

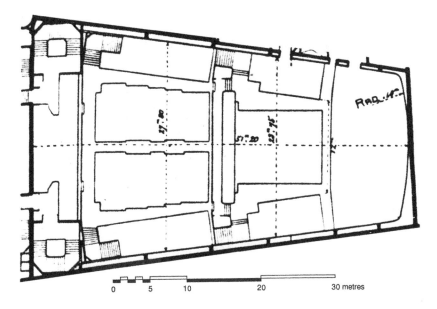

Figure 4.16 Plan of the Salle Pleyel, Paris

reflected out in a parallel beam (hence the use of the parabola for searchlights and large telescopes). Already from Figure 4.15 one senses that the design is aimed at a single source position, a lecturer not an orchestra. It was a design scheme which was also proposed by Le Corbusier for the 2600 seat Debating Chamber of the League of Nations, Geneva, in 1927 (Figure 4.17).

Unfortunately such concave ceilings are not without their problems. But first let us return to the experiments on speech transmission. Though reflections are important, the two examples quoted overstate their value. Unaided speech can be projected further than 11 m (section 7.2), so background noise is likely to have influenced that result, while the long-distance propagation over water must have been enhanced by vertical air temperature gradient effects. An early reflection in a room only doubles the energy of the direct sound.

What the design of the Salle Pleyel achieved was to project nearly all the energy which hits the ceiling onto absorbent audience. This offers good transmission to the listener, but the performer suffers because sound travels in both directions.

Initially the performers were frustrated by a strong echo off the ceiling and rear wall. After this had been reduced by absorbent on the rear auditorium wall, the performers found that so much audience noise was focused on them they could barely hear each other. After rebuilding in 1928 following a fire, high zigzag screens were placed around the orchestra as a response to the early experience.

The acoustic character of the Salle Pleyel was highly non-diffuse, almost the opposite extreme to the classical hall. A loud clear sound was achieved but at the expense of most other aspects considered important for music listening. The hall was renovated in 1981, 1994 and 2006. Following the 1994 work (Xu, 1995), the hall seated 2400, with a volume of 19 200 m³ and occupied reverberation time around 1.6 seconds. The most recent refurbishment in 2006 by Artec Consultants of New York was substantial, involving major revision of the stage area, extending balconies along the side walls, reducing the seat capacity by a further 500 seats and raising the ceiling.

The Salle Pleyel was probably the most extreme example of a hall suffering from what McKean (1996)

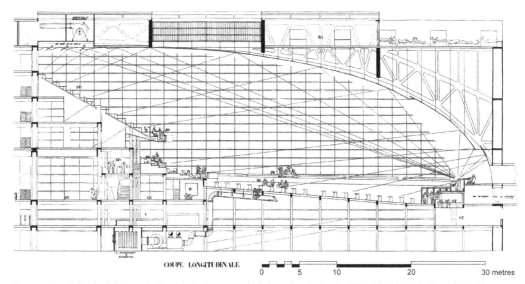

Figure 4.17 Salle des Nations design for the League of Nations, Le Corbusier(©FLC/ADAGP, Paris and DACS, London 2009)

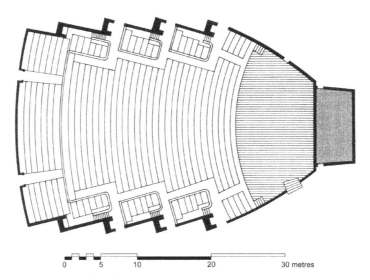

Figure 4.18 Plan of Gothenburg Konserthus

has charmingly called the 'megaphone fallacy'. It was based on the belief that 'absolute clarity' of directly received sound was needed. We now know that sound arriving from many directions is appropriate for music as well as a reverberant response. Two famous concert halls from the inter-war years used a megaphone profile in plan: the 1935

Konserthus of Gothenburg, Sweden, and the Kleinhans Music Hall in Buffalo, USA, of 1940; both are reviewed in Beranek (2004). These halls are in fact very different in size, the Gothenburg hall (Figure 4.18) has 1286 seats and no balcony, whereas the Buffalo hall has a large audience capacity of 2839 and a substantial single balcony. Both halls suffer from short reverberation times of around 1.6 and 1.3 seconds but also from a lack of spatial sound. The fan-shaped plan hall is another auditorium form with similar problems.

An even more bizarre philosophy was proposed by F.R. Watson (1923). Watson had concluded that nearly all acoustic defects were caused by long delayed acoustic reflections, while nearby reflections are beneficial. He proposed that auditoria should be designed with outdoor acoustics, with absorbent liberally employed in the auditorium but with reflecting surfaces on stage. The Severance Hall, Cleveland, Ohio, of 1930 was based on Watson's ideas. Acoustic design had taken a serious step backwards.

4.5 The fan-shaped hall

A major misconception of many of the designs in the first half of the twentieth century was the (tacit?) assumption that there was one 'perfect acoustics', that there was no distinction between the appropriate conditions for the spoken voice and instrumental music. In view of the totally different forms developed historically for drama theatres and concert halls, this was a blinkered view. In the 1920s a new art form developed with its own auditorium type: the cinema. The acoustic requirements for cinema are not particularly stringent and in order to maximize the audience size the obvious plan form is the fan shape, originally developed so successfully by the ancient Greeks for outdoor theatre. Inevitably the fan-shape plan was also adopted for concert halls, however during the last thirty years or so this plan form has become anathema to many acousticians. There is more than one reason why the fan shape is unflattering to concert performance.

The most obvious problem with the fan shape is that the rear auditorium wall is automatically generated as a concave curved surface, which produces a focused echo back to the stage. There is a simple remedy for this in tilting the rear wall to reflect sound down onto the audience. The alternatives of fragmenting the rear wall surface to make it scattering or placing absorbent on it have often

been used but if the degree of focusing is too great, echoes may still be audible. Of these remedies, only by rendering the surface scattering is the rear wall retained as reflecting. As a major bounding surface it is desirable that it should not absorb acoustic energy.

Other acoustic problems of the fan-shape plan include the following. The extreme width at the rear of the hall tends to leave seats in the centre rear of the stalls with few early reflections. This is particularly the case for reflections from the side. As seen in Figure 3.13, the angle of arrival for side wall reflections is small in the fan shape. One of the acoustic characteristics therefore of this plan form is a limited degree of source broadening. The acoustic imperfections of the fan shape also extend to the later part of the received sound. Whereas in the rectangular plan there is vigorous interreflection of sound between the parallel side walls, the potential for multiple reflection is much reduced in the fan shape (Figure 4.19). This compromises the sense of feeling surrounded by sound. It also leaves a low level of late sound towards the rear of the hall, where the sense of reverberation is lower. Krokstad, Strøm and Sørsdal (1968) report from computer studies a lack of mid-period reflections (40–180 ms) in fan-shaped designs. This may contribute to an early decay time shorter than the reverberation time, again contributing to a less reverberant sensation. The degree to which a fan-shape plan suffers these various deficiencies is predictably influenced by the angle of fan; larger angles of fan have more extreme acoustics.

Subjectively therefore the fan-shaped hall tends towards, though is not as extreme as, the directed sound hall discussed in the previous section. The sound is more frontal than lateral and the sense of reverberation is diminished by the limited degree of diffusion. If the ceiling reflections are strong, the sound quality can become harsh due to tone colouration effects. Within these halls pronounced variations in quality can often be observed, with a tendency for a dull sound towards the rear of the stalls seating and a lesser sense of reverberation at seats distant from the stage. Two extreme cases

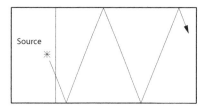

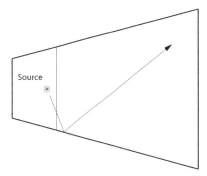

Figure 4.19 Multiple reflections from the walls of a rectangular plan and a fan-shape plan

with a semicircular plan are discussed in sections 11.1 and 11.2.

As an example of a more realistic fan shape, the 2700-seat **Alberta Jubilee Halls**, Canada, are of interest (Figure 4.20). These virtually identical halls in Calgary and Edmonton bear some resemblance to the earlier Kleinhans Music Hall of 1940 in Buffalo. Opening in 1957, they were designed by the Alberta Department of Public Works, not in fact as pure concert halls but rather as multi-purpose spaces with stage flying facilities, though concerts were a major use employing an orchestral shell on stage. Among design details which deserve note is the elaborate diffusing structure on the rear wall which successfully suppresses an echo (Northwood and Stevens, 1958). The front sections of the ceiling were made convex to render less intense the ceiling reflections to the front stalls with delays of 40–50 ms.

The auditorium volume (21 500 m³) was selected as intermediate between the requirements for speech and music, giving an occupied reverberation time of 1.4 seconds. That this was short for music

use was noted by many of the listeners at a test concert. Objectively the seating area with minimum sound level in the hall was found to occur in the middle of the ground floor, testifying to the hole-in-the-middle reflection problem with this plan form. On the other hand at 'the rear of the top balcony ... the sound had brilliance and an incisive quality of attack that would please the "hi-fi" enthusiast, but [with] much less large-hall effect'. This proves to be due to the facetted ceiling design which concentrates reflections onto the top balcony. O'Keefe has provided an incisive analysis of the acoustics of these Jubilee and other post-war halls, which suggests that a major issue for concert acoustics is the height–width ratio (O'Keefe, 2002). He found that wide, low halls suffered not only from poor spatial sound but also poor reverberance and sound level.

As a modern comment on these designs, both cities have subsequently commissioned new concert halls following classical rectangular lines: the Jack Singer Concert Hall in Calgary of 1985 (Forsyth, 1987) and the Winspear Centre in Edmonton of 1997. The Alberta halls themselves underwent major refurbishment in 2005 to improve their acoustics; a major modification was to raise seating sections next to the side walls, thereby creating reverse-splay seating areas rather than fan-shaped (Jordan and Rindel, 2006).

In Europe, the Konserthus in Oslo of 1977 with 1700 seats is a further example of a fan-shape plan that has yet to be seriously modified (Jordan, 1980). The architect best known for using almost exclusively the fan-shape plan for auditorium designs was the Finnish architect, Alvar Aalto (1898–1976). Most notable are two halls in Helsinki: the Kulttuuritalo of 1958 (1500 seats) and the Finlandia Concert Hall of 1971 with 1750 seats (Forsyth, 1985). For the Finlandia Hall, Aalto acted as his own acoustic consultant, yet subsequently the hall's acoustics have been rated as disappointing. At the time of writing there is some electronic enhancement in the hall and the city of Helsinki is building a new concert hall. Even in the hands of a famous architect, the fan-shape plan can offer inferior acoustics for symphony concerts.

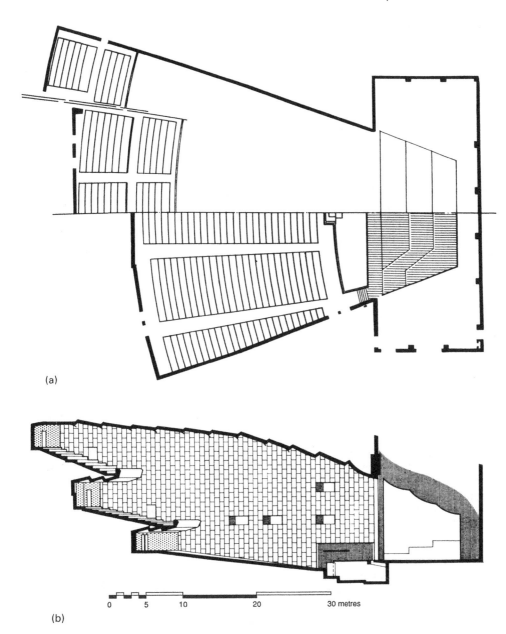

(a)

(b)

0   5   10   20   30 metres

Figure 4.20 (a) Plan and (b) long section of the Alberta Jubilee Halls, Canada

4.6 The synthesis in British post-war halls

Following the massive bombardment of European cities during the Second World War, Britain was the first country to build new concert halls to replace war damage. The major acoustician in Britain from the 1930s was Hope Bagenal (1888–1979) – see Trevor-Jones (2001) for a brief biography. His book with Wood *Planning for good acoustics* (1931) was remarkable for its time, though his views continued to develop (see especially Bagenal, 1950). For the design of the Royal Festival Hall, London, and its 1951 contemporaries in Bristol and Manchester, the designers had experience of two models: the classical rectangular plan and the more recent focused designs, often following a fan-shaped plan form. With demands for an audience capacity approaching 3000, they knew they were unable to simply scale up the classical designs. Yet on balance it was felt that the sound quality associated with the rectangular plan was to be preferred. In the rectangular hall, not only are there no concave surfaces likely to create serious echoes but 'in addition [it] has the possible advantage that there is more cross-reflection between parallel walls which may give added "fullness"' (Parkin *et al.*, 1953). The solutions involved a synthesis of the rectangular plan with long sections responding to more recent trends. The choice of ceiling profile was partially directed in the case of the Royal Festival Hall (Figure 4.21), slightly concave in Bristol and diffusing in Manchester.

The three British 1951 halls are all treated in some detail in the next chapter, so it is appropriate here to consider philosophies rather than detailed achievements. All three halls are rectangular in plan, though with widths significantly larger than their classical forebears: 32 m in the Festival Hall, for example. Care was exercised in the design of the side balconies to give adequate sightlines, which are often unsatisfactory in earlier rectangular halls. In the case of the London hall, there was an interesting attempt at overcoming the limitations of a fixed reverberation time. Bagenal hoped to avoid the perceived conflict between requirements for choral

and instrumental music. With an optimum reverberation time for instrumental music, the acoustics are too dry for choral works, while with the optimum time for choral music the sound would be too reverberant for the orchestra alone. By designing for a long reverberation time but profiling surfaces near the source (in this case the ceiling) to give enhanced early reflections, it was hoped to have both clarity and reverberance. It was a bold proposal which would have worked had the reverberation time been long enough. Before long, others were attempting to pull off the same trick, but for slightly different reasons (section 4.8).

In the event, the Royal Festival Hall and to a lesser extent its British contemporaries all suffer from inadequate reverberation. The designers were wise to rely on the rectangular plan, but it is clear that they were obliged in each hall to make compromises to accommodate high audience numbers.

4.7 The diffuse solution

While the British designers in 1951 had retained from the classical halls the rectangular plan form in the (correct) belief that this was a salient feature of their acoustic success, the Göttingen acoustics group in the 1950s isolated the scattering character of the wall and ceiling surfaces of the nineteenth-century halls as crucial to their success. There were persuasive logical reasons behind their argument. Fundamental to a good environment for hearing music is an acoustical 'sense of space'. The listener should feel that the concert hall is responding to the sound produced by the musicians and hear reflected sound from all directions. This spatial sensation is of course particularly marked in large churches. It had long been assumed among acousticians that a full 'sense of space' was produced by a sound field which is uniform in directional terms, or in other words is fully diffuse (Kuttruff, 2000, p. 223). At least for the later sound, theory suggests that a diffuse sound field can be created in a space with highly scattering boundary surfaces. The plaster mouldings, niches, statuettes, coffered ceilings etc.

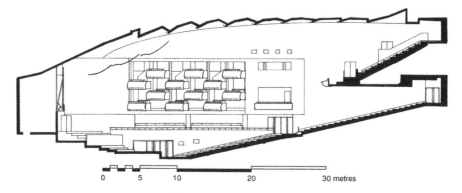

Figure 4.21 Long section of the Royal Festival Hall, London

in classical halls provided such highly scattering surfaces. This line of thought was also tempting, since a diffuse sound field is a fundamental point of reference in acoustic theory.

The most radical expression of this theory is to be found in the **Beethovenhalle** in **Bonn** of 1959 (architect: S. Wolske). The consultants, Meyer and Kuttruff (1959), justify their advice as follows: 'To avoid corner reflections and sound focusing, it was natural to place diffusing elements on either the whole or part of the ceiling or walls'. Not only is the

ceiling covered with a dense combination of hemispheres, pyramids and truncated cylinders projecting about 300 mm, but the forward side walls are also highly diffusing with vertical cylinders behind acoustically transparent screens (Figure 4.22). Certainly they had reason to be concerned about the concave curvature of the ceiling, but their justification is surely an understatement. The Beethovenhalle contains one of the most explicit instances of substantial acoustic scattering treatment. The hall seats a modest 1420 with an internal volume of 16

Figure 4.22 Beethovenhalle, Bonn

000 m³. Beranek (1962) rates the acoustics as merely 'Good'.

Returning to the question of scattering treatment, research studies at Göttingen eight years later showed that our ears require a much less extreme directional distribution to feel a full 'sense of space'. Subjective experiments by Damaske (1967) indicated that sounds from only four distributed directions were adequate to provide a

sensation of arrival from nearly all directions. This particular experiment related to the later reverberant sound. The design implication is that surfaces should be available to reflect sound from the major directions to the side of and behind the listener, rather than it being necessary to achieve a uniform spatial distribution.

Concern for a 'sense of space' is certainly appropriate and there are locations in many halls where

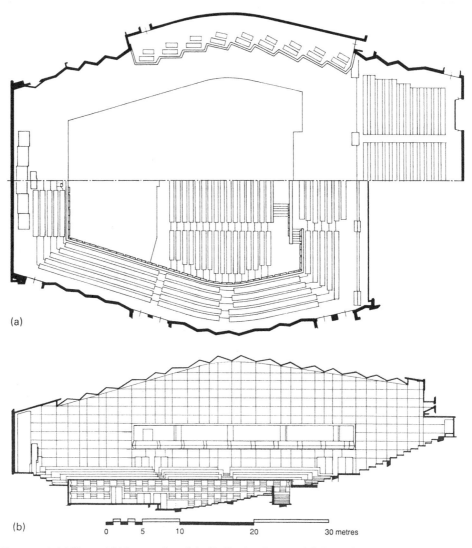

(a)

(b)

Figure 4.23 (a) Plan and (b) long section of the De Doelen Concert Hall, Rotterdam

Figure 4.24 De Doelen Concert Hall, Rotterdam

there is a disturbing lack of a sense of sound from behind. The diffuse hall solution can have its drawbacks and is not recommended as a substitute for a suitable form. For instance, early lateral reflections will be stronger and thus produce greater source broadening from a surface with the correct orientation than a surface pointing in the wrong direction but profiled to make it scattering to sound. The diffusion question remains unresolved; it is considered further in section 3.6.

The **De Doelen** concert hall in **Rotterdam** of 1966 provides an example of a hall with scattering walls and ceiling but inspired reflection design (Figures 4.23 and 4.24). The overall shape with an elongated hexagonal plan was chosen by the architects, Kraaijvanger and Fledderus. The acoustic consultants, Kosten and de Lange (1965), introduced a secondary hexagon around the lower audience close to the stage for additional local early reflections. The hall seats 2230 with a volume of 27 000 m^3 and reverberation time of 2.1 seconds. With a substantial maximum width of 40 m, no overhangs and ample seating rakes, more listeners are closer to the stage and with good sightlines than in a rectangular hall. Of particular interest in this hall is that the plan can be generated from logical consideration of provision of reflections within a certain delay, as discussed in section 3.5.

4.8 A subdivided acoustic space?

In the mid-1950s, the city of New York decided to build the Lincoln Center, housing ballet, opera, theatre, the Juilliard School of Music and a new concert hall. The **Philharmonic Hall** was to replace Carnegie Hall, which was then scheduled for demolition, a fate averted by the efforts of the violinist Isaac Stern. The acoustic consultants, Bolt, Beranek and Newman, were engaged to work with the architect M. Abramovitz. Before this commission, Bolt, Beranek and Newman had begun a comprehensive survey of 54 of the world's concert halls and opera

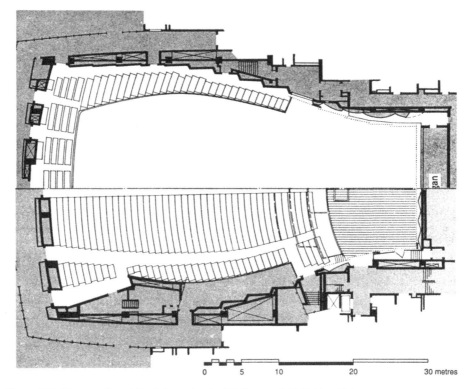

Figure 4.25 Plan at stalls and first balcony level, of Philharmonic Hall, New York

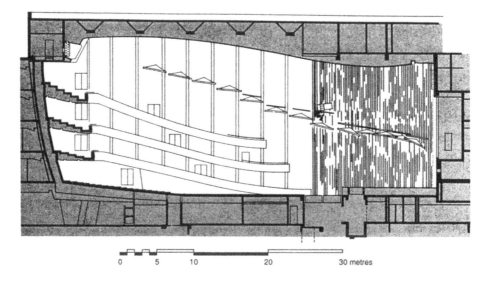

Figure 4.26 Long section of Philharmonic Hall, New York

houses. The study was accelerated to provide data to design a hall whose acoustics were intended 'to assume a place among the best halls of the world'. The invaluable product of Beranek's survey of world halls, the book *Music, Acoustics and Architecture*, was published in 1962 in the same year as Philharmonic Hall opened.

A comprehensive discussion of the design and the experiences during the tuning week are contained in the final chapter of Beranek's book. With minor adjustments the promised acoustic goal appeared to have been reached. One has no intimation of the storms which were to rage after the opening. Other prestigious consultants were called in to rectify the problems, over two million dollars were spent on modifications, until with a new private grant the whole of the concert hall envelope was removed to be replaced in 1976 by a completely new design named after the benefactor, Avery Fisher. Philharmonic Hall rates as the most publicized acoustic disaster of the twentieth century.

What had gone wrong? When a problem proves to be so intractable, it is perhaps predictable that there is no simple answer and that no one involved has attempted to provide a full assessment. The puzzle is especially perplexing since the reverberation time was close to 2 seconds, often considered an optimum value, and the gross form was not outrageous by acoustical standards (Figures 4.25 and 4.26).

Beranek concluded from his survey of world concert halls that several aspects were important for the best acoustics. His analysis indicated that the most important property was a sense of acoustic 'intimacy'. He generalized from the observation that halls with a good sense of intimacy have surfaces not too distant from the audience. This suggested a crucial design parameter: the initial-time-delay-gap, that is the delay of the first reflection (Figure 3.4). Beranek considered that for the best acoustics the delay gap should not exceed 20 ms. A small delay gap is characteristic of the classical rectangular halls with their narrow plans. But how could this criterion be satisfied at the larger scale necessary for modern concert halls?

With regard to Philharmonic Hall, Beranek (1962, p. 515) states the 'large concert hall problem' (section 3.4) as follows:

> An absolute limit of 24 000 m³ was recommended for the size of the hall, a limit that exceeds by about 40 per cent the median of four excellent [European classical] halls. The compromise on cubic volume was accepted in the expectation that the application of new principles and techniques could bring about the intimacy heretofore so closely dependent on small size.

The most significant new technique was to install a suspended reflector array substantially below the auditorium ceiling to provide reflections with an initial-time-delay-gap of less than 23 ms at the centre of the main floor. The reflecting panels were to cover more than 50 per cent of the projected area where they hung, and were placed over the stage and front half of the auditorium. What Beranek had done was to replace lateral reflections in classical halls by overhead reflections in the new hall.

The evidence (Beranek *et al.*, 1964) that suspended arrays could be beneficial was limited to the experience in the 6000 seat Music Shed at Tanglewood, now named after Koussevitzky, plus some much smaller halls. The acoustics of the wide fan-shaped plan Tanglewood Shed, at Lenox, Massachusetts, had been remarkably improved in 1959 by the installation by Bolt, Beranek and Newman of a diffusing stage enclosure and reflector array, to the point where in Beranek's view (1962) 'it is the only place that houses a very large audience, 6000 listeners, under acoustical conditions that rival the best in America'.

By mid-1959 a design was adopted for Philharmonic Hall which was basically rectangular both in plan and section, with three balcony levels, a modest suspended reflector array and sound scattering walls (Fantel, 1976). Apart from the reflector array, the form owed much to Boston Symphony Hall. The final design of the hall involved numerous changes to increase seat capacity and for cost reasons; the latter ruled out the scattering wall treatment, for instance. Beranek (2008) records a series of modifications which were made without adequate consultation between acoustic consultants and architects, particularly during the later design period. To accommodate some design changes of which they were aware, Beranek recommended substantially extending the coverage of the suspended array into the hall. Philharmonic Hall seated an audience of 2646.

Of the opening concert, the *New York Times* critic, Harold Schonberg, described the sound in the stalls as 'clear, a little dry, with not much reverberation and a decided lack of bass'. Fantel (1976) refers to a 'steely hardness, the fiddles sounded harsh, and the orchestra's sections failed to blend, as if invisible walls stood between strings, woodwinds, and brasses'. Schroeder *et al.* (1966) list as faults:

> a poor frequency response affecting audibility of cellos and double basses, a lack of subjectively felt reverberation, echoes from the rear, inadequate sound diffusion and poor hearing conditions for musicians on stage.

Comments on the subjective character of the hall relate, it seems, nearly always to the stalls or lower balcony seating. Schroeder *et al.* (1966) note

the subjective superiority of the higher balconies. Schonberg is more forthright about the top terrace, where he found that:

> the tonal characteristics are altogether different. Here the bass can definitely be heard ... so full in sound that it almost appears amplified and too live. It is exciting, though, and in climactic movements of the *Gloria* the effect lifts one off his seat.

Examination of the plan and section reveals several anomalies. The main floor rake is remarkably shallow, the balcony overhangs are excessive by Beranek's own standards (section 3.7) and the rear wall is concave both in plan and section, which must have encouraged echoes back to the stage. The curvature on the side walls means that not only is there some focusing at the rear of the stalls, but also that many stalls seats receive no lateral reflections (Figure 4.27). Finally a peculiarity of the stage design is its remarkable size, enough for about 170 musicians, which probably left orchestras too widely spaced. But many of these features are found elsewhere. A flat stalls floor was characteristic of all nineteenth-century designs. Even in combination

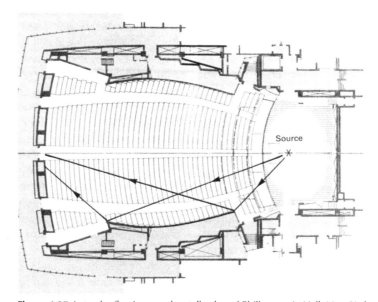

Figure 4.27 Lateral reflections on the stalls plan of Philharmonic Hall, New York

they do not seem sufficient to produce such disappointing acoustics beyond redemption.

A fault, which was soon isolated, related to the behaviour of the suspended reflectors, or 'clouds' as they were often called. Owing to the small size and regularity of the reflector panels, as well as the use of a double layer of reflectors, it was found that the array did not reflect bass sound (Watters *et al.*, 1963; Meyer and Kuttruff, 1963). The designers were doubly unlucky because this experience uncovered a phenomenon found in all halls which no one had suspected: that low-frequency sound is attenuated as it passes at grazing incidence over seating (section 2.6.1). Without either low-frequency direct sound or low-frequency early reflections, it was not surprising that the bass sounded deficient. In the event, reorienting the clouds failed to quell all the criticisms (Lanier, 1963). But the scale of the effects was certainly gross. Measurements in the hall's original condition by Schroeder (1984) of the 750 Hz total sound level in the stalls (Figure 4.28) show a rate of decrease with distance of 4 dB per 10 m in Philharmonic Hall, compared with a mean of 1.5 db per 10 m in British halls (with a lowest value of 2.4 dB/10 m). (To produce Figure 4.28 two assumptions were necessary: the source position was not stated so a position 3 m from the stage front was taken and the 750 Hz level at 9 m has been assumed to be equal to theory. Neither assumption affects the comments here however.) Also in Figure 4.28, the level of bass relative to mid-frequency sound approaches –20 dB, while the criterion is nearer –2 dB. No wonder the critics were vociferous. Further interesting objective results for the orignal hall are to be found in Schultz (1965).

In conclusion it seems likely that an extreme redistribution of acoustic energy in the hall was associated with the excessive 'cloud' array. The effect of the clouds extended beyond a failure to provide a first reflection with its full complement of bass energy. The low percentage open area, of less than 50 per cent in Philharmonic Hall, would create a poorly coupled acoustic space above the array, whose behaviour varied with frequency. The poor coupling would inhibit reverberation in the whole

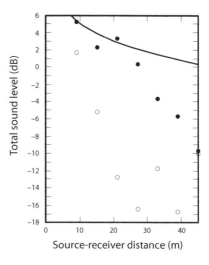

Figure 4.28 Total sound level versus distance in Philharmonic Hall, New York (after Schroeder, 1984). •, measurements in band around 750 Hz; o, mean of measurements in bands around 125 Hz and 250 Hz. Solid curve according to revised theory. See text regarding assumptions

space; Cremer and Müller (1982, p. 112) suggest a minimum open area of 70 per cent. In the stalls the clouds were responsible for inadequate bass and a poor sense of reverberation. Shankland (1963) provided an early suggestion that a predominance in this hall of overhead compared with lateral early reflections was also undesirable (see section 3.2). In addition, a harshness of string tone is commonly perceived in spaces with strong overhead reflections.

By 1969 under advice from H. Keilholz the clouds were totally removed, and a completely new stepped ceiling was installed with diffusion added to the side walls (Young, 1980). The audience seemed reasonably content but the musicians were still unhappy with conditions on stage. This should hardly have necessitated the demolition of a concert hall, but in a climate where the orchestras were sensing that other venues did better justice to their efforts, the radical solution was taken. The present Avery Fisher Hall, by acoustic consultant Cyril Harris, is essentially rectangular in plan with three layers of balconies (Bliven, 1976). The design bears an uncanny resemblance to the original 1959 proposal (Fantel, 1976), but it is not obscured by clouds!

4.9 The terraced concert hall

Of all the halls built as part of post-war reconstruction, the **Berlin Philharmonie** of 1963 stands as the most innovative. The 'lovers of harmony' in former West Berlin were certainly more fortunate than their New York counterparts. Of particular interest in its design was the interaction of two masters of their crafts, the architect Hans Scharoun (1893–1972) and acoustician Lothar Cremer (1905–1990). Cremer had consulted on the Liederhalle, Stuttgart (1956), Sender Freies Berlin (1959 – both discussed by Beranek, 1962) and the Deutsche Oper, Berlin of 1961 (section 9.4). In each case there is attention to early reflections and each offers an interesting example of successful acoustic design. Scharoun's architecture, though often called expressionist, is more meaningfully described as organic, in which the form develops from the inside outwards, in which space acts as a positive force upon the life it contains and vice versa. His earlier theatre schemes had experimented with novel stage–audience relationships (Blundell Jones, 1995). Scharoun had noted that 'people always gather in circles when listening to music informally' and wished to adopt this natural arrangement for a concert hall. The design of the Philharmonie was developed between architect and consultant from the beginning and won the design competition in 1956.

With an arena form there are two major acoustic concerns: that many instruments are directional and that surfaces are required to provide early reflections. Initially Scharoun had proposed a fully central stage but was persuaded on acoustical grounds to introduce a directional bias in the plan so that the majority of the audience was in front of the stage. Beyond this measure, there was not much in a hall of this size (compared with that of the neighbouring Chamber Music Hall, section 6.3) that could be done to mitigate the directional effects. There are of course many halls in which audience is placed in

choir seats behind the stage for purely instrumental concerts, so this is a matter of degree rather than a new departure. The directional problem does not exist for instruments by themselves with small radiating surfaces, such as woodwind, nor for tympani for instance with horizontal radiating surfaces. And while brass instruments, especially trumpets, become highly directional in their high registers, the lack of brilliance in other directions is not normally severe. The most serious problem is shadowing produced by the players' bodies, especially for string players and singers. This affects balance but the listener may well accept the bias if it conforms to his expectations. Of all musical 'instruments', the directivity problems of solo singers seem the most severe.

It was in the disposition of the audience of 2230 (Figures 4.29 and 4.30) that the interplay of architect and consultant was so productive (Frampton, 1965). The use of terraces had already been tried in 1956 in the 800-seat Mozartsaal in the Stuttgart Liederhalle (section 6.3). In the Philharmonie, the surfaces bounding the individual 'vineyard terraces' (as they are usually called) are used to create a fascinating three-dimensional space, while offering planes able to create acoustic reflections. Frequently these surfaces are inclined to the vertical to direct sound down onto neighbouring seating blocks (Figure 4.31). In the long section, there are steps in what is basically continuous seating. As well as dividing up the multitude, this serves two acoustic functions: to provide reflections from behind to the back rows of the seating blocks (though admittedly from low surfaces) and to reduce the attenuation effect for sound travelling at grazing incidence over extensive audience areas. A prow, at the back of the hall facing the stage, serves to subdivide the seating further and provides a surface for late side reflections.

Cremer (1964, 1965) was particularly concerned that the acoustic conditions for the orchestra should closely resemble those of other halls. Tall surfaces, as much as 3 m high, surround the stage, while the extreme ceiling height above the stage is compensated by the use of suspended reflecting panels. The fact that von Karajan, as resident

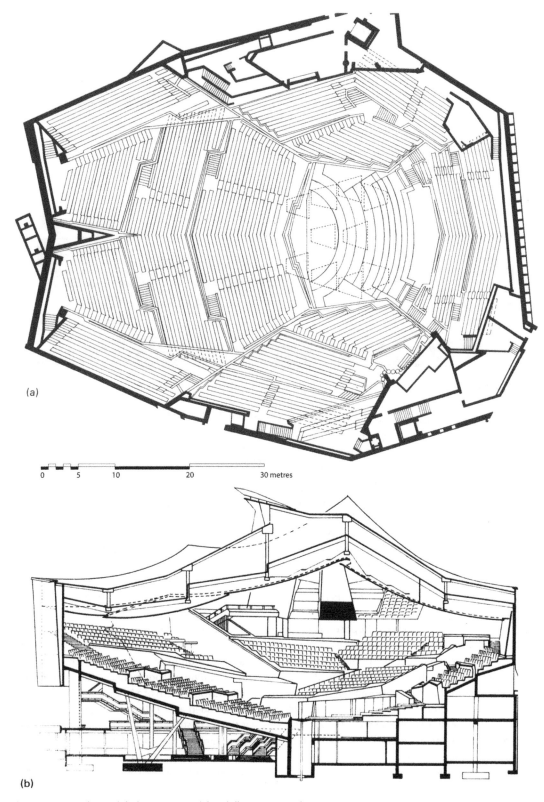

(a)

0   5   10   20   30 metres

(b)

Figure 4.29 (a) Plan and (b) long section of the Philharmonie, Berlin

Figure 4.30 Philharmonie, Berlin

conductor, preferred the panels to be high, suggests that the stage acoustics are good even without the maximum number of reflections available. The stage area of 330 m² contains space for full choir.

Provision of adequate early reflections throughout the hall was the other main concern and this aspect was tested in an acoustic model. It was found necessary to use the overhead orchestral reflectors to provide reflections to the front stalls as well. For the roof, Cremer wished to have a low ceiling round the perimeter to provide good reflections to remote seats. A ceiling at this level running horizontally across the whole hall would have provided too small a volume for reverberation purposes, while a concave ceiling would have presented serious risks of focusing. The chosen tent-like profile of convex surfaces encloses a substantial volume of 25 000 m³, while its form has the advantage of assisting diffusion. A large number of pyramidal diffusers are placed on lower sections of the ceiling. The pyramids also have slits in them in order to act as Helmholtz resonator absorbers and limit low-frequency reverberation. The measured reverberation time with full

occupancy is 1.9 seconds at mid-frequencies, rising to 2.1 seconds at 125 Hz.

Scharoun described his conception in the following terms:

> Music as the focal point. This was the keynote from the very beginning. This dominating thought not only gave shape to the auditorium of Berlin's new Philharmonie but also ensured its undisputed priority within the entire building scheme. The orchestra and conductor stand spatially and optically in the very middle of things; if not at the mathematical centre then certainly completely enveloped by their audience. Here you will find no segregation of 'producers' and 'consumers', but rather a community of listeners grouped around an orchestra in the most natural of all seating arrangements. ... The construction follows the pattern of a landscape, with the auditorium seen as a valley, and there at its bottom is the orchestra surrounded by a sprawling vineyard climbing the sides of its neighbouring hills. The ceiling, resembling a tent, encounters this 'landscape' like a 'skyscape'. Convex in character,

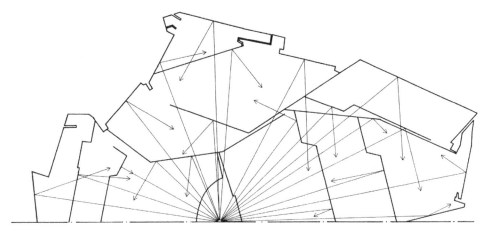

Figure 4.31 Reflections in the Philharmonie, Berlin (reproduced by permission of Pennsylvania State University Libraries)

the tent-like ceiling is very much linked with the acoustics.

Subjectively, the sound is intimate at most seats, providing clarity within a strong sense of reverberation. In front of the stage, the acoustics in many locations are certainly very good. This hall was built before there was any suggestion that lateral reflections were beneficial, so that some seating areas are better served in this respect than others (Figure 4.31). The reverberant sound is however highly diffuse, which compensates spatially. One has to admit that as predicted the social advantages are at the expense of some acoustic uniformity, particularly with regard to balance. On one side of the stage the violins are weak, while on the other the double basses become difficult to hear. Balance problems are less severe behind the stage. Yet as a concert experience, the Philharmonie is indeed remarkable.

In terms of design for early reflections, the Philharmonie established a major shift in approach. Instead of relying on reflecting surfaces in the stage area (as in the Salle Pleyel, Paris, and the Royal Festival Hall), the subdivision of audience seating generates surfaces close to the audience able to provide reflections, as shown in Figure 3.15. With overhead reflections becoming unacceptable, the subdivided audience offers for this reason a valuable solution to the problems of large concert hall design.

Cremer's design schemes since the Philharmonie continued to employ intriguing ways of providing reflections in large halls (Cremer, 1989). The early reflections in these designs tended to be lateral. Two 'theoretical' schemes, discussed in section 3.5, have both been exploited in real halls: the hexagonal and trapezium terraced halls. These halls inevitably rely on reflections from quite shallow surfaces between seating blocks, from which low frequencies are unlikely to be reflected. Cremer was less concerned than some who consider low frequencies crucial to the sense of source broadening. This matter has yet to be conclusively resolved by research.

It was inevitable that true symphonic music-in-the-round would have to be tried. The **Muziekcentrum Vredenburg** in **Utrecht** of 1979 by Hertzberger seats 1550 in a volume of 17 000 m³ with a gross plan very close to a regular octagon (Figure 4.32). The stage occupies roughly one half of the central arena. The consultants, de Lange and Booy, insisted on vertical planes at different heights in the audience seating, aiming in particular to provide lateral reflections (de Lange, 1980; Padovan, 1980). The reverberation time occupied is 1.9 seconds. This scheme was also tested in an acoustic scale model.

Figure 4.32 Muziekcentrum Vredenburg, Utrecht

4.10 The lateral directed reflection sequence hall

When asked in the mid-1960s to assess the competition for and subsequently to advise on the design of a new hall for Christchurch, New Zealand, Marshall (1979a, 1979b) was dismayed to discover the lack of existing recommendations regarding the appropriate shape for a concert hall enclosure. In considering this problem, he developed a new theory proposing that 'lateral reflections were the most important single component of the early reflection sequence' (section 3.2). This proposal provided a further explanation for the subjective superiority of classical concert halls. The **Christchurch Town Hall** of 1972 was the first product of this hypothesis (Marshall and Barron, 2001). It was not an example of timid acoustic design.

The architects, Warren and Mahoney, had proposed a near-elliptical plan with a cantilevered gallery. The brief specified a flat central stalls floor. The elliptical plan has the particular visual virtue that from opposite the stage one tends to 'convert' the form into a circular one, providing the desirable illusion of proximity to the stage. In spite of the large audience of 2338 plus 324 in choir seating, none are more than 28 m from the stage front (Figure 4.33). With a ceiling as much as 21 m above the floor there is a sensation of vastness, while the arena form imparts a vivid sense of occasion. The reflectors also contribute markedly to the visual impression (Figure 4.34).

Fourteen vertical wall elements define the elliptical plan and generate adjacent seating areas. The primary aim of the acoustical design was 'the provision of unmasked lateral reflections'. The architects and consultants, A.H. Marshall and W.A. Allen, carefully developed a design with individual seating groups in the gallery for each vertical wall element. Figure 4.35 shows how each seating group

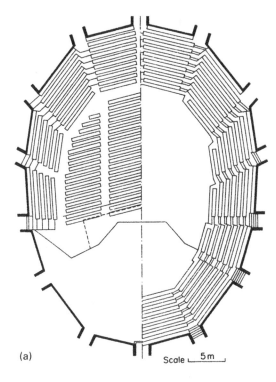

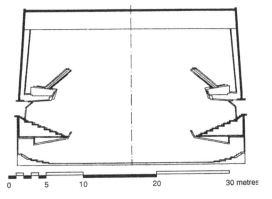

Figure 4.33 (a) Plan and (b) cross-section of the Town Hall, Christchurch, New Zealand

(a)   Scale ⌊___5 m___⌋

Figure 4.34 Town Hall, Christchurch, New Zealand

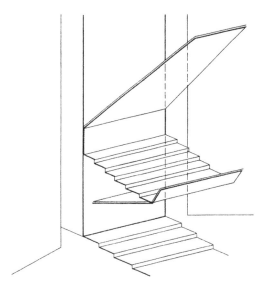

Figure 4.35 Individual seating group engendering reflecting surfaces in the Town Hall, Christchurch, New Zealand

engenders three reflecting surfaces: an inclined balcony front to reflect sound into the central stalls, a balcony soffit to enhance reflections to the overhung seats and, most significant, a large suspended reflector providing lateral reflections both into adjacent gallery seating areas and into the stalls. Two of these elements, the balcony soffit and the suspended reflector, also serve an essential masking function to avoid focusing by the elliptical plan. At stalls level the seat rake and soffit design sufficiently limit the exposed wall height. Immediately above balcony seating but below the reflectors the surfaces are treated with absorbent. Orientation of the gallery reflectors was refined with both an acoustic scale model and a computer model. Opposite the stage it was found necessary to use dihedral reflectors in order to achieve reflections from the side. The volume above the reflectors is substantial; the total volume of 20 700 m³ generates a reverberation time of 2.3 seconds when the hall is occupied with audience.

Subjective response to the Christchurch Town Hall has, in the main, been highly favourable, though for some its marked character is not optimal. As a personal view, the most remarkable feature is not the spatial character of the sound but the sense of intimacy and identification with the performance. With a high degree of clarity, it is possible to listen with ease to individual musical lines. Yet there is also a rich sense of reverberation; the long reverberation time has seldom led to criticisms of excess. In the flat stalls seating, there can be problems hearing the woodwind due to obscured line-of-sight. Those who criticize the acoustics are probably unenthusiastic about reflections from large plane surfaces. The successor to this design in Wellington, New Zealand, aimed to overcome these shortcomings.

The objective behaviour of Christchurch Town Hall is interesting, showing sizable deviations from average behaviour. The results given in Appendix C were measured in 1983 (and are thought to be more reliable than those quoted by Marshall, 1979a, owing to a superior sound source). Particularly marked is the difference between the early decay time (EDT) and reverberation time, with the EDT being only 82 per cent of the latter at mid-frequencies. In subjective terms this explains why there is no sense of excessive reverberation. It probably occurs because a high proportion of sound leaving the source is reflected onto absorbent seating. This suggests that in designs of this nature a long reverberation time is not a luxury but is essential. In line with the shorter relative EDT, there is a high value for the objective clarity. Analysis shows that this is mainly caused by a low level of late sound relative to theory. Late sound energy may well be screened by the reflectors. The measured total sound level is less than theory, but with such a wide discrepancy between reverberation time and early decay time, the validity of the theory is somewhat compromised. Total sound values are all however above the 0 db criterion. The measured objective source broadening is typical rather than exceptional. Also of interest is the observation that all gallery seats receive a reflection within 20 ms of the direct sound.

The new hall for **Wellington** of 1983, known as the **Michael Fowler Centre**, had to be designed in a mere six weeks, so it was natural to use the

Figure 4.36 Michael Fowler Centre, Wellington, New Zealand

Christchurch design as its starting point (Marshall and Hyde, 1979). Warren and Mahoney were again the architects, while Marshall was joined by Hyde as acoustic consultants. The Wellington plan form is virtually identical but in this case the stalls flooring is raked (Figure 4.36). The total seat capacity is 2566 including choir. As a design criterion, research since the Christchurch design had indicated less stringent requirements for the creation of source broadening. Strong lateral reflections on paths remote from the audience seating planes were now the aim, with less emphasis on lateral reflection arrival time. Experience with the Christchurch hall had indicated two risks associated with specular reflections from plane surfaces: false localization and tone colouration. With reflection from a highly scattering surface, these subjective risks disappear. Traditionally the effectiveness of scattering surfaces had been a hit-and-miss affair until Schroeder proposed a series of slot diffuser designs with predictable behaviour (section 2.6.4). At the time the quadratic residue diffuser (QRD) was the most promising to give deterministic scattering over a specific frequency range. The Michael Fowler Centre was the first concert hall to use such diffusers.

Below some frequency, scattering surfaces are no longer effective. To maintain 'warmth', Marshall wished the reflections to be specular over the range of the audience attenuation dip, and subjective experiments suggested that the scattering action should be effective above about 350 Hz (QRD design frequency 500 Hz). At high frequencies there was concern that the diffuser fins would reflect sound back towards the source. This backscattering behaviour had been predicted by H.W. Strube of Göttingen and was indeed experienced in subjective recordings in a 1:10 scale model of the hall. But as well as the change from specular to scattering surfaces, the reflectors became independent of the room boundaries. While in Christchurch, reflectors direct sound to the same side of the auditorium, in Wellington the QRD surfaces reflect to the opposite side. In addition to the diffusing reflectors in Wellington, lower convex surfaces direct sound to the neighbouring side wall before reaching the listener. Model tests also confirmed the need for some additional plane reflecting surfaces located behind the major reflectors to serve adjacent seating areas. A further difference with Christchurch is that all but the rear gallery seating blocks have an opening behind them to allow late reflected sound to reach overhung seats. This last feature works most effectively.

In subjective character the Michael Fowler Centre shares many characteristics with its Christchurch forebear. Again there is intimacy, envelopment and a remarkable degree of transparency, enabling individual musical lines to be followed. The objective measurements (Appendix C) point to the minor failing of a slight lack of sense of reverberation for the Romantic repertoire. The early decay time (EDT) is 83 per cent of the reverberation time at mid-frequencies, but with a conventional reverberation time of 2 seconds (occupied) the EDT becomes less than optimal. Presumably the reason for the shorter reverberation time than in Christchurch is incidental absorption by the greater area of exposed reflector (the volume is in fact larger at 22 700 m³).

A third design based on the Christchurch model opened in the Hong Kong Cultural Centre in 1989 (Marshall, Nielsen and Halstead, 1998). This uses yet another reflector scheme, but again with QRD surfaces. Marshall's design with Hyde and Paoletti of the Segerstrom Hall in Orange County, California, is a further radical departure (section 10.6).

4.11 Concert hall design in the 1980s

During the 1980s acoustic consultancy became more international, with for instance some American consultants being invited to advise on halls in Europe. The American experience during this period turned out to be a strong influence on the worldwide situation in the 1990s and beyond. Many halls from this period are included in Hoffman *et al*. (2003).

Few consultants in America had been untouched by the ill-fated New York Philharmonic Hall. The original consultants, Bolt, Beranek and Newman, remained enthusiastic about the use of suspended panel arrays, but developed a more sophisticated design procedure (Beranek and Schultz, 1965). Having realized that bass sound is attenuated as it passes over audience (section 2.6.1), but that in

most halls this is not perceptible as a lack of warmth, they concluded and confirmed to their satisfaction from simple subjective experiments that bass energy only needs to be present in the later reverberant sound. On the other hand, their experience with suspended overhead reflectors had shown that only the mid- and high frequencies were associated with the sense of intimacy and clarity. In design terms this is particularly attractive because it allows small reflector panels to be used. Indeed, according to their argument large panels are inappropriate since they starve the reverberant field of bass sound. Beranek and Schultz also suggested an additional measure beyond those listed by Beranek (1962): the ratio of early-to-late energy, referred to here as objective clarity; they considered it to affect the sense of clarity and intimacy. They found that the acceptable range of this quantity was small and concluded that it was too delicate a matter to be left to chance: in large halls some acoustical adjustment (with for instance movable reflectors) must be built in. Adjustability of reverberation time through the use of absorbent banners was also used. Thus the tunable concert hall was born.

Many of the Bolt, Beranek and Newman designs in the two decades after 1962 are illustrated in Talaske *et al*. (1982). Though some of these halls had large capacities, they were generally not very high-profile facilities by national standards; many were located on university campuses. But between 1980 and 1982, Bolt, Beranek and Newman were responsible for four major full-scale concert halls, each the home of a significant orchestra: Louise M. Davies Symphony Hall, San Francisco (1980, 3000 seats), Joseph Meyerhoff Symphony Hall, Baltimore (1982, 2467 seats), Victorian Arts Centre Concert Hall, Melbourne (1982, 2600 seats, now Hamer Hall, Melbourne Arts Centre) and Roy Thomson Hall, Toronto (1982, 2812 seats). (Accounts of these halls can be found in Beranek, 2004, and Lord and Templeton, 1986.) In each case the acoustic design was predominantly directed by the late T.J. Schultz. All four halls were inspired by the design of Massey Hall in Toronto of 1894; not surprisingly they share many common features. They each have two balconies, which in the case of the three North American halls are subdivided into individual small seating sections. The walls and ceilings are highly profiled to produce scattered sound reflections. In plan the hall widths next to the stage are modest, but beyond the stage front the width increases substantially to a large maximum value. Each hall has suspended reflecting panels but the coverage is now limited to above and a little beyond the stage, much less than in Philharmonic Hall. Adjustable banners control reverberation in each hall.

Roy Thomson Hall, Toronto, designed by Erickson, is a highly photogenic and stunning visual experience. The gross plan is roughly circular with a 'bicycle wheel' construction supporting the ceiling, based on a 'hub' above the stage front, Figures 4.37 and 4.38. Such a concave plan is liable

to produce severe focusing problems, which have been avoided here by substantial segmentation of the walls. At higher levels the hall envelope consists of 26 vertical convex surfaces. The advantage of this plan form is that it allows all seats to be within 33 m of the stage, though the penalty is a substantial maximum width of around 39 m. Indeed this has been suggested as a cause of disappointing sound in some seats and in 1987 additional tilted reflectors were installed around the side walls to improve early lateral reflections. The internal volume at 28 300 m³ is substantial.

The overstage panels were originally 2.1 m diameter convex circular 'saucers' made of clear acrylic plastic, covering 40 per cent of the stage area. These reflecting panels were oriented to serve the stage, main floor and first balcony. Additional smaller

Figure 4.37 Roy Thomson Hall, Toronto

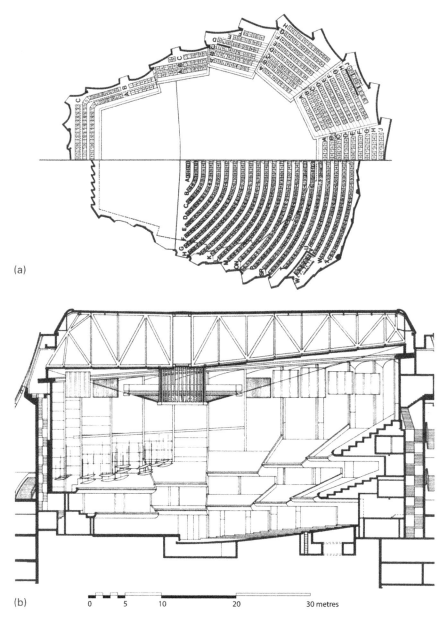

(a)

(b)

0   5   10   20   30 metres

Figure 4.38 (a) Plan at stalls and mezzanine level and (b) long section of Roy Thomson Hall, Toronto

panels improved coupling between the stage and choir seating. The reverberation time in the fully occupied condition of close to 1.8 seconds could be reduced to 1.4 seconds (for the classical repertoire, for instance) by an elaborate array of absorbent cylinders and banners. There was however a slight penalty of reduced sound level when they were lowered into the hall. Schultz (1986), the original principal consultant, summarized his philosophy as follows:

> Drawing the audience close to the performers in a large concert hall, and compensating for the resulting increased width of the hall by a very flexible array of sound reflecting panels above the stage has made it possible to achieve very fine acoustics while preserving a great sense of theater.

There have been criticisms of the acoustics of each of the 1980–82 halls. Significant changes have been made at Davies Symphony Hall (1992), the Meyerhoff Hall (2001) and Roy Thomson Hall (2002). Among the issues addressed, there has been concern to improve communication between musicians on stage, to provide a greater sense of reverberance and to enhance early lateral reflections for the audience.

While Bolt, Beranek and Newman elaborated on solutions proposed in Beranek's book of 1962 by invoking the notion of an optimum ratio of early-to-late sound, an alternative interpretation of Beranek's overall subjective ratings was published by West

(1966). West found a high correlation between the subjective ratings and the ratio of height-to-width of halls. This matched the ideas of Marshall concerning the importance of early lateral reflections (section 3.2).

The North American consultant who responded most enthusiastically to the idea of lateral reflections was Russell Johnson (1923–2007). Between 1954 and 1970 Johnson worked with Bolt, Beranek and Newman and was involved with several multi-purpose auditoria which employed the 'suspended array' philosophy. He subsequently established an independent consultancy, now called Artec Consultants. He moved completely away from suspended arrays to continuous canopies, suspended in free space above the stage platform and over the front rows of audience seated in the stalls. The **Centre in the Square** in **Kitchener**, Ontario, of 1980 (Figure 4.39) is one of a line of such designs. Many of the halls at that time were multi-purpose but concert requirements were little compromised. The Kitchener hall with a volume of 15 300 m³ seats 1920, and has a flytower over the stage. Two balconies at the rear link with shallow box-balconies running down

Figure 4.39 Centre in the Square, Kitchener, Ontario

the side walls. Balcony soffit reflections and parallel side walls enhance early reflections. The rear ceiling is profiled to redistribute ceiling reflections. A substantial continuous canopy, fully adjustable in height, is 'the chief instrument for clarity, intelligibility and articulation' (Johnson, 1981). The novel feature of this design was the use of towers on air castors, which can either hold audience, or work as lighting perches or provide an acoustic shell for orchestral use. By horizontally closing off the flytower, the void behind the towers acts as a reverberation chamber to provide secondary reverberation for concert performance. Another hallmark of Artec designs is massive construction to minimize bass absorption and maximize bass sound level.

In **Pikes Peak Center, Colorado Springs**, of 1982 (1955 seats, volume 16 950 m³) the Kitchener approach was developed one stage further with tilted upper side wall surfaces to provide additional early lateral reflections (Figure 4.40). These surfaces are also slightly tilted in plan to direct reflections more towards the stage. Although this is at the expense of more remote balcony seats, these latter areas are generally more than well served. Light

models were used to optimize the reflector design. A reverberant void behind movable stage towers was again used, which in the designer's words 'results in excellent clarity and envelopment in the presence of satisfying reverberance' (Johnson, Essert and Walsh, 1986; see also Edwards, 1985). A similar cross-section shape was also used for the concert hall in Nottingham, England (section 5.12). The subsequent Festival Hall for Tampa, Florida (1986), in addition uses reverse-fan shapes in plan to enhance lateral reflections. Towards the end of the 1980s however, Johnson and Artec made an abrupt change of design philosophy. They abandoned schemes with room surfaces inclined specifically to provide lateral reflections and adopted the parallel-sided solution for virtually all their new concert auditoria, discussed further below.

Response to Scharoun's revolutionary 1963 design for the Berlin Philharmonie was slow in coming. Somewhat surprisingly, the first completed hall inspired by the vineyard terraced audience arrangement was the **Boettcher Hall** in **Denver**, Colorado (Schmertz and Jaffe, 1979). Boettcher Hall, which opened in 1978, holds a large audience of

Figure 4.40 Pikes Peak Centre, Colorado Springs

Figure 4.41 Neues Gewandhaus (1981), Leipzig

2750 in a huge 37 200 m³ volume. It is more 'music-in-the-round' than the Philharmonie, with about 20 per cent of the audience to the side of or behind the orchestra, compared with a figure of 10 per cent for the Berlin Hall. A consequence of this seating arrangement is that the furthest seat is only 26 m from the stage.

In Europe, the city of Leipzig finally replaced its famous hall in 1981 with a new **Neues Gewandhaus** (Skoda, 1985; Winkler and Tennhardt, 1988). The design has been labelled as inspired by the Berlin Philharmonie, whereas it is much less wayward than the Berlin model (Figure 4.41). The most obvious stimulus is the concept formalized by Cremer (1986) as a terraced trapezium design (Figure 3.20). It can be described as a stepped reverse-splay hall. Interesting details include inclined balcony fronts and a diffusing ceiling of convex panels. It holds an audience of 1905 and has a volume of 21 000 m³ with an occupied reverberation time of 2 seconds. The design was elaborated by extensive scale model testing, checking among other aspects the responses caused by the directional nature of different instrument groups (Tennhardt, 1984). Responses to the hall's acoustics have been favourable.

Another concert hall influenced by the Philharmonie is the **Suntory Hall** in **Tokyo** of 1986. In this case the seating arrangement is a fairly conventional response to the Philharmonie (Figure 4.42). The roof profile in long section is like a tent as found in Berlin, while being slightly convex in cross-section to direct sound towards the side walls. The seat capacity is 2006, with a volume of 21 000 m³ and a mid-frequency reverberation time of 2.0 seconds; the acoustic consultant was Nagata Acoustics.

A further well-received hall in Japan has been the **Osaka Symphony Hall** of 1982, for which Prof. Ishii was consultant. The hall is roughly rectangular in plan, with 1702 seats within a volume of 17 800 m³; the occupied reverberation time is 1.9 seconds. Of particular interest is the design of the ceiling with four tent-like convex surfaces linking to a central horizontal roof (Figure 4.43). Not only is this ceiling likely to promote diffusion but the two convex surfaces running along the hall are probably also beneficial for earlier lateral reflections. A further feature of interest is that though two levels of balcony run along the side walls, there is hardly any overhang of stalls seating by the side section of balcony.

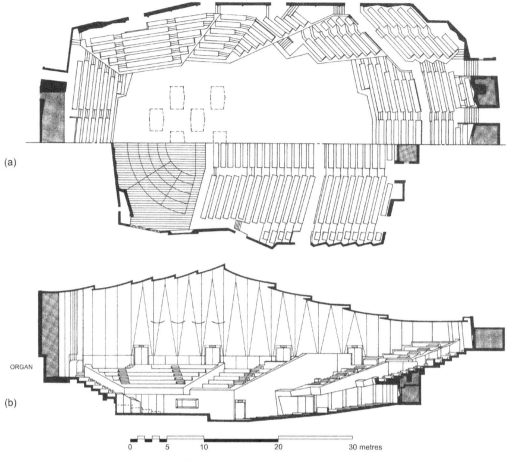

(a)

(b)

ORGAN

| 0 | 5 | 10 | | 20 | | 30 metres |

Figure 4.42 Plan and section of Suntory Hall, Tokyo

Figure 4.43 Osaka Symphony Hall, Japan

4.12   Return to precedents

Though many novel designs for concert halls have been tried, the rectangular or shoebox concert hall from the nineteenth century has remained a continual point of reference. Halls of this form have continued to be built throughout the last 50 years, yet around the turn of the millennium the rectangular plan hall has become particularly popular. The reasons for this have much to do with the changing nature of concert-going during 50 years. The expectations of audiences have risen sharply, stimulated in part by the remarkable progress in the quality of recorded and broadcast sound. And concerts now have to compete with a wide range of entertainment possibilities available in modern cities. For musicians, modern travel exposes them to a broader range of performing environments and comparisons are inevitably made. The outcome is that the acoustics of new concert halls is critical not only for their reputation but also for their financial viability. Clients, hall managers and musicians have responded by becoming less willing to take risks. The rectangular or more precisely parallel-sided hall has since around 1990 become the design of choice for many.

Two American consultants cite a further reason for the shift to parallel-sided halls (Scarbrough and Jaffe, 1999). Between the 1950s and '70s, acoustic consultancy in America had been dominated by the firm of Bolt, Beranek and Newman. Their association with the unfortunate New York Philharmonic Hall had dented their reputation but not affected their status. However the response to their four major halls between 1980 and 1982 (see previous section) was to have a more serious impact:

> The most unfortunate legacy of these [four] halls was to feed a growing perception among symphony orchestra conductors, musicians, managers and audiences that the acoustical profession could not be counted upon to deliver results in new facilities. In parallel with this was a growing impression that only the traditional shoebox shape offered a viable model for new concert halls (despite the existence of more than a few

shoebox shaped halls whose acoustics range from merely mediocre to positively abysmal).

Of the rectangular halls built in the United States in the years between 1950 and 1980, many were designed by Cyril Harris. For his earlier large halls the design of Boston Symphony Hall was the stimulus. As in Boston, these halls have in each case a stage enclosure and three balconies that extend along the side walls (compared with two in Boston). The concert hall in the Kennedy Center of 1971 in Washington DC holds 2759. Harris was famously called upon in 1976 to consult on Avery Fisher Hall, to replace the ill-fated New York Philharmonic Hall. A recent example of his design was the 2500-seat **Benaroya Hall** for **Seattle** completed in 1998. In this last case the walls and ceiling have been heavily profiled compared with the lighter surface treatment of earlier halls by Harris (Harris, 2001).

In Europe the then East German authorities decided in 1986 to build a 'new' nineteenth-century rectangular concert hall in East Berlin. The hall even has classical columns and chandeliers as well as the surface decoration on the walls and ceiling characteristic of its precedents. The **Konzerthaus** (formerly the Schauspielhaus) has a modest capacity of up to 1677 with one surprising feature: a flat rather than raked main floor (Beranek, 2004).

A modern interpretation of the rectangular hall was built in **The Hague** in 1987 (Metkemeijer *et al.*, 1988). Known as the **Dr Anton Philips Hall**, this 1900-seat auditorium with a single balcony which wraps round all four walls is of interest as a low-budget solution. The ceiling is of concrete with the roof structure exposed below it. The walls are made of damped steel panels profiled to provide scattering; the profiling consists of different-depth slots in both a vertical and horizontal direction.

A close copy of the Vienna Musikvereinssaal was chosen for the **Seiji Ozawa Hall** to complement the Koussevitzky Music Shed at **Tanglewood**, Lenox, Massachusetts. This was opened in 1994 with Kirkegaard Associates as acoustic consultants. Though the capacity of 1180 is modest, the rear wall of the hall can be completely opened to allow audience to listen on the lawn outside.

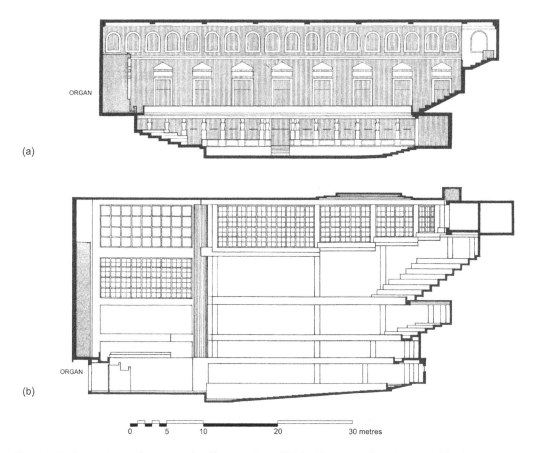

(a)

(b)

0   5   10   20   30 metres

Figure 4.44 Comparison to the same scale of long sections of (a) the Vienna Musikvereinssaal and (b) the McDermott Concert Hall in Dallas

While many designers have stuck closely to earlier precedents, two consultants have made developments to parallel-sided halls beyond the traditional shoebox proportions. Artec Consultants discarded their approach based on providing strong lateral reflections, which had generated highly moulded forms (section 4.11). The **McDermott Concert Hall** in **Dallas** of 1989 was the first of a continuing series of parallel-sided halls designed by Artec (Cavanaugh and Wilkes, 1999, p. 278).

To accommodate just over 2000 seats at a modern seating standard with a sufficiently large hall volume for reverberation purposes, it was realized that some dimension(s) would need to be extended compared with the nineteenth-century shoebox halls. With the distance from the stage front to the furthest audience seat limited to 40 m, extending the hall length was not possible. Substantially increasing the hall width on the other hand would be detrimental to lateral reflections. This left the height, which was made significantly higher than in the nineteenth-century halls (Figure 4.44).

The Dallas hall was soon followed by Birmingham Symphony Hall in England, discussed in detail in section 5.14. The concert hall in the Lucerne Cultural and Congress Centre of 1998 also belongs to the same family of designs. Each of these halls has seating opposite the stage organized very much in the manner of a traditional opera house with a curved balcony front to three or four balconies. The balconies continue along the side walls in narrow horizontal strips with no more than two rows of seating each. A novel feature of these halls is the inclusion of reverberation chambers; in the case of the Dallas hall the chamber has a volume of 7200 m^3, 30 per cent of the auditorium volume. The intention is that sound will enter the chamber, reverberate around it and leak back into the auditorium. In this way it can be expected to extend the terminal reverberation (section 2.8.1).

In Tokyo, Beranek with Takenaka R&D Institute acted as consultants for the 1632-seat **Takemitsu Memorial Hall** in **Tokyo Opera City** (Hidaka, Beranek *et al.*, 2000). While the auditorium plan is very traditional and two horizontal balconies run round the hall, the ceiling is pyramid shaped with its peak at 28 m above the main floor (Figure 4.45).

The inside surface of the pyramid is highly scattering including quadratic residue diffusers (QRD).

Who would have predicted that the concert hall form favoured around 1900 would again become so popular a century later? It is almost as if the science of acoustics has had nothing to offer the design of concert halls. Science has provided quantification of both the subjective experience of listening to music and the objective acoustic behaviour. In most important respects, we are now able to predict the performance of an auditorium before it is built using either scale or computer models. It seems as if 100 years' experience has just brought us round in a circle.

There are acoustic consultants who consider that the parallel-sided hall is the only viable solution and that the science of acoustics beyond reverberation time still has little to offer. Many other consultants keep a more open mind. The alternatives to parallel-sided halls with good reputations include the lateral

Figure 4.45 Takemitsu Memorial Hall, Tokyo Opera City

directed reflection sequence hall (section 4.10). This can offer exciting acoustics but has not been picked up by other designers following Marshall; successful design with this auditorium form is probably more demanding than other options.

The terraced concert hall has now been tried in several locations and if designed with care seems capable of offering acoustic conditions as good as any parallel-sided hall. Of recent examples, the 1997 **Kitara Concert Hall** in **Sapporo**, Japan, by Nagata Acoustics with 2000 seats is an interesting development of the terraced hall, in that it uses a lot of large convex surfaces (Beranek, 2004). Compared with parallel-sided halls, the terraced concert hall has benefits in performance terms: it can provide an exciting relationship between performers and audience giving a strong sense of shared experience as opposed to the formality in shoebox halls of performers facing audience directly. It has two further advantages. The first is flexibility of design: design details can be modified for individual seating areas usually without repercussions for other seating. This is often not the case in parallel-sided halls, where if conditions for instance under a balcony overhang are unsatisfactory, increasing the height under the soffit is likely to have major effects elsewhere in the hall.

The second advantage of the terraced hall concerns seat numbers. The maximum seat capacity in parallel-sided halls is around 2200, whereas several successful concert halls of different forms have been built with larger capacities up to 3000 seats. In this context, it is particularly interesting to note that, for the parallel-sided Lucerne concert hall, the Lucerne authorities were persuaded that it would be in their best interests to reduce the proposed seat count from 2000 to 1840 for acoustic reasons. The smaller number was considered the 'optimum figure for mass clarity' (Ryan, 1998).

Rather than the broad-brush approach used here to compare the acoustic merits of different gross auditorium forms, the next chapter deals with 16 British concert halls in detail. The discussion will be based on both subjective test results from listeners who completed questionnaires and objective acoustic measurements.

References

General

Bagenal, H. and Wood, A. (1931) *Planning for good acoustics*, Methuen, London.

Barron, M. (1992) Precedents in concert hall form. *Proceedings of the Institute of Acoustics*, **14**, Part 2, 147–156.

Beranek, L.L. (1962) *Music, acoustics and architecture*, John Wiley, New York.

Beranek, L.L. (2004) *Concert and opera houses: Music, acoustics and architecture*, 2nd edn, Springer, New York.

Forsyth, M. (1985) *Buildings for music*, Cambridge University Press, England and MIT Press, Cambridge, MA.

Forsyth, M. (1987) *Auditoria – designing for the performing arts*, Mitchell, London.

References by section

Section 4.2

Elkin, R. (1955) *The old concert rooms of London*, Edward Arnold, London.

Meyer, J. (1978) Raumakustik und Orchesterklang in den Konzertsälen Joseph Haydns. *Acustica*, **41**, 145–162.

Section 4.3

Allen, W.A. (1969) Acoustics twenty years after the Festival Hall. *Royal Institute of British Architects Journal*, February, pp. 62–67.

Beranek, L.L. (1977) The notebooks of Wallace C. Sabine. *Journal of the Acoustical Society of America*, **61**, 629–639.

Beranek, L.L. (1979) The acoustical design of Boston Symphony Hall. *Journal of the Acoustical Society of America*, **66**, 1220–1221.

Beranek, L.L. (1988) Boston Symphony Hall: an acoustician's tour. *Journal of the Audio Engineering Society*, **36**, 918–930.

Bradley, J.S. (1991) A comparison of three classical concert halls. *Journal of the Acoustical Society of America*, **89**, 1176–1191.

Clements, P.A. (1998) The interrelationship of musical excellence and acoustical excellence:

a case study of the Gewandhaus, Leipzig, 1880–1900. *Proceedings of the 16th International Congress on Acoustics, Seattle*, Vol. IV, 2455–2456.

Clements, P.A. (1999) Reflections on an ideal: tradition and change at the Grosser Musikvereinssaal, Vienna. *Proceedings of the Institute of Acoustics*, **21**, Part 6, 5–14.

Elkin, R. (1944) *Queen's Hall, 1893–1941*, Rider, London.

Gropius, M. and Schmieden, H. (1887) *Das neue Gewandhaus in Leipzig*, Von Ernst und Korn, Berlin (included as appendix in Skoda, 1985).

Meyer, E. and Cremer, L. (1933) Uber die Hörsamkeit holzausgekleideter Räume. *Zeitschrift für technische Physik*, **11**, 500–507.

Müller, H.A. (1992) The simple design of shoebox concert halls and their shortcomings. *Proceedings of the Institute of Acoustics*, **14**, Part 2, 9–16.

Parkin, P.H., Scholes, W.E. and Derbyshire, A.G. (1952) The reverberation times of ten British concert halls. *Acustica*, **2**, 97–100.

Sabine, W.C. (1922) *Collected papers on acoustics*, Harvard University Press (reprinted 1964, Dover, New York).

Skoda, R. (1985) *Neues Gewandhaus Leipzig*, VEB Verlag fur Bauwesen, Berlin.

Section 4.4

Andrade, E.N. da C. (1932) The Salle Pleyel, Paris, and architectural acoustics. *Nature*, **130**, 332–333.

Building (1928) New concert hall in Paris. February, pp. 62–64.

McKean, J. (1996) Musical heirs, from Epidauros to Manchester. *RIBA Profile*, **103/10** October, 7–9.

Watson, F.R. (1923) *Acoustics of Buildings*, Wiley, New York.

Xu, A.Y. (1995) Acoustic problems of the Salle Pleyel (Paris) and the modifications in 1994. *Proceedings of the Institute of Acoustics*, **17**, Part 1, 65–71.

Section 4.5

Jordan, V.L. (1980) *Acoustical design of concert halls and theatres*, Applied Science Publishers, London.

Jordan, N.V. and Rindel, J.H. (2006) The Alberta Jubilee Halls reborn with up-to-date acoustics. *Proceedings of the Institute of Acoustics*, **28**, Part 2, 297–304.

Krokstad, A., Strøm, S. and Sørsdal, S. (1968) Calculating the acoustical room response by use of a ray tracing technique. *Journal of Sound and Vibration*, **8**, 118–125.

Northwood, T.D. and Stevens, E.J. (1958) Acoustical design of the Alberta Jubilee auditoria. *Journal of the Acoustical Society of America*, **30**, 507–516.

O'Keefe, J. (2002) Acoustical problems in large post-war auditoria. *Proceedings of the Institute of Acoustics*, **24**, Part 4.

Section 4.6

Bagenal, H. (1950) Concert halls. *Royal Institute of British Architects Journal*, January, 83–93.

Parkin, P.H., Allen, W.A., Purkis, H.J. and Scholes, W.E. (1953) The acoustics of the Royal Festival Hall, London. *Acustica*, **3**, 1–21.

Trevor-Jones, D. (2001) Hope Bagenal and the Royal Festival Hall. *Acoustics Bulletin*, **26**, No. 3, 18–21.

Section 4.7

Damaske, P. (1967) Subjektive Untersuchungen von Schallfeldern. *Acustica*, **19**, 199–213.

Kosten, C.W. and de Lange, P.A. (1965) The new Rotterdam Concert Hall – some aspects of the acoustic design. *Proceedings of the 5th International Congress on Acoustics, Liège*, Paper G43.

Kuttruff, H. (2000) *Room Acoustics*, 4th edn, Spon Press, London.

Meyer, E. and Kuttruff, H. (1959) Zur akustischen Gestaltung der neuerbauten Beethovenhalle in Bonn. *Acustica*, **9**, 465–468.

Section 4.8

Beranek, L.L. (2008) *Riding the waves: a life in sound, science and industry*. MIT Press, Cambridge, MA.

Beranek, L.L., Johnson, F.R., Schultz, T.J. and Watters, B.G. (1964) Acoustics of Philharmonic Hall, New York, during its first season. *Journal of the Acoustical Society of America*, **36**, 1247–1262.

Bliven, B. (1976) Annals of architecture – a better sound. *New Yorker*, November 8, 51–135.

Cremer, L. and Müller, H.A. (trans. T.J. Schultz) (1982) *Principles and applications of room acoustics*, Vol. 1, Applied Science, London.

Fantel, H. (1976). Back to square one for Avery Fisher Hall. *High Fidelity*, October.

Lanier, R.S. (1963) Acoustics – what happened at Philharmonic Hall? *Architectural Forum*, **119**, December, 118–123.

Meyer, E. and Kuttruff, H. (1963) Reflexions-eigenschaften durchbrochener Decken. *Acustica*, **13**, 183–186.

Schroeder, M.R. (1984) Progress in architectural acoustics and artificial reverberation: Concert hall acoustics and number theory. *Journal of the Audio Engineering Society*, **32**, 194–203.

Schroeder, M.R., Atal, B.S., Sessler, G.M. and West, J.E. (1966) Acoustical measurements in Philharmonic Hall (New York). *Journal of the Acoustical Society of America*, **40**, 434–440.

Schultz, T.J. (1965) Acoustics of the concert hall. *IEEE Spectrum*, June, 56–67.

Shankland, R.S. (1963) Acoustics of the New York Philharmonic Hall. *Journal of the Acoustical Society of America*, **35**, 725–726.

Watters, B.G., Beranek, L.L., Johnson, F.R. and Dyer, I. (1963) Reflectivity of panel arrays in concert halls. *Sound – Its Uses and Control*, **2**, 26–30.

Young, E.B. (1980) *Lincoln Center – the building of an institution*, New York University Press.

Section 4.9

Blundell Jones, P. (1995) *Hans Scharoun*, Phaidon, London.

Cremer, L. (1964) Die raum- und bauakustischen Massnahmen beim Wiederaufbau der Berliner Philharmonie. *Die Schalltechnik*, **57**, 1–11.

Cremer, L. (1965) Die akustischen Gegebenheiten in der neuen Berliner Philharmonie. *Deutsche Bauzeitung*, **10**, 850–862.

Cremer, L. (1989) Early lateral reflections in some modern concert halls. *Journal of the Acoustical Society of America*, **85**, 1213–1225.

Frampton, K. (1965) Genesis of the Philharmonie. *Architectural Design*, March, **35**, 111–128.

de Lange, P.A. (1980) The acoustics of the new 'surround' concert hall of Utrecht, The Netherlands. *Proceedings of the 10th International Congress on Acoustics, Sydney*. Paper E-1.1.

Padovan, R. (1980) Music Centre, Utrecht, Holland. *Architectural Review*, February, **167**, 79–87.

Section 4.10

Marshall, A.H. (1979a) Aspects of the acoustical design and properties of Christchurch Town Hall, New Zealand. *Journal of Sound and Vibration*, **62**, 181–194.

Marshall, A.H. (1979b) Acoustical design and evaluation of Christchurch Town Hall, New Zealand. *Journal of the Acoustical Society of America*, **65**, 951–957.

Marshall, A.H. and Barron, M. (2001) Spatial responsiveness in concert halls and the origins of spatial impression. *Applied Acoustics* **62**, 91–108.

Marshall, A.H. and Hyde, J.R. (1979) Some preliminary acoustical considerations in the design for the proposed Wellington (New Zealand) Town Hall. *Journal of Sound and Vibration*, **63**, 201–211.

Marshall, A.H., Nielsen, J.L. and Halstead, M.M. (1998) The Hong Kong Cultural Centre Halls – acoustical design and measurements. *Proceedings of the 16th International Congress on Acoustics, Seattle*, Vol. IV, 2447–2448.

Section 4.11

Beranek, L.L. and Schultz, T.J. (1965) Some recent experiences in the design and testing of concert halls with suspended panel arrays. *Akustische Beihefte*, **1**, 307–316.

Cremer, L. (1986) Der Trapezterrassenraum. *Acustica*, **61**, 144–148.

Edwards, N. (1985) Design methods in auditorium acoustics. *Proceedings of the Institute of Acoustics*, **7**, Part 1, 73–79.

Hoffman, I.B., Storch, C.A. and Foulkes, T.J. (2003) *Halls for music performance: another two decades of experience 1982–2002*, Acoustical Society of America, Melville, New York.

Johnson, F.R. (1981) An answer to the enigma of flexibility for music and theater. *Architectural Record*, mid-August, 68–73.

Johnson, R., Essert, R. and Walsh, J. (1986) Coordination of acoustics and theatre planning in the Pikes Peak Center. *Proceedings of the Vancouver Symposium on Acoustics and Theatre Planning for the Performing Arts*, August, 45–48.

Lord, P. and Templeton, D. (1986) *The architecture of sound*, Architectural Press, London.

Schmertz, M.F. and Jaffe, J.C. (1979) Denver's Boettcher Concert Hall. *Architectural Record*, March, 99–110.

Schultz, T.J. (1986) Room acoustics in the design and use of large contemporary concert halls. *Proceedings of the Vancouver Symposium on Acoustics and Theatre Planning for the Performing Arts*, August, 7–12.

Skoda, R. (1985) *Neues Gewandhaus Leipzig*, VEB Verlag fur Bauwesen, Berlin.

Talaske, R.H., Wetherill, E.A. and Cavanaugh, W.J. (eds) (1982) *Halls for music performance, two decades of experience: 1962–1982*. American Institute of Physics, New York.

Tennhardt, H.-P. (1984) Modellmessverfahren für Balanceuntersuchungen bei Musikdarbietungen am Beispiel der Projektierung des Grossen Saales im Neuen Gewandhaus Leipzig. *Acustica*, **56**, 126–135.

West, J.E. (1966) Possible subjective significance of the ratio of height to width in concert halls. *Journal of the Acoustical Society of America*, **40**, 1245.

Winkler, H. and Tennhardt, H.-P. (1988) Die Semperoper Dresden, das Neue Gewandhaus Leipzig und das Schauspielhaus Berlin und ihre Akustik. *Fortschritte der Akustik DAGA '88*, 43–56.

Section 4.12

Cavanaugh, W.J. and Wilkes, J.A. (1999) *Architectural acoustics: principles and practice*, John Wiley, New York.

Harris, C.M. (2001) Acoustical design of Benaroya Hall, Seattle. *Journal of the Acoustical Society of America*, **110**, 2841–2844.

Hidaka, T., Beranek, L.L., Masuda, S., Nishihara, N. and Okano, T. (2000) Acoustical design of the Tokyo Opera City (TOC) concert hall, Japan. *Journal of the Acoustical Society of America*, **107**, 340–354.

Metkemeijer, R.A., Heringa, P.H. and Peutz, V.M.A. (1988) The Dr Anton Philips Hall in the Hague. *Proceedings of the Institute of Acoustics* **10**, Part 2, 271–280.

Ryan R. (1998) Lakeside spectacular (Cultural and Conference Centre, Lucerne). *Architectural Review*, October, 38–43.

Scarbrough, P. and Jaffe, J.C. (1999) Breaking out of the (shoe) box: perspectives on non-traditional architectural geometries in symphonic concert hall design over the past half-century. *Proceedings of the Institute of Acoustics*, **21**, Part 6, 27–34.

5 British concert halls and conclusions for concert hall acoustics

Traditionally, data presented about concert halls consisted of its size and reverberation time plus a plan and long section. In some cases, this may be adequate: a rectangular hall with a respectable reverberation time and sound scattering walls is likely to have predictable acoustic characteristics. But for the majority of halls, this limited data is far from sufficient. With at least five subjective dimensions, one objective acoustic quantity will not suffice, while the effect of form and size on the acoustics is rarely direct.

In the previous chapter it was mostly necessary to rely on conventional information to categorize many of the existing halls. The chance to make detailed investigations of 16 halls, both at the subjective and objective level, shows just how diverse are the peculiarities of existing halls. In addition the study throws some light on the crucial problem for designers: the nature of the link between architectural form and acoustic behaviour. The arguments here are often based on the behaviour of the new objective quantities in addition to reverberation time, though the conclusions can usually be stated in simple design terms.

Subjective testing was conducted at public concerts performed by reputable symphony orchestras. The listeners were mainly acoustic consultants, though before concluding that this must be a biased group, it transpires that they were not wholly unanimous in their views. The questionnaire, shown in Figure 3.7, has major scales of clarity, reverberance, envelopment, intimacy and loudness. Analysis of the results indicated that the questionnaire

technique was reproducible. The initial subjective exercise involving 11 halls is described in Barron (1988a). The results averaged by hall from this exercise are shown in Figure 5.1.

For objective measurements, the five major quantities discussed in sections 3.3 and 3.10 were used: reverberation time, early decay time, the early-to-late sound index (called here 'objective clarity'), the early lateral energy fraction (called here 'objective source broadening') and total sound level. Measurements were made with a single source position (generally 3 m from the stage front) and at 11 or more receiver positions. Since each of these halls is basically symmetrical about the main axis, only positions on one side of the hall needed to be tested. The measured results for each hall are presented diagrammatically, as explained in section 3.10.6. As an objective measure of the strength of bass sound, mean bass level balance results are given in Figure 5.73.

For each hall there are two photographs, a scaled plan and long section plus a figure with measured objective results. Except for the first hall reviewed, the figure numbers for the photos and plans/sections are generally omitted from the text. In addition, some general references about the individual halls are only listed under references at the end of the chapter for readers who may wish to study the written record further.

The 16 British concert halls should be considered as case studies. They were measured and tested subjectively during two separate periods: for halls up the Nottingham Royal Concert Hall in section

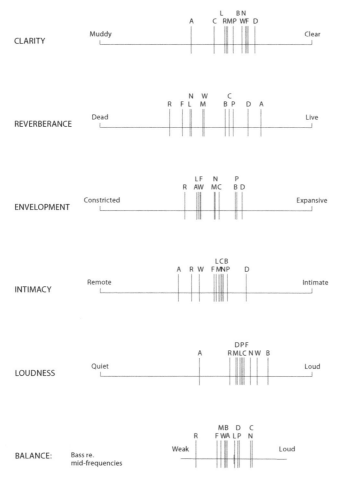

Figure 5.1 Mean subjective responses by hall on the questionnaire scales.
Labels: A, Royal Albert Hall, London; B, Bristol Colston Hall; C, Croydon Fairfield Hall; D, St David's Hall, Cardiff;
F, Royal Festival Hall, London; L, Liverpool Philharmonic Hall; M, Manchester Free Trade Hall; N, Nottingham Royal
Concert Hall; P, Poole Lighthouse Concert Hall; R, Barbican Concert Hall, London; W, Watford Colosseum.

5.12, measurements were made between 1982–4, whilst the remainder were measured during the 1990s. Measurement dates are recorded with other data at the end of each section. The text in each case relates to the hall around the time of measurement. Several halls have subsequently undergone significant modifications and in the case of the Free Trade Hall in Manchester, section 5.7, the hall no longer exists.

The history of concert hall design in Britain extends over three centuries and is significant on the world scale. London in the eighteenth century could claim to be the musical capital of the world, but unfortunately nothing of significant size remains from this period. A further regrettable absence is the lack of a large so-called classical rectangular hall in Britain. The last of these, St Andrew's Hall in Glasgow, tragically burnt down in 1962 (see Beranek, 1962). Of the British halls now in existence, they can be divided into three groups. In the first of these, we have two halls designed before the First World War, before auditorium acoustics became a

Table 5.1 Basic details of 16 British concert halls

Hall	Date	Seats	Auditorium volume (m³)	Reverb. time (s)	Acoustic consultant
Royal Albert Hall, London	1871	5090	86 650	2.4	–
Usher Hall, Edinburgh	1914	2217 + 333	16 000	1.7	–
Philharmonic Hall, Liverpool	1939	1767 + 184	13 560	1.55	H. Bagenal
Watford Colosseum	1940	1586	11 600	1.45	H. Bagenal
Royal Festival Hall, London	1951	2645 + 256	21 950	1.45	H. Bagenal, P.H. Parkin and W.A. Allen
Colston Hall, Bristol	1951	1940 + 182	13 450	1.7	H.R. Humphreys, P.H. Parkin and W.A. Allen
Free Trade Hall, Manchester	1951–1996	2529	15 430	1.55	H. Bagenal
Fairfield Hall, Croydon	1962	1539 + 250	15 400	1.65*	H. Bagenal
Lighthouse Concert Hall, Poole	1978	1473 + 120	12 430	1.55*	P.H. Parkin
Barbican Concert Hall, London	1982	2026	17 750	1.6	H. Creighton
St David's Hall, Cardiff	1982	1687 + 270	22 000	1.95	Sandy Brown Associates
Royal Concert Hall, Nottingham	1982	2315 + 186	17 510	1.75*	Artec Consultants Inc.
Glasgow Royal Concert Hall	1990	2195 + 263	28 700	1.75	Fleming & Barron with Sandy Brown Associates
Symphony Hall, Birmingham	1991	1990 + 221	25 000	1.85	Artec Consultants Inc.
Bridgewater Hall, Manchester	1996	2127 + 276	25 050	2.00	Arup Acoustics
Waterfront Hall, Belfast	1997	2039 + 195	30 800	1.95	Sandy Brown Associates

The quoted reverberation times are mean occupied values at 500/1000 Hz (* = predicted from unoccupied measurement). Seat numbers in halls which include choir seating are shown as 'auditorium + choir'.

science. In both cases we find the now obvious fault of focusing problems due to concave reflecting surfaces. The second group from the mid-1930s to the mid-1960s are all halls either designed acoustically or strongly influenced by one consultant, Hope Bagenal. Central among these halls is the Royal Festival Hall in London. The final group of halls represents relatively diverse design philosophies inspired by acoustic advances since the 1950s. Discussion of these halls will be made chronologically; their basic details are given in Table 5.1. Following individual sections on the 16 halls, the conclusions for concert hall design are considered.

5.1 Royal Albert Hall, London

For a building as quintessentially Victorian as the Royal Albert Hall in both style and scale, it is surprising to discover the tortuous road to its completion in 1871. On more than one occasion the Queen's support had to be elicited by gentle reminders of her beloved Prince Consort's interest in the scheme (even though his own enthusiasm fell noticeably short on commitment) and the appropriateness of this memorial to him. In the absence of generous support from wealthy bodies or government, a novel method was finally used to collect funds to

Figure 5.2 Royal Albert Hall, London

Figure 5.3 The 'flying saucers' in the Royal Albert Hall, London

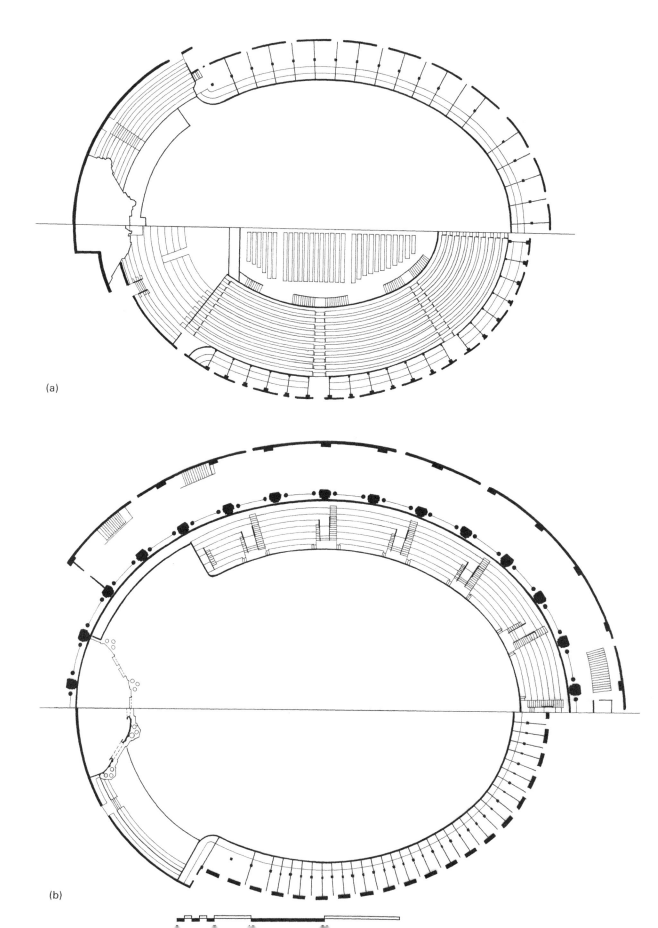

(a)

(b)

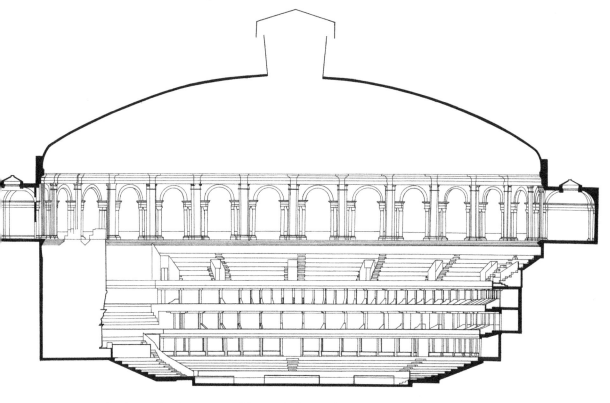

Figure 5.4 (a), (b) Plans and (c) long section of the Royal Albert Hall, London

finance the building: to sell seats for £100 for free use 'in perpetuity'. The private ownership of seats and boxes persists to this day, even though the ownership is not now as free as those early investors must have been led to believe.

The story begins with the Great Exhibition of 1851 in Hyde Park, which produced not only Paxton's remarkable Crystal Palace but also a very substantial profit for the Commissioners of the Exhibition. Some of this money was wisely invested in purchasing the land on which museums and later the Albert Hall were built. The guiding entrepreneur behind the hall, to be devoted to Arts and Sciences, was a Mr Henry Cole. Inspired by the Roman arenas he saw on a continental voyage in 1858, he proposed a 'modern' equivalent to hold 30 000 spectators. As time progressed and financial limitations

forced some realism onto the project, the numbers shrank first to 12 000 and then to the 6000 or so in the hall which exists today. The design was much criticized: *The Engineer* magazine considered it 'wrong for anything except gladiatorial combat'! Nevertheless the elliptical plan with layers of boxes surrounding it survived, Figures 5.2–5.4. To cover the immense area, the largest unsupported span roof (at the time) was designed and constructed of glass on an iron frame. What might have remained the largest anachronism of its age has surprisingly found a healthy niche for itself in London's musical life.

For a hall 'erected for the advancement of the Arts and Sciences' it can claim to have provided its lessons for the science of acoustics. The problem was already perceived at the opening address by the Prince of Wales (later King Edward VII). As *The Times* reported on the next day: 'The address was slowly and distinctly read by his Royal Highness, but the reading was somewhat marred by an echo which seemed to be suddenly awoke from the organ or picture gallery, and repeated the words with a mocking emphasis which at another time would have been amusing'. To Col. Scott, the designer of the huge roof structure, this acoustic problem was a mere irritant which could be solved by a velarium of cloth (weighing $1\frac{1}{4}$ tons) stretched underneath the roof. But in spite of experiments with the height of the velarium, the echo persisted owing to the strength of focusing by the concave dome. Mr Wentworth Cole in the 1890s had wires stretched across the hall with rabbit-netting suspended from each wire. He pronounced that 'there does not appear any doubt that these wires have proved effective in diminishing in a marked degree, if not altogether getting rid of the echo, and in effect bringing the sound of instruments, and of the voice, markedly nearer to the listener'. We are now able to prove that such enthusiasm for the effectiveness of stretched wires must have been simple self-delusion! By the 1920s there was concern that

the echo might have become even worse. One suggested cause was the current men's fashion which involved an 'increase in boiled shirt fronts'. Yet the recommended treatment at the time of 4000 m² of suspended felt was deemed too expensive.

The event which spelt a brighter future both for music in the hall and for its acoustics was the otherwise much lamented destruction of the Queen's Hall in May 1941, which left the Albert Hall as the only large hall in the capital. The 1941 summer Promenade Concert season was scheduled to be held in the hall and during preparations it was found that lowering the velarium by about 10 m down to the level of the Gallery floor almost cured the echo. Though the risk of shattered glass accelerating and cutting through the velarium precluded the adoption of this solution during wartime, a substantial reflector over the orchestra was installed which obscured many of the echo paths. After the war, the velarium was removed and a fluted aluminium inner dome with a perforated lower skin backed by mineral wool was installed inside the dome. This was effective as an acoustic absorber though not quite over the whole frequency range. Secondary echo paths remained, such as off the perimeter of

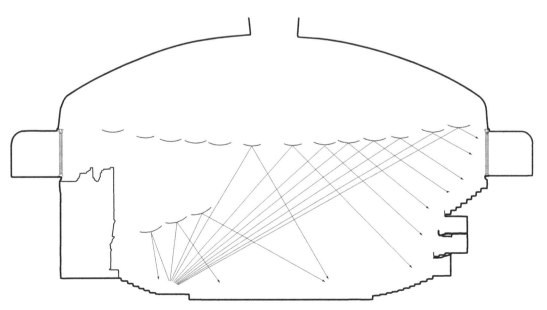

Figure 5.5 Longitudinal section through the Royal Albert Hall showing sound paths for reflections off suspended reflectors

the dome, and it was only in 1968–70 that the echo problem was finally suppressed by suspending 134 'flying saucer' diffusers at the level of the Gallery ceiling.

In spite of its problems and its immense size, more than twice the volume of any other auditorium considered here, the Royal Albert Hall manages annually to house London's vibrant 'Prom' concerts during eight weeks between July and September. In the 1990s the average attendance at 60 concerts was nearly 4000 per concert – a remarkable achievement. This success owes little to the acoustics but is likely to be affected by the peculiar power of the arena form. This power to stimulate a strong sense of the shared experience has not escaped the notice of many organizations, including the British Union of Fascists in the 1930s. The hall has always been used for a very wide range of events from sports, in particular boxing, to exhibitions, reunions, balls and of course its musical uses. It is regularly used for large-scale performances, an ideal location for such spectaculars as Tchaikovsky's 1812 Overture. In this and grand choral works the space really comes alive, but at the smaller scale the impact can be dull. Several singers have however gained the reputation of being able to 'fill the Albert Hall', the most famous being the contralto Clara Butt (1873–1936). It is likely that her particularity was not just a matter of greater sound volume (see section 9.2.1).

Until the modifications of the late 1960s, the problem of the echo dominated comments about the acoustics. With a hall of this size though, there are inevitably additional problems. Two aspects were also dealt with during the 1968–70 period. The enclosed volume of the Albert Hall is larger than necessary for an appropriate reverberation time. The reverberation time was particularly long at mid-frequencies, so by placing absorbent which was most efficient at these frequencies on the upper side of the saucers, the reverberation time characteristic was rendered more uniform with frequency. This location for absorbent is particularly apt because it increases the effect of the saucers by absorbing some remaining sound reflected from the dome. Another problem with a large space is

maintaining sufficient early reflections when some seats are inevitably remote from useful surfaces. The reflections from the suspended saucers are very valuable at the upper seating levels (Figure 5.5). A reflector above the stage, as first used in 1941, has been retained to provide reflections particularly to the Arena and Stalls seating.

Subjective characteristics

For such an extreme design of hall as the Royal Albert Hall, it is no surprise that the hall is found to be subjectively extreme too. It was judged as the least clear, the most reverberant, the least intimate and the quietest British hall. Spatial response was also poor. Predictably the overall judgement was not enthusiastic. Nevertheless there are some compensating features which deserve mention. Although the loudness of sound is inadequate, the dynamic response is particularly good: the transition during *crescendos* is extremely smooth with no sense of constraint. This is aided by a high degree of blend in the orchestral sound with a reasonable tonal balance.

However, the dominant response is to the quietness of the sound. Compared with other British halls, the high sense of reverberance, which is generally thought desirable, is inadequate compensation for this quiet sound. Interestingly the listeners were in difficulties about judging the clarity and in the stalls the sense of reverberation is less than one expects in such a voluminous space. Reverberance is significantly higher in the balcony; surrounded as it is by the hard surfaces of the Gallery; the reverberation there provides a rich spatial experience.

A hall of this size without detailed acoustic design must have problems of uniformity, yet it is a tribute to the modifications of the late 1960s that the variation in acoustic quality is not larger. The boxes will of course sound much better at their mouths than in the rear (see section 9.2.6). Minor echo problems still exist due mainly to the elliptical plan form.

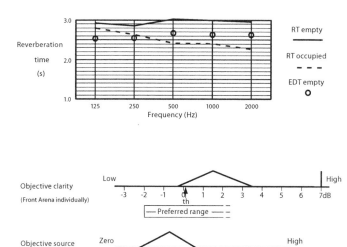

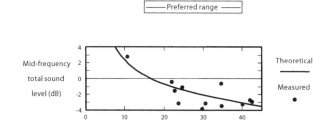

Figure 5.6 Royal Albert Hall: objective characteristics

Objective characteristics

Measured objective data for the Albert Hall are summarised in Figure 5.6. One notices two marked characteristics: a long reverberation time and a low total sound level. The occupied reverberation time is rather long at mid-frequencies (2.4 seconds) rising healthily in the bass. However, the quantity which is thought to be more relevant for subjective appreciation, the early decay time, is sufficiently shorter and probably within the optimum range. Yet in spite of the long reverberation time, we find high objective clarity, higher than expected. What appears to be happening is that the reflector directs a lot of sound energy onto the audience which is then no longer available for later reverberant sound. Further evidence of this is found in the measured sound decays, which at most seats are curved, or 'sagging'.

This may explain the difficulties which listeners had in judging clarity.

When one looks at the total sound level situation, the measured values are surprisingly close to the predicted according to the revised theory (Barron and Lee, 1988). In a space as large as this we might expect quieter sound at seats remote from walls. The uniformity found here is probably due in part to the careful orientation of the stage reflector. Theory says that the total sound level is a function of the total acoustic absorption. In the Albert Hall this absorption is due to the very large audience as well as the additional absorption incorporated to control the echo from the dome, so these two elements are responsible for the quiet sound, with levels at all but one position below the 0 db criterion.

In some seats, before the 1968–70 'flying saucer' treatment, the level of the echo was 3 db greater than the direct sound with a delay of around 1/6th of a second – certainly a serious echo. The echoes which now remain are much less prominent and are only likely to be obvious with more directional musical instruments, such as trumpets in their higher registers.

Conclusions

Several clear lessons emerge from the experiences of the Royal Albert Hall. Extensive concave surfaces create unacceptable conditions if the curvature is such that it causes focusing at audience positions. The enormous dome in the Albert Hall falls into this category and only substantial modifications have cured the problem. Since the enclosed volume of the hall is also too large for orchestral music use, control of both echo and reverberation by acoustic absorbent was sensible. The history of this famous echo also shows that more than simple absorption on a concave surface can be necessary. The addition of suspended reflectors has now substantially solved the echo problem and by providing earlier reflections has further improved listening conditions. This modification, together with the reflector over the orchestra, has provided reasonable uniformity and sound with enough clarity to enable successful concerts to be held.

The most serious remaining problem, which is unavoidable in a hall seating so many, is the quietness of the sound. This problem has been compounded though by the excessive hall volume and additional absorbent added in the region of the dome. This extra absorption, though valuable in controlling the reverberation time, is probably responsible for a sound level decrease of about 1.5 db which cannot be afforded in a hall of this capacity. As it stands, electroacoustic assistance is the only way by which sound levels could be increased. Immediately after the opening in 1871, *The Times'* critic concluded with remarkable prescience that 'happily there exist choral works by genuine masters which may be advantageously brought into request; and these, at any rate for a time, would most fairly try the capabilities of the Albert Hall'. The echo has gone

Royal Albert Hall, London: Acoustic and building details

Volume = 86 650 m³

Total hall length = 67 m

Number of seats = 5090

Volume/seat = 17.0 m³

True seating area = 2116 m² plus 564 box seats

Stage area = 104 m²

Acoustic seating area = 2986 m²

True area/seat = 0.47 m²

Volume/acoustic seating area = 29.0 m

Mean occupied reverberation time (125–2000 Hz) = 2.5 seconds (measured by K. Shearer, 1970, from music)

Unoccupied objective data measured February 1982 (stage partially occupied with chairs); subjective assessments in September 1983 and August 1985

Owned by Royal Albert Hall

Building opened in March 1871

Designed by Capt. F. Fowke and subsequently LtCol. H.Y.D. Scott

Uses: concerts, sports events, meetings, exhibitions, balls etc.

Construction:

Floor – timber except in balcony which is concrete

Walls – generally plaster on masonry. Back of boxes is thin wood

Ceiling – plaster round periphery of dome, see text for aluminium dome treatment

Suspended saucers – glass-reinforced plastic dishes with 38 mm fibreglass above (covered with 0.04 mm plastic film)

Stage reflectors – glass reinforced plastic

Stage – thick timber over air space

Organ – behind choir seating

Seating – upholstered movable and fixed seats, tip-up in the balcony

but large-scale choral works still 'try the capabilities' the most effectively.

In 1996 a major redevelopment began at the Albert Hall, conducted so as not to interfere with the normal running of the venue. The most substantial changes have been underground with new truck access and loading docks and much improved backstage facilities for performers. The acoustics were studied in a 1:12 scale acoustic model by acoustic consultants Peutz & Associates (Metkemeijer, 1997, 2002). The work was completed in 2003, with the major visible change a revised array of the saucers suspended below the dome.

5.2   Usher Hall, Edinburgh

Private bequests to build public concert halls or theatres are much more rare in Europe than on the other side of the Atlantic. It is probably a reflection of Britain's historical standing in the world that this lone example dates from the turn of the century. In 1896 Mr Andrew Usher donated £100 000 for a City Hall in the hope 'that the opportunities afforded by the Hall and its adjuncts might promote and extend the cultivation of, and taste for, music, not only in Edinburgh, but throughout the country'. Selection of a site for the hall was the cause of such controversy and delays that Mr Usher did not live to admire the fruits of his generosity. The hall was completed in 1914 and greeted by the *Architects' and Builders' Journal* (20 May 1914) 'as one of the most successful buildings of the monumental type that have recently been erected in the kingdom'. The exterior itself is in the classical mould with high-level sculptures representing 'Municipal Beneficence', 'The Soul of Music' and the like. The interior attempts to impose classical details onto the form of the usually more florid Victorian and Edwardian theatres. Though in design terms this involves several uncomfortable compromises, the hall is successful in use and is a crucial venue for the annual International Edinburgh Festival. Mr Usher would surely be pleased that music in his hall inspires an international as well as a broad British audience.

Up to the arrival of the Modern Movement in architecture, design owed much to precedent. This process led to distinct forms for the different auditorium types. Yet there were frequent failures during the last century when designers ignored the significance of differences between the evolved forms of theatres, concert halls and opera houses. The design of the Usher Hall is an adaptation of the theatre form for music use.

The hall conforms to the horseshoe plan at all three seating levels. A distinct proscenium frame divides the auditorium from the stage, with fluted Ionic columns on each side and with a drop of ceiling height over the stage and choir/audience seating behind. When viewed on axis the area behind the proscenium is reminiscent of the perspective scene with a raked 'perspective' floor which was the norm for eighteenth-century theatre stages. In the *Architectural Review* in 1914 we find that since the building was 'intended expressly for the hearing of good music, great attention had to be devoted to the acoustics. To this end the hall was given a flat ceiling, which is treated with broad deep ribs'. The significance of this comment relates to the fact that the interior ceiling does not follow the shape of the exterior domed roof. It seems that at least for its section a lesson had been learnt from the Royal Albert Hall in London. In the original building the central ceiling panel was glazed to provide daylight, as were additional clerestory windows. The glazed ceiling panel is now replaced by an ugly perforated saucer containing ventilation outlets and the windows have been covered over with curtains.

The concept of reverberation time dates from the work of Sabine at the turn of the century. However, it seems improbable that Sabine's work was known to the anonymous adviser(s) on the acoustics of the Usher Hall. It is surprising therefore that the hall volume should have been built large enough to provide a respectable reverberation time. In theatres the reverberation time is generally about half the value found in this hall, so we can only speculate as to why the hall has this desirable large volume. It is however the gross form of the hall which gives greatest cause for concern.

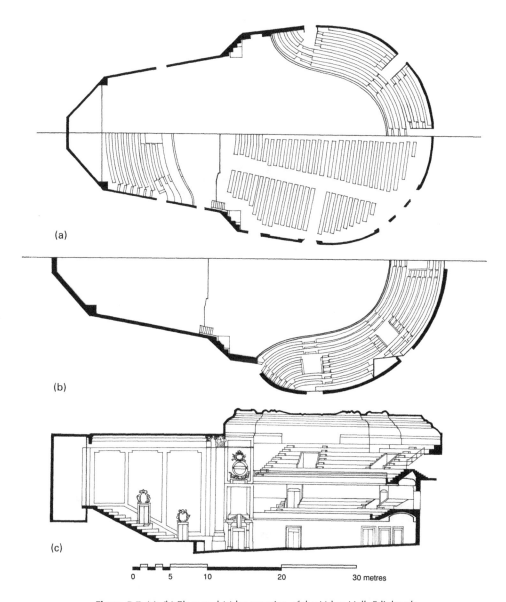

Figure 5.7 (a), (b) Plans and (c) long section of the Usher Hall, Edinburgh

Subjective characteristics

The characteristic of the Usher Hall which distinguishes it from many halls discussed here is the variability in the acoustics at different seat locations.

In the major part of the stalls, the sound quality is reasonable but without any pronounced positive features. As one moves towards the rear of the stalls the sound becomes quieter and less impressive. In many seats in the rear stalls there is also

Figure 5.8   Usher Hall, Edinburgh

Figure 5.9   Usher Hall, Edinburgh

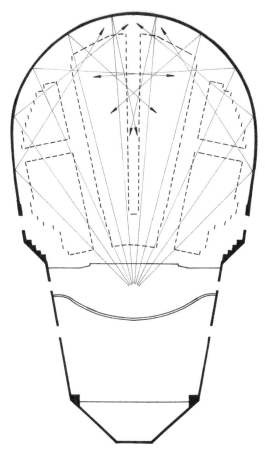

Figure 5.10 Sound paths for reflections from the curved wall at stalls level in the Usher Hall, Edinburgh

his enthusiasm 'unquestionably' for the Upper Circle where he finds the clarity restored though music still sounds subdued in level there.

Unfortunately, owing to its geographical location it has not been possible to conduct the normal subjective analysis in the Usher Hall. Results which are to hand support Beranek's views, with comments such as 'lifeless and lacking in brilliance' in the Grand Circle. As well as the Upper Circle, a further seating area with its champions is the choir seating behind the stage. The sound there is predictably judged as loud and clear with only the sense of reverberance judged by some as deficient. In seats behind the orchestra the balance between instruments is obviously not ideal but the excitement of the experience is a rich compensation.

Objective characteristics

The reverberation time of the Usher Hall is reasonable, at the bottom of the optimum range (Figure 5.11). The change with occupancy is larger than normal due to there being little upholstery on seats in the Gallery. Concerning the other objective measures, the early decay time is unusual in that it is higher than the reverberation time at mid-frequencies. The measured values of the objective clarity are not remarkable but the objective source broadening contains some very high values. As plotted here the total sound level shows evidence of non-uniform behaviour, with some positions louder than expected and some quieter. However from a superficial view of the results in Figure 5.11, nothing obviously suggests disappointing acoustics.

Yet the high values of objective source broadening do require explanation; this concert hall has the highest average value measured. One would first think that the focusing by the concave wall surfaces might cause the effect, but one sees in Figure 5.10 that few seats can benefit from strong lateral sound coming from the curved walls. Since objective source broadening is the fraction of early energy which arrives from the side, an alternative though unusual explanation might be that the high values are caused by a lack of early frontal sound.

the distraction of a focused reflection from behind (Figure 5.10). While the delay of this strong reflection is too short for any echo to be heard, there can be a strong sense of false localization here with individual instrument groups appearing stronger from behind than in front.

The major surprise and disappointment in the hall is the situation in the Grand Circle. Seats at the first balcony level would generally be considered the best and this is reflected in the pricing policy in this hall. Beranek (1962) describes the sound here as lacking in clarity and brilliance, without the reverberation of the hall being apparent either. He also comments that the sound is weak. Beranek reserves

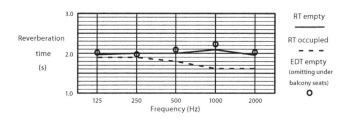

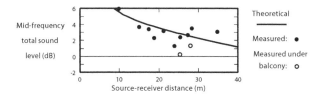

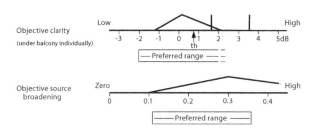

Figure 5.11 Usher Hall, Edinburgh: objective characteristics

To investigate this, it is possible from the measurements to calculate the early frontal sound and compare measured values with predictions (based on revised theory, with an assumed value for the fraction of lateral energy of 0.15). One finds that in only a quarter of the measurement positions does the measured frontal sound correspond with predictions. Areas where there is adequate early frontal sound are at the front of the stalls and in the Upper Circle opposite the stage. In the remaining three-quarters of the hall the frontal sound is deficient (by between 1 and 4 dB). Quiet total sound also occurs at several of these seats, due to the same cause. Another cause of quiet sound is lack of late sound underneath balcony overhangs.

This interpretation of the major acoustic problem in the Usher Hall is something of a heresy according to modern thinking. A high proportion of early

sound from the side is generally associated with the best concert hall acoustics. Here we find a hall where most seats have a high proportion of lateral sound yet the sound is judged as lifeless. In the Usher Hall the lateral reflections are at the expense of other reflections and the result is not judged enthusiastically. The cause of lack of frontal sound is easy to appreciate from the view of the hall looking towards the stage (Figure 5.8). The curved side walls constitute about the only surface at stalls level which can provide useful early reflections and from Figure 5.10 these are rarely frontal. The seating behind the stage will be absorbent when occupied, the stage walls are not angled to project sound into the auditorium and the wall profile immediately in front of the proscenium where the wall 'steps back' at the sides negates a possible reflection there. At higher seating levels even the chance of reflection off the curved walls is limited owing to balconies extending round the sides of the hall and obscuring 'rising' sound.

Subjectively a lack of sound from the front may make it difficult for listeners to identify with the orchestral source. In the Upper Circle, listeners will receive a strong reflection from the ceiling and the acoustics there are found to work well. For seating behind the stage, the opposite conditions prevail: a reverse-splay plan provides a high reflection density and a loud sound. It is not surprising that for some people these are the best seats in the house.

Since the reported measurements, the hall was renovated in 1999–2000. Of acoustic significance was the replacement of audience seating and removal of carpet in aisles. The unoccupied reverberation time has been increased by about 0.3 seconds at 500/1000 Hz. A brighter sound than before is reported.

Conclusions

The Usher Hall is designed in the monumental manner but without attention to the successful acoustic tradition of rectangular concert halls. The auditorium form has its origins in theatre designs but the scale has been made appropriate for concert use. Fortunately this has meant a large enough hall volume to provide a reasonable reverberation time. However, the theatre form with its concave walls results in unpleasant focused reflections at seats close to that wall, whereas elsewhere there is a lack of early sound, particularly from the front. The evidence of this hall suggests that this too is judged as detrimental acoustically. Where strong reflections do arrive, in the Upper Circle and behind the stage, the acoustics work well. Refurbishment in the year 2000 involved modifying the seating arrangement in the stalls, removing carpet and changing the ventilation into a displacement system, with Sandy Brown Associates as acoustic consultants

The Usher Hall clearly illustrates the problems of taking a building form appropriate to speech and applying it for music.

Usher Hall, Edinburgh: Acoustic and building details

Volume = 16 000 m³

Total hall length = 52.2 m

Number of seats = 2215 plus 287 choir

Volume/seat = 6.3 m³

True seating area = 1040 m²

Stage area = 120 m²

Acoustic seating area = 1234 m²

True area/seat = 0.41 m²

Volume/acoustic seating area = 13.0 m

Mean occupied reverberation time (125–2000 Hz) = 1.75 seconds (measured by Building Research Station, 1951, from music signal)

Unoccupied objective data measured August 1982 (unoccupied stage); limited subjective assessment in 1982

Owned by City of Edinburgh

Building opened in 1914

Designed by Stockdale Harrison and Sons and H.H. Thomson

Acoustic consultant (1999–2000): Sandy Brown Associates

Uses: principally popular and orchestral music

Construction:

Floor – timber (aisles of stalls and Grand Circle carpeted prior to 2000)

Walls – plaster on lath over air space

Ceiling – plaster on lath with thick plaster profiling

Stage and choir enclosure – plaster on wooden lath

Stage – timber over large air space

Organ – located behind choir seating

Clerestory windows – covered with curtains

Seating – stalls and Grand Circle, upholstered tip-up with unperforated bases

Upper Circle, light upholstered with solid backs

Choir, solid timber seats without upholstery

5.3 Philharmonic Hall, Liverpool

In the middle of Liverpool's golden era, two substantial halls intended for music performance were built in the city within five years of each other. The original Philharmonic Hall was commissioned by the Liverpool Philharmonic Society and had a good acoustic reputation. It opened in 1849 with an audience capacity of 2100 within a rectangular plan form. St George's Hall also followed the rectangular form but on a sumptuous scale in the grand classical manner. It is reputed to be the largest town hall in England. It opened in 1854 and survives to this day but is now little used for music (see Forsyth, 1985, for discussion of both halls). The Philharmonic

Hall was destroyed by fire in 1933, to be replaced by the present building on the same site in 1939. The present hall is the only example in Britain to be run by its resident orchestra, the Royal Liverpool Philharmonic.

During the 1920s an acoustic design concept developed which was at odds with both the traditional rectangular form and the ideas of Sabine. It was felt that room surfaces should reflect sound directly onto the audience. The purest example of a so-called 'directed sound' hall is the Salle Pleyel, Paris (section 4.4). The Philharmonic Hall is the closest British representative of this design philosophy, though much less extreme than its French counterpart. In plan the front half of the hall is

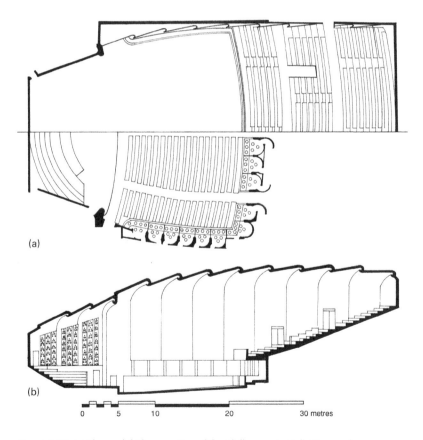

(a)

(b)

0	5	10	20	30 metres

Figure 5.12 (a) Plan and (b) long section of the Philharmonic Hall, Liverpool

Figure 5.13 Philharmonic Hall, Liverpool

Figure 5.14 Philharmonic Hall, Liverpool

fan-shaped (with approximately 40° apex angle), but at the rear the side walls become virtually parallel. In this way one of the major hazards of the fan shape has been avoided: adequate reflections are likely to exist in the rear of the hall at the centre. The ceiling follows a curve which has the effect of enhancing the strength of the ceiling reflection at the rear. In gross terms the form of the Philharmonic Hall can be expected to be less propitious for producing diffuse sound, with fewer possible directions from which sound can arrive at listeners when compared with the rectangular form. The wall and ceiling construction is of a series of shells of plaster on expanded metal, which are arranged in echelon fashion. The front edge of each shell is rolled back to allow lighting to be concealed. The wall/ceiling profile will provide a degree of diffusion in the plane of the long section. The same wall/ceiling scheme continues into the stage area with a vestigial proscenium arch separating the two. Grills in the stage walls conceal an organ.

The design is clearly influenced by contemporary cinema forms with detailing characteristic of the 1930s, but how far the acoustic consultant, Hope Bagenal, influenced the design is not known. In his book of 1931 we find the now surprising suggestion (p. 50) that 'much failure and inconvenience would be avoided in acoustics if the fan shape could be used in preference to square and oblong'. Not long after this time his view changed definitively in favour of the rectangular plan form for concert halls. Two details provide clear evidence of acoustic advice: the absorbent treatment on the rear wall and the upward inclination of the balcony front. Both features will minimize focused reflections back to the stage. In the late 1940s with post-war reconstruction underway, this hall was the subject of close study as it offered the chance to investigate the relative merits of the 'directed sound' and rectangular concert hall. Somerville (1949), using the Liverpool Philharmonic Hall as his example of the former, compared it with the (then) best British rectangular hall: St Andrew's Hall, Glasgow.

Subjective characteristics

There are several sources for subjective comments on this hall. Somerville (1949) mentions that 'the definition was good' but that the deadness of the hall results 'in a hard tone in loud passages, which robs *crescendo* passages of their brilliance'. He found a hardness in the string tone and poor balance between instruments at the front and rear of the stage, instruments at the rear being more prominent. In their studies they also observed colouration at around 110 Hz. This last peculiarity is probably not noticed explicitly by many listeners.

Parkin *et al.* (1952) conducted a survey, also during 1949, of the views of music critics, professors of music and composers. This group considered the Philharmonic Hall to have the best acoustics of halls then existing in Britain. St Andrew's Hall, Glasgow, was voted second. In an unpublished Building Research Station report, six listeners (mainly acousticians) commented on the acoustics during a concert in 1949. They felt that there was inadequate mid- and high frequency reverberation, but 'an excessive low frequency reverberation ... (which) might have been a resonance at a particular frequency'. (Somerville in 1949 found evidence of resonance around 110 Hz by all sections of the wall and ceiling, which coloured low-frequency sound.) Sound was perceived as weaker in the balcony. Overall there was 'broad agreement that the hall was definitely not very good acoustically'. Beranek (1962) echoes these views but he was obviously impressed by the clarity of the music. He suggests that the lack of reverberation makes the acoustics less favourable for Romantic symphonies than for music of an earlier period.

The responses to this author's questionnaire survey were average on most scales with the exception of the reverberance judgement, which was consistently judged as low. Surprisingly the clarity was not judged as correspondingly high. Most listeners preferred the stalls to the balcony. One criticism of the situation in the balcony is the limited angle from which sound is perceived; the experience there is principally frontal. No adverse comments were made about the dynamic response of the hall, which seemed quite natural, though it does not 'sound like a large space'. Overall the assessment was of acoustics that were 'Good', which when compared with other halls places this hall in a group below the best visited.

The dominant characteristic in this hall is lack of reverberance but this is compensated by an ability to hear individual parts with a pleasant brilliance to the sound. This characteristic may explain why Parkin *et al.*'s group liked the hall while others did not. Musicians tend above all else to seek high clarity, which enables them to hear fine musical detail.

Objective characteristics

With such a well-documented lack of sense of reverberation one expects to find a short reverberation time and the measured mid-frequency value of only 1.55 seconds in the occupied condition ties in with the subjective reactions (Figure 5.15). Somerville (1949) suggested that the measured reverberation time is shorter than predicted by Sabine's formula because of the room form. However a simple reverberation time analysis using equivalent absorption coefficients (section 2.8.6) for both the unoccupied and occupied conditions does not support Somerville's suggestion. The mid-frequency reverberation time is close to prediction, the low measured value occurs because of the small hall volume caused by inadequate ceiling height.

One peculiarity in the measured results is the high objective clarity in seats at the rear of the balcony. Objectively this is a consequence of a deficiency of later reverberant sound at these seats, similar to the situation which occurs under a balcony overhang. Yet the ceiling height above the

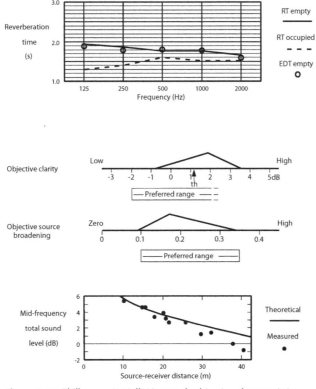

Figure 5.15 Philharmonic Hall, Liverpool: objective characteristics

balcony is not especially low. The directed nature of the hall surfaces as opposed to a design with more scattering may be the cause of this behaviour. The deficient late sound at rear balcony seats is linked to the low total sound levels in the graph in Figure 5.15. Sound at these seats will be perceived as lacking both reverberance and loudness, which would explain listeners' preference for the stalls seating. The objective source broadening, however, is better than one expects in fan-shaped halls; the fact that the rear side walls become parallel to each other is likely to help in this regard.

Conclusions

The Philharmonic Hall is a valuable example of a hall following a design philosophy now discredited by most acousticians. Yet the major fault of the hall is simply to have inadequate volume and hence too short a reverberation time. This makes the hall more suitable for music of the Classical and Baroque eras than for the subsequent grand Romantic repertoire. Although many comments about the hall are unambiguously critical, the overall assessment in this study is of good acoustics though not the best. The major compensation in its acoustic quality is the high degree of definition and brilliance, which allows musical detail to be clearly followed. This is a characteristic much appreciated by musicians.

The detailed acoustic analysis of this hall has indicated certain trends associated with the directed ceiling form. The subjective effects of these probably make the hall sound even less reverberant than a rectangular hall with the same reverberation time. The degree of directedness in the design and hence the subjective effects are much less severe than in a more obviously directed-sound design like the original Salle Pleyel, Paris.

Philharmonic Hall was refurbished in 1995, principally because rust had been found in the steel structure supporting the internal walls and ceiling. On advice from Kirkegaard Associates, the plaster of the internal walls and ceiling was replaced with concrete, the platform was modified, the ventilation rerouted and the carpet removed,

---

Philharmonic Hall, Liverpool: Acoustic and building details

Volume = 13 560 m³

Total hall length = 50.0 m

Number of seats = 1767 plus 184 choir

Volume/seat = 7.0 m³

True seating area = 836 m² plus 134 box seats

Stage area = 148 m²

Acoustic seating area = 1150 m²

True area/seat = 0.46 m²

Volume/acoustic seating area = 11.8 m

Mean occupied reverberation time (125–2000 Hz) = 1.45 seconds (measured by Building Research Station, 1949, from music signal)

Unoccupied objective data measured April 1982 (stage occupied with chairs and music stands); subjective assessment in November 1984

Owned by Royal Liverpool Philharmonic Society.

Building opened in 1939

Designed by H.J. Rowse

Acoustic consultant: H. Bagenal

Uses: concerts, meetings etc.

Construction:

Floor – carpet on concrete

Walls – plaster on metal lath. Rear wall of balcony has applied acoustic tile

Boxes – of wood with timber panelling behind

Ceiling – plaster on metal lath

Stage and choir enclosure – plaster on metal lath, the sides perforated with organ behind. Wall behind choir is timber panelled

Stage – timber boards on joists

Organ – behind perforated screen on each side of stage

Seating – upholstered tip-up with unperforated bases

5.4 Watford Colosseum

Three town hall auditoriums built in the interwar years in London's hinterland were known affectionately as the three Ws: Walthamstow and Wembley (now Brent) Town Halls are also rectangular in plan. Previously known as the Watford Town Hall Assembly Hall, it was renamed in 1994 as the Colosseum, a strange choice perhaps for a rectangular plan hall. A particular interest with this hall is the involvement of Hope Bagenal as acoustic consultant. Though his name is also associated with Liverpool's Philharmonic Hall, the Watford Town Hall of 1940 represents the first hall discussed here in which his influence is unmistakable. The architect, C.C. Voysey (son of the architect C.F.A. Voysey, famous for his Arts and Crafts houses), won the design competition and produced a complex with 'chaste neo-Georgian forms' (Pevsner, 1953). For the Assembly Hall there

was apparently some doubt in the client's mind regarding the uses for the hall. This ambivalence may explain why the interior volume of the hall is somewhat small for concert use.

This hall has gross proportions close to the often-quoted shoebox form of halls famous for their acoustics. Stalls seating is removable and on a flat floor, which is carpeted to protect a dancing surface. There is a single balcony with a flat balcony front. The walls contain windows which unfortunately are now covered by curtains. These will introduce extra acoustic absorption and have the effect of depressing the reverberation time. The lower walls are timber panelled; together with the ceiling cornice detail they were chosen on acoustic grounds.

In his major work on architectural acoustics of 1931 (Bagenal and Wood), the consultant appears to have been somewhat ambivalent regarding the relative merits of the fan shape and rectangular

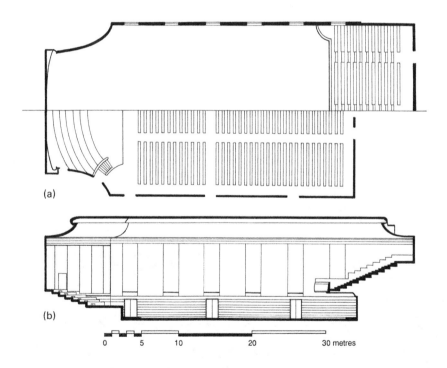

(a)

(b)

0   5   10   20   30 metres

Figure 5.16 (a) Plan and (b) long section of the Watford Colosseum

Figure 5.17 Watford Colosseum

Figure 5.18 Watford Colosseum

plan form. From non-acoustic points of view the fan shape has obvious advantages by bringing audience as close as possible to the stage. Its use as a common form for cinemas at the time must have been persuasive. However, the concert halls with the best reputations were rectangular in plan. By the time of the design of this hall, Bagenal had concluded that the rectangular or 'oblong plan is preferable to the fan'. In his subsequent book of 1942 we find the principal recipe for the design of the Watford hall. The convex splays at the stage end as well as the convex cornice treatment on the junction of walls and ceiling derive directly from the Queen's Hall, London (section 4.3), which 'was in many ways so satisfactory for general musical purposes as to set a standard and leave a tradition' (Bagenal, 1942). The convex splays at the stage end were thought to distribute sound uniformly over the hall. Likewise the convex cornice treatment was felt to have advantages of distributing sound and the cornice on the rear wall obscures surfaces which would otherwise be responsible for a reflection returning back to the stage. Large areas of wood panelling were also thought desirable at this time. From his book of 1942, it would seem that this hall design involved few compromises compared with Bagenal's preferred concert hall scheme at the time. A modern appraisal places different interpretations on the critical acoustic characteristics of this design.

Subjective characteristics

The subjective response to this auditorium can be simply stated. The clarity is good but this is at the expense of mediocre reverberance. Source broadening and intimacy are both smaller than one expects in a rectangular hall of this modest size. Though the sound was judged as loud, uniform throughout the hall and well balanced, the overall judgement was of only 'Reasonable' acoustics. This requires explanation because in traditional thinking a narrow rectangular hall of this sort should score well.

The major deficiency would seem to be a lack of response from the room, due to a lack of both

---

Watford Colosseum: Acoustic and building details

Volume = 11 600 m³

Total hall length = 50.0 m

Number of seats = 1586

Volume/seat = 7.3 m³

True seating area = 660 m²

Stage area = 166 m²

Acoustic seating area = 822 m²

True area/seat = 0.42 m²

Volume/acoustic seating area = 14.1 m

Mean occupied reverberation time (125–2000 Hz) = 1.45 seconds (measured by Building Research Station, 1949, from music)

Unoccupied objective data measured March 1982 (stage occupied with chairs); subjective assessment in January 1985

Owned by Borough of Watford.

Building opened in 1940

Designed by C.C. Voysey

Acoustic consultant: H. Bagenal

Uses: concerts, recordings, dances, banquets, boxing etc.

Construction:

Floor – carpet covering on timber boarding (stalls) and concrete (Gallery)

Walls – plaster on brick pierced with windows, recesses covered with curtain. At stalls level timber panelling over airspace

Ceiling – fibrous plaster

Stage and choir enclosure – plaster and perforated organ screen

Stage – timber boarding

Organ – located on either side of stage behind transparent screens

Seating – upholstered tip-up, bases perforated in the stalls only

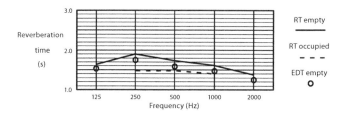

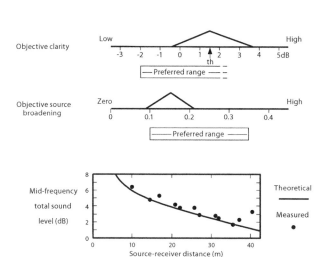

Figure 5.19 Watford Colosseum: objective characteristics

reverberance and source broadening. The sound was 'looked at', rather than there being any sense of surround and without a real sense of being in a large space. A couple of comments of false localization in the stalls and of strange sound quality in the balcony were also made.

Objective characteristics

The major peculiarity here, the short reverberation time in Figure 5.19, explains the major criticism of its sound quality. The occupied mid-frequency value below 1.5 seconds is too short for orchestral purposes and the rise in reverberation time in the bass is too modest to be useful acoustically. Unless a major use was for unassisted speech, one would not nowadays install so much timber panelling which limits the reverberation time rise in the bass. The

perceptually more important early decay time (EDT) is also shorter than the reverberation time, implying an even lower sense of reverberance. Objective clarity and the early decay time follow closely the expectations from the reverberation time, so in this hall the traditional acoustic measure explains much of its acoustic character.

For the case of source broadening, objective and subjective results are in agreement: values are less than one expects for a narrow hall of this sort. The presence of curtains on sections of the side walls will certainly be detrimental to the crucial early lateral reflections and the presence of low-frequency absorbing panelling at lower wall levels is also undesirable for the same reason. Bagenal's concern for good sound distribution has however been achieved, at least in terms of level. The sound levels at the rear of the hall prove to be particularly high, caused no doubt by the close proximity of reflecting ceiling surfaces.

Conclusions

This hall is used for many purposes, including banquets and dances as well as concerts and orchestral recording. Because of this it is probably unfair to judge it by the standards of a purpose-built concert hall. But with so many elements of acoustical origin, it behaves somewhat disappointingly. This re-analysis of his design suggests that several of the specific elements included by Bagenal for acoustic reasons are probably not crucial.

The main reason for disappointment here is the short reverberation time. A significant improvement should be achievable by removing the curtains and blocking off the window splays with slightly recessed hard surfaces. The predicted effect of this measure would be to increase the occupied mid-frequency reverberation time to 1.6 seconds. This would be worthwhile in itself, with the additional bonus that it would enhance early side reflections and hence subjective source broadening. Yet there still remains a mystery about this hall. When it is compared with what was, before the war, the hall with the best acoustic reputation, the Leipzig Neues Gewandhaus, we find they had the same volume, the same number of seats and both were rectangular with short reverberation times. Why do they appear to behave so differently? The only obvious difference is greater diffusing surfaces in the Leipzig hall, yet some very good halls have little surface decoration.

5.5 Royal Festival Hall, London

By 1951 Britain had endured peace-time food rationing for as many years as the war itself had lasted. A 'Tonic to the Nation' was prescribed: the Festival of Britain. The date of the Festival also marked the centenary of the Great Exhibition (linked with the Royal Albert Hall). The main exhibition site was an area of land on the south side of the River Thames opposite Charing Cross railway station at the point where the river curves sharply, providing excellent views in both directions. The site, formerly that of a brewery, was obviously ripe for redevelopment in order to link the South Bank with the traditional centres of the city. The Royal Festival Hall was the only permanent building among the exhibition structures; it was built as London's prime concert space to replace the Queen's Hall, destroyed in 1941.

The design of this hall was based on an extensive consideration of the current state of the art as well as tests on many materials of then unknown acoustic characteristics. Remarkably the whole design and construction of the hall was completed in only 2½ years. For those involved, the Royal Festival Hall offered a long-awaited opportunity to put into practice years of frustrated speculation, becoming a formative experience of their professional lives. Probably because of this, there is an extensive literature published both before and after its completion.

Few concert halls had been built in the inter-war years. Two main experiments had been made in terms of room form. The first of these involved profiling the ceiling to direct reflected sound onto the audience. The Salle Pleyel, Paris, designed according to this principle, was not considered a success. With the rise of popular entertainment in the cinema, it was inevitable that the fan-shaped plan, so appropriate for visual reasons, should also be tried for concert conditions. The relative merits of the fan-shaped plan have been discussed in section 4.5 and under the Philharmonic Hall, Liverpool. The consultants of the Festival Hall weighed the fan shape against the traditional rectangular plan form and decided to use the rectangular plan, bearing in mind its good, unchallenged reputation.

The classical rectangular halls are characterized by high ceilings, narrow cross-sections and balconies along the side walls and across the rear of the hall. For a hall the size of the Festival Hall with 2900 seats, it would not be possible to simply scale up the classical design. The distance to the furthest seat should not significantly exceed 40 m, for both visual and acoustic reasons. Balconies along side walls have poor sightlines for all but the front row of seats. Thus in the design of this hall many experiments had to

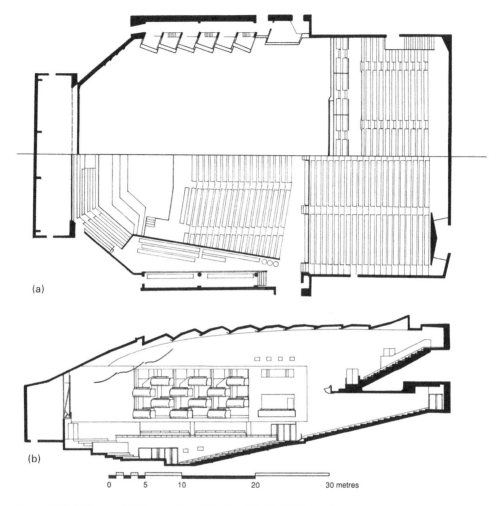

Figure 5.20 (a) Plan and (b) long section of the Royal Festival Hall, London

be run for the first time. To enlarge seating capacity the width was increased compared with classical halls to 32 m. Since the stage could not extend the full width of the hall, the walls bounding the lower stalls splay out from the stage. Side balconies were omitted and a single deep balcony was used facing the stage. Suspended boxes on the upper side walls provide additional seats. But if its plan can be seen as a development of the classical rectangular plan, in its long section the Festival Hall owes some debt to the experiments of the preceding decades.

The advantage of the rectangular plan was considered to be that 'there is more cross-reflection between parallel walls which may give added "fullness"' (Parkin *et al.*, 1953). Bagenal, the principal acoustic consultant, provided a comprehensive account of his design philosophy in the year before the hall opened (Bagenal, 1950) and again immediately after the opening (Bagenal, 1951). A particular concern for Bagenal was the risk of echo, which as he says obviously increases as the auditorium size gets larger. Areas of absorbing material were

Figure 5.21 Royal Festival Hall, London

Figure 5.22 Royal Festival Hall, London

included to quell possible echoes, whereas nowadays one would judge the policy as being over-cautious. Another concern was to limit low-frequency reverberation with substantial areas of the side walls as timber panel over airspace.

The most intriguing aspect in Bagenal's earlier paper (1950) concerns the question of clarity and sense of reverberation. He distinguished two schools of preference, those that draw upon choral tone for a standard and those that draw on instrumental tone. If the reverberation time is the only parameter at one's disposal, the former school would demand a longer reverberation time than the latter. Bagenal realized (correctly as we now know) that both these tastes can be satisfied by directing additional reflections onto the audience. In his words: 'if you can get a good registering on the ear of that first fraction-of-a-second you can stand a longer reverberation time and still follow rapid passages'. In this way you can achieve both adequate clarity and rich reverberance. In the Royal Festival Hall the clarity in the Grand Tier (i.e. balcony) was thought to be most at risk and the ceiling is profiled to enhance early reflections onto that region. Unfortunately this inspired scheme was undermined because the eventual reverberation time was too short even for instrumental music, so that no enhancement of the clarity was necessary (Barron, 1988b). The resulting lack of reverberance has been the continual criticism of the acoustics of the Royal Festival Hall, particularly in the bass.

The reverberation time problem in the Royal Festival Hall stimulated Parkin to invent the first electronic reverberation enhancement system (see section 10.5 and Parkin and Morgan, 1965, 1970). In the early days of 'assisted resonance', as it was known, some critics responded to the new 'richness' they perceived in the sound. Performers were more consistently enthusiastic, with conductors finding it much easier to obtain a homogeneous blending and a more resonant and warmer tone. Objections to the introduction of electronics to a concert hall were expected, but none were serious. For listeners, the assisted resonance system was reckoned to improve the warmth of the sound. The system

was used continuously until December 1998, when instead of upgrading or replacing it, the decision was taken to seek a solution without electronics.

In several respects the Royal Festival Hall was seminal in establishing standard features for large concert halls, even though in each case it may not have been the first example. A cavity wall construction was used for the auditorium envelope to guarantee a quiet interior on a noisy urban site (Figure 2.27). To ensure that all instruments can be heard by all members of the audience, both the stage and audience seating are well raked. Rather than use any form of proscenium frame, the stage is exposed with choir seating behind and audience seating to the sides. The views and preferences of the performers were also elicited and among other things this resulted in a sound-reflecting canopy to assist their ability to hear themselves and their colleagues. Observant visitors to the hall may notice a less seminal feature on the floor between the stage and the first row of stalls seats. Slate bedded into concrete was used here to reflect sound from the front row of instruments to the rear of the hall.

The acoustic character of the Royal Festival Hall proved to be far removed from the qualities one associates with the old classical halls. Its 'modern sound' demanded adaptation both on the part of audiences to appreciate its virtues fully and on the part of orchestral musicians whose individual performances were exposed as never before. The hall has many champions for whom good acoustics and those of the Festival Hall are synonymous. This underlines a crucial conclusion of recent subjective acoustic research: that individuals have different preferences even though they are probably capable of perceiving the same effects.

Subjective characteristics

During the first 18 months of the opened hall, a conscientious survey of responses was undertaken (Parkin *et al.*, 1953). In the national press the views ranged from 'perfect' to 'sickening' together with enigmas like 'Chaucerian'. There was wide favourable comment on the sense of definition or clarity;

balance and blend were more often complimented than criticized. Likewise, surprisingly, there were more favourable than unfavourable comments on the 'fullness of tone or resonance' (for which the expression reverberance is used here). Questionnaires were also completed at a concert by 18 foreign scientists working in acoustics; half did not know a better hall, but a third wanted more 'fullness'. The general opinion of a group of eminent musicians was that the definition and clarity were 'outstandingly good and there were several comments that this had led to an improvement in orchestral performance'.

The main dissenting voice in the early days came from the BBC (Somerville and Gilford, 1957). Both these authors and Parkin *et al.* comment that the hall worked best for chamber music and small orchestras. But Somerville and Gilford were not enthusiastic about larger-scale performance and endorsed the comments of the critic, D. Shawe-Taylor, who said that he could 'think of no famous auditorium with the peculiar dryness of tone which is to be felt at every big climax in the Festival Hall. There seems to be a growing consensus of opinion about this fault ... '. History bears out this consensus but the diversity of views serves to highlight different individual tastes. For those with a musical background, the ability to hear individual instruments and individual musical lines in the Festival Hall can be a magical quality which compensates for the lack of a rich sense of reverberation. By the standards of the classical rectangular halls, the acoustics of the hall are dry and for some the clarity might seem 'clinical'. The subsequent introduction of assisted resonance in the hall lessened the criticisms of dryness and particularly improved the warmth of the bass.

In the author's survey made when the assisted resonance system was in operation, the judgements of high clarity and low reverberance were predictable. On other attributes the judgements of source broadening were slightly below average, but average for intimacy and loudness. A lack of bass sound was also noted. Overall the acoustics were rated as 'Reasonable', which must be lower than the initial reactions to the hall. It should be added that the author's listeners did not contain anyone for whom clarity was an overriding concern. A difference in intimacy between the Terrace seat (not overhung) and a seat in the Grand Tier (with the lower intimacy value) was observed. Seats below the large overhang were not comprehensively sampled but they are known for their distinctly remote sound and lack of intimacy.

A preoccupation of the BBC in the 1950s and '60s, inspired no doubt in part by the experience of the Royal Festival Hall, was concern for the detrimental effects of strong overhead reflections (Somerville *et al.*, 1966). Three listeners in the author's survey made comments of harsh, unnatural sound in the Grand Tier but it seems to be something that many people can adapt to. Interestingly, the ticket-pricing policy in the hall suggests that the sound in the Grand Tier is appreciated less; less even than seats immediately below them under the overhang. Some of the author's listeners were convinced in their preference for the front of the Terrace (that is just behind the gangway dividing the lower level of seating), but on average the difference in overall assessment relative to the Grand Tier was not significant.

Objective characteristics

The original design goal for the reverberation time in the Festival Hall was 2.2 seconds (Parkin *et al.*, 1953) whereas in reality before electronic assistance the value turned out to be 1.45 seconds (Figure 5.23). The reasons for this large discrepancy appear to be several. Firstly the absorption figures used for audience seating at the time were serious underestimates (see section 2.8.3 and Barron, 1988b) and this accounts for about two-thirds of the error. The remedy would have been to increase the volume of the hall. The remaining discrepancy is likely to be associated with absorbent material or details which effectively absorb. There was until recently 270 m^2 of leather 'cushion' stuffed with glass wool on the rear surfaces in the hall and behind the seating below the boxes. The box arrangement also provides a system by which

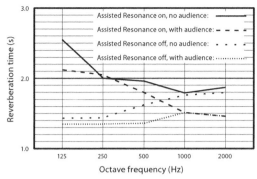

Figure 5.23 Measured reverberation times in the Royal Festival Hall, London with and without assisted resonance, occupied and unoccupied. Occupied hall values after Parkin and Morgan (1970)

sound has great difficulty escaping, particularly since they have drapes hanging behind them; the 'open area' of the boxes (i.e. excluding balcony fronts) is 90 m^2. The organ will also provide some absorption. Taken together these elements are likely to be sufficient to explain why the reverberation time in this hall was below the criterion. Those involved in the hall were especially concerned about the suspended ceiling which was built much thinner than intended. Though this is unlikely to be very absorbent at mid-frequencies, it is probably the major cause in combination with the elm wall panels of the short low-frequency reverberation time.

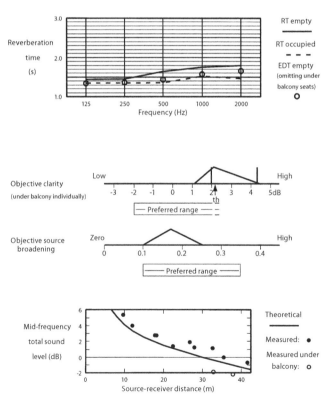

Figure 5.24 Royal Festival Hall, London: objective characteristics without assisted resonance

The effect of assisted resonance on the measured reverberation time is shown in Figure 5.23. But at least as important subjectively as the reverberation time is the loudness of the sound and in this respect the changes due to the system were less impressive. With the opportunity to measure sound levels with the system turned on and off, the mean increase due to the system in the 125 Hz octave was found to be only 1.4 dB, with lower values at higher frequencies. This probably explains why the system did not fully quell the criticisms of 'dryness' in the acoustics.

The values in Figure 5.24 refer to the situation with the assisted resonance turned off. One notices at mid-frequencies that the early decay time is shorter than the reverberation time. This is a characteristic of halls in which early reflections are directed at audience. Bagenal's intention to use the ceiling shape to enhance early reflections could produce this objective effect, especially for the Grand Tier seating. The two points in Figure 5.24 above the line for the theoretical total sound level (at 27 and 32 m) bear testimony to the fact that the level has indeed been increased at the front of that seating area. On the negative side, the acoustic shielding effect by the balcony is also unmistakable in the plot of total sound level. Overall though we find that many of the objective characteristics are determined by the measured reverberation time.

Conclusions

The Royal Festival Hall is a landmark in concert hall design after a long period when little was built. The acoustic consultants attempted to achieve a synthesis of the traditional rectangular design and the directed designs of the 1920s and '30s. Their task was particularly demanding given that the audience size was 2900 seats. Yet they were not afraid to experiment and the attempt at providing both clarity and reverberance is one that was probably first realized only 20 years later with enhanced reflections from the side rather than from overhead. Unfortunately they ended with a short reverberation time which made such subtleties irrelevant.

Royal Festival Hall, London: Acoustic and building details

Volume = 21950 m³

Total hall length = 52.0 m

Number of seats = 2645 plus 256 choir

Volume/seat = 7.6 m³

True seating area = 1181 m² plus 120 box seats

Stage area = 163 m²

Acoustic seating area = 1587 m²

True area/seat = 0.42 m²

Volume/acoustic seating area = 13.8 m

Mean occupied reverberation time (125–2000 Hz, assisted resonance off) = 1.40 seconds (measured by Parkin and Morgan, 1970, from music)

Unoccupied objective data measured February 1982 (stage occupied with chairs, organ exposed); subjective assessments in March 1983 and March 1986

Owned by Arts Council of Great Britain

Building opened in May 1951

Designed by London County Council Architects' Dept.: R.H. Matthews, J.L. Martin with E. Williams and P.M. Moro

Acoustic consultants: H. Bagenal, P.H. Parkin and W.A. Allen

Uses: concerts plus summer ballet season

Construction:

Floor – cork tile on concrete, transverse aisles carpeted

Walls – parallel walls of elm panels over air space. Stalls splay walls of ribbed timber. Rear walls and side walls below boxes covered with leather over glass wool

Ceiling – 10–20 mm fibrous plaster with added vermiculite plaster above to make up to 50 mm

Stage and choir enclosure – plywood panels

Stage reflector – 50 mm polished wood

Stage – timber boards over airspace

Organ – behind choir seating with openable shutter door

Seating – upholstered tip-up with perforated bases

The hall has an acoustic quality characteristic of many mid-twentieth century designs as opposed to traditional halls but not as extreme as the fan shape. With its novel solution of the architectural auditorium problem of the time, it proved to be an influential design. Its several successors are discussed on following pages.

It would be a mistake to dismiss this hall only for its short reverberation time. The sense of occasion has changed little from what it was within a year of its opening. It remains as Bagenal said (Martin *et al.*, 1952) 'a home for the mind, a home for the imagination ... Music lovers go there because they like it. It is spacious and hospitable; and compared to the old halls of my day, it is really a kind of paradise'. We now have more 'paradises' but few that are so exacting for the performers and rewarding for the listener in projecting the musical detail. As the 440 Hz tuning 'A' sounds in the subdued acoustics of the foyers, beckoning latecomers to their seats, we know we can expect the excitement of clear sound in a large space.

A major refurbishment of the Royal Festival Hall was undertaken between 2005 and 2007 (Pearman *et al.*, 2007). In the auditorium there were two main acoustic goals for the acoustic consultant Kirkegaard Associates: to increase by natural means the reverberation time and to rebuild the stage area. The stage geometry has been modified and the original timber overstage reflector replaced by an adjustable one which employs a special reflective fabric. To enhance reverberation time, sound absorbing surfaces were removed and reset to render them as sound reflective as possible; the suspended ceiling has been replaced with a thicker one. Small increases in auditorium volume were also possible. The resulting occupied reverberation time at 500/1000 Hz is now 1.65 seconds, with a 21 per cent rise at 125 Hz (Hartman, 2007).

5.6   Colston Hall, Bristol

Fire, the continual enemy of auditoria, claimed two previous rectangular Colston Halls in less than a century. Remarkably the third Colston Hall, of traditional design with two balconies extending round three sides of the hall, had survived intact the heavy wartime bombing of Bristol only to be destroyed by fire in February 1945. The new hall was completed in 1951 as part of the Festival of Britain. Regarding its architecture, the brochure produced for the reopening stated: 'An effort has been made towards a straightforward and direct architectural style embodying the free and flowing lines of contemporary architecture, combined with a subdued and dignified colour scheme'. The existing walls of the old hall were reused which imposed limitations on the roof construction and its height. However, this did mean a rectangular design in plan, in line with the preferences of the acoustic consultants. The balcony was completely redesigned. In order to accommodate the audience of 2000 it extends right up to the stage, stepping down to optimize sight-lines. The front stalls floor is horizontal and extends underneath removable staging for the lower sections of the raked seating. The hall can thus also be used for dancing and exhibitions.

By present-day standards this hall, like the Festival Hall, has surprisingly large areas of absorbent installed for acoustic reasons. At mid-frequencies the absorbent areas are restricted to the rear wall and the cornices between the ceiling and the side walls; in each case this was included to prevent echoes. Efforts were made to limit the rise of low-frequency reverberation time with low-frequency absorbent treatment applied to the walls below the balcony and with panels on the side walls above the balcony. Fortunately these areas of absorbent have left a satisfactory reverberation characteristic.

The Colston Hall design shares many similarities with the two other British halls of the same year. Of particular interest are the acoustic effects of the grosser elements. Compared with the Festival Hall, the stalls floor is much flatter, but the stage is raked steeply to maintain sight-lines from the stalls, and there is a high wall separating the orchestra from the choir seating behind. The different balcony scheme has already been mentioned, though what is not immediately obvious is that the hall width is larger at the balcony level. The ceilings are different.

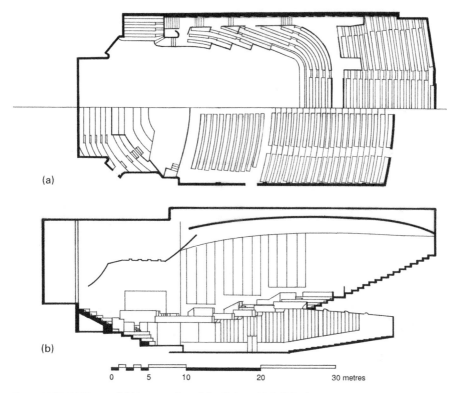

Figure 5.25 (a) Plan and (b) long section of the Colston Hall, Bristol

Subjective characteristics

Whereas the Royal Festival Hall ceiling is profiled in the longitudinal direction and the Manchester Free Trade Hall had a highly scattering ceiling, that of the Colston Hall is, from an acoustic point of view, basically plane. No doubt serious consideration was given to the ceiling design in each case. All three halls had a reflector over the stage.

Subjective characteristics

Beranek (1962) quotes and obviously concurs with the then contemporary view that the acoustics of the Colston Hall are the best of the 'Festival of Britain' halls. He found it 'almost excellent', with clarity, brilliance, warmth and adequate liveness. He raises the problem of poor balance, which can arise with imprudent brass and timpani placed in front of the rear wall of the stage. Somerville and Gilford

(1957) remark on the fuller tonal quality which has been achieved without echoes and on the 'definition which is good but not, perhaps, exceptional'.

In this author's survey, the Colston Hall also emerges as the most successful of the 1951 halls. Clarity and source broadening were judged as good, reverberance and intimacy reasonable and the sound as loud, in fact the loudest of the pre-1990 British halls in the survey. At the concert we attended, there appeared to be no particular balance problem associated with over-dominant brass. But there was a weakness associated with the woodwind, especially in the stalls. The sense of reverberation is still slightly lacking, with an absence of perceived sound from above and behind. This minor deficiency seems to occur both in the stalls and balcony, though it must become severe well behind the overhang in the rear stalls. One risk associated

Figure 5.26 Colston Hall, Bristol

Figure 5.27 Colston Hall, Bristol

with frontal reflections off large plane surfaces, in this case the ceiling, is a possible harsh tonal quality. Though there were comments about a certain edginess to the sound quality in the balcony, this was not perceived by the majority.

Objective characteristics

A major contribution to the relative success of the Colston Hall is probably associated with the reverberation time, which approaches the optimum value (Figure 5.28). Nevertheless the fact that the empty early decay time (EDT) is shorter than the equivalent reverberation time at mid-frequencies means that the sense of reverberance may yet be below optimum. A low EDT often occurs with high values for objective clarity, and this is the case here.

The high values of the objective clarity (the early-to-late index) occur predominantly due to a lack of later sound especially in the stalls. A plot of late sound vs. source–receiver distance is particularly revealing (Figure 3.33). There is a clear subset of results for the stalls and only the values in the balcony follow theory. The extended balcony appears to subdivide the space acoustically, mainly to the detriment of the stalls seating. The particularly low values for the overhung seats of the late and total sound point to poor conditions there, with quiet unreverberant sound.

The objective source broadening is reasonable however. The overhead and ceiling reflections are obviously strong in this hall but the width is also small, providing spatially balanced early reflections. The measured total sound is close to theory, with the exception of the overhung seats already mentioned.

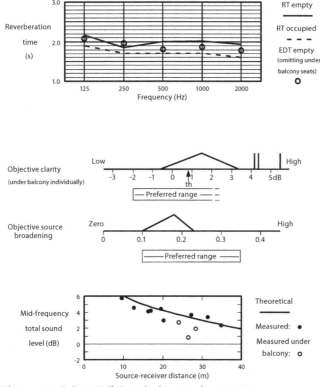

Figure 5.28 Colston Hall, Bristol: objective characteristics

Yet it was judged as being particularly loud. One can suggest the relatively small size of the hall and the reasonable reverberation time as contributory factors, but in addition there appears to be a tendency for listeners to judge sound as loud when there are strong early reflections. This may be one more effect associated with the large plane surfaces in this hall.

Conclusions

The designers of the Colston Hall were very constrained in their options, having to build within existing walls but accommodate a large audience with, no doubt, a more comfortable seating standard than before. The result was a well-proportioned space with well-balanced acoustics. Its success compared with its contemporaries is probably helped by its capacity of 2000 which places it in the

'difficult' acoustics category, but avoids the extreme problems of even larger halls.

In spite of a limitation on the ceiling height (the walls could not be extended higher) the reverberation time is satisfactory and this is the only hall of the period in which this occurs. The balcony design was probably an inevitable consequence of the building constraints and produces the major acoustic deficiency of the design: poor conditions for the more remote overhung seats. There is also evidence of the balcony subdividing the space acoustically, at the expense of conditions in the stalls.

A particularly interesting feature of the acoustic design of this hall is the use of large plane surfaces. The question of surface treatment, of whether to install scattering elements or not, has been a preoccupation among acousticians for many years (see, for example, Somerville and Gilford, 1957). It

**Colston Hall, Bristol:
Acoustic and building details**

Volume = 13 450 m³

Total hall length = 47.7 m

Number of seats = 1940 plus 182 choir

Volume/seat = 6.3 m³

True seating area = 870 m²

Stage area = 111 m²

Acoustic seating area = 1037 m²

True area/seat = 0.41 m²

Volume/acoustic seating area = 13.0 m

Mean occupied reverberation time (125–2000 Hz) = 1.70 seconds (measured by Building Research Station, 1951, with pistol shots)

Unoccupied objective data measured June 1982 (with empty stage); subjective assessments in April 1983 and January 1986

Owned by the City of Bristol.

Building opened in July 1951

Designed by City Architect: J. Nelson Meredith

Acoustic consultant: Building Research Station (H.R. Humphreys, P.H. Parkin and W.A. Allen)

Uses: classical concerts, popular music, wrestling etc.

Construction:

Floor – timber on flat stalls, linoleum on concrete on raked sections

Walls – at stalls level: timber panel over airspace up to about 1.5 m, above which are solid convex wooden panels separated by slots with Rockwool behind. At balcony level the walls are plaster on masonry with applied panels of 13 mm timber board in echelon fashion over mineral wool of mean depth 125 mm. Rear walls are acoustic tile over 38 mm airspace

Ceiling – 19 mm plaster on metal lath with 38 mm vermiculite concrete above

Ceiling cornices – exposed woodwool

Stage and choir enclosure – timber mainly over airspace

Stage reflector – plaster on metal lath

Stage – hardwood boarding with large airspace below

Organ – behind grill beyond choir seating, covered with curtain

Seating – upholstered tip-up with plywood base

is a question which remains far from resolved today since diffusion affects so many different attributes. In the 1950s the major concern was the effect of poor diffusion on the later sound decay; uneven and ragged decays were scorned. Today it is clear that the gross shape can have influences on the decay which are at least as marked. The more obvious subjective effects caused by strong reflections from nearby plane surfaces are a tonal harshness and a loud sound. Both these characteristics have been perceived in the Colston Hall, but neither is sufficiently prominent to undermine its good acoustic reputation, particularly in the balcony seating area.

While the acoustics of the Colston Hall can be rated as 'good for their time', the hall is very deficient in front- and back-stage accommodation. An architectural competition was held for a replacement concert hall on a dramatic waterside site in the city centre. A design team appointed in 1996 were developing an exciting scheme that would have become a landmark for the city (Blundell Jones, 1998). Then in July 1998 having previously welcomed the scheme, the Arts Council of England turned down their lottery-funding, which effectively killed the project. Thus for the moment, the Colston Hall remains the prime concert venue for Bristol.

5.7 Free Trade Hall, Manchester

In the mid-nineteenth century, Manchester was famous for its school of political and economic liberalism which was at the forefront of the 'free trade' movement opposing the Corn Laws. The first Free Trade Hall was intended for political meetings, but from about 1843 the replacement hall began to be used for musical events. In all, four halls were built on the same site. The third hall of 1856 had a capacity of about 2200 and was considered by Bagenal to be 'one of the best of the Victorian halls' (see Bagenal and Wood, 1931, and Bagenal, 1952). This hall was destroyed by bombing in 1940 and replaced by the new hall in 1951, again within the context of the Festival of Britain. This Free Trade Hall was used as the main concert venue in Manchester

for 45 years until 1996 when it was replaced by the new Bridgewater Hall (section 5.16). As an auditorium it suffered from a small area per seat, excessive balcony overhangs and a failure to accommodate the choral and organ traditions of the city. However the quality of sound in the hall was acceptable and by itself the acoustics would probably not have been sufficient cause for a rebuild. The major failure of the Free Trade Hall was linked to the elements in auditoria which have changed most since the nineteenth century: the front- and back-of-house provision. The Free Trade Hall has now been converted into a hotel; it is included here as an interesting acoustic case study.

The 1951 Free Trade Hall had to be built within the remaining boundary walls and accommodate a large audience of 2500. Two balcony levels were used with the height of the hall raised to accommodate the upper balcony. This hall like its predecessor was destined for multiple use; in particular the stalls floor was for many years polished and protected by a canvas cloth for dancing. The slight stalls floor rake of the earlier hall was retained but this required a high stage level to give acceptable sightlines. Rather than extend the rear Circle away from the stage (and raise the second balcony to give a suitable height for the overhang), much of the Circle seating was located on the sides, leaving a substantial part of stalls seating shaded by the Circle.

One of Bagenal's main concerns was whether the 'remotest seats at these extremes (could) be made to hear the *pianissimo*' (Bagenal, 1952). To ensure this, convex over-stage reflectors directed sound to most of the seating area, but crucially from his viewpoint to the rear seating. Again there was concern about echoes returning to the stage, but rather than just applying absorbent material a more enlightened approach was used. At the rear of the stalls the soffit slopes down to enhance the ceiling reflection to the rear seats while eliminating the possible echo back to the stage. The reverse splay surfaces at the rear of the stalls and Circle were likewise stepped to avoid echoes. Coffering 0.9 m deep, picked out in bright colours, was used for the ceiling to prevent troublesome echoes to the floor seating.

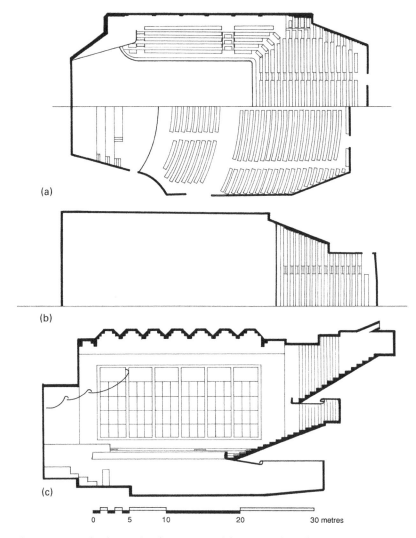

(a)

(b)

(c)

0   5   10   20   30 metres

Figure 5.29 (a), (b) Plans and (c) long section of the Free Trade Hall, Manchester

Efforts were also made to minimize absorption at both low and mid-frequencies.

Subjective characteristics

For a hall so drastically subdivided by its main balcony, which overhangs more than half the stalls seating, and with a second balcony added as an uncomfortable appendage, the acoustic characteristics must surely vary between different seating areas. Somerville and Gilford (1957) stated succinctly that 'hearing conditions vary somewhat throughout the auditorium. Because of screening, hearing is only fair at the back of stalls, circle and balcony'.

Figure 5.30 Free Trade Hall, Manchester

Figure 5.31 Free Trade Hall, Manchester

They are happier though about the tonal quality which they felt was less harsh than in its contemporaries, owing to the diffusing ceiling. Beranek (1962) commented on how loud the sound appeared, which does not concur with our experience. He also found the reverberation unreal and was worried by the balance between instruments, favouring the brass in particular.

In the present survey of five audience positions, the obvious preferred seat is at the front of the circle. It was judged as the more reverberant and loudest seat with a 'Good' assessment overall. The main criticism was of a lack of reverberance but otherwise the sound had a pleasant natural quality about it, which was dynamically alive. The front stalls and rear circle were judged as intermediate, with the rear stalls and mid-balcony the least liked. These last were judged particularly as lacking in intimacy and loudness. With such substantial overhangs, the sense of reverberation also tends to diminish markedly in extreme positions.

An inevitable problem in this design was with balance. The stage was very steeply raked, but this could not compensate for masking for seats in the stalls of sound from the centre of the orchestra. On the other hand in the circle the woodwind tended to be over-exposed. The situation at the edge of the stage where the scrolls of the double basses seemed almost to touch the balcony soffit was simply bizarre.

But the surprise about the Free Trade Hall was that the disadvantaged seats did not sound worse. For instance at the rear of the stalls, the acoustical response was better than the very unsatisfactory visual situation. One could even distinguish individual musical parts with some success there. The big compensation for the hall's acoustics may have been related to spatial effects. Envelopment was judged, unusually, as varying throughout the hall with a particularly high value in the front side stalls. Overall this hall performed better with envelopment than expected for its width. There was an obvious design feature which may explain this.

Objective characteristics

In line with other British halls of this era, the reverberation time was too short, which, worse still, decreased in the bass, as can be seen in Figure 5.32. The designers realized that the hall volume was borderline (Bagenal, 1952) but since the hall was fully contemporary with the Royal Festival Hall, the same anomalous seat absorption figures would have caused the same calculation error resulting in inadequate volume for the hall.

The objective clarity, though centred on the theoretical value, shows a wide spread. While this was to be expected for a hall as subdivided as this, the total sound level remained surprisingly close to predictions even at the overhung seats. It is objective source broadening that was particularly high in the Free Trade Hall. The region with the highest values was the stalls and the reasons are not hard to find. With the main balcony extending the full length of the side wall on each side, listeners received lateral reflections from both the side walls and the side wall/balcony soffits. This is considered one of the merits of the classical concert hall design (Figure 4.5). The other propitious design feature was the reverse splay plan at the rear of each seating level. This form is especially effective at providing strong lateral reflections and maintaining good sound level at distant seats.

In Bagenal's discussion (1952) of the hall, he gave no indication that he felt that the reverse-splay plan form contributed towards improving the acoustics. Its virtues were only proposed 17 years later. The reason for its presence in the Free Trade Hall may have been a fortunate coincidence resulting from having to squeeze a large hall onto an existing site. The orientation of the reverse-splay walls derived from the angle of the axis of the hall to the line of the main building facade; by narrowing the hall at the rear, the foyers could be made of constant width. The net acoustic result was that listening conditions below the overhangs were better than expected.

A virtually unavoidable consequence of large overhangs is that the late sound decreases substantially at remote seats, which causes the subjective effect of a diminished sense of reverberation. In this

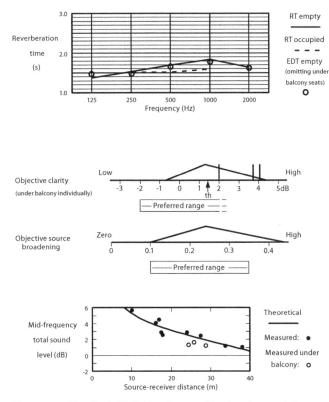

Figure 5.32 Free Trade Hall, Manchester: objective characteristics

hall there was the interesting compensation of a high proportion of early lateral sound, such that for some listeners the absence of reverberant sound may have been compensated by source broadening effects due to the early reflections. The acceptability of this trade-off would though be a matter of personal taste.

Conclusions

More than any other large concert hall discussed here, the Free Trade Hall was an attempt, and a rather uncomfortable one (for tall people literally) to squeeze a quart into a pint pot. The resulting short reverberation time was characteristic for its period and can be blamed on the inadequate absorption data of the time. The extensive subdivision by balconies made this hall an example of a design pushed beyond the limits of acceptability. For performers in

this hall there was no possibility of providing good balance between instruments for all seating areas.

The surprise about the acoustics is that they were not worse at the disadvantaged seats. The use of a highly diffusing ceiling appeared to be acceptable here, given that the walls were able to provide good reflections. A further positive feature of the design, which appears to have been accidental, was the use of the reverse-splay form at the rear of each seating level. This enhanced reflections, particularly lateral ones, so that sound levels did not deteriorate seriously at remote seats.

Bagenal was concerned about audibility of *pianissimi* at these remote seats and paid particular attention at the stage end with splay walls in plan and a reflector above. In the event it would seem that his concern was satisfied at least as much by the details surrounding the listeners at the remote seats.

Free Trade Hall, Manchester: Acoustic and building details

Volume = 15 430 m³

Total hall length = 47.5 m

Number of seats = 2529

Volume/seat = 6.1 m³

True seating area = 938 m²

Stage area = 173 m²

Acoustic seating area = 1181 m²

True area/seat = 0.37 m²

Volume/acoustic seating area = 13.0 m

Mean occupied reverberation time (125–2000 Hz) = 1.5 seconds (measured by Building Research Station, 1951, from music)

Unoccupied objective data measured April 1982 (stage occupied with chairs and stands); subjective assessments in July 1984 and March 1985

Owned by Manchester City Council

Building opened in Nov. 1951, closed 1996

Designed by City Architect: L. Howitt

Acoustic consultant: H. Bagenal

Uses: orchestral concerts, meetings, popular music, occasional boxing

Construction:

Floor – timber boarding in stalls; cork tile at higher levels

Walls – stalls: timber panels, mainly flush on masonry, stepped section is over airspace. Main side walls masonry, Circle side wall panels of hardboard over rockwool plus airspace

Ceiling – fibrous plaster coffering with screeded woodwool at top of coffer

Stage and choir enclosure – mainly timber panelling mounted without airspace

Stage – timber boards on timber beams

Stage reflector – solid wood

Seating – upholstered tip-up with perforated bases

5.8 Fairfield Hall, Croydon

To anyone who knows the Royal Festival Hall, the Fairfield Hall in Croydon, south London, feels like familiar territory. Not only are the two exteriors similar but many interior details, often not consciously noticed, are powerfully reminiscent. The foyers are similarly disposed in the two buildings and the same circular pillars rise to the underbellies of the stalls floors. Acoustically the hall is also the descendant of the Festival Hall; indeed after the disappointments of London's major concert hall, the consultant Hope Bagenal viewed the Fairfield Hall as an opportunity to remedy the faults inherent in the design of the earlier auditorium. It was, in fact, his last major concert hall and offers an interesting example of the fruits of his long experience.

The Fairfield Halls complex comprises the concert hall, a 763-seat theatre (the Ashcroft Theatre) and exhibition spaces. Both the theatre (a traditional proscenium design) and the concert hall have that over-stage reflector which is a hallmark of many of Bagenal's designs. The concert hall seats 1539 plus 250 choir, which is only 62 per cent of the capacity of the Festival Hall. This is already a characteristic in the Fairfield Hall's favour. Flexibility has been incorporated in the design with the stage on lifts offering the possibility of lowering the front section to produce an orchestra pit; the rear stalls seating area can also be screened off with a fibreglass curtain. The concert hall is used for music of all sorts, classical, popular and jazz, as well as films and sporting events. In spite of this multi-purpose use the hall exhibits few compromises in its design as a symphony concert hall.

In common with nearly all Bagenal's designs, the plan shape is rectangular, but with this seating number the width is a comfortably small 26 m. Whereas in the Festival Hall the central stalls taper towards the stage, this detail is not required in this narrower hall. There is again a single balcony with a straight balcony front. The long section of the Fairfield Hall illustrates the designers' continual concern with getting reflections to the most distant seats: firstly with the aid of the reflector above the stage and also with tilted panels immediately above

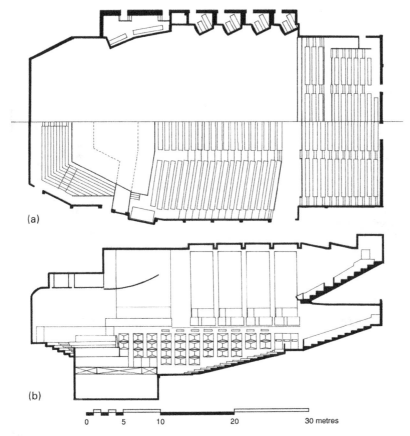

Figure 5.33 (a) Plan and (b) long section of the Fairfield Hall, Croydon

seating in the balcony. The wall treatment bounding the sides of the front stalls is timber panel over an airspace (to provide some low-frequency absorption) and is profiled to scatter reflections from these surfaces. Two design details are unique to the Fairfield Hall among the halls discussed here: the use of deep (0.75 m) transverse ceiling beams and the use of fins separating the boxes which extend in height from the level of the boxes to the ceiling.

Subjective characteristics

The acoustics of the hall were sampled at four seat positions at a concert by a major London orchestra.

Subjectively the sound quality was liked with an overall judgement of 'Good'. The best judged seat though provides a surprise: it was the seat beneath the balcony overhang. And the seat in the balcony almost immediately above it was the least liked position, whereas often it is the most liked. A clue to the reason for this reversal compared with expectations can be found in the observation that the intimacy judged in the balcony was significantly less than elsewhere in the hall. Judgements in the main stalls were less extreme but they fall well short of the excellence that the designers may have hoped for with this design developed progressively from the Leipzig Gewandhaus model. The clarity in this

Figure 5.34 Fairfield Hall, Croydon

Figure 5.35 Fairfield Hall, Croydon

hall was not judged as high, in clear contrast to the Royal Festival Hall. Reverberance and envelopment were both judged reasonable. Likewise intimacy and loudness were viewed as reasonable on average, but for both these qualities judgements decreased as one moved towards the rear of the hall. Conditions in the stalls could allow for exciting experiences but somehow the sound appears rather constrained. The unexpected enthusiasm for the position below the balcony clearly indicates good balcony overhang design, but demands explanation.

The Fairfield Hall rates well in one further characteristic, at least by British standards: the bass was judged as loud relative to other frequencies with a response second only to the Royal Concert Hall, Nottingham in the pre-1990 halls surveyed. Tradi-

tionally this would be associated with a bass rise in the reverberation time characteristic.

Objective characteristics

As shown in Figure 5.36, the reverberation time does indeed rise in the bass in the Fairfield Hall but though the unoccupied mid-frequency value is a reasonable 1.9 seconds, the predicted occupied reverberation time is as low as 1.6 seconds. In spite of this, the subjective reverberance judgement was reasonable, probably due to the bass reverberation time rise, which will also occur with an audience present. The objective clarity requires no special comment overall but the measured value at the front row of the balcony is the lowest in the hall and very low in itself. The lowest value for the

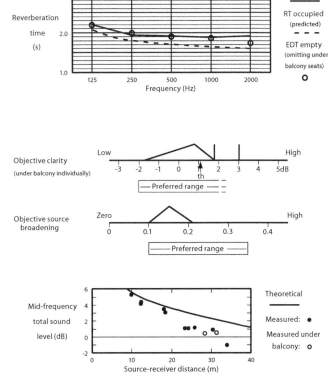

Figure 5.36 Fairfield Hall, Croydon: objective characteristics

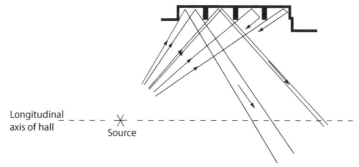

Longitudinal
axis of hall
Source

Figure 5.37 Ray diagram, presented in plan, for reflections off the wall area above the boxes in the Fairfield Hall. The source is located on stage

source broadening measure occurs at the same seat. In most concert halls the front of the balcony is considered as the best listening area, but here we have the suggestion that early reflections and particularly early lateral reflections (from the side walls) may be missing at this position.

The other unusual observation from the objective measurements concerns the measured sound levels. Comparing measured and predicted sound levels, the agreement is good close to the stage but there is a consistent deviation in the rear half of the hall. From a subjective point of view, this explains the lack of perceived intimacy in the balcony. Objectively it points to reflections not reaching the rear seats. The measured total sound level at the furthest balcony seat is less than the criterion of 0 dB. Examination of the behaviour of early and late sound shows that in the balcony it is the early sound which is most deficient. This behaviour has also been found in the Barbican Concert Hall and St David's Hall, Cardiff; in each case they have highly sound scattering ceilings. Our suspicions therefore turn to the transverse ceiling beams and, perhaps even more significantly, the fins between the boxes in the Fairfield Hall. Figure 5.37 shows how the effect of these fins is to return more sound to the stage than is reflected to the rear of the hall. The transverse ceiling beams will have the same effect and both will influence sound in the balcony. The problem at the furthest seats is particularly interesting in view of the specific measures taken to avoid quiet sound. In fact the stage reflector does not provide a reflection to these distant seats, at least for

performers towards the front of the stage. But the inclined ceiling panels above the balcony, though surely valuable, cannot provide sufficient compensating energy.

Finally the situation below the balcony requires discussion. Compared with other under-balcony situations, the sound is loud here, in particular the early sound. It is likely that both the over-stage reflector and the balcony soffit contribute reflections for these seats, providing a clear sound which was liked by our listeners.

Conclusions

The Fairfield Hall possesses several characteristics which promise good acoustic conditions: it has parallel side walls and a cross-section with a modest width, the seating capacity is moderate and the balcony overhang is likewise modest. The hall was Hope Bagenal's last major design and, with the exception of the seating capacity, all these characteristics were major concerns of his. As an attempt to avoid the failings of the Royal Festival Hall, the Fairfield Hall can be considered a success: there is no criticism of 'clinical clarity' here, the bass sound is warm and the conditions do not deteriorate below the balcony overhang. Though judged better than the other London halls, measurements highlight two shortcomings in the acoustic design. The reverberation time is again too short due to inadequate hall volume; a 15 per cent increase in volume would improve conditions for symphony music (though it

Fairfield Hall, Croydon: Acoustic and building details

Volume = 15 400 m³

Total hall length = 47.6 m

Number of seats = 1539 plus 250 choir

Volume/seat = 8.6 m³

True seating area = 864 m² plus 61 box seats

Stage area = 165 m²

Acoustic seating area = 1138 m²

True area/seat = 0.50 m²

Volume/acoustic seating area = 13.5 m

Mean occupied reverberation time (predicted 125–2000 Hz) = 1.75 seconds

Unoccupied objective data measured December 1980 and October 1982 (stage occupied with some chairs and stands); subjective assessment in October 1984

Owned by London Borough of Croydon

Building opened in Nov. 1962

Designed by Robert Atkinson and Partners (partner-in-charge D.H. Beaty-Ponnall)

Acoustic consultant: H. Bagenal

Uses: orchestral and popular music, jazz, films and sporting events

Construction:

Floor – cork tile on concrete except in front of stage where it is woodstrip on battens

Walls – generally plaster on masonry. To the sides of the stalls is timber panel over airspace. Upper sections of back walls behind rear stalls and balcony seating are acoustic tile

Ceiling – plaster on concrete slabs

Stage and choir enclosure – timber panel over airspace

Stage – hardwood strip on timber joists

Organ – freestanding adjacent to choir seating

Seating – upholstered tip-up with perforated bases

might perhaps compromise other uses). The second point is more novel and relates to the treatment of the upper side walls and ceiling. Both the fins separating the box slots and the transverse ceiling beams prevent reflections from these surfaces reaching the audience. The acoustic effect of this is particularly noticeable in the balcony which sounds more remote than it should. Remedial measures in the form of panels between the fins and ceiling beams should produce noticeable improvements. The lesson from the Fairfield Hall is that the character of the major surfaces in an auditorium generally has acoustic implications.

5.9 The Lighthouse Concert Hall, Poole

This hall was originally known as the Wessex Hall in the Poole Arts Centre. From an acoustical point of view, the Centre offers two instructive case studies: here of the concert hall and of a theatre (section 8.4.6). It is convenient for the discussion that this hall follows chronologically on the Fairfield Hall, though in the British context the long gap of 16 years was filled with two more original auditoria, which are treated here as recital halls. The design of the Lighthouse Hall can be seen as a postscript to the elaboration of the rectangular plan hall developed by Hope Bagenal.

The acoustic consultant for the Lighthouse Concert Hall was Peter Parkin, a reticent man whose long career at the Building Research Establishment had involved offering advice on many of the auditoria discussed here. It is therefore surprising to many people to learn that this was his only sole consultancy. Parkin's formative concert hall experience was the Royal Festival Hall and we find in the Lighthouse Hall clear echoes of the details of the London hall, firstly in the plan form but most pointedly in the ceiling profile. The ceiling is uniform in the cross-section direction but scattering in the long section just as the suspended ceiling in the Festival Hall. The seating numbers in this hall are obviously much smaller (1473 plus 120 choir), which allows for a minimal overhang by the balcony. Fortunately

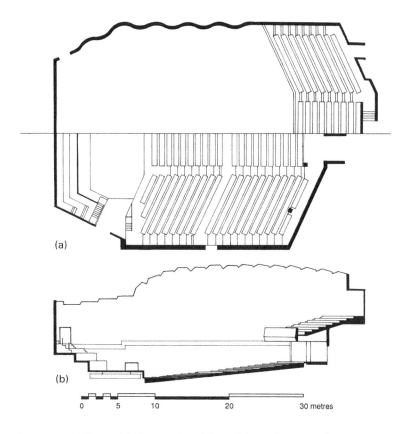

Figure 5.38 (a) Plan and (b) long section of the Lighthouse Concert Hall, Poole

the side walls here were left clear; though they are mildly scattering there is nothing to prevent them providing good lateral reflections.

Inevitably for a hall in a provincial city, multiple use is essential. It is ingeniously catered for in this hall. The whole stalls floor is built as a raft supported on screw jacks which enable the floor to be lowered from its normal raked position to allow for the seating to be either rolled away for storage below the stage or raised to the upper stage level to provide a large flat level floor, suitable for dancing, exhibitions, banquets, etc. The observant visitor can spot some truncated doors in the side walls as evidence of this feature. Though this floor scheme places minor constraints on sightlines, it offers flexibility with very few compromises for concert use.

Subjective characteristics

The acoustics of the Lighthouse Concert Hall were judged favourably. At the two seating positions, in the rear stalls and in the rear Circle, the sound was rated as 'Good' (towards the boundary with 'Very Good'). Though clarity and loudness were judged as average, for the three other major characteristics of reverberance, envelopment and intimacy the hall was rated highly. Indeed the envelopment judgement was the highest recorded in the eleven pre-1990 halls. A further favourable judgement

Figure 5.39 Lighthouse Concert Hall, Poole

Figure 5.40 Lighthouse Concert Hall, Poole

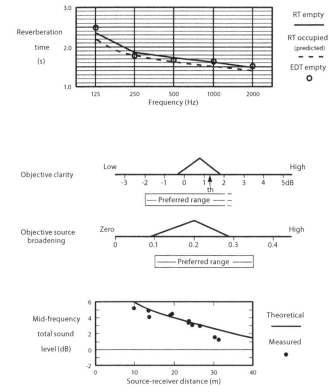

Figure 5.41 Lighthouse Concert Hall, Poole: objective characteristics

recorded on the questionnaires was a positive but modest bass balance. The listening position in the Circle was only three rows from the back, which predictably was not judged so well (the constricted ceiling height is a concern here). Yet the perceived differences between the two seats were very small, which suggests good uniformity in the hall.

The sound in the Lighthouse Hall has a very natural quality about it, with a well-developed response from the space as a whole. Dynamically the hall responds well. To the author's ears the sound in the rear stalls of the solo violin (in a Mozart concerto) verged on the magical, for which the outdated expression 'singing tone' came to mind. The most consistent, if minor, criticism concerned a slight lack of brilliance. Some comments on poor balance between different orchestral groups were also voiced.

Objective characteristics

For people familiar with reverberation time, it is very surprising to see the measured reverberation time characteristic in Figure 5.41 after the comments above. How does one explain a subjective judgement of high reverberance with a mid-frequency reverberation time of only 1.5 seconds? Of eleven halls tested subjectively (Barron, 1988a), the Lighthouse Hall comes third on the reverberance judgement but ninth in the order of mid-frequency reverberation times. The early decay time and reverberation time are identical at these frequencies so the explanation does not lie there.

The reverberation time at mid-frequencies is short but it does rise in the bass. This leads to the idea that the bass must also be contributing to the sense of reverberation. The sound here is well diffused which may also contribute to the sense of reverberance. If both the bass and mid-frequency reverberation time (RT) contribute to reverberance, it has significant implications for concert hall design. If it is possible to trade off between the bass and mid-frequencies, one can either have a reasonably flat RT characteristic within the optimum range (probably greater than 1.8 seconds) or have a shorter mid-frequency RT and a longer bass RT. In construction terms the first option requires a large hall volume but allows for lighter weight materials whereas the second can be achieved with a smaller volume but needs more massive construction. This design option is probably only available for halls with scattering wall and ceiling surfaces (see the Royal Concert Hall, Nottingham as a contrary example).

Before leaving the reverberation time, it can be noted that the frequency characteristic is falling at 2 kHz, which suggests more than normal absorption at high frequencies. This absorption is likely to explain the perceived lack of brilliance. As regards the location of the absorption, the culprit may be the reconstructed stone-faced wall blocks.

As far as the remaining objective measures are concerned, the objective clarity and total sound both behave as expected from the reverberation time value. The behaviour of the early and late energy in the Lighthouse Concert Hall was illustrated in Figures 3.30 and 3.31; it is typical of the average hall. Values of the source broadening measure are high, which fits in with the subjective judgement. These objective values are higher than those in the Fairfield Hall even though the Lighthouse Hall is the wider of the two. This underlines the importance of having walls that are clear of obstructions to provide strong lateral reflections.

Conclusions

The example of the Lighthouse Concert Hall shows that good subjective reverberance, envelopment and intimacy can be achieved in a rectangular hall seating 1500. Among British halls these three characteristics prove to be the most closely related to overall preference and probably account for this hall

being assessed as among the best three of pre-1990 British halls. The big surprise revealed by our study is that in a diffuse situation a reverberation time which is short at mid-frequencies but rises in the bass can be acceptable. More predictably, this hall demonstrates the desirability of continuous reflecting side walls in creating a good sense of spatial impression.

Lighthouse Concert Hall, Poole: Acoustic and building details

Volume = 12 430 m³

Total hall length = 41.4 m

Number of seats = 1473 plus 120 choir

Volume/seat = 7.8 m³

True seating area = 809 m²

Stage area = 158 m²

Acoustic seating area = 980 m²

True area/seat = 0.48 m²

Volume/acoustic seating area = 12.7 m

Mean occupied reverberation time (predicted 125–2000 Hz) = 1.7 seconds

Unoccupied objective data measured May 1982 (stage occupied with chairs); subjective assessment in November 1984

Owned by Poole Borough Council

Building opened in April 1978

Designed by the Borough Architect: G. Hopkinson

Acoustic consultant: P.H. Parkin

Uses: popular and orchestral music, variety, boxing etc.

Construction:

Floor – woodstrip on framing (stalls), woodblock on concrete (Circle)

Walls – lower level: ribbed timber panel over airspace; upper level: purpose-made reconstructed Portland stone block; upper rear wall: plastered masonry

Ceiling – sprayed fibrous concrete (approx. 50 mm)

Stage – woodstrip on steel joists

Seating – upholstered tip-up with unperforated bases

At the time of the name change of the Centre in 2001/2, the acoustics of the concert hall were also addressed by Arup Acoustics. The stone blocks on the side walls were sealed with epoxy and the seating replaced with less sound-absorbing seats. The unoccupied reverberation time at 500/1000 Hz has risen from 1.7 to 2.2 seconds, a change which is surely noticeable. The total sound level has increased by an impressive 1.5 db on average.

5.10 Barbican Concert Hall, Barbican Centre, London

Three British concert halls opened in 1982; they make an interesting trio. A different design philosophy inspired each of them but in each case the link with the parallel-sided descendants of the Royal Festival Hall has been severed. The Barbican Concert Hall belongs to the Barbican Arts Centre, to be found within walking distance to the north of St Paul's Cathedral.

The term 'barbican' refers to a gatehouse or watchtower which stood in Roman times in the city wall adjacent to the present location. What before the Second World War had been a mainly industrial site devoted to textiles and printing was devastated by bombing in December 1940. An elaborate development programme for the area was conceived in the 1950s to include housing, schools, the Guildhall School of Music and Drama and finally a Centre of Arts and Conferences. The Centre contains a theatre (section 8.6.2), exhibition spaces, cinemas, a library and art galleries. At the time of its opening it was claimed to be the largest centre of its kind in Western Europe. The price for this diversity within the limiting space remaining on the site has been a labyrinthine structure with cramped conditions, particularly for the concert hall. Few visitors fail to get lost on their first encounter!

The Barbican development is unusual in that it was designed over a 20-year period by a single architectural practice. This guaranteed a stylistic uniformity but meant that the Centre preserved the ideals of 1960s architecture already 20 years

out of date at its opening – perhaps the last of the 'concrete culture palaces'. Architecturally it has been criticized as being excessively serious, with an inappropriate solemnity 'for the simple enjoyment of ordinary people'. Its massive concrete columns would certainly be the pride of any Roman builder! As an Arts Centre, it has become a busy new complement to the South Bank complex and with it the City of London has added greatly to the capital's cultural life.

The concert hall itself is mainly below ground level, totally concealed from the exterior. It sits immediately below a sculpture court framed by a U-shaped block of flats. This constraint has proved to be crucial to its design and acoustics. Since excessive cost prevented excavation for a lower floor level, the height and consequently the volume is small by concert hall standards. Further, the necessary structural requirements for the roof/sculpture court have necessitated substantial beams within the hall. Several other halls discussed in this book have open truss structural support within the acoustic volume. In this case, pairs of solid beams 3.7 m deep run both across and along the hall to provide routes for ventilation ducting as well as structural support. Substantial solid transverse beams are cited in several acoustic texts for creating dead spots in the seating area. To mitigate against this problem, diffusing coffering is included at ceiling level and around 2000 diffusing spheres were hung, as shown in Figure 5.45.

The other major use for the hall apart from music is for conferences but few compromises have been made to accommodate this requirement. The audience seating is on three levels with minimal overhangs. Given a need to limit the distance to the furthest seat, the hall has acquired a substantial width of 43 m, again a potentially undesirable acoustic feature. The stage is similar in form to an enclosed space beyond a proscenium opening but contains inadequate space for large choral performance with orchestra.

When the hall opened the reviews of the acoustics were surprisingly diverse but there was general approval among the critics for the generous seating

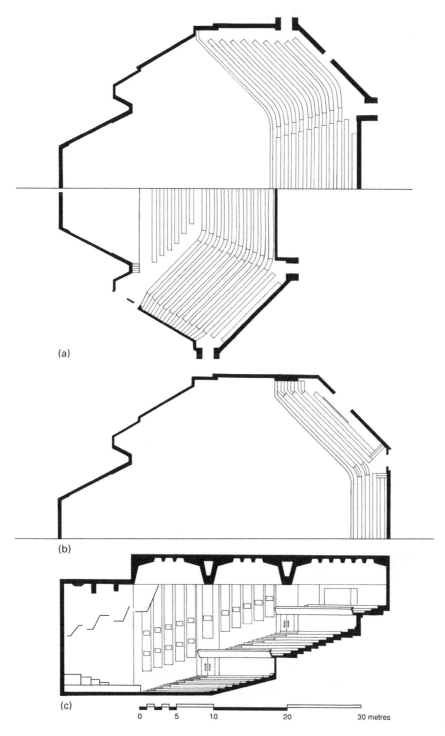

Figure 5.42 (a), (b) Plans and (c) long section of the Barbican Concert Hall, Barbican Centre, London

Figure 5.43 Barbican Concert Hall, Barbican Centre, London

Figure 5.44 Barbican Concert Hall, Barbican Centre, London

standard. Subsequently the Barbican Concert Hall has been labelled as having acoustic problems just as did its fellow London halls before it: the Royal Albert Hall dogged by an echo and the Royal Festival Hall characterized by clinical clarity and inadequate warmth. Attempts to improve the acoustics have been going on since the hall opened, the most obvious of early modifications being removal of all the diffusing spheres. The assessments here were made in 1984 after initial modifications.

The acoustic situation in the Barbican Concert Hall is complicated since several problems coexist. At an early stage it was clear that there was a problem at low frequencies; the cellos and basses were frequently inaudible in *tutti* passages. Objectively the bass reverberation time was much shorter than anticipated and this suggested that there was unexpected bass absorption somewhere in the hall. This was certainly surprising since the hall is constructed of solid concrete and although many wall finishes are timber they are mounted flush on the concrete surface. Investigations showed that the most likely cause of the problem was the seats. Meanwhile all additional bass absorption was eliminated as far as possible (for example, the side wall elements looking like gross organ pipes were converted from double-panel absorbers to single non-absorbent panels). However, it is often very difficult to determine the details of low-frequency behaviour. Reverberation chamber measurements on the seats indicated that their bass absorption was reduced by installing a hardboard panel beneath the upholstery. This was then done to half the seats in the hall and a small increase in bass reverberation time (RT) resulted. So the remaining seats were similarly treated but no further RT change occurred! (It has also been suggested that the excess absorption may be due to resonant behaviour in the ventilation extract ducting below the seats, but there is no measured evidence that this is occurring.) The suspended spheres were clearly easy to remove and there are always those who are antipathetic towards such objects – among them the conductor Claudio Abbado. The spheres had been introduced for scattering purposes, and there is evidence that they

produced more uniform conditions. They were open at both ends which minimized resonant behaviour, though some incidental absorption was associated with them. Removing the spheres has increased the reverberation time somewhat but, as found in an acoustic model of the hall, without spheres the RT does not follow theory in this hall. Overall only small objective changes have been achieved by all these modifications, but what appears to be the main problem of excessive low-frequency seat absorption remains to be solved.

Subjective characteristics

A concert by a leading conductor and orchestra was attended in the hall as it stood without spheres at ceiling level. Though the clarity was reasonable, both the reverberance and envelopment were judged as lower than elsewhere. Only the Royal Albert Hall was judged as less intimate and less loud, with the overall judgement disappointing. The perception of the bass, judged under bass balance, was also the worst of the halls tested before 1990.

The subjective experience of listening to a symphony orchestra does lack some excitement in this hall. It is considered to work better with smaller orchestral forces. The listener can find difficulty in relating to the performance and sensing the response of the room. Inability to hear the bass is also frustrating quite apart from a lack of warmth in the sound. Subjective responses throughout the space were fairly uniform except for particularly low judgements of intimacy towards the rear. This and the judgement of background noise were both significantly worse at the upper seating positions. Especially in the Second Tier, the sound appears remote as if seen through a window, and the high noise level there does not help.

Objective characteristics

The reverberation time for the occupied hall was certainly short for concert hall purposes, with a mid-frequency value of 1.6 seconds and no bass rise, Figure 5.46. This short mid-frequency reverberation

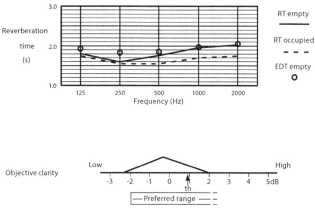

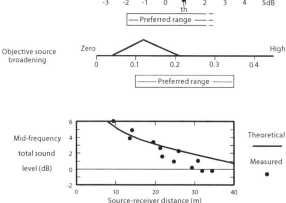

Figure 5.46 Barbican Concert Hall, London: objective characteristics

time was mainly due to inadequate interior volume in this hall. Traditionally this would be considered a basic cause for disappointing sound quality, except that in the author's survey other halls with shorter reverberation times were judged as more reverberant. Nowadays the early decay time (EDT) is thought to be better related to reverberance but the EDT value in the Barbican Hall is not low enough to fully explain the subjective response; there is no obvious reason for this.

Among the other measurements, objective clarity was not exceptional, objective source broadening was low and the measured sound level dropped off more than expected with increasing distance. This last behaviour would explain the lack of intimacy in the Second Tier. For most of these measures, the behaviour is in the right direction relative to typical values in order to explain subjective

observations but the differences compared with typical values are not especially large. It is possible that because there were not many early reflections in this hall, the effect of introducing an audience on direct and other sound travelling at grazing incidence to seating was particularly significant. In other words the difference between occupied and unoccupied conditions may have been larger in the Barbican Hall than elsewhere.

One difference which does deserve mention was that between theoretical and measured objective clarity. Though poor clarity was not detected subjectively, this disparity may point to one of the problems in this hall. The explanation for both the difference between measured and theoretical objective clarity and the low sound levels is a deficiency of early sound (comprising the early reflections), particularly at the rear of the hall, as illustrated in Figure

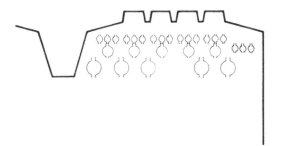

Figure 5.45 Cross-section through a lateral ceiling bay of the Barbican Concert Hall, showing the location of spheres present when the hall opened

3.32. With bass sound deficient as well, this lack of early sound ties in with the intimacy judgements. The cause of inadequate early sound must surely be the poor reflective qualities of the ceiling (similar to that off walls in the Fairfield Hall, Figure 5.37) as well as the substantial width of the hall. In the Upper Tier the limited angle from which sound can arrive would further exacerbate the situation there.

The short bass reverberation time is almost certainly due to high bass absorption but subjectively the reverberation time in itself is unlikely to be the sole cause of a poor bass judgement. The level of bass relative to mid-frequency sound is probably more important for the subjective judgement and the measured values of bass level balance proved to be consistently bad throughout the hall (Figure 5.73).

Conclusions

The Barbican Concert Hall is a lavishly executed auditorium but disappointing in its acoustics like the other major London concert halls. The most obvious failure is associated with the bass and this has been tracked down to a probable anomalous absorption by the seating. Additional characteristics are probably due to the height limitation imposed on the design. The low hall volume causes a short mid-frequency reverberation time. The substantial width and unusual ceiling design produce a lack of early reflected energy which is not immediately clear from objective measures. The experience of this hall suggests that listeners are very sensitive

Barbican Concert Hall, London: Acoustic and building details

Volume = 17 750 m³

Total hall length = 44.2 m

Number of seats = 2026

Volume/seat = 8.8 m³

True seating area = 1207 m²

Stage area = 157 m²

Acoustic seating area = 1369 m²

True area/seat = 0.60 m²

Volume/acoustic seating area = 13.0 m

Mean occupied reverberation time (125–2000 Hz) = 1.65 seconds (measured by Arup Acoustics, 1983, from pistol shots)

Unoccupied objective data measured May 1984 (empty stage with rear tiers raised); subjective assessment in July 1984

Owned by City of London

Building opened in March 1982

Designed by Chamberlain, Powell and Bon

Acoustic consultant: H. Creighton

Uses: concerts and conferences

Construction:

Floor – endgrain woodblock

Walls – timber facing on concrete. Wall panels are of single-layer plywood on timber frames

Ceiling – solid concrete

Stage enclosure – relief perforated timber sculpture with alternate areas of porous absorber, panel absorber and concrete behind

Stage – individual stage lifts with timber boarding

Seating – fixed upholstered with ventilation extract below. Seat backs are solid timber as a continuation of the floor

to this early reflected sound. The additional absence of bass sound probably compounds the problem.

It is surely sad when concert halls are built so infrequently that the crucial decision to build with inadequate ceiling height was taken for reasons which were more self-imposed than intractable. The Barbican Concert Hall was an expensive hall on a per-seat basis, so decisions were a matter of priorities rather than straight economics. Kirkegaard Associates were employed over the period 1994–2001 to improve, over three phases, the acoustics of the hall. They extended the platform, replaced the overstage canopy with a fully adjustable one and introduced about 550 m² of reflectors in the ceiling area above the audience to give stronger reflections than were provided by the original diffusing ceiling.

5.11 St David's Hall, Cardiff

One only has to compare Scharoun's 1963 Berlin Philharmonie design with its contemporaries to appreciate how far ahead of its time it was (section 4.9). Yet few architects have confronted the challenge of developing Scharoun's ideas further; Hertzberger's Utrecht Muziekcentrum of 1977 is the most obvious descendant. St David's Hall in Cardiff owes an obvious debt to Scharoun's masterpiece, though it lacks some of the subtlety of its predecessor. The Philharmonie was developed from a reassessment of the relationship between the performers and the audience. The design incorporated two major breaks with tradition: the stage was sited much more centrally in the hall and the audience was subdivided into terraces which are comparable in size with the orchestra. In St David's Hall only the second of these design features is used; the performing end of the hall preserves the traditional orchestral stage arrangement with seating behind for the choir. The St David's Hall audience is seated on a shallow stalls area and twelve independent tiers. The orientation of the tiers is simpler than Scharoun's; the expression 'vineyard terraces' is less appropriate to the Cardiff hall, not least because the latter has many overhangs. Most surfaces bounding the audience tiers in St David's Hall are oriented radially to the stage. This could have acoustic implications since these surfaces are for this reason unavailable to direct acoustic first reflections.

St David's Hall has several other peculiarities as a concert hall, not least the process of its design. Cardiff, as the capital city of Wales, lacked a good performing facility. To help defray costs, the hall was included in a commercial development; it sits above two levels of shops. The same architects were responsible for both elements. However, the design contains a much larger input from the consultants than would normally be the case, in particular from Martin Carr, the theatre consultant. It is a considerable credit to the design team that the whole is so harmonious and successful. Inevitably multi-purpose needs had to be catered for, particularly popular music and sporting events. The hall was also designed to provide a large conference venue to enhance an already lucrative conference business in the city. The use of the hall for speech is handled with a public address system but all non-musical uses benefit from keeping the distance of audience as short as possible from the stage. A maximum distance of 33 m from the stage front has been achieved here.

The seating capacity of just under 2000 is very similar to that of the Barbican Concert Hall; they also both have the same maximum width of around 43 m. But in St David's Hall the gross plan form is an elongated hexagon with the width tapering towards the rear. By subdividing the seating into tiers, each tier can be arranged to face the stage and tiers can be positioned to the side of the orchestra. The listener in a tier feels himself a member of a group with its own identity rather than feeling submerged into a mass. This is surely no less democratic than a continuous multitude. Acoustically the tiered system provides a host of surfaces able to provide reflections, which on the whole are not true mirror-reflections.

The sightline situation has received considerable attention in this hall. With such small blocks of seating, each tier can be individually raked and positioned, as opposed to continuous balconies where row heights are determined by sightlines for the worst case. Indeed sightlines have been further optimized by making rows non-horizontal in the side upper tiers, a feature also present in the Berlin Philharmonie. The seating rakes are particularly steep in this hall and the seating standard is also generous, giving the listener good views and comfortable conditions. Commentators tend to ascribe

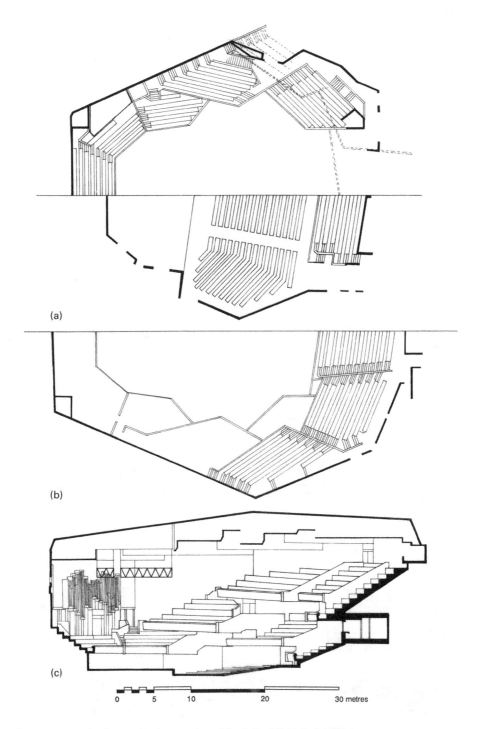

(a)

(b)

(c)

0 5 10 20 30 metres

Figure 5.47 (a), (b) Plans and (c) long section of the St David's Hall, Cardiff

Figure 5.48 St David's Hall, Cardiff

Figure 5.49 St David's Hall, Cardiff

the high reflective backs of the seats (found at higher levels) to some ingenious acoustic purpose. In reality the reason is more mundane: in response to the steep rakes they prevent listeners falling into the row below!

With two levels of shops below and a large volume requirement for the hall for acoustic reasons, the height of the building was likely to contravene city planning regulations. The solution has been to include both structural support for the roof and ventilation ducting within the acoustic volume. The open-truss beams and ducting are concealed behind an acoustically transparent ceiling. However, this suspended ceiling has some acoustic absorbency of its own, demanding a yet larger volume to maintain the reverberation time. Given that the seating standard is also generous, it is inevitable that this hall had a high volume per seat; at 11.2 m^3/seat this is the fifth largest volume per seat of the British halls considered in this chapter.

The acoustician has admitted that his priorities for the acoustic design were traditional. The primary concern was to achieve 'a reverberation time of 2 seconds with a diffuse well-distributed sound field'. In view of the need to accommodate speech, no rise in bass reverberation time was intended. The additional bass absorption necessary to achieve this was provided incidentally by the lightweight roof, which in a city like Cardiff with no aeroplane flyovers is an obvious economical solution. To minimize other incidental absorption, thermal lagging of the ductwork had to be included inside the ducts. In some seating areas, listeners receive early side reflections from neighbouring walls. It was intended to install a number of reflecting surfaces above the suspended ceiling to provide additional lateral reflections, where they were deficient. In the event the plethora of reflecting surfaces above listeners has rendered this unnecessary.

In the design of St David's Hall a conventional acoustic approach has been taken to an unconventional design. Its major precedent, the Berlin Philharmonie, has some disappointing areas though in general it works well acoustically. The

St David's design was tested in a 1:50 scale model. While the test results were encouraging, two concerns remained about which very little information existed to suggest their acoustic significance: whether the suspended ceiling and highly diffusing elements above would obscure overhead reflections which would be missed subjectively and secondly, with many bounding surfaces to the tiers running radially from the stage, would the proportion of early side reflections be adequate?

Subjective characteristics

In a subjective survey of eleven British concert halls (Barron, 1988a), St David's Hall has proved to be among the best liked, an assessment which has been corroborated by two visits. In detail the clarity was found to be typical, the reverberance high, envelopment and intimacy good, with loudness and bass balance average. In other words the hall scores well on all scales and this proves to be characteristic of the halls which were liked best. In agreement with the model results, subjective conditions were perceived as uniform; differences exist of course but they are not gross.

It is always more difficult to establish the reasons for success than failure in acoustics, because deficiencies can normally be isolated. The reasons for the acoustics of classical rectangular halls being appreciated are no less enigmatic than those of St David's Hall. A dominant characteristic of its acoustics is the diffuse reverberation, which provides a pleasing response from the room without undermining acoustic clarity. The good visual intimacy, due to very good sightlines and proximity to the stage, is well complemented acoustically. The sound has a natural quality with good dynamic response. Individuals will have their own preferences for the best seats. The acoustic spatial experience is predominantly reverberant rather than one of source broadening. For the author the most exciting listening positions contain both these spatial elements, which in this hall means locations with the strongest lateral reflections, such as towards the rear of the stalls. Subjective reactions to seats below overhangs will also vary between individuals, but differences are unlikely to be serious enough to affect enjoyment of a concert.

Objective characteristics

The objective measurements in St David's Hall (Figure 5.50) show a respectable reverberation time, close to the design value. As intended, at low frequencies the value does not rise. The mean early decay time agrees with the reverberation time at mid-frequencies, which points to a well-diffused sound field. The objective source broadening is probably in line with subjective reactions. But there were no subjective responses here to match surprising behaviour in both the objective clarity and total sound level. At distant seats, each measure shows differences relative to predictions. In both this hall

and the Barbican Concert Hall there are these same objective characteristics caused by a deficiency of early sound at the rear; this has been ascribed to the highly diffusing nature of the ceilings. But why is this perceived in the Barbican and not in St David's Hall?

There are two factors which may contribute to the different perceptions of low total sound level in the two halls, though this remains speculative. In St David's Hall the bass sound does not drop off with anything like the steepness of the mid-frequency level plotted in Figure 5.50. The lower mid-frequency levels may not therefore be detected as such. The second possibility is that a quiet sound level may be more acceptable if the sound is more diffuse: the low ceiling height and structural beams in the Barbican Hall must produce a less diffuse sound in the Upper Tier than the higher roof line in St David's.

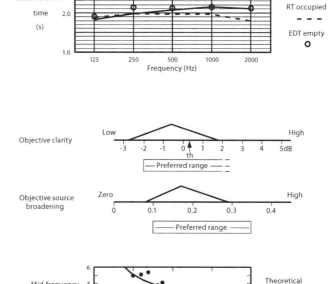

Figure 5.50 St David's Hall, Cardiff: objective characteristics

The boundary of acceptability in these circumstances is probably a fine one.

Conclusions

Much of the discussion above has been concerned with possible reasons for this hall being so highly appreciated. Two obvious features in its favour are the optimum reverberation time and the diffuse nature of its sound field. While this survey has shown that the reverberation time criterion is still relevant (though it should rather be applied to the early decay time), the question of diffuse or not has to remain an open one, the example of the Colston Hall in Bristol providing the opposite view. The extent to which a sound field dominated by components from specific directions is liked or not must remain a matter of personal preference.

However, the experience of St David's Hall shows that a highly diffuse sound field is liked. And with both a diffusing ceiling and many reflecting surfaces among the acoustically absorbent seating, this design scheme probably creates conditions that are more diffuse than is found in a simple geometrical space with a multitude of protrusions on the walls. The aspect of this design which could have been further pursued was that of early lateral reflections; while it clearly complicates the design, the Berlin Philharmonie shows that tier side walls which are not radial relative to the stage can be accommodated to provide these reflections.

St David's Hall offers a novel auditorium design scheme which can surely act as a precedent for other new halls. It indicates that structural support of the roof within the acoustic space is not inimical to good acoustics. Subdivision into many seating tiers offers large flexibility for architects and allows design for good sight lines with less compromise than occurs with large seating blocks. The minor risks of low sound levels should be amenable to local acoustic design with specifically oriented reflectors. The Belfast Waterfront Hall (section 5.16) was designed by the same key players as were responsible for St David's Hall.

St David's Hall, Cardiff: Acoustic and building details

Volume = 22 000 m³

Total hall length = 47.8 m

Number of seats = 1687 plus 270 choir

Volume/seat = 11.2 m³

True seating area = 1193 m²

Stage area = 179 m²

Acoustic seating area = 1367 m²

True area/seat = 0.61 m²

Volume/acoustic seating area = 16.1 m

Mean occupied reverberation time (125–2000 Hz) = 1.9 seconds (measured by Sandy Brown Associates, 1983, from music)

Unoccupied objective data measured August 1982 (empty stage); subjective assessments in May 1984 and October 1985

Owned by City of Cardiff

Building opened in Sept. 1982

Designed by J. Seymour Harris Partnership

Acoustic consultant: Sandy Brown Associates (A.N. Burd)

Theatre consultant: Carr and Angier

Uses: concerts, conferences and sporting events

Construction:

Floor – timber

Walls – plastered concrete

Ceiling – 200 mm reinforced aerated concrete slabs painted to seal pores

Balcony soffits – plaster

Suspended ceiling – 100 mm lattice of compressed chipboard (Formalux). Vertical sections of plywood

Stage – maple strip on timber beams on lifts

Organ – located in front of structural wall

Seating – upholstered tip-up with solid base. At upper levels plywood seat backs extend to head height

5.12 Royal Concert Hall, Nottingham

The City of Nottingham can boast a rich artistic life which must be the envy of many provincial cities. Its star element is the Royal Centre containing a renovated theatre and this new concert hall. The two auditoria sit adjacent to one another on a modest detached site in the heart of the city. The 1865 Theatre Royal, by Phipps and later Matcham, was comprehensively renovated in 1978 (Glasstone, 1978) by the same architectural practice that designed the concert hall. Rather than attempt to imitate nineteenth-century style for the concert hall, they chose confrontation with the modern, a contrast which extends to the two foyer spaces. The concert hall exterior has a jazzy Art Deco air to it, with the foyers fully exposed after dusk to passers-by through extensive reflective glazing. The hall itself seats 2500, though the designers are at pains to point out that this number was imposed by the client and necessitated compromises in order to fit onto a tight site with height limitations as well.

The gross form of the hall is the result of protracted development between the architects and acousticians. The latter were the American firm, Artec Consultants, with considerable experience in North America. The hall contains several features not found in earlier British concert halls. The most obvious of these are the massive wall and ceiling constructions and the three-dimensional faceting of the ceiling of the hall. This faceting also lends a 1930s flavour to the design and produces a sense of it having been carved from a solid mass. The surfaces provide both acoustical reflections and act as 'effective modulators of architectural space and scale'. The subjective characteristic being sought by the acousticians was for enveloping sound. It is perhaps no coincidence that the hall interior has a womb-like character, particularly under the orange lighting at their disposal.

The acoustical design philosophy applied by the acoustic consultants is in stark contrast to that of others a generation earlier. Their principal consideration has been to provide strong early lateral reflections rather than to be much concerned about reverberation time. Their concern for early lateral reflections has been expressed in two design principles: 'to adopt a room shape that engenders strong lateral energy and adopt narrow, tall room proportions to support multiple lateral reflections' (Edwards, 1985). These criteria derived originally from comparisons between classical rectangular halls and more modern fan-shaped halls; early lateral reflections are a strong contender to explain the preferred acoustic quality of the former. Design forms which encourage early lateral reflections include reverse-splay fan shapes in plan and angled ceiling profiles in cross-section. For this hall the appropriate shapes were designed with the help of light models. The reverse-splay form has been incorporated here by replacing rear corners with 45° walls, so that the hall width tapers at the rear of each seating level. A typical cross-section is shown in Figure 5.54: the surface A projects a reflection towards the centre of the seating rows, whereas towards the perimeter second-order reflections off the wall C and soffit B provide local reflections. Virtually all room surfaces have been angled where possible to provide appropriate reflections.

The design of this hall deviates from the classical not only in its form but also with respect to its surface structures. There is no equivalent here to the surface decoration of the classical halls; the

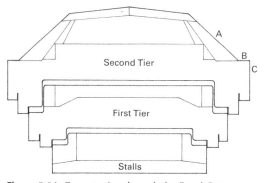

Figure 5.51 Cross-section through the Royal Concert Hall, Nottingham, looking towards the rear seating. Upper surfaces responsible for lateral reflections are labelled A, B and C (see text)

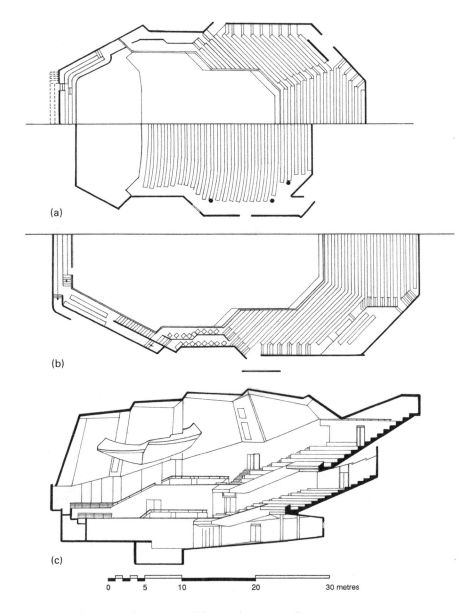

Figure 5.52 (a), (b) Plans and (c) long section of the Royal Concert Hall, Nottingham

Figure 5.53 Royal Concert Hall, Nottingham

Figure 5.54 Royal Concert Hall, Nottingham

design is completely at the other extreme to the highly scattering schemes discussed in section 4.7. The consultants also believe in providing a rich bass sound and have insisted on a massive envelope for the hall, with all perforations for lighting etc. sealed, in order to avoid incidental bass sound absorption. The hall includes a massive 32-tonne canopy which contains lighting and loudspeakers as well as performing an acoustic function; the height and tilt of the canopy are adjustable. The acoustic aim was stated as being to provide reflections both for musicians and for the front few rows in the stalls. Many would consider that a less massive object could achieve the same goal at far less than 3.5 per cent of the total building cost.

With a theatre already within the complex, multi-purpose needs for the hall are limited to music in all its forms, with films occasionally being shown. Nevertheless requirements for popular music are at odds with those for a symphony orchestra. For popular music, 15 absorbent banners can be extended down the front ceiling surfaces guided by wires. The total banner area is inadequate to produce much noticeable effect on the reverberation time but their location prevents disturbing reflections for sound from the loudspeakers. An orchestral pit can be exposed by lifts at the stage front.

Subjective characteristics

The acoustics of the Royal Concert Hall have been sampled at four seat positions during two separate concerts with results well reproduced. The judgements are especially interesting since the acoustics score well on several scales, yet only for the minority of listeners did this correspond with a 'Very Good' overall assessment. The sound character here is one which exposes individual preferences.

The clarity of the sound was judged as the highest of the halls tested pre-1990 but the sense of reverberance was correspondingly low (with only the Barbican Hall as less reverberant). These two responses would normally be associated with a short reverberation time, but the situation proves

not to be as simple as this. The envelopment and intimacy were judged as reasonable, the loudness as high and overall the judgement was 'Good'. The bass balance was the highest of the halls tested. Verbal comments were more overtly critical: several listeners referred to the harsh quality of the sound, with brash, shrill and blasty used as synonyms. The loudness of the sound was surprising enough for some to mention it specifically; one listener labelled it as tiring! There were also several comments about poor orchestral balance with the brass being too prominent and an odd mention of false localization.

The majority of these responses were consistent throughout the space. The sense of reverberance was however judged significantly higher right at the rear in the Second Tier compared with the other seats. Overall the sound shows good uniformity in spite of restrictions placed on the hall design, but for many the sound character was slightly displeasing.

Objective characteristics

The measured reverberation time in Figure 5.55 without audience is close to 2 seconds at mid-frequencies rising substantially in the bass; the predicted occupied value falls to 1.7 seconds at middle frequencies. The early decay time (EDT) is however significantly shorter than the measured reverberation time and at mid-frequencies the EDT occupied may be as low as 1.5 seconds. This low value probably accounts for the low perceived reverberance. Objective clarity is high in line with the subjective response; tellingly the mean value is clearly higher than the theoretical. The objective source broadening is also high, and uniformly so, testifying to consistent presence of good early lateral reflections. A further agreement between objective and subjective occurs for the bass sound; the mean level balance between bass and mid-frequencies is second highest in the Royal Concert Hall (Figure 5.73). Lastly the total sound level is well maintained throughout the hall, with little drop-off below balcony overhangs.

In this hall therefore, more than in some others, objective behaviour matches closely the subjective

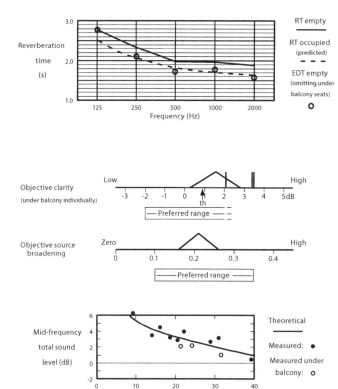

Figure 5.55 Royal Concert Hall, Nottingham: objective characteristics

response. The objective behaviour can be considered a consequence of the acoustic design. The room surfaces have been specifically oriented to direct sound from the source onto the audience. The orientation of these surfaces is such that a high proportion of this sound arrives at the listener from the side. This creates a rich spatial sense. It also maintains a uniformly loud sound throughout the hall and due to the solid auditorium envelope the bass sound is also strong (and these two components further contribute to a high degree of subjective envelopment). However there is a penalty inherent in this design procedure. The early sound level is raised by directing so much early sound onto the audience and consequently the early-to-late energy index (i.e. objective clarity) becomes high. This in itself is acceptable but the EDT will fall for the same reason and this is perceived as low reverberance.

The necessary antidote to prevent subjective lack of reverberance is to have a longer reverberation time by having a larger hall volume. In this way, though the EDT remains shorter than the reverberation time (RT), raising the RT would also increase the EDT towards the preferred range (probably greater than 1.8 seconds). The Christchurch Town Hall, New Zealand, provides an example of this (section 4.10). This hall therefore demonstrates how the reverberation time, or more specifically the EDT, cannot be ignored. Unfortunately a larger volume was not an option with this design.

The other major criticism was of harsh sound quality. This is generally referred to as 'tone colouration' and is a well-known occurrence where there are strong early reflections. Though it is an effect which is most pronounced for overhead reflections, it can also occur for lateral reflections off large plane surfaces (section 3.2). In addition, it is possible that it is perceived more strongly when the sound is loud. This hall demonstrates, more clearly than any other considered here, that large plane reflecting surfaces have to be treated with respect, and that surface modulation, otherwise known as scattering treatment, is not an outmoded delusion of the past.

Conclusions

With a restricted site on which to accommodate a 2500 audience, the designers selected a two-balcony scheme with some affinities with Edwardian theatre designs. Acoustically the design conforms with a precise philosophy to incorporate multiple early reflections arriving laterally at the listeners. In this respect the design is very successful; the objective measure for lateral reflections was found to be consistently high. Although the balcony overhangs are larger than the designers would have wished, the sound quality is surprisingly uniform. The hall has characteristics missing in several other halls: the sound is clear, intimate and loud with a rich bass. Yet overall the acoustic quality was not judged as favourably as these characteristics might suggest. The extensive manipulation of early reflections has as a by-product a tendency to destroy the sense of reverberation, in this case to its detriment. The absence of any acoustic scattering treatment both in the body of the hall and around the stage is perceptible to many ears and tends to cause a harsh sound quality and poor orchestral balance.

Strong early lateral reflections have been hailed by some as the magic ingredient of the classical halls which were missing from many designs from the middle of the twentieth century. While this may be true, the experience of this hall suggests that traditional concerns for reverberation time (or, better, early decay time) and some scattering treatment cannot be ignored. Good lateral reflections are a complement to other requirements, not a substitute.

Royal Concert Hall, Nottingham: Acoustic and building details

Volume = 17 510 m³

Total hall length = 49.8 m

Number of seats = 2315 plus 186 choir

Volume/seat = 7.0 m³

True seating area = 1273 m² plus 72 box seats

Stage area = 155 m²

Acoustic seating area = 1475 m²

True area/seat = 0.52 m²

Volume/acoustic seating area = 11.9 m

Mean occupied reverberation time (predicted 125–2000 Hz) = 1.95 seconds

Unoccupied objective data measured August 1983 (empty stage, 2/3 of floor covered with theatre velour); subjective assessments in June 1983 and April 1984

Owned by City of Nottingham   Building opened in Nov. 1982

Designed by RHWL Partnership

Acoustic consultant: Artec Consultants Inc.

Uses: music of all forms, films and conventions

Construction:

Floor – studded rubber on concrete

Walls – plaster on solid masonry

Ceiling – 100 mm sprayed concrete, plastered. Balcony soffits of fibrous plaster

Stage – timber boarding. Front sections on lifts to create orchestral pit

Seating – upholstered tip-up with unperforated bases

5.13 Glasgow Royal Concert Hall

In St Andrew's Hall, built in 1877, Glasgow had a concert hall with fine acoustics from the golden era of concert hall design. The hall latterly held 2133 (Beranek, 1962) in a 22.5 m wide rectangular plan space with a single balcony; the occupied reverberation time was about 1.9 seconds. St Andrew's was felt by many (Somerville, 1949), but not all (Parkin *et al.*, 1952), to have the best acoustics in Britain. Tragically St Andrew's Hall burnt down in 1962 and many years elapsed before a replacement was built.

In 1968 Leslie Martin's architectural practice began design of an arts complex including not only a replacement concert hall but also a drama theatre and accommodation for the Royal Scottish Academy of Music and Drama (Martin, 1983). Early schemes included a large auditorium suitable for concerts with a flytower above the stage for dramatic performance. However no solid commitment to actually build it was made until the mid-1980s. In the meantime, Scottish Opera had moved into the renovated Glasgow Theatre Royal and the Royal Scottish Academy had commissioned purpose-built headquarters.

Despite changes in the brief, the site for the development had remained constant. The design for a large concert hall, produced in 1984, took the old St Andrew's Hall as its starting point, with a single balcony running round the sides and rear. The plan form was developed by pulling out the sides at their centre point to turn a parallel-sided plan into an elongated hexagon. This plan was subsequently modified to become parallel-sided at the stage end with a reverse-fan plan towards the rear. The resulting plan has the form of an arena, which contributes to the sense of shared experience between performers and audience.

The architect was keen to restrict the overall height of the building and made an early decision to include structural support of the auditorium ceiling and roof within the auditorium. As this structure was covered in perforated cladding, some ventilation ducting could be included within the auditorium volume without being visible from below. To further limit the building height, but provide sufficient volume for adequate reverberation, the auditorium width increases at higher levels. This solution had implications in terms of the provision of early acoustic reflections. Acoustic scale models at 1:50 scale were used to check and develop the design.

Serious progress towards realization of the scheme occurred in early 1987 when Glasgow learnt that it had succeeded in its nomination as the European City of Culture for 1990. Design now proceeded in earnest with RMJM Architects (Edinburgh) taking over design as executive architects. The hall opened in October 1990, ready to contribute to the city's year of culture.

Balconies in classical halls can introduce several problems, among them that of sightlines from side balconies to the stage and the effects of balcony overhangs. 'Pulling out' the sides of the plan offers sightline improvements. By having a hall width that increases with height, the degree of overhang can be reduced. Maximum widths of the Royal Concert Hall are 31 m at stalls level, 42 m at balcony level and 44.5 m at ceiling level. A second advantage of such a cross-section is that it offers a large auditorium volume while maintaining a modest hall width at stage level.

The primary acoustic concerns for this hall were provision of adequate reverberation and sufficient early reflections at all seating areas. Early reflections for the stalls seating were promoted by the reverse-splay form in plan and balcony fronts which tilted down. The surface at the rear of the stalls facing the stage was also tilted down to prevent a possible echo back to the stage. To compensate for the substantial hall width at ceiling level, reflectors were suspended above balcony level to provide lateral reflections for the seating below. Their design drew on the experience of Marshall in 'lateral directed reflection sequence halls' (section 4.10). Four pairs of single-element quadratic residue diffusers (QRDs) were used on each side of the hall to provide scattered reflections (Figures 2.21 and A.4).

Overstage reflectors were also provided to assist the orchestral players in hearing themselves and one another. This feature was progressively

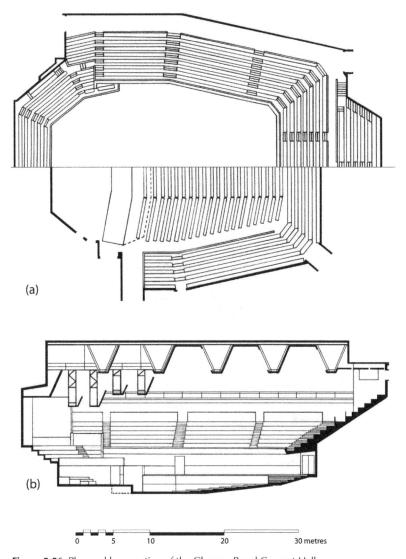

(a)

(b)

```
0    5    10          20          30 metres
```

Figure 5.56 Plan and long section of the Glasgow Royal Concert Hall

modified during the early years of the hall. Five substantial V-shaped trusses running across the hall support the ceiling and roof. Timber boards perforated with slots provide a visual screen on both sides of the trusses; at the time of design a 36 per cent perforation was felt adequate.

All large modern auditoria have to cope with multiple use. In the Royal Concert Hall eleven rows

of raked stalls seating can be removed, using an air-castor system. By adjustment of the heights of stage platforms, an extensive flat floor area can be created for arena-format productions.

As well as the site being noisy as regards road traffic, it also has an underground railway line passing immediately beneath it at the stage end. To exclude vibration induced by the railway, both the

Figure 5.57 Glasgow Royal Concert Hall (in original form)

Figure 5.58 Glasgow Royal Concert Hall (in original form)

auditorium and some surrounding accommodation are resiliently mounted on pads of natural rubber. The noise from the ventilation system has proved to be very quiet, at approximately NR10.

Subjective characteristics

The Glasgow Royal Concert Hall has not been tested reliably by subjective questionnaire study. Before the hall opened, a group of acousticians was invited to attend the first orchestral rehearsal in the hall. Their responses suggest basically uniform acoustic conditions with a slight difference in character between the stalls and circle level. The clarity, envelopment, intimacy and loudness were judged as average. The sense of reverberation was also judged as average by the standards of British concert halls in 1990, but that group contains several halls with low reverberation times.

Subsequent experience in the hall has shown that some orchestras have had difficulty determining the best arrangement on the stage. The high wall around the platform has also made orchestral balance problematic on occasions. Comments in the press around the time of the opening were mixed, but have generally improved. Some very positive comments have been made, while one critic continued to comment about 'unflattering string tone'. It is possible that this hall is more demanding of the musicians than some but that for the audience it offers a high degree of transparency.

In the (somewhat biased) view of this author, the sound is clear, intimate, enveloping and sufficiently loud. In the stalls the sound tends to appear more immediate than in the circle, where the sound is more reverberant. The principle acoustic criticism is a slight lack of reverberance.

Objective characteristics

On the crude measure of volume per seat, this hall scores well with a value of 11.7 m³/person. This volume gave a predicted reverberation time for the hall of around 2 seconds. Yet the measured time at mid-frequencies (500/1000 Hz) in the occupied hall is 1.75 seconds (Figure 5.59). According to a Sabine equation calculation, this implies 14 per cent more absorption than expected. During construction some gaps were left open at ceiling level which would account for some of the discrepancy, but a careful study found that 8 per cent of the absorption measured in the occupied hall cannot be accounted for. The best explanation to date is that this is caused by the details of the construction at ceiling level with its many suspended elements, including the perforated panels which provide a visual screen for the roof trusses. Reverberation time is influenced by the number of reflections that sound 'rays' experience; the number of impacts between sound rays and objects will be high for sound that enters the ceiling region.

A mid-frequency reverberation time of 1.75 seconds is however only just outside the recommended range of 1.8 – 2.2 seconds. But in halls with surfaces inclined to promote early reflections, the early decay time (EDT) tends to become shorter. In the unoccupied Glasgow hall, the ratio of EDT to reverberation time is 0.90 (Barron, 1995b). This suggests an occupied EDT of around 1.6 seconds. The sense of reverberation is now considered to be more closely related to the early decay time and a value for the EDT of 1.6 seconds corresponds well with subjective comments.

Turning to the other objective measures, objective clarity is in line with the measured reverberation time, though some high values are observed at the overhung seat locations. Objective source broadening is uniformly high and indeed the mean value is one of the highest measured in British halls. The total sound level has some quite high values in the front part of the hall and some low values at distant seats. The low values of level, found in the Upper Circle opposite the stage, were not predicted. They have not been a source of complaints by audience.

Conclusions

Readers will notice that the author contributed to the acoustic design of the Glasgow Royal Concert Hall, firstly as an unofficial advisor to Leslie Martin and then working jointly with Sandy Brown

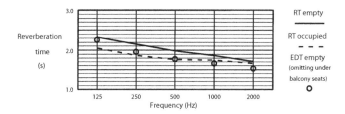

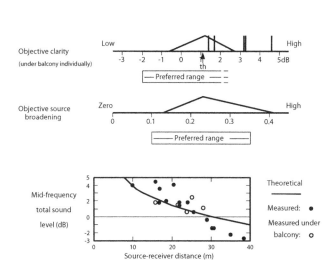

Figure 5.59 Glasgow Royal Concert Hall: objective characteristics

Associates during the executive phase. Comments above are thus well informed but may be biased. The Glasgow hall does not follow closely a recognized precedent but offers a novel version of an arena plan form. This solution offers the advantages of that form with regards to the close relationship between performers and audience but also has a few disadvantages. In retrospect, the overhangs are somewhat excessive and the perforated cladding used to conceal the open-truss beams should have been made more acoustically transparent. Another interesting observation is that this stalls design with its reverse-splay plan generates perhaps rather too many early reflections for the stalls seating. The overall aim of the acoustic design of the concert hall was to provide a uniform, intimate and enveloping sound with a suitable balance between clarity and sense of reverberation. These aims were substantially achieved.

The account here considers the original design and experience of the first five years of the hall. During this time the only modifications made were to reflectors above the stage in order to improve conditions for musicians. In 1996 a new director of the hall was appointed who called in Kirkegaard Associates to advise on the acoustics. The principal modification has been the unnecessary removal of the QRD reflectors along the sides of the hall at high level.

Glasgow Royal Concert Hall: Acoustic and building details

Volume = 28 700 m³

Total hall length = 49.6 m

Number of seats = 2195 plus 263 choir

Volume/seat = 11.7 m³

True seating area = 1300 m²

Stage area (medium format) = 176 m²

Acoustic seating area = 1542 m²

True area/seat = 0.53 m²

Volume/acoustic seating area = 18.6 m

Mean occupied reverberation time (125–2000 Hz) = 1.8 seconds (measured by Sandy Brown Associates, 1990, from pistol shots)

Unoccupied objective data measured August 1990; subjective assessment September 1990

Owned by Glasgow Cultural Enterprises Ltd. Building opened in October 1990

Architects: Sir Leslie Martin and RMJM (Edinburgh)

Acoustic consultants: Fleming & Barron with Sandy Brown Associates

Uses: classical music, popular music, occasional arena events and conventions

Construction:

Floor – timber boarding on plywood on concrete

Walls – plaster on masonry and timber bonded to concrete

Ceiling – precast concrete with plastic cover

Balcony soffits – three layers of 50 mm plasterboard

Stage – timber boarding on plywood

Stage reflector – 114 mm timber on steel frame

Seating – upholstered with solid bases, wooden arms and backs

5.14 Birmingham Symphony Hall

As England's second city, Birmingham clearly needed a purpose-built concert hall. Its principal orchestra, the City of Birmingham Symphony Orchestra, had grown since 1980 under the baton of Sir Simon Rattle into a world-class orchestra yet they only had a modest Town Hall to perform in. By the time this concert hall was being considered, the city authorities were beginning to appreciate the value of their orchestra. As Rattle said: 'Birmingham has realized that the arts are an incredible growth industry'. In consequence, little expense was spared for the new concert hall. The hall opened amid a blaze of publicity in 1991 with claims that it would offer 'perhaps the best acoustics in Europe'.

Birmingham Symphony Hall forms part of the International Convention Centre comprising 11 halls in all, ranging in capacities from 120 to about 3000. Symphony Hall was designed for multiple uses but the concert requirement was definitely over-riding. With acousticians, Artec Consultants of New York, the hall's design was acoustic led; joint design development with the architects Percy Thomas Partnership continued throughout the project.

Artec Consultants have a reputation for being uncompromising about their acoustic requirements. However this hall represents a radical change of direction for Artec. Prior to the mid-1980s their emphasis had been on provision of early lateral reflections; in most of their designs this resulted in pitched ceiling profiles (see section 4.11). Plan forms varied but generally there was a splay in front of the stage, a section with parallel side walls and a reverse splay at the rear. In 1989 in Dallas (McDermott Concert Hall, Meyerson Symphony Centre) and in the Birmingham Symphony Hall two years later, Artec chose a plan form that relates directly back to the nineteenth-century parallel-sided halls with basically horizontal ceilings. In fact, the Dallas (Beranek, 2004) and Birmingham halls have many similarities.

The classical halls (section 4.3) had proportions of roughly a double cube, with the width and height similar and the length about twice the width. The

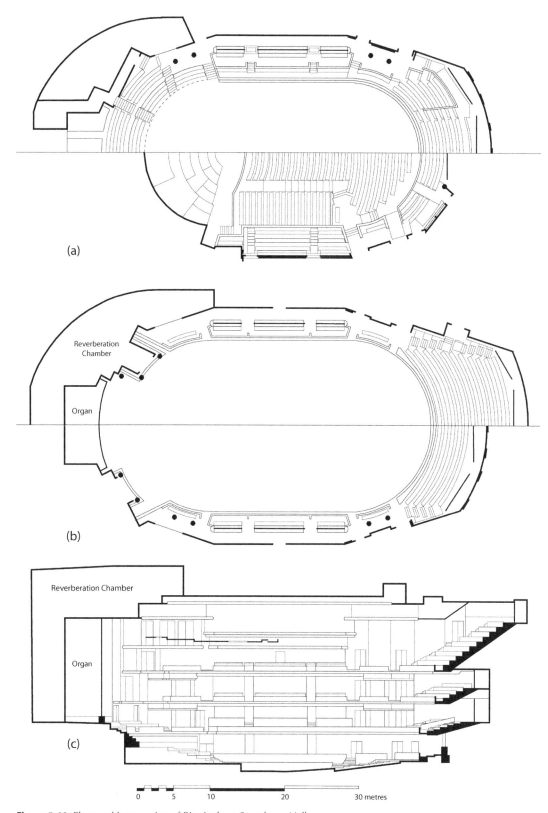

(a)

(b)

Reverberation
Chamber

Organ

(c)

Reverberation Chamber

Organ

0   5   10   20   30 metres

Figure 5.60 Plans and long section of Birmingham Symphony Hall

principal dimensions of Symphony Hall are width 27 m, height 23 m and length 57 m. The ratio of width to height in Birmingham at 1.17 is in fact intermediate between the ratios in the Vienna Musikvereinssaal (1.11) and Boston Symphony Hall (1.23). Both the width and height of Birmingham Symphony Hall are significantly more than in the key nineteenth-century halls (Table 4.2). The height of 23 m in Birmingham is large by concert hall standards and is significantly more than the heights of around 18 m in the Boston and Vienna halls, for instance (see Figure 4.44). The other major departure concerns the plan shape; a century earlier the plan was generally rectangular, whereas in Birmingham the rear seating follows a semicircular curve as you might find in a traditional opera house.

In nineteenth-century halls, not only were seating numbers usually modest but seating density was high. Artec generally insist on a seat count that is well short of the presumed maximum for concert halls of 3000. The 2211 seat capacity in Birmingham is modest relative to this maximum but large by the standards of many earlier shoebox-shape halls. It is therefore not surprising that a considerable height was necessary with four levels of seating.

Except in the largest world cities, individual auditoria have to accommodate more than just use for classical concerts. There are three variable acoustic elements in the hall: the stage canopy, a reverberation chamber and movable acoustically absorbent panels. The 42-tonne stage canopy is mainly horizontal and provides reflections both back to the orchestra and to audience in the front stalls. The canopy height can be varied over a very wide range to provide appropriate acoustic support and appropriate visual conditions. The standard symphony orchestra position for the canopy is 14 m above stage level.

A reverberation chamber was a new element for British concert halls. The chamber wraps round the stage end of the hall from the base level of the organ to a height of 4 m above the hall ceiling. It has a volume of 10 300 m^3 which compared with the auditorium volume of 25 000 m^3 offers a 40 per cent increase. Twenty motorized doors with a total area of 195 m^2 link the chamber to the auditorium; the doors are constructed of 150 mm thick concrete. The rationale behind the chamber is that sound entering the chamber reverberates within it and leeches back into the auditorium. By adjusting the position of the motorized doors, the degree of reverberation enhancement can be varied.

Two interesting possibilities are offered by the chamber. The first is the simultaneous presence of clarity and reverberance. Secondly with the chamber linked to the auditorium, there is the possibility that more appropriate conditions might be available for choral music, for which the standard 2 second reverberation time may be a little short. Acoustical critics however have pointed out that normally one is concerned about running reverberation (section 2.8.1), whereas this chamber will probably only influence the terminal reverberation. In other words the effect of the chamber is only likely to be audible when the music stops. In fact, Artec has a somewhat ambivalent view about reverberation. They claimed neither to assess it nor to consider it of value for development of the design and expressed dismay at the emphasis British consultants appear to place on reverberation time (Charles and Allen, 1989). And yet their design contains a major investment to enhance that very reverberation.

The third variable feature is intended for events such as amplified music or speech, not for classical music. Acoustically absorbent screens suspended from tracks can be placed either in front of or behind walls, which are themselves behind seating both at the sides and rear of the auditorium. This has been used here as an alternative to the more common suspended banners, to be found for instance in the Nottingham Royal Concert Hall. The total screen area is 625 m^2.

Another long-time hallmark of Artec designs is massive construction to minimize low-frequency absorption. Hard surfaces are used throughout with walls for instance of granite or plaster.

There was a further major departure in the design of this hall, namely the very low background noise level. A criterion of 'perfect silence' was mentioned. Considerable care was taken over the design

Figure 5.61 Birmingham Symphony Hall

Figure 5.62 Birmingham Symphony Hall

of the ventilation system with lower than normal air velocities. The resulting background noise level is a very quiet NR10 and this offers the potential for a larger dynamic range than in other halls. Detractors complain that this makes audience noise more obvious, that one's pleasure in hearing music can very easily be frustrated by a single member of the audience coughing. What is particularly interesting is that the audience in Birmingham has learnt to be quieter than audiences elsewhere (Newton and James, 1992); evidence suggests they may be over 5 db quieter. The strict background noise criterion required all potential noise sources to be sufficiently quietened. A special cause for concern was the proximity of the busiest railway tunnel in the UK, which carries mainline trains. The hall was located as far as possible from the tunnel but its nearest point was only 35 m away. The whole hall has been mounted on rubber mounts with considerable attention to detailing; the isolation renders trains inaudible in the hall (Cowell, 1991).

Subjective characteristics

At its opening in 1991, Symphony Hall was greeted by the press with considerable enthusiasm. Not untypical was the comment that 'the opening marks a watershed in the history of British music-making'. Nearly all critics commented with surprise at the virtual absence of background noise (other than from the audience), which allows 'extraordinary *pianissimos*'. A typical comment from another critic was that 'no-one could have predicted the richness and variety of sound' in the hall.

Two concerts have been visited with listeners completing questionnaires. Useful results are available for two locations: in the rear half of the stalls and around the fifth row in the Grand Tier opposite the stage; the Grand Tier is the highest seating level. Overall the acoustics were rated as 'Good' at both locations. Responses in the rear stalls show substantial perceived envelopment and a very high rating for loudness. Yet the intimacy at this position, which we would expect to be high with a loud sound, is only average and judgements of

clarity and reverberance are very variable, some listeners detected low clarity or reverberance, some high. Two listeners commented that the sound was bright but with a slightly brash quality about it.

In the Grand Tier, the loudness is more average, the envelopment and intimacy are relatively low but again the spread of judgements for both clarity and reverberance are large. Comments were made that the sound quality was better for lower sound levels than it was in loud *tutti*.

For all listeners in all locations the ventilation noise was inaudible. The low background noise presumably allows not only weaker sounds but also longer reverberant tails to be heard; perhaps this contributed to the confusions in judgements of clarity and reverberance. Yet the difference between the views of music critics and the listeners who were mainly acoustic consultants comes as a surprise.

Objective characteristics

Measurements have been made in the unoccupied hall at 20 audience positions with the doors to the reverberation chamber closed. During the same measurement session, measurements were also made with the doors open at seven of the 20 positions. Despite Artec's apparent indifference to reverberation time, the measured values are typical for symphony concert halls (Figure 5.63). The occupied value, measured by Kimura *et al.* (1992), appears to be about 1.85 seconds at mid-frequencies, though measurement conditions for that were not ideal. One can observe that there is a bass rise in the reverberation time (at 125 Hz) though the rise is substantially less than is found in another Artec design, the Nottingham Royal Concert Hall. The unoccupied early decay time is only slightly less than the unoccupied reverberation time with a ratio of 94 per cent between them; this is a respectable value indicating reasonable diffusion.

As already mentioned, measurements were made which allowed the effect of opening the doors to the reverberation chamber to be assessed accurately. The quantity that the chamber is likely to most effect is reverberation time. Yet the measured change in reverberation time proved to be small, probably too small for many listeners to notice. With an acoustically coupled space, one expects the decay of sound to be double sloped, the first slope related to the auditorium and the second to the more reverberant chamber, but here this effect could not be found. Changes in the other measured quantities are not quoted here since they are very small.

Measurements of the objective clarity show it to be in the preferred range but the spread of measured values is wide. It appears that the main cause of this spread is that some locations receive a lot of early reflections whereas others have just an average number. Values of objective clarity measured in seats overhung by balconies are likewise rather spread out.

Objective source broadening behaves in a rather similar way to clarity. Measured values are more or less within the preferred range but the spread is wide. Some particularly high values were measured in the stalls.

The fact that conditions are rather variable within Symphony Hall can be seen most clearly in the total sound level graph in Figure 5.63. Within 27 m of the source, there are several positions with sound levels louder than one would expect. Beyond 27 m the trend is for sound levels to be quieter than expected. At remote seats in the highest seating level, the Grand Tier, sound levels are below the criterion of 0 dB. One reason for the quiet sound in the Grand Tier can be seen in the long section of the hall; ceiling reflections to the Grand Tier are blocked by a ledge included for follow-spot lights.

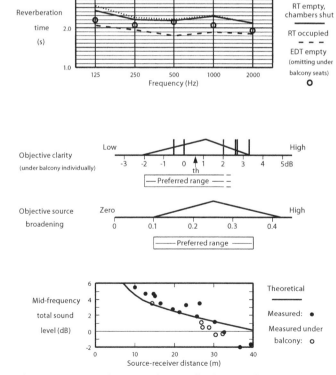

Figure 5.63 Birmingham Symphony Hall: objective characteristics

Conclusions

Birmingham Symphony Hall opened in 1991 amid grand claims about its acoustics. The press responded enthusiastically, in some cases hailing the hall as a new departure for Britain. With the passing of time, it is possible to review the real status of this concert hall. What aspects of the design are likely to be copied in the future?

In earlier concert hall designs by the consultants of this hall, Artec, the emphasis had been on providing strong lateral reflections. A major departure in the design of this hall was to revert to the plan exploited in the nineteenth century with parallel side walls. To accommodate over 2000 people to modern seating standards, this hall is substantially higher than those of the previous century but this does not appear to present problems. Parallel-sided plans are also good for strong lateral reflections and the reverse-splay surfaces towards the rear of seating areas also help in this respect.

The hall has two novel characteristics. The first of these, the reverberation chamber, appears from measurements made in the hall not to have fulfilled its initial promise. The second feature is a low ventilation noise level; this has definitely been noticed, mainly favourably, and the question of background noise has been brought to the fore whereas previously it tended to be taken for granted.

In terms of subjective response, there are surprising differences between responses by music critics and acousticians. Of course, there are other agendas involved for both these groups but consistent views usually contain a grain of truth in them. The media continue to refer to the acoustical excellence of the hall, while acousticians remain less enthusiastic. However the evidence of objective measurements in the hall may offer an explanation for the difference of view.

The hall performs well in terms of temporal quantities such as reverberation time, early decay time and objective clarity. Objective source broadening, which is associated with early lateral reflections, is high in this hall. But examination of the total sound situation shows that for most of the

Birmingham Symphony Hall: Acoustic and building details

Volume = 25 000 m³

Total hall length = 57.0 m

Reverberation chamber volume = 10 300 m³

Number of seats = 1990 plus 221 choir

Volume/seat = 11.3 m³

True seating area = 1115 m²

Stage area = 200 m²

Acoustic seating area = 1450 m²

True area/seat = 0.50 m²

Volume/acoustic seating area = 17.2 m

Mean occupied reverberation time (125–2000 Hz) = 1.9 seconds (measured by H. Kimura *et al.* ca. 1992, from music)

Unoccupied objective data measured February 1999; subjective assessments June 1991 and May 1992

Owned by Symphony Hall (Birmingham) Ltd.

Building opened in April 1991

Architects: Percy Thomas Partnership

Acoustic consultant: Artec Consultants Inc.

Uses: classical music, popular music, occasional dance and conventions

Construction:

Floor – timber boarding on plywood on concrete

Walls – granite, plaster on masonry and timber bonded to concrete

Ceiling – precast concrete with plastic cover

Balcony soffits – three layers of 50 mm plasterboard

Stage – timber boarding on plywood

Stage reflector – 114 mm timber on steel frame

Absorbent screens for amplified music/speech – 75 mm mineral wool in timber frames

Seating – upholstered with solid bases, wooden arms and backs

seats where critics are likely to sit the sound level is higher than one would expect in a more diffuse space. These high levels are associated with the early part of the sound, in other words there are more than the average number of early reflections, offering the acoustics of a space smaller than the actual one. This can provide an exciting effect but to those filling in the questionnaires there were some subtle penalties to pay in terms of sound quality.

While the closer seats to the stage are favoured, the more remote ones suffer from lack of sound level. To what extent one is the consequence of the other is difficult to say; there is also the question of the unfortunate ceiling line.

This hall is interesting as a modern response to the parallel-sided design. It has also been responsible for posing the question again of how quiet the background noise should be. The acoustic consultants of this hall have continued to develop the parallel-sided solution in their more recent designs.

5.15 Bridgewater Hall, Manchester

Rather unusually for the 1990s, the city of Manchester decided to use a new concert hall as a major element of urban regeneration in a city which had suffered serious industrial decline. Remarkably it can claim to be the first large auditorium to be built in Britain as a free-standing structure since the Royal Festival Hall of 1951. With the experienced team of RHWL as architects (particularly Nick Thompson) and Arup Acoustics, the aim was to provide a significant new building for the city and a new home for the local Hallé Orchestra with facilities far superior to its predecessor, the Free Trade Hall (section 5.7).

The hall exterior uses red sandstone and metal cladding to provide the appearance of solidity, which is cut through by glazed foyers, appearing light and free against the major mass (Thompson, 1999). Contrasts extend to the interior, where the foyers are visually bright whereas the auditorium is in subdued browns, whites and greys.

As in Birmingham Symphony Hall, an uncompromising approach to background noise levels

and noise intrusion was applied here; the background noise in the auditorium is close to the limit of audibility at PNC15 and the whole building is mounted on springs to exclude vibration from the nearby urban railway. A further feature which benefits low noise design is the location of all major plant in a tower detached from the main building. For the auditorium envelope, massive construction was used to minimize sound absorption at low frequencies.

To achieve suitable concert hall acoustics, a considerable auditorium volume is needed and this generally results in high ceilings. This can leave a rather awkward empty upper space. A common remedy to this visual problem is to include some of the structure supporting the roof within the auditorium. This solution can have the additional advantage of keeping down the overall height of the building exterior. In Bridgewater Hall cast steel vertical members are joined by a lattice of steel rods within the hall. Lines of low voltage lights are suspended at the lower level of the structure. They are one element among many which contribute to the architects' aim of an 'appropriate balance between aural and visual intimacy'.

Two plan forms for concert halls have emerged from earlier decades: the parallel-sided hall and the terraced hall. Each has its limitations. For the parallel-sided hall the limitation is in seating numbers, with the width limited on acoustic grounds the auditorium length and height can become too large. The terraced hall normally has audience seating on either side of the stage, from which the orchestral balance is markedly skewed.

The novel feature of Bridgewater Hall is the synthesis between these two traditional forms. The front half of the hall follows a basically parallel-sided design scheme with three levels of balcony. The traditional rectangular shape offers listeners in the stalls reflections from the side walls and 'cue-ball' reflections off the side walls and side balcony soffits. In the rear half of the hall the seating blocks follow a 'vineyard' arrangement with steeply raked tiers. At the three levels opposite the stage there are reverse-splay surfaces in plan which can be

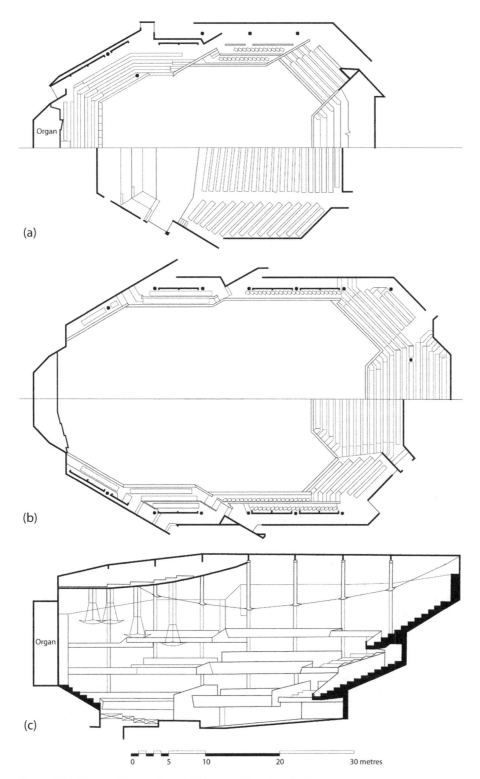

(a)

(b)

(c)

Organ

0 5 10 20 30 metres

Figure 5.64 Plans and long section of Bridgewater Hall, Manchester

Figure 5.65 Bridgewater Hall, Manchester

Figure 5.66 Bridgewater Hall, Manchester

expected to provide strong sound reflections from the side. However accommodating the full audience of 2400 must have created some difficulties since the balcony overhangs at both stalls and Circle level are rather deeper than one would like, with low heights at the openings.

Choir seating wraps round the platform. The strong tradition in northern England for organ music meant that the 5000-pipe organ, rather than being left until future funds became available, was included in the original construction. It is placed in the conventional position behind the choir seating, providing a visual focal point for the audience.

The acoustics of the auditorium was checked both with a computer simulation model and a 1:50 scale model. Timber screens behind the side balcony seating have clearly been designed to optimize the reflection situation in nearby seats. It is no secret that the architect of the hall prefers a classic modernist style with clean lines and plane surfaces. This approach can create difficulties when the acoustic consultant wants surfaces to be acoustically scattering. Shallow grooves are found on the side balcony fronts and ribs have been added to the upper sections of some of the timber screens; both these details will provide a little diffusion, but only at high frequencies. The majority of the ceiling is stepped in cross-section which will also scatter reflections. Is this enough?

Subjective characteristics

When a new concert hall opens, the music critics suddenly turn their attention from the subtleties of individual musical interpretations to the acoustics of halls. Their response is necessarily often influenced by the quality of the music played. In the case of the Bridgewater Hall, the reaction in the press to the opening concerts was mainly of disappointment. With time, the reputation of the hall's acoustics has risen; this often occurs as musicians adapt to their new acoustic environment.

Careful reading of the various initial reactions does highlight one consistent comment: that in musical climaxes the definition tended to disappear among congested textures. This lack of smoothness over the full dynamic range is one of the comments made by listeners completing questionnaires. Several listeners also mentioned hearing image shifts, with sound appearing to come from directions different to the true direction.

On the questionnaire scales themselves, the overall judgement was of good acoustics. Listeners found the sound to be particularly live and reverberant and offering a particularly high degree of source broadening. This last judgement was incidentally higher than in any other British hall, including Birmingham Symphony Hall. But though the source sounded broad, two listeners commented that they did not feel enveloped by sound. On the other scales, the clarity and intimacy was average and the sound was considered loud. There is some evidence that acoustic quality is better in the stalls than at the upper balcony levels.

A telling piece of evidence is the lack of agreement about the degree of intimacy; there was a large range of individual responses on this scale for the same locations in the hall. Is the sound loud but in character not ideal? If something is missing in the acoustics of this hall, it is difficult to be specific about what it is that is actually lacking.

Objective characteristics

Objectively the situation in Bridgewater Hall proves to be straightforward (Figure 5.67). The reverberation time of the unoccupied hall has a high value around 2.4 seconds, while in the occupied hall the mid-frequency reverberation time is a thoroughly respectable 2.0 seconds. Both empty and occupied there is a significant rise in reverberation time in the bass, no doubt as a result of massive construction. The early decay time (EDT) is close to the reverberation time.

Objective clarity is in the preferred range, symmetrical about the theoretical mean. Clarity is higher in two overhung seat positions opposite the stage. Objective source broadening is high and consistent. The only obvious criticism relates to the measured sound level in the rear third of the hall. The level values here are slightly below expectations

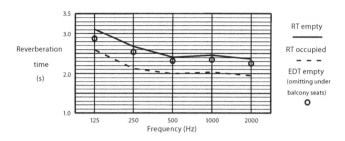

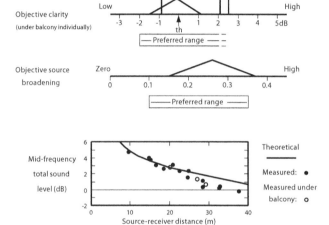

Figure 5.67 Bridgewater Hall, Manchester: objective characteristics

and in several positions close to the minimum criterion of 0 dB. The fact that sound levels drop more than expected in the rear third of the hall may help explain why the quality of sound is judged as less successful in this area.

In objective terms, this hall conforms as one of the closest to proposed criteria for the various objective measures. It is likely that the careful testing of the acoustics during the design of the hall with computer and scale models contributed to the good objective performance.

Conclusions

A continual discussion in concert hall acoustics concerns the optimum plan form. There are those who consider that only the rectangular plan or shoebox

hall is capable of excellent acoustics. The 'vineyard' terrace plan has meanwhile gained a good reputation, particularly from the halls in Berlin and Cardiff. Exploiting the virtues of each plan by combining a parallel-sided front half with a terraced rear half of the hall is a very attractive idea.

The designers of Bridgewater Hall have shown that good objective performance can be realized with this geometry. For instance, high values of objective source broadening, which are found in narrow rectangular halls, have been achieved here, probably by careful local design. The balcony fronts direct sound down onto audience and likewise many panels behind the side balcony seating also direct sound down. These details may contribute to the drop off of sound level towards the rear of the auditorium.

Bridgewater Hall, Manchester: Acoustic and building details

Volume = 25 050 m³

Total hall length = 53.0 m

Number of seats = 2127 plus 276 choir

Volume/seat = 10.4 m³

True seating area = 1156 m²

Stage area = 227 m²

Acoustic seating area = 1424 m²

True area/seat = 0.48 m²

Volume/acoustic seating area = 17.6 m

Mean occupied reverberation time (125–2000 Hz) = 2.1 seconds (measured from impulses by Arup Acoustics in June 1996 and corrected for small modifications undertaken in August 1996)

Measurement date for unoccupied objective data: 22 July 1999

Owned by Manchester City Council

Building opened in September 1996

Designed by RHWL Partnership

Acoustic consultant: Arup Acoustics

Uses: classical music (including symphonic, choral and organ music), light popular music

Construction:

Floor – timber boarding embedded onto concrete

Walls – limestone, plaster on masonry and concrete

Ceiling – precast concrete panels on concrete beams

Screens behind side balcony seating – veneered timber on MDF

Stage – timber boarding over air space

Balcony fronts – 30 mm glass reinforced gypsum

Ensemble reflectors – curved glass 2.6 × 2.6 m

Seating – upholstered with slotted bases, wooden arms and backs

Some critical comments have been made of the subjective acoustic quality. Among the minor modifications to the hall between the first test concerts and the opening was to introduce more scattering on some surfaces. It is difficult to be confident about something as enigmatic as diffusion, but the nature of the critical comments does point to the need for greater scattering treatment. The acoustic consultants have also concluded that 'the hall might benefit from further high frequency sound diffusion' (Harris, 1997). But this is in the context of a hall which has been greeted with great enthusiasm by performers and audiences.

5.16 Waterfront Hall, Belfast

It took 18 years from the initiation of a study of arts provision for Belfast City to the completion of the Waterfront Hall in January 1997. The persistence of the City Council finally won through to produce a world-class facility, designed as a 2234-seat concert hall but with considerable flexibility of use. The principal members of the design team were the Belfast-based architects Robinson & McIlwaine, the theatre consultants Carr & Angier and acoustic consultants Sandy Brown Associates. The last two had worked together on St David's Hall in Cardiff and this hall was clearly the precedent for the Waterfront Hall. The basic auditorium design of St David's Hall owed much to Martin Carr of Carr & Angier, who set himself the challenge of introducing even more flexibility for the Belfast hall.

Waterfront Hall is at a prominent position in the city next to the River Lagan; 43 per cent of the exterior walls of the building are glazed to exploit the views the site offers. The exterior of the building is circular with a shallow dome roof. There is no hint from outside that the auditorium follows the 'vineyard terrace' scheme pioneered by Scharoun in the Berlin Philharmonie in 1963.

A vineyard terrace hall appears to offer all the acoustic advantages of the parallel-sided or shoebox hall: good balance between clarity and reverberance, the possibility of early reflections from the side and a diffuse sound field. To achieve

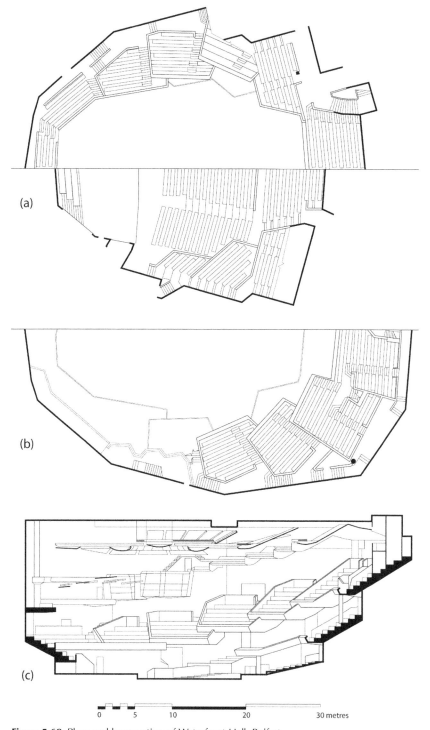

(a)

(b)

(c)

0   5   10   20   30 metres

Figure 5.68 Plans and long section of Waterfront Hall, Belfast

these aims requires the surfaces separating individual seating blocks to be used to provide suitable acoustic reflections. The advantages of the terrace design include the possibility of having a larger maximum hall width and having an audience arrangement that surrounds the performing platform. This greater degree of surround can offer a more intimate performer–audience arrangement. A further big advantage for designers of the terrace scheme is much greater flexibility; a detail in one part of the hall can generally be altered independent of the rest of the auditorium. The parallel-sided hall and the vineyard terrace have become the two preferred solutions for large concert hall design.

The Waterfront Hall has a significantly larger seat capacity than the 1957 seats of St David's Hall in Cardiff. To accommodate these extra numbers, two levels of tiers run round the rear half of the hall with a total of three levels of seating opposite the stage. The approach to design of the tiers is very similar in the Cardiff and Belfast halls with the surfaces separating them running basically radially from the stage but the vertical height of these upstands is on average smaller in the Belfast Hall. Both halls are used for conferences; numerous gangways and circulation routes allow easy access between tiers.

In the Cardiff hall the reverberation time had been around 2 seconds with a volume per seat of 11.2 m^3/seat. The starting point for the acoustic design of Waterfront Hall was a long natural reverberation time with a design aim of 2.3 seconds (Burd, Haslam and Stringer, 1997). To achieve this a large volume would be needed; the ultimate volume per seat in Belfast is 13.8 m^3/seat. The other major difference between the two halls occurs in the ceiling area. In the Cardiff hall the upper volume is obscured by an acoustically transparent suspended ceiling, which conceals roof structure and some ventilation ducting. In Waterfront Hall the visual ceiling consists of rings of reflecting surfaces with a convex profile and acoustically transparent elements made of timber slats with 50 per cent open area. The volume above is principally occupied by technical equipment, the largest element being a full theatrical suspension grid which extends over both the stage and arena (stalls) seating area.

Flexibility of use includes possible modifications both to the audience area and the stage area (Carr, 1997). The arena seating is on wagons, which can be moved on air castors into storage below the auditorium. The arena floor is on elevators, so that a large flat area including the stage can be created or alternatively a promenade floor giving a total capacity of about 3000 for events such as pop concerts. The front stage elevator can be lowered to produce an orchestra pit for opera, dance etc. The suspension grid allows extensive possibilities for hanging lights, scenery, loudspeakers etc. With some reluctance from the theatre consultant, an optional proscenium arch facility has been included which gives acceptable sightlines for about 1900 spectators. For amplified events, some acoustical variability is available from retractable absorbing drapes located in the ceiling space out of view.

The stage itself is reasonably conventional in shape but contains some removable choir seating at the rear, in addition to the fixed choir seating behind. Acoustical support for musicians on stage is provided by reflections from the soffits and adjacent walls of galleries which extend around the stage end of the hall. For reflections from above, there are circular reflectors with a convex profile whose height is fully adjustable.

As has now become the norm, auditorium ventilation is via a displacement system with supply from plenums below the floor. The background noise level achieved is below NR20.

With the basic design of the hall complete in 1992, the acoustic consultants asked the author to test a 1:50 scale model of the hall. The tests revealed that the early decay time (EDT) was lower than expected, with a ratio between EDT and RT of 0.83. This ratio tends to respond to the degree of diffusion (Barron, 1995b), which is otherwise very difficult to measure. The ratio in St David's Hall is much higher at 1.03; the Cardiff hall is considered to offer diffuse sound whereas the new hall appeared not to do so. A more diffusing array of suspended elements was recommended for the ceiling space.

Figure 5.69 Waterfront Hall, Belfast, with black suspended cloth at rear of stage

Figure 5.70 Waterfront Hall, Belfast

Other suggested modifications included increasing the reverse-splay for surfaces to the side of the seating opposite the stage at stalls level and raising the soffit above seating at the highest level opposite the stage. In building projects which extend over very long periods, it often occurs that when the decision to build is finally made, progress on the design has to be extremely rapid! For Waterfront Hall just eleven months were available before tenders were invited. Results from the first acoustic model tests appeared at this time and it was fortunate that the architects were willing to undertake changes in the design at this late stage. A second model test was conducted on the revised design (essentially the scheme actually built) which showed significant improvements in all the issues detected earlier.

Subjective characteristics

A newspaper critic reviewing the opening concert in January 1997 in Waterfront Hall described the acoustics as 'balanced, warm and natural' (*Daily Telegraph*, 20 January 1997). Enthusiasm for the sound in the hall seems to be the general response.

A listening exercise with questionnaires was conducted at a concert during the first year of operation of the hall. The clarity of sound was judged as high and the sense of reverberance a little above average. On the other scales the responses are average for British concert halls. Two locations were tested, one in the Lower Terrace opposite the stage, the other in the Upper Terrace also opposite the stage. In concert halls in general more remote seats are liked less, but in this hall there was no obvious preference between the two locations. (One significant feature here may be that at the Lower Terrace location the measured proportion of early lateral sound is the lowest in the hall. A lesser spatial sense here may counteract the advantages associated with its closer position.) There were some comments about orchestral balance, but with a visiting orchestra who were new to the hall, minor balance problems are to be expected.

Overall, listeners completing questionnaires made an average judgement of Waterfront Hall between 'Good' and 'Very Good'. Tonal balance was judged as even.

Objective characteristics

From Figure 5.71 containing objective values measured in Waterfront Hall, we find a respectable mid-frequency reverberation time (500/1000 Hz) of 2.2 seconds unoccupied and 2.0 seconds occupied. The occupied reverberation time was measured by Sandy Brown Associates, the acoustic consultants, with an audience of about 60 per cent capacity. The values given in the figure are an estimate for full occupancy. There is slight rise in reverberation time at low frequencies, helpful for acoustical warmth, but at 2 kHz the reverberation time drops more than is usual. One presumes that this high frequency drop is associated with extra absorption somewhere, but its location is not obvious.

The reverberation time achieved in the hall is notably shorter than the design aim of 2.3 seconds. As hinted at in the acoustic model tests, though all suspended elements in the ceiling are acoustically hard, their acoustic influence appears to be to reduce the reverberation time (due to reduction of the mean free path). The occupied value of 2.0 seconds that has been achieved is perfectly respectable, but it is clear that the larger auditorium volume was necessary to compensate for the effects of the multitude of suspended items in the ceiling.

The occupied early decay time (EDT) will also be acceptably within the criterion of greater than 1.8 seconds. While discussing scale model tests above, the ratio of EDT to RT was mentioned as a measure of diffusion. The measured ratio in the full-size hall is 0.96, close to that in the second model of 0.94. Both these values suggest good diffuse conditions, whereas with the first model a value of 0.83 did not.

No particular comment is needed about objective clarity other than that values measured under balcony overhangs are no higher than measured elsewhere. Sound levels at overhung seats can also be seen in the figure; these levels are not lower than those in exposed seats. Both results suggest good balcony overhang design in this hall.

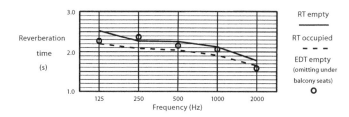

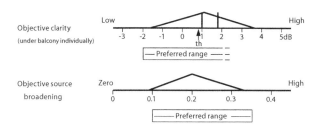

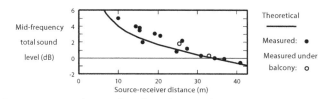

Figure 5.71 Waterfront Hall, Belfast: objective characteristics

Objective source broadening is within the preferred range in Waterfront Hall, but average values are significantly lower than in the recent Birmingham and Manchester concert halls. With bounding walls to individual seating blocks running roughly radial to the stage, it is perhaps surprising that sufficient early reflections do in fact arrive from the side in Waterfront Hall.

With the total sound level, one sees that several positions have louder levels than expected according to theory. This is mainly due to strong early reflections. Agreement with theory is good at distant seats but theoretical levels dip below the 0 db criterion beyond 33 m. This is a consequence of the total acoustic absorption in the hall, including no doubt contributions from the extensive gangways and the many suspended elements in the ceiling space.

On the question of tonal balance, the spectrum of sound is probably more important for listeners

than the reverberation time characteristic which has been considered in the past. With regard to the balance between bass and mid-frequency sound levels, the situation in this hall is healthy. The received sound level at 2 kHz relative to lower frequencies is also typical by concert hall standards. This objective evidence therefore accords with subjective judgements of neutral tonal balance.

Conclusions

Since 1963 when Scharoun's Berlin Philharmonie opened, few concert hall designers have taken up the challenge of the 'vineyard terrace' scheme. The Waterfront Hall demonstrates yet again how in careful hands the terrace solution can provide acoustics at least as good as many parallel-sided halls. Though the geometry of a vineyard terrace hall is more demanding of the designer, it offers much less constrained architectural solutions to problems of sightlines and accommodating particular seat numbers. For instance, Waterfront Hall has very modest overhangs by balconies; overhangs in a parallel-sided hall of the same capacity are generally much deeper.

Another significant feature of Waterfront Hall is the flexibility of use built into the auditorium: standard concert format, open arena, proscenium arch and all with extensive flying facilities. The range of possible formats here is the largest to be found within the UK; it is probably the most versatile hall in Europe. Much of this flexibility comes from physical support options contained in the ceiling space. One interesting result revealed from this project is that the extensive suspended elements including a large suspension grid do result in a lowering of the reverberation time. Luckily since a generous auditorium volume was used, there was scope for some reduction of reverberation time relative to what was predicted according to standard procedures.

In one respect there is scope for further improvement in the acoustic characteristics of a vineyard terrace hall compared with both St David's Hall, Cardiff, and Waterfront Hall. Other recent British halls have included strong early reflections from the side which contribute to source broadening and probably also a sense of envelopment. By having vertical surfaces on each side of individual seating tiers which are oriented radially from the stage, they are not capable of supplying first order reflections to listeners nearby. Lateral reflections can be provided by having reverse-splay plans for individual seating blocks. This presents a more demanding, but definitely worthwhile, architectural challenge than the radial arrangement.

Waterfront Hall, Belfast: Acoustic and building details

Volume = 30 800 m^3

Total hall length = 52.2 m

Number of seats = 2039 plus 195 choir

Volume/seat = 13.8 m^3

True seating area = 1090 m^2

Stage area = 179 m^2

Acoustic seating area = 1315 m^2

True area/seat = 0.49 m^2

Volume/acoustic seating area = 23.0 m

Mean occupied reverberation time (125–2000 Hz) = 2.0 seconds (measured by Sandy Brown Associates 1997 and corrected to full occupancy, from noise decays)

Unoccupied objective data measured July 1999

Owned by Belfast City Council

Building opened in January 1997

Architects: Robinson & McIlwaine

Acoustic consultant: Sandy Brown Associates

Uses: classical music, popular music, conferences, proscenium stage events including opera, commercial events

Construction:

Floor – timber aisles, carpet between seating rows

Walls – plaster on masonry

Ceiling – grouted concrete planks

Balcony soffits – plasterboard under metal studs

Stage – timber boarding on timber joists

Ceiling reflectors – 12 mm plywood on steel frames

Seating – upholstered including an under-seat pad with cloth covering, wooden arms and backs

5.17 Acoustical design of concert halls in subjective and objective terms

As well as providing detailed discussion of the individual halls, it is hoped that the above can contribute to an understanding of both subjective and objective behaviour in concert halls. The background to subjective interpretation was given in section 3.2. A detailed study has been made of results, both subjective and objective, of eleven of the British halls built before 1983 (Barron, 1988a). Major results of this study are presented here and compared with earlier work by others.

Firstly on the subjective side, it is clear that with a little practice perceptive listeners can respond consistently to an acoustic questionnaire with five or more scales. It also seems that when different listeners are responding on a particular questionnaire scale they are generally responding to the same attribute. However, when they make an overall judgement on the acoustics, listeners differ in their preferences.

Two surveys have provided interesting evidence about preference. In the author's 1988 study, the listeners subdivided into two groups: those that preferred intimacy and those that preferred reverberance. In the earlier Berlin study (section 3.2) there was a subdivision between clarity and loudness. These two different results require reconciliation. Firstly there is in fact agreement in the selection of intimacy and loudness. Both intimacy and loudness are well related to the objective sound level (which in the case of loudness is as expected). Also, in subjective terms, in this survey intimacy and loudness are highly intercorrelated. Clarity and reverberance can be less easily reconciled. In extreme cases the two aspects are inversely related (with a long reverberation time, the clarity is low and the reverberance is high and vice versa). But for concert halls it is found that clarity and reverberance can be independent and must therefore be treated separately. In the Berlin study, their particular group preferred high clarity, whereas in the author's survey the relevant listeners preferred high reverberance. These

must therefore be two separate groups. High clarity seems to be preferred by those with a musical background; their championing of the Royal Festival Hall has already been mentioned. By chance, this survey contained no-one in this group but there seems little doubt that it exists. We thus conclude that listeners subdivide into at least three groups regarding their preference: for clarity, reverberance and intimacy.

In this study, the halls which were liked best were those which scored well on the attributes of reverberance, envelopment (or rather source broadening) and intimacy. For the same reasons as above, clarity should be added to this list. If listeners fall into different groups, only by satisfying each group will a hall be good for all. This is a demanding and humbling challenge. It means of course that an individual designer should not rely on his ears alone. What then are the objective requirements for high clarity, reverberance, envelopment and intimacy?

The question is best tackled by answering it initially in technical terms: what objective measure correlates best with each subjective quality? The relationship for the case of intimacy is shown in Figure 5.72; the best correlation with intimacy is with the measured sound level (averaged over frequency). However intimacy is also found to be well correlated with source–receiver distance; seats closer to the orchestra are judged as more intimate.

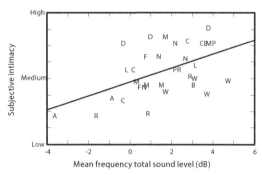

Figure 5.72 Relationship between subjective intimacy and measured total sound level (averaged over 125–2000 Hz) for 34 positions in 11 halls. Points are labelled by hall as in Figure 5.1. Line is best fit regression line (r = 0.53)

Table 5.2 Objective measures corresponding to subjective attributes

Subjective attribute	Objective measure
Clarity	Mid-frequency early-to-late index
Reverberance	Early decay time
Source broadening	Early lateral energy fraction + mean frequency total sound level
Intimacy	Total sound level
Loudness	Mean frequency total sound level and source–receiver distance
Warmth	Bass level balance

But since sound level and source–receiver distance are themselves related, are listeners responding to an acoustical or a visual stimulus? Simple statistics on the results of this survey cannot answer this question, but experience of a wider range of halls provides strong evidence that acoustic sound level is important for intimacy.

The objective measures which most closely match six of the subjective attributes considered here are listed in Table 5.2. The results in the table are based on the consensus of recent research work.

The table also includes 'Warmth' as a subjective attribute. This term is commonly used to describe the sensation of a rich bass sound, though it has been little used here as the questionnaire used a scale of bass balance instead. The objective total sound level at bass relative to mid-frequencies, called bass level balance, is listed in Table 5.2. The relationship between subjective and objective is not as well substantiated for warmth as for other attributes. Figure 5.73 shows the mean bass level

balance values measured in the 16 British halls. Extreme high or low values for the level balance correspond with rich or weak perceived bass. For example, the Manchester Bridgewater Hall, the Belfast and Nottingham halls (T, S and N) all scored well subjectively for bass sound in the questionnaire studies, while the Barbican concert hall (R) scored poorly. An investigation has shown that the design features responsible for influencing bass level balance (Barron, 1995a) include both geometrical features as well as the ratio of bass to mid-frequency reverberation time.

If we are aiming at good clarity, reverberance, envelopment and intimacy, then the implications for concert hall design are as follows.

1. Provide sufficient early reflections to maintain good clarity.
2. Design for adequate early decay time (probably at least 1.8 seconds occupied with audience) to provide reverberance. This may require a longer than normal reverberation time in halls with directed reflections (section 3.5).
3. Ensure that a significant proportion of early reflections arrive at listeners from the side.
4. Design for adequate sound levels with sufficient early reflections and later sound in order to maintain intimacy. (A level of 0 db is the minimum criterion.) This task is eased with shorter distances to the most remote seats.

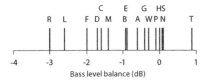

Figure 5.73 Mean bass level balance (i.e. bass vs. mid-frequency total sound level) by hall for occupied halls. More positive implies more bass energy. Labels as per Figure 5.1 (plus E, Edinburgh Usher Hall; G, Glasgow Royal Concert Hall; H, Birmingham Symphony Hall; S, Belfast Waterfront Hall and T, Manchester Bridgewater Hall)

There has been no mention here of listener envelopment (section 3.10.3), because to a certain extent it introduces no new criteria. With attention to three objective quantities: early lateral energy, sound level and reverberation time, it is likely that listener envelopment will be good (Barron, 2001).

A further omission concerns diffusion and the need for scattering surfaces. This is an unresolved area in concert hall acoustics; the experience of the Manchester Bridgewater Hall (section 5.15) is instructive.

The proposal here is that clarity, reverberance, source broadening and intimacy carry roughly equal weight for good acoustics. Two alternative opinions should be mentioned. Ando (1985) suggests that four objective quantities should be optimized for the best acoustic conditions: the delay of the first reflection, a quantity linked to the reverberation time, the interaural cross-correlation coefficient (IACC) and relative total sound level. The last three of these are similar to measures used here, with IACC as an alternative to objective source broadening. The delay of the first reflection, or initial-time-delay gap, has been critically reviewed in section 3.2.

An second alternative view is proposed by Hidaka, Beranek and Okano (1995) also found in Okano, Beranek and Hidaka (1998), who suggest that source broadening, as measured by the IACC, is most closely related to acoustic quality. While there is evidence for the importance of source broadening among British halls, there are some halls with high values of objective source broadening (similar to IACC) which are not judged as having the best acoustics overall (Marshall and Barron, 2001). The following halls have high mean objective source broadening but not the highest overall reputations: the Usher Hall, Edinburgh, and the Royal Concert Halls in Nottingham and Glasgow (sections 5.2, 5.12 and 5.13).

Returning to the search above for the one measurable quantity which matches acoustic quality most closely, surely this is to ignore the nature of subjective response? Several independent subjective quantities are important and people have their own personal bias in terms of what is for them most important. This implies that there is no single quantity that is most important. Rather, several measurable quantities are important and in

a well-designed hall values for each quantity need to be within acceptable limits throughout the auditorium. The concert hall experience is definitely multi-dimensional.

Recent surveys of concert halls have sought, among other things, to clarify the degree to which objective measures can describe the subjective response to concert acoustics. If these objective measures are useful, we need to be able to predict them for designs of the future. In the case of reverberation time, this can usually be done from drawings plus acoustic absorption data. For other measures, either computer or acoustic scale models can be used (section 3.9). It has been argued above that measurement of the quantities listed in Table 5.2 considerably enlarges the scope for objective description of concert hall acoustics. These objective measures can greatly enhance the confidence of predictions about the quality of untried designs. Yet they should not be seen as the total answer: the science cannot replace the art.

From a conceptual standpoint, these newer objective measures are still quite crude. They have been selected by subjective experiment with its inevitable imprecision. The human hearing system is much too complex for it to be possible to use physiology as our starting point. One very obvious limitation of the objective measures is the crude assumption about the directivity of the source. A violin radiates more sound energy at right angles to the belly of the instrument, particularly at the higher frequencies. But this directional characteristic is complicated, and with 60 or more different instruments each with their own directional behaviour, it is no surprise that the simplest assumption is taken: that the orchestra as a whole radiates sound equally in all directions. What is certainly clear is that objective description of concert halls could be further evolved, probably using computers, to accommodate such complications.

The new objective measures provide a valuable guide, they can help to avoid serious failures, and they can assist in the appraisal of basic design forms. But they are not sufficiently refined yet to be able to guarantee excellence as opposed to just good acoustics. And there are several aspects of almost equal importance which cannot as yet be expressed numerically, such as sensations relating to diffusion.

5.18 Solutions to the large concert hall problem

As was discussed in Chapter 3, the difficulty of achieving good acoustics in a concert hall rises sharply when seating numbers exceed 1500. Halls with seating capacities in excess of 2000 which have very good acoustics are rare, even in the world-wide context. There is no single reason for this size barrier. An optimum reverberation time can in principle always be achieved, as long as an unrestricted hall volume is at the disposal of the designers. It is the other aspects which are not automatically satisfied in the large hall with the appropriate reverberation time. Foremost among these other aspects are likely to be the early reflection situation and the total sound level.

To discuss the total sound level, it is best to consider separately the early reflections and the later reverberant sound. Dealing with the later sound first, this is normally well behaved, unless the directions from which it arrives at the listener become restricted. Inauspicious design forms can exhibit this trait; the semicircular plan is one of these, discussed in detail in section 11.1. In general terms, a design should not have appendages nor should angles of view for remote listeners be less than of the whole hall. A common situation with inadequate late sound is underneath a substantial balcony overhang, where little sound energy can arrive from above or behind. The depth of the overhang in plan must be restricted and if possible the soffit height raised significantly above the standard (section 3.7).

It is the early sound situation which is most likely to suffer in the large hall. For both the early and the late sound, a unified compact design is desirable, but with a large seating capacity this tends to leave some seats remote from reflecting surfaces. The problem of ensuring adequate early reflections at all seats has been called the large concert hall problem here. This problem undermines the use of simple plan forms for large concert halls. The situation of seats being remote from reflecting surfaces is particularly marked in the case of the fan-shape plan; this simple plan form is now considered inappropriate for concert use.

With the rectangular plan the situation is much better behaved. The 'solution' in this case to the large concert hall problem involves restricting the hall width, but the penalty here is that halls of this form are only possible up to a certain seat count. The example of the Royal Festival Hall, with for instance its excessive balcony overhang, demonstrates the necessary compromises required in the large rectangular hall. In that hall early reflections were enhanced from above by the suspended reflector, but this is now a discredited approach. Today the trend is to enhance reflections from the side, yet once this step is taken there seems no reason to remain tied to the simple plan form. Individual seating areas can be considered independently in terms of their early reflection requirements and remote seating groups can desirably be accommodated in reverse-splay enclosures.

Three design solutions are current contenders for promising acoustics in large halls. The first solution addresses the large concert hall problem most directly and offers what has been a frequent goal for acoustic designers: the possibility of achieving both high clarity and high reverberance. This can be done by designing for a long reverberation time and enhancing the early reflections from lateral directions. The early reflections can be directed either by profiling the walls and ceilings (e.g. the Royal Concert Hall, Nottingham) or by introducing suspended reflectors (as in Christchurch Town Hall, section 4.10). The challenge is to provide subjective diffuseness by ensuring that sound arrives at all listeners from a wide range of directions. False localization and harsh tonal quality associated with strong discrete reflections must certainly be avoided, probably by modulating reflecting surfaces and avoiding large unbroken plane surfaces. This design scheme, called here the directed reflection approach, offers considerable flexibility in design and the opportunity to deal with the requirements for different seating areas individually.

The subdivided audience solution, often called the vineyard terrace design, exemplified by the Berlin Philharmonie (section 4.9) and St David's Hall, Cardiff, tends to produce a highly diffuse sound which proves to be well liked. Early reflections are provided by acoustically reflective bounding walls and balcony fronts; with care in the design of these surfaces, a good degree of acoustic uniformity is possible. Architecturally the subdivided audience scheme has the virtue of offering a high degree of flexibility to the design.

As described in section 4.12, the parallel-sided solution has made a somewhat surprising return to favour. The side walls, which must not be too far apart, provide a serious limit to design flexibility, so that the maximum seat capacity for parallel-sided halls is less than for the other two solutions. With pressure on seat numbers, it is difficult with parallel-sided halls to maintain good acoustic conditions throughout the audience area, particularly under balcony overhangs. On the other hand, the presence of balconies to the sides of the stage platform is likely to furnish valuable reflections for the performers.

The location of the stage and acoustic conditions for the musicians also require serious consideration. The ability of players to hear themselves and each other will clearly influence their performance, a crucial issue. Conditions for the performers are also pertinent since they as a group often influence the reputations of halls. The stage position surrounded by audience with choir seating behind is used in many successful halls and works well for the listener. Careful consideration should be taken of vertical surfaces around the performing area. The alternative stage arrangement with the orchestra surrounded by some form of shell is common in North America but elsewhere it is generally restricted to multi-purpose halls with flying facilities. An orchestral shell that provides generous space above and if possible around the orchestra seems desirable acoustically.

Concert hall design has come a long way since the hit-and-miss days of the nineteenth century. Much of this progress has occurred since 1960, inspired by research which was undertaken in the knowledge that a rigid scientific approach was inappropriate. Concert hall acoustics has now reached an exciting point where promising approaches to the acoustic problems exist. These now wait to be further exploited by architects and acousticians willing to accept the challenge of working within subtle constraints. The three solutions presented here to the large hall problem of the directed reflection approach, the subdivided audience and the parallel-sided hall offer a wealth of possible interpretations in design.

References

General

Ando, Y. (1985) *Concert hall acoustics,* Springer-Verlag, Berlin.

Barron, M. (1988a) Subjective study of British symphony concert halls. *Acustica,* **66**, 1–14.

Barron, M. (1995a) Bass sound in concert auditoria. *Journal of the Acoustical Society of America,* **97**, 1088–1098.

Barron, M. (1995b) Interpretation of early decay times in concert auditoria. *Acustica,* **81**, 320–331.

Barron, M. (2001) Late lateral energy fractions and the envelopment question in concert halls. *Applied Acoustics,* **62**, 185–202.

Barron, M. and Lee, L.-J. (1988) Energy relations in concert auditoriums, I. *Journal of the Acoustical Society of America,* **84**, 618–628.

Beranek, L.L. (1962) *Music, acoustics and architecture,* John Wiley, New York.

Beranek, L.L. (2004) *Concert and opera houses: Music, acoustics and architecture,* 2nd edn, Springer, New York.

Hidaka, T., Beranek, L.L. and Okano, T. (1995) Interaural cross-correlation, lateral fraction, and low- and high-frequency sound levels as measures of acoustical quality in concert halls. *Journal of the Acoustical Society of America,* **98**, 968–1007.

Marshall, A.H. and Barron, M. (2001) Spatial responsiveness in concert halls and the origins of spatial impression. *Applied Acoustics,* **62**, 91–108.

Okano, T., Beranek, L.L. and Hidaka T. (1998) Relations among interaural cross-correlation coefficient (IACCE), lateral fraction (LFE), and apparent source width (ASW) in concert halls. *Journal of the Acoustical Society of America*, **104**, 255–265.

References by section

Section 5.1

Bagenal, H. (1941) *Concert music in the Albert Hall.* Royal Institute of British Architects Journal, August, 169–171.

Beranek, L.L. (1962) *Music, acoustics and architecture*, John Wiley, New York.

Clark, R.W. (1958) *The Royal Albert Hall*, Hamish Hamilton, London.

Izenour, G. (1977) *Theater Design*, McGraw Hill, New York.

Metkemeijer, R.A. (1997) The Royal Albert Hall, past, present and future. *Proceedings of the Institute of Acoustics*, **19**, Part 3, 57–66.

Metkemeijer, R.A. (2002) The acoustics of the auditorium of the Royal Albert Hall before and after redevelopment. *Proceedings of the Institute of Acoustics*, **24**, Part 4.

Shearer, K. (1970) The acoustics of the Royal Albert Hall. *British Kinematography Sound and Television*, February, 32–36.

Section 5.2

Architectural Review (1914) **36**, July, 18–20.

Beranek, L.L. (1962) *Music, acoustics and architecture*, John Wiley, New York.

Section 5.3

Architects' Journal (1939) Philharmonic Hall, Liverpool. 24 August, 273–275.

Bagenal, H. and A. Wood, A. (1931) *Planning for good acoustics*, Methuen, London.

Beranek, L.L. (1962) *Music, acoustics and architecture*, John Wiley, New York.

Forsyth, M. (1985) *Buildings for music*, Cambridge University Press.

Parkin, P.H., Scholes, W.E. and Derbyshire, A.G. (1952) The reverberation times of ten British concert halls. *Acustica*, **2**, 97–100.

Somerville, T. (1949) A comparison of the acoustics of the Philharmonic Hall, Liverpool and St Andrew's Grand Hall, Glasgow. *BBC Quarterly*, **4**, 41–54.

Section 5.4

Bagenal, H. (1942) *Practical acoustics and planning against noise*, Methuen, London.

Bagenal, H. and Wood, A. (1931) *Planning for good acoustics*, Methuen, London.

Pevsner, N. (1953) *The buildings of England: Hertfordshire*, Penguin, London.

Section 5.5

Architectural Review (1951) Special issue: Royal Festival Hall. June, **109**, 335–394.

Bagenal, H. (1950) Concert halls. *Royal Institute of British Architects Journal*, January, 83–93.

Bagenal, H. (1951) Musical taste and concert hall design. *Proceedings of the Royal Musical Association*, 1951/2, 11–29.

Barron, M. (1988b) The Royal Festival Hall acoustics revisited. *Applied Acoustics*, **24**, 255–273.

Beranek, L.L. (1962) *Music, acoustics and architecture*, John Wiley, New York.

Cox, T.J. and Shield, B.M. (1999) Audience questionnaire survey of the acoustics of the Royal Festival Hall, London, England. *Acustica/Acta Acustica*, **85**, 547–559.

Hartman, H. (2007) Tuning up the Royal Festival Hall. *Architects' Journal*, 13 September, 38–40.

Martin, J.L. *et al.* (1952) Science and the design of the Royal Festival Hall. *Royal Institute of British Architects Journal*, April, 196–204.

McKean, J. (1991) Royal Festival Hall. *Architects' Journal*, 9 October, 22–47.

McKean, J. (1992) *Royal Festival Hall*, Phaidon, London.

Parkin, P.H. and Morgan, K. (1965) 'Assisted resonance' in the Royal Festival Hall, London. *Journal of Sound and Vibration*, **2**, 74–85.

Parkin, P.H. and Morgan, K. (1970) 'Assisted resonance' in the Royal Festival Hall, London: 1965–1969. *Journal of the Acoustical Society of America*, **48**, 1025–1035.

Parkin, P.H., Allen, W.A., Purkis, H.J. and Scholes, W.E. (1953) The acoustics of the Royal Festival Hall, London. *Acustica*, **3**, 1–21.

Pearman, H. *et al.* (2007) Back by popular demand. *RIBA Journal*, June, 28–52.

Shield, B. (2001) The acoustics of the Royal Festival Hall. *Acoustics Bulletin*, **26**, No. 3, 12–17.

Somerville, T. and Gilford, C.L.S. (1957) Acoustics of large orchestral studios and concert halls. *Proceedings of the IEE*, **104**, 85–97.

Somerville, T., Gilford, C.L.S., Spring, N.F. and Negus, R.D.M. (1966) Recent work on the effects of reflections in concert halls and music studios. *Journal of Sound and Vibration*, **3**, 127–134.

Trevor-Jones, D. (2001) Hope Bagenal and the Royal Festival Hall. *Acoustics Bulletin*, **26**, No. 3, 18–21.

Section 5.6

Allen, W.A. and Humphreys, H.R. (1951) Acoustics in the new concert halls. *Royal Institute of British Architects Journal*, December, 39–41.

Beranek, L.L. (1962) *Music, acoustics and architecture*, John Wiley, New York.

Blundell Jones, P. (1998) Harbour master: concert hall, Bristol, England. *Architectural Review*, December, 41–45.

Nelson Meredith, J. (1951) The Colston Hall, Bristol. *Royal Institute of British Architects Journal*, December, 44–46.

Somerville, T. and Gilford, C.L.S. (1957) Acoustics of large orchestral studios and concert halls. *Proceedings of the IEE*, **104**, 85–97.

Section 5.7

The Architect and Building News (1952) Free Trade Hall, Manchester. **201**, January, 7–17.

Bagenal, H. (1952) The auditorium of the Free Trade Hall. *Royal Institute of British Architects Journal*, March, 180–182.

Beranek, L.L. (1962) *Music, acoustics and architecture*, John Wiley, New York.

Somerville, T. and Gilford, C.L.S. (1957) Acoustics of large orchestral studios and concert halls. *Proceedings of the IEE*, **104**, 85–97.

Section 5.10

Beranek, L.L. (2004) *Concert and opera houses: Music, acoustics and architecture*, 2nd edn, Springer, New York.

Knobel, L. (1981) *Architectural Review*, October, 238–254.

Long, K. (2001) Caruso sings. *Building Design*, 12 October, 14–19.

Section 5.11

Beranek, L.L. (2004) *Concert and opera houses: Music, acoustics and architecture*, 2nd edn, Springer, New York.

Burd, A.N. (1982) St David's Hall, Cardiff. *Proceedings of the Institute of Acoustics, Auditorium Acoustics and Electro-acoustics meeting*, Edinburgh, September.

Forsyth, M. (1983) Building study: Cardiff in concert. *Architects' Journal*, 2 March, 53–70.

Section 5.12

Edwards, N. (1985) Design methods in auditorium acoustics. *Proceedings of the Institute of Acoustics*, **7**, Part 1, 73–79.

Forsyth, M. (1983) Building study: right Royal Concert Hall. *Architects' Journal*, 23 February, 55–68.

Glasstone, V. (1978) Royal pastiche. *Architectural Review*, May, **163**, 277–287

Thompson, N. (1982) Nottingham Concert Hall and Derngate Centre, Northampton. *Proceedings of the Institute of Acoustics, Auditorium Acoustics and Electro-acoustics meeting*, Edinburgh, September.

Trombley, S. (1982) Harmonic structure: RHWL in Nottingham. *Royal Institute of British Architects Journal*, May, 35–42.

Section 5.13

Barron, M. and Burd, A.N. (1992) The acoustics of Glasgow Royal Concert Hall. *Proceedings of the Institute of Acoustics*, **14**, Part 2, 21–29.

Beranek, L.L. (2004) *Concert and opera houses: Music, acoustics and architecture*, 2nd edn, Springer, New York.

Carolin, P. and Dannatt, T. (1996) *Architecture, education and research, the work of Leslie Martin:*

papers and selected articles. Academy Editions, London.

Martin, J.L. (1983) *Buildings and ideas, 1933–83; from the studio of Leslie Martin and his associates*, Cambridge University Press.

Martin, L. and Bakker, G. (1990) Urban design – Leslie Martin: the Glasgow Halls project. *Architecture Today*, **8**, 34–38.

Parkin, P.H., Scholes, W.E. and Derbyshire, A.G. (1952) The reverberation times of ten British concert halls. *Acustica*, **2**, 97–100.

Somerville, T. (1949) A comparison of the acoustics of the Philharmonic Hall, Liverpool and St Andrew's Grand Hall, Glasgow. *BBC Quarterly*, **4**, 41–54.

Swan, R. (1990) Concerted effort. *Concrete Quarterly*, No. 167, Winter 1990, 10–17.

Section 5.14

Beranek, L.L. (2004) *Concert and opera houses: Music, acoustics and architecture*, 2nd edn, Springer, New York.

Charles, J. and Allen, W.A. (1989) Symphonic Variations. *Royal Institute of British Architects Journal*, November, 52–55.

Collier, T. (1991) The International Convention Centre Birmingham. *Architecture Today*, April 1991, **17**, 52–68.

Cowell, R. (1991) Railway vibration isolation at the Birmingham International Convention Centre. *Acoustics Bulletin*, **16**, No. 5, 17–19.

Graham, R.N. (1992) Symphony Hall, Birmingham: a fusion of architecture and acoustics. *Proceedings of the Institute of Acoustics*, **14**, Part 2, 73–75.

Kimura, H., Matsueda, K., Shirai H. and Katsuragawa J. (1992) Research of the reverberation time of European concert halls using actual concert sound. *Proceedings of the 1992 Spring meeting of the Acoustical Society of Japan*, **2**, 765–766.

Newton, J.P. and James, A.W. (1992) Audience noise – how low can you get? *Proceedings of the Institute of Acoustics*, **14**, Part 2, 65–72.

Reid, F. (1991) Rattle and Brum. *Architects' Journal*, 1 May, 24–27.

Section 5.15

Anderson, D. *et al*. (1997) Bridgewater Hall, Manchester. *Arup Journal*, **32**, No.1, 15–20.

Harris, R. (1997) The acoustic design of the Bridgewater Hall, Manchester. *Proceedings of the Institute of Acoustics*, **19**, Part 3, 129–135.

Thompson, N. (1999) The architecture of the Bridgewater Hall, Manchester. *Proceedings of the Institute of Acoustics*, **21**, Part 6, 1–4.

Welsh, J. (1996) The city hall. *RIBA Profile*, **103/10** October 1996, 16–29.

Section 5.16

Burd, A. (1997) Belfast's Waterfront Hall. *Acoustics Bulletin*, **22**, No. 2, 21–24.

Burd, A., Haslam, L. and Stringer, S. (1997) Through tiers to smiles – Belfast Waterfront Hall. *Proceedings of the Institute of Acoustics*, **19**, Part 3, 87–94.

Carr, M. (1997) War and peace over 18 years. *Proceedings of the Institute of Acoustics*, **19**, Part 3, 81–86.

Fawcett, P. *et al*. (1997) Building study: On the waterfront. *Architects' Journal*, 6 March, 35–43.

Robinson, V. *et al*. (1997) Bravo Belfast Waterfront Hall. *Perspective (The Journal of the Royal Society of Ulster Architects)*, January.

6 Chamber music and recital halls

6.1 Introduction

Over the years, visual style for chamber music halls has ranged from the florid baroque of Mozart and Haydn's day to some very austere essays in concrete of more recent years. The visual impinges on a fundamental consideration for all auditoria: the relationship between performers and listeners. In chamber music and recital halls an essential starting point is the creation of a sense of shared experience. Intimacy, both visual and acoustic, must surely be the goal for the design of chamber music halls. A major concern is thus to limit seat capacity, ideally to around 500–600 for a hall used for string quartets, for instance.

Acoustically the situation for chamber music halls proves to be less critical than in the full-size concert hall. The risks of acoustic disaster are smaller, though many challenges remain. More than in symphonic music, one needs to be able to hear the individual musical lines. One should also hear a response from the room, but not such that it reduces clarity. The loudness of music has to be maintained, bearing in mind that a string quartet produces only one hundredth of the sound power of a full orchestra (Meyer, 1990). The number of musicians performing in a particular hall will also often vary, at the extremes from the soloist to the chamber orchestra of 25 musicians or more.

The early history of the chamber music hall has in the main already been covered in section 4.2 under the origins of the concert hall. The seminal halls such as the Altes Gewandhaus in Leipzig and Haydn's courtly halls are discussed there. Particularly in the eighteenth century the acoustic character of the performing space affected composition. The nature of chamber music was influenced by whether the music was intended for courtly performance, amateur domestic use or for a public auditorium. The acoustic character of the early performing spaces was accidental but for larger spaces with audiences the acoustic needs become more stringent.

While referring to the eighteenth century, it is well to record the oldest surviving public recital hall: the **Holywell Music Room** in **Oxford**, England of 1748 (Figure 6.1). It can boast visits by Handel, Mozart and Haydn in its early years. This charming hall was designed by Dr T. Caplin specifically for the performance of Handel oratorios (Mee,

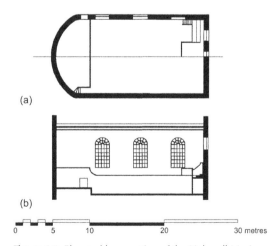

(a)

(b)

0 5 10 20 30 metres

Figure 6.1 Plan and long section of the Holywell Music Room, Oxford, of 1748

Table 6.1 Basic details of four British recital halls

Hall	Date	Seats	Auditorium volume (m³)	Reverberation time (s)	Acoustic consultant
Wigmore Hall, London	1901	544	2,900	1.5	–
Queen Elizabeth Hall, London	1967	1106	9,600	1.8*	H. Creighton
Maltings Concert Hall, Snape	1967	824	7,590	1.6*	Arup Associates (D. Sugden)
Music School Hall, Cambridge	1978	496	4,100	1.5*	H. Creighton

The quoted reverberation times are estimates of mid-frequency occupied values.
* Chamber concert with exposed curtain around stage.

1911; Bagenal and Wood, 1931) and is still used for regular concerts. The hall is rectangular in plan with a coved performing area at one end; it holds 300 in a volume of 1660 m³. Bagenal calculated the reverberation time occupied as 1.5 seconds, but this may be a slight overestimate. A parallel-sided hall of this modest size is acoustically reliable.

For acoustic design, reverberation time remains a prime consideration. Historically values between 1.0 and 1.5 seconds have prevailed, with for instance the calculated value for the 1781 Altes Gewandhaus in Leipzig being 1.2 seconds. To illustrate the discussion of chamber music halls, four British halls are considered in detail below, with characteristics summarized in Table 6.1. Three European halls are also briefly discussed in section 6.3. A sequence of three books chart examples of auditoria, both large and more modest in size, over more than 50 years: Beranek (1962), Talaske *et al.* (1982) and Hoffman *et al.* (2003).

6.2 Reverberation time and loudness

The choice of appropriate reverberation time for a recital hall is at the same time more difficult but less critical than for a full symphony concert hall. In a small hall, reflections arrive earlier and this means that maintaining satisfactory clarity should be less of a concern. For the largest symphony concert halls, the loudness of sound at distant seats is at the limit of acceptability; poor designs for 3000 seats can often be criticized for inadequate loudness. Recital halls are usually not built with the maximum

seating numbers acceptable on acoustic grounds. Visual demands and the need for a sense of intimacy appropriate to the nature of chamber music should prevent this. The suitable choice of reverberation time, as it affects loudness, should therefore be less stringent in the smaller hall.

For symphony concert halls, the recommended reverberation time is a function of programme only. For smaller halls there has been a long-standing tradition of relating the optimum time to the hall volume as well. Different sources in the literature give different recommendations and the final selected values should be influenced by experience of individual halls, as well as the acoustic intentions of the designers. A shorter reverberation time will enhance musical definition at the expense of the sense of reverberation. A long reverberation time will give a more sumptuous sound with better blend but less clarity.

The recommended reverberation time for chamber music in Table 2.3 was 1.4 – 1.7 seconds, shorter than for symphony music where blending of the orchestral sound is desirable. One could further argue that different times were needed for solo or accompanied performing as opposed to chamber orchestras. But the range of programme in a hall will depend on its overall size and the size of its stage. Larger halls will be used for larger performing groups, but probably also for solo recitals as well. Which brings one back to considering the reverberation time as a function of volume. Historically there has been much discussion of the optimum relationship. Watson in his book of 1923 made a major contribution in which he suggested, with some mathematical justification, that reverberation time should depend on the cube root of the volume ($V^{1/3}$). No simple theory will however account for all the aspects involved and experience shows that a weaker dependence on hall size is appropriate.

Cremer and Müller (1982, p. 613) suggest a relationship based on the seventh root of the volume ($V^{1/7}$), which they justify on the basis of the smallest audible changes of reverberation time and sound level. Their recommendation is plotted in Figure 6.2; lower and upper limits are presented rather than 'optimal' values to account for taste. Values in Figure 6.2 would seem to offer the best current recommendations. However when developing a design, one does not normally begin with a hall volume but with a seat count. The seat count can be used to determine the total acoustic absorption (*A*). With the relationship in Figure 6.2 between reverberation time (*T*) and volume (*V*) and the Sabine equation between *T*, *V* and *A*, a relationship between *A* and *T* can be derived, plotted in Figure 6.3.

To use Figure 6.3, one would ideally begin with a scaled seating plan, from which the acoustic seating area can be calculated (section 2.8.4). The total acoustic absorption is estimated by adding the stage area to the acoustic seating area and multiplying this total area by the Kosten equivalent absorption coefficient (= 1.14 for occupied seating at mid-frequencies, section 2.8.6). The appropriate reverberation time from Figure 6.3 can then be inserted into the Sabine equation to determine the required auditorium volume. When a seating plan is unavailable, the seating standard (i.e. true area per seat) is used to establish the true seating area. This is multiplied by a factor to determine the acoustic seating area; a factor of 1.21 is the mean of measured values in Britain. For calculations prior to the seating being selected, a mean measured acoustic area per seat of 0.58 m² could be used for **average-size seating**. In this case, the acoustic absorption per seat is 0.66 m² (= 0.58 × 1.14).

Also included in Figure 6.2 are the points for the five recital halls discussed in detail below. However, only in the case of the Berlin hall is there

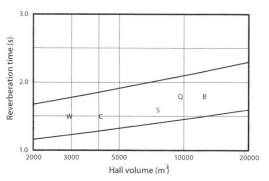

Figure 6.2 Recommended mid-frequency reverberation times for music performance as a function of hall volume (after Cremer and Müller, 1982). (Equation of the mean time is: $\log T = 0.138 \log V - 0.306$.) Individual points: **B**, Berlin Chamber Music Hall; **C**, Cambridge Music School; **Q**, Queen Elizabeth Hall, London; **S**, Snape Maltings Concert Hall; **W**, Wigmore Hall, London

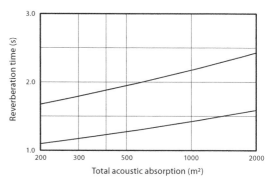

Figure 6.3 Recommended mid-frequency reverberation times for music performance as a function of total acoustic absorption. (Equation for the mean time is: $\log T = 0.160 \log A - 0.229$.)

the likelihood that the relationship was causal. The values have been inserted to allow comments on subjective behaviour to be assessed relative to Cremer and Müller's recommendations.

The reverberation time characteristic with frequency is also a necessary concern in order to maintain tonal balance. A high-frequency reverberation time close to the mid-frequency value should encourage a brilliant tone. The low-frequency time will determine the sense of acoustic warmth. The recommendations regarding frequency variation found in section 2.8.1 for larger music halls could also be applied for chamber music. Some designers, however, recommend a flat frequency characteristic in the bass for chamber conditions.

A further problem can arise in recital halls, when the stage occupancy is highly variable. Halls with large stages find themselves being used at one extreme for solo performers and at the other for full symphony orchestras. In a small hall, this can cause the reverberation time to fall considerably with a full orchestra, in the very situation in which one would like it to be longer. The difficulties of achieving variable reverberation time to compensate for this are discussed in detail in section 10.4, though in the case of smaller halls control by variable absorption is more realistic than in larger spaces. Retractable curtains are an obvious possibility, yet the location of them involves the same problems that introduction of acoustic absorbent poses in any auditorium: does the absorbent influence the acoustics on stage or does it obscure early reflections? And if so, will it matter?

Introducing absorbent also reduces the total sound level and loudness, and these must always be considered in auditoria. Limits for symphony concert halls have been established by subjective experiment, with a limiting sound level of 0dB relative to the direct sound level at 10 m from the source. What is required is the equivalent figure for recital halls of the minimum acceptable level relative to this datum, as a function of programme. In the case of recital halls, the situation is less well researched than in symphony halls, though much could be done. For instance, Meyer (1990) has provided comprehensive data on sound power from individual orchestral instruments, which it would be interesting to use to elucidate this matter. The following are rough guidelines only: a minimum level of +4 db at mid-frequencies in chamber music halls, and a level of +6 db in solo recital halls.

Excessive sound level is also possible in smaller halls, particularly when medium-size to large professional orchestras play in them (amateur orchestras tend to be quieter than professional ones). If large numbers of performers are expected on a regular basis, variable absorption especially around the stage is recommended; thick curtains are an obvious option.

As regards the other measures identified for concert halls, they will also be used here for chamber music halls; numerical differences between symphony and chamber music halls are mainly obvious. Clarity should be no problem given a suitable choice of reverberation time. A sense of source broadening is probably of lesser significance for small performing groups, but provision of good lateral reflections is a wise strategy. Acoustic intimacy depends on overall loudness and proximity to the source. One can summarize the situation in small spaces by saying that the early reflections usually look after themselves and that reverberation time is the dominant determinant of behaviour in smaller halls. Yet even here the tolerance for short or long reverberation times is greater than in large concert halls.

A valuable source of measured data has been published by Hidaka and Nishihara (2004). They present results of measurements in nine predominantly nineteenth-century European chamber music halls and nine much more recent Japanese halls. The authors also make a case for lighter, flexible stage floors that reradiate cello and double bass sound.

6.3 Room form for chamber music

As with all auditoria, careful design of the performing platform is called for to ensure that musicians can hear both themselves and their colleagues to achieve good ensemble and balance. Small chamber groups, such as quartets or solo performers, will usually prefer a small stage to a large one. But the large stage can usefully be modified by moveable reflective screens. An interesting study by Meyer (1998) concludes that moveable reflectors on the stage placed only a few metres from the players are the optimum solution for chamber music.

Regarding overall room form, many concerns remain the same for both small and large spaces for music. Echoes should be avoided though they are less likely to be clearly audible in a smaller space, where reflection delays will be shorter. Flutter echoes between parallel surfaces remain undesirable. Even colouration due to widely spaced resonant frequencies in the bass is unlikely, since minimum dimensions of chamber halls are still large compared with half the wavelength of musical frequencies.

The arrangement of audience seating contributes considerably to the character of the space. Performers have a need to feel that they are communicating to a unified audience. Good sightlines are obviously desirable, with suitably raked seating. The rectangular plan hall, though admirable in many respects, fails to create a sense of focus when it has straight rows, whereas where visual contact exists between members of the audience, this heightens the shared emotional experience.

Whether room surfaces should be plane or modulated in order to scatter reflections is an ongoing point of discussion among designers. Resolving the question is frustrated by the subtle nature of the subjective differences and the considerable difficulty of measuring diffusion. Jones (1972), an obvious champion of scattering, makes some interesting comments in a discussion about renovations to a university hall of about 450 seats. He suggests two hypotheses regarding the virtues of a diffuse space: that it sounds more live than a less diffuse space with the same reverberation time and that in diffuse halls the low frequency reverberation time need not exceed the mid-frequency value. The first hypothesis has been mentioned before, such as in the discussion of classical rectangular halls (section 4.3). In this multi-purpose university hall, a shorter reverberation time was preferable since the hall is also used for speech, though with a high-quality reinforcement system. Concerning low-frequency reverberation time, Jones was in the fortunate position of making progressive modifications. Finding that there was 'a confused noise within', additional low-frequency absorption was added resulting in a nearly flat reverberation time characteristic.

The case for scattering of early reflections is more obvious, particularly for overhead reflections. Discrete early reflections from above tend to create a shrill sound quality (due to interference effects). This can be particularly prominent with small string ensembles, with the additional risk that strong overhead reflections can enhance bowing noise. High ceilings to ensure that overhead reflections do not arrive first are therefore necessary, while rendering them scattering as well should be considered. The disadvantages of scattering treatment, as listed in section 3.10.5, are unlikely to be of sufficient magnitude to matter in a smaller hall.

The preference for good lateral reflections remains in chamber music halls. The traditional rectangular hall is in this respect near optimal for chamber use. The high ceiling, narrow hall is again appropriate to provide reflections from preferred directions. Parallel side walls encourage a high reflection density. Ratios of the basic room proportions are not critical in arithmetic terms but the gross ratios are significant. The double cube proportions of the classical concert hall may also have virtues at smaller scale.

Just as with the larger symphony hall, the fan-shape plan has little to recommend it acoustically. In larger chamber music halls, there is however a strong argument for employing a splay at the performing end, in order to bridge between the narrow stage and broader auditorium. While perhaps not ideal acoustically, its disadvantages can be mitigated either by reducing the width at the auditorium rear (giving an overall elongated hexagonal plan) or by extensive scattering treatment on the walls. Both solutions were used by the consultants Kosten and de Lange in the **De Doelen Chamber Music Hall** in **Rotterdam** of 1966 (Talaske *et al.*, 1982). The architects for the complex were Kraaijvanger, Kraaijvanger and Fledderus (Figures 6.4 and 6.5). A grid underneath the ceiling also stimulates diffusion. This hall seats 604 in a volume of 4040 m³; the reverberation time is short at 1.2 seconds.

Moving further away from the rectangular plan, the 800-seat **Mozartsaal** in the **Stuttgart Liederhalle** of 1956 is of interest (volume 5500 m³,

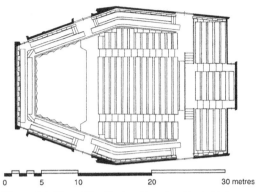

Figure 6.4 Plan of De Doelen Chamber Music Hall, Rotterdam

Figure 6.5 De Doelen Chamber Music Hall, Rotterdam

occupied reverberation time about 1.7 seconds). The plan, developed between the architects Abel and Gutbrod and acoustic consultants Cremer and colleagues, is five-sided (Figure 6.6). This avoids parallel walls and encourages spatial distribution of the late reflections (Cremer, Keidel and Müller, 1956). The particular interest of this hall is that it was the first occasion on which 'vineyard terraces' were used (section 4.9). As well as enhancing early sound reflections, it has an interesting characteristic that the hall seems large from the stage, while for the audience the performers appear very close.

An extreme in auditorium design is in the round. This form had been proposed by Scharoun for the Berlin Philharmonie, but his consultant Cremer had dissuaded him owing to the directional nature of musical instruments. For the adjacent **Chamber Music Hall of the Philharmonie**, the design was developed posthumously by Wisniewsky based on a casual sketch by Scharoun with a fully central stage (Blundell Jones, 1988). The difficulties this posed the acoustic consultants, Cremer and Fütterer, have been recorded in detail by Fütterer (1988). The Chamber Music Hall, opened in 1987, has 1064 seats with a volume of 12 500 m³ (Figures 6.7 and 6.8).

There are two major problems caused by a central stage: how to accommodate the directional nature of musical instruments and how to provide

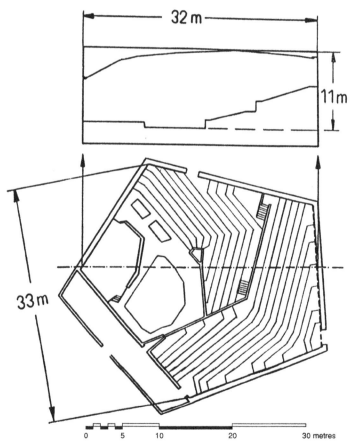

Figure 6.6 Long section and plan of the Mozartsaal of the Liederhalle, Stuttgart (after Cremer, Keidel and Müller, 1956)

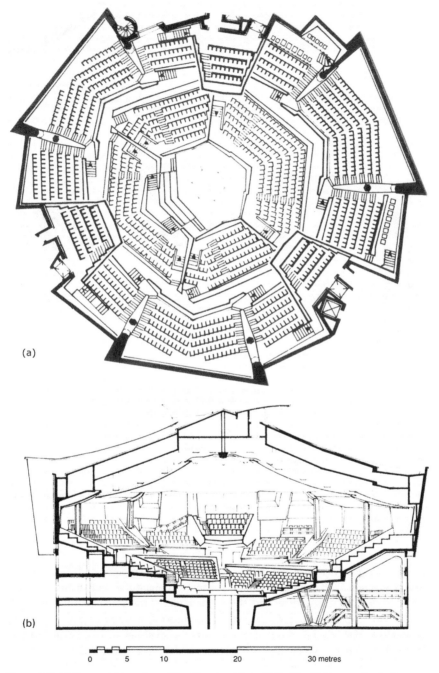

Figure 6.7 (a) Plan and (b) section of the Chamber Music Hall of the Philharmonie, Berlin

Figure 6.8 The Chamber Music Hall of the Philharmonie, Berlin

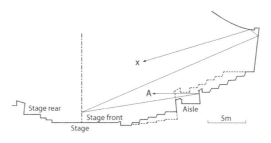

Figure 6.9 Section through the Chamber Music Hall of the Philharmonie, Berlin (solid line). Reflection A to listeners behind the performers comes off a surface of limited height, whereas the dotted proposal without an aisle offered a much larger reflecting surface. Reflection X arrives too late to be useful and is diffused by a convex soffit

early lateral reflections. Experience of the Philharmonie has shown directional difficulties, particularly with strings, which tend to radiate away from the instrument body, and especially with solo singers. In the Chamber Music Hall, though the performing area is fully in the centre of a nominally circular plan, the stage possesses a clear orientation. The smaller scale of a chamber music hall provides the possibility of some compensation for instrument directivity: reflecting surfaces placed in front of the performers can direct early reflections to audience seated behind the stage (Figure 6.9). However, the architect of the Berlin Chamber Music Hall insisted on an aisle half way up the seating, which leaves only a small height of vertical surface capable of providing such reflections.

The question of early lateral reflections in the round is more intractable. With a continuous circular seating arrangement and circular perimeter wall, no lateral reflections exist. Subdivision into vineyard terraces provides additional vertical surfaces, but it is difficult in a basically circular plan to achieve lateral reflections to most of the seats; most surfaces are either radial or tangential to a circle. Only subdivision into separate seating areas, each bounded by their own walls, is likely to succeed. Provision of lateral reflections from the ceiling is likewise frustrated by the problem that reflectors can easily obscure reflections from other reflectors. The acousticians for the Philharmonie Chamber Music Hall decided to abandon attempts at uniform early lateral reflections and chose to rely for spatial effects mostly on the later reverberation, which is rendered diffuse by the subdivision of the audience, and by scatterers and suspended reflectors in the ceiling.

Both problems of a central stage mentioned above, of instrument directivity and lateral reflections, can be clearly heard in the different seats around the stage. The reverberation time has been selected as a long 1.8 seconds, which contributes to a live, diffuse sound. As with the neighbouring concert hall, the tent-like roof provides strong reflections to remote seats and improves uniformity. The hall has been received enthusiastically for ensembles ranging from string quartets to chamber orchestras. It creates the shared experience intended by Scharoun.

6.4 Wigmore Hall, London

The interior of the Wigmore Hall displays the opulence of its age, known in Britain as the Edwardian era. Designed in the Renaissance style by T.E. Collcutt in 1901, the hall was originally built by the Bechstein piano company, and bore their name. Details include a frieze of red Verona marble running the entire length of the hall and Caribbean mahogany panelling. The platform is located in an alcove surmounted by a cupola with a coloured bas relief of 'Ideal inspiring the art world'. The hall was sold owing to bankruptcy in 1916 during the First World War (Bechstein being a German company), with the new name taken from Wigmore Street on which it lies. A visit to the Green Room offers a nostalgic experience, with its signed photographs of numerous famous artists, many of whom display touching affection for the hall in which they made their London debuts. Busoni, Rubinstein and Solomon among many others performed there in its early years.

The auditorium plan form is rectangular with a shallow balcony opposite the stage. Yet in two respects the design differs from the 'classical' model, as epitomized by the Vienna Musikvereinssaal. The ceiling is barrel-vaulted rather than flat with a curvature capable of causing focusing. Secondly the

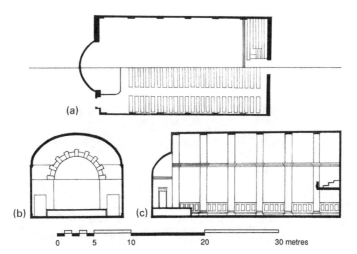

Figure 6.10 (a) Plan and (b), (c) sections of the Wigmore Hall, London

Figure 6.11 The Wigmore Hall, London

Figure 6.12 The Wigmore Hall, London

wall surfaces are not particularly scattering. The seating capacity is 544, while the stage is only 33 m² in area. This restricts use to recital situations; performances with more than six players are rare. Plans and photos are to be found in Figures 6.10–6.12.

Subjective characteristics

At a recital in 1984 for solo piano, flute and piano and finally string quartet, the sound was uniformly judged as loud and intimate. Intimacy is particularly germane to recital halls; the Wigmore Hall possesses this quality and allows the listener to savour the sound of individual instruments. Clarity was also judged as high (except at a seat position towards the rear of the stalls; though not overhung, the reason for this is not obvious). Reverberance was judged as mid-way between 'Live' and 'Dead', which may imply an optimum value. Source broadening was likewise judged as reasonable, but not

exceptional. Overall acoustic quality was rated as 'Good'. As regards background noise, underground trains are unfortunately audible in quiet passages here. Several listeners also mentioned the hiss of gas lamps which were then still in place at the rear of the hall!

Though not revealed explicitly from statistical analysis of the questionnaire results, the sound character in the balcony of this hall is distinct. Whereas judgements of intimacy normally diminish towards the rear of a hall, the balcony location in the Wigmore Hall was judged as the most intimate. For the listeners as a group, this was the preferred seat overall, though some individuals favoured the stalls.

Objective characteristics

With a measured reverberation time of 1.6 seconds, and predicted value of 1.5 seconds occupied, the

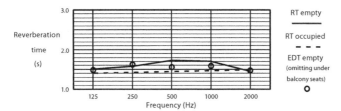

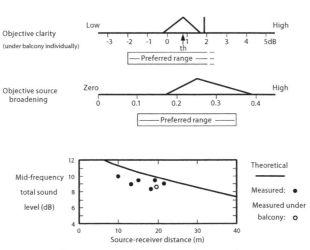

Figure 6.13 The Wigmore Hall, London – objective characteristics

volume of the hall appears to be well chosen for an appropriate balance between clarity and reverberance. Variation of reverberation time with frequency is small (Figure 6.13). Average objective clarity is as expected, but high values occur in the balcony (and more predictably below it). The reason for high clarity in the balcony is obvious from measured impulse responses: these balcony positions receive intense early reflections which are louder than the direct sound. Focusing by the barrel-vaulted ceiling is clearly the cause, but with reflection delays of 30 ms or less no echo is apparent. Within the listener group, only one member seemed worried by the harsh sound quality which can be perceived in these circumstances.

Total sound is a little less than expected, for reasons that are not clear. (Perhaps the concave ceiling focuses more sound onto the absorbent audience than normally occurs, thereby starving the later reverberant sound.)

Conclusions

The Wigmore Hall was built as a commercial showcase but now functions as London's prime recital hall. Being built before Sabine's work was known, the choice of hall volume, and therefore reverberation time, was either a fortunate coincidence or inspired by comparable precedents. The hall is rightly well loved as an instrument for the soloist and small chamber group. In rational acoustic design terms one would criticize the concave ceiling and indicate a preference for some more scattering treatment on the room surfaces. Yet the subjective implications of these deficiencies are minor in a space of this modest size.

6.5 Queen Elizabeth Hall, London

The generous foyer spaces under the belly of the Royal Festival Hall auditorium bear witness to the original intention to place a second smaller hall within the same building. This plan was however abandoned when the site to the side of the Festival Hall became available. The Queen Elizabeth Hall and Purcell Room (a 372-seat recital hall) were completed in 1967 to designs by the then Greater London Council. The contrast between old and new is startling. Peter Moro summarized the distinction in external character (*Architects' Journal*, 1967a): 'The old concert hall is an extrovert structure, more transparent than solid: the new one is solid and introspective'. Perhaps it is necessary in life for extremes to be tried before they are recognized for what they are. The 'brutalist' architecture of the Queen Elizabeth Hall was 'architects' architecture' in 1967 but it has found little favour among laymen (C. Jencks, 1968; P. Moro *et al.*, 1968). Thirty years later the architecture has few defenders and one option in the on-going saga of redevelopment of the South Bank during the 1980s and '90s was demolition. Yet the choice of this particular architectural style does not have many acoustic ramifications, apart from the difficulty of including scattering surfaces when surface decoration is viewed as superficial excess.

The Queen Elizabeth Hall seats 1106 and was intended as a chamber music hall to complement the Royal Festival Hall. After the trials with reverberation time in the older hall, the designers were determined to ensure the same mistake was not repeated. Thus somewhat paradoxically the reverberation time in this chamber music hall is longer than in its neighbour, the symphony concert hall. Seating in the auditorium is on one continuously raked block. In front of the main cross-aisle, the plan is fan-shape (apex angle 28°) and parallel-sided beyond it. In an attempt to mitigate against the assumed disadvantages of the fan shape (section 4.5), the inclined walls have stepped parallel facades, though this only begins above head height. The ceiling is nomi-

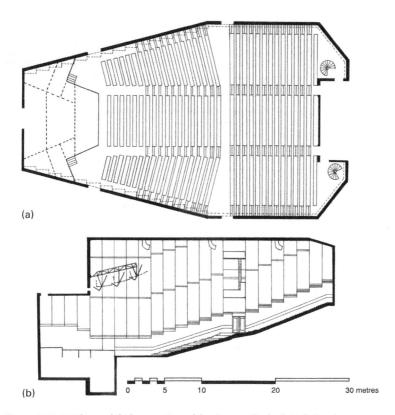

(a)

(b)

Figure 6.14 (a) Plan and (b) long section of the Queen Elizabeth Hall, London

nally flat and horizontal with great ventilation inlets, intended to provide some acoustic scattering.

A major point of discussion following the opening of the Festival Hall and its contemporaries was the acoustic merits of an overstage reflector to reflect sound onto the audience (section 5.5). Their introduction was due to concern about inadequate clarity at the rear of the new large concert halls, but some acousticians criticized the harsh quality they imparted to the sound. In the Queen Elizabeth Hall the reflector is movable between a reflecting and non-reflecting position. Subjective tests conducted at the time of opening suggested that the audible effect of the reflector was small but that in general the preference was for the vertical, no-reflection condition. For solo piano however the reflector was favoured in its down position. (The fact that the subjective effect of this reflector proved to be small is

not wholly surprising given the relatively small size and narrow width of the hall, which creates other reflections of similar delay to that of the reflector.)

With a massive auditorium shell (the walls here are 375 mm thick concrete), the bass reverberation time will rise. For symphony concerts many consider this to be highly desirable but for chamber music the argument for a bass rise is less clear. Low-frequency absorption is provided in the Queen Elizabeth Hall by 2300 Helmholtz resonators built into the boxes lining the walls (Figure 6.17). The internal resonator volumes are all the same but the holes linking them to the auditorium volume are of four sizes, providing four frequencies of maximum absorption. The resonators are well damped to spread their absorption (in technical parlance they have low Q); they are effective absorbers at frequencies below 250 Hz.

Figure 6.15 The Queen Elizabeth Hall, London

Figure 6.16 The Queen Elizabeth Hall, London

Figure 6.17 Helmholtz resonator absorbers in the Queen Elizabeth Hall, London

Subjective characteristics

At a concert in 1984 by a chamber orchestra tested in four locations, the judgements overall were 'Good'. But the results were distinctive in having marked perceived differences between seats with regard to clarity and intimacy. Such subjective non-uniformity is unusual among the concert halls tested in the survey. Predictably the position at the front was judged as the most clear and intimate, and these seats (in the ninth row) were judged as 'Very Good'. The position at the rear of the hall was least liked. Overall reverberance and loudness were judged as somewhat above average, while source broadening was rated as near average.

It is clear from listeners' comments that many responses are individual (one listener greatly preferred the rear location). To the author's ears, there is at the front a sense of presence verging on

transparency which can be very exciting. At the rear the clarity seemed inadequate for fast passages, with excessive reverberance. Several listeners when sitting at the front commented that the reverberation came from behind.

Objective characteristics

A long reverberation time has certainly been achieved in this hall: 2.1 seconds measured unoccupied with 120 m² of curtain around the stage (Figure 6.18, measured in 1982); 2.4 seconds as quoted in 1967 (*Architects' Journal*, 1967a, presumably without the curtains extended). These unoccupied values are longer than prediction, which may be due to limited absorption by the leather-covered seats. An alternative explanation is inadequate diffusion, but the available data do not allow confirmation of this. The occupied mid-frequency value was measured in 1967 as 1.9 seconds (*Architects' Journal*, 1967a), though the curtain will presumably reduce it further (hence the figure of 1.8 seconds is used in Figure 6.18). The marked dip in the unoccupied characteristic at 250 Hz appears to be associated with the seating, since it is not evident in the 1967 occupied curve.

Of the other results shown in Figure 6.18, only the total sound behaviour draws comment. Measured total sound is on average 1.7 db less than expected and the cause is very likely to be the presence of the absorbing curtain around the stage. But while the late sound is uniformly reduced, the early sound is weakest around and behind the main cross aisle. This is a known characteristic of fan-shaped halls, but it is surprising to encounter it here with only a shallow angle of fan and stepped upper side walls. The early sound level near to the stage is much closer to expectations, which may go some way to explaining the varied subjective response at different locations in this hall.

Measurements have also been made of the effect of lowering the reflector. As expected the presence of an extra overhead reflection increases objective clarity and sound level. The magnitude of the change is unexceptional with a maximum of about 1 db in each measure at any seat.

Conclusions

Investigation of the acoustics of the Queen Elizabeth Hall throws up two interesting questions: is there adequate diffusion of the sound in this hall and does the part-fan shape in plan, modest though it is, have acoustic implications? Both questions are significant for acoustic design but neither can be answered without further study. One might be more categorical about the stage design. The choice of concrete around a stage is particularly unfortunate visually and ever since it opened a curtain has usually covered the rear wall. What

should have been a variable element for acoustic purposes has become fixed. A less austere-looking stage enclosure which included more scattering surfaces would be desirable.

Though the architecture of the Queen Elizabeth Hall has received much criticism, this has never extended to its acoustics, which certainly after the Festival Hall are responsive and kinder to performers. Successfully operating the hall commercially has been more problematic and one response since 1988 has been to use the hall for more operatic performances. Two massive lighting bridges now hang from the ceiling and more absorbing drapes have crept into the platform area. These additions are unlikely to have improved the acoustics.

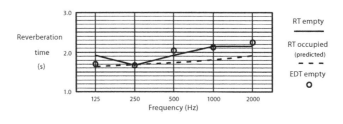

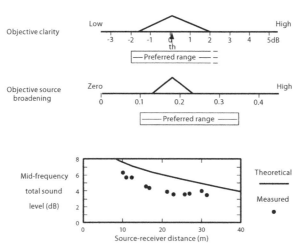

Figure 6.18 The Queen Elizabeth Hall, London – objective characteristics with the reflector in the upward position. Predicted RT with 30 players on stage

6.6 The Maltings Concert Hall, Snape, Suffolk

Set next to the River Alde, overlooking reed marshes which stretch to the coast, the Snape Maltings must surely rank as having the most beautiful natural location of any auditorium in the UK. It owes its position to the Aldeburgh Festival, begun in 1948 by Benjamin Britten and Peter Pears. The Festival had grown from an originally intimate affair to larger attendances and more ambitious productions. Neighbouring churches and the small Jubilee Hall in Aldeburgh were becoming unsuitable, especially for secular concert works. The then disused Maltings forms part of a large complex of mid-nineteenth-century buildings. Its conversion into a concert hall was sensitively undertaken by Arup Associates in 1967 (it was re-opened in 1970 after being gutted by fire). The simple lines and straightforward finishes in generous-sized spaces

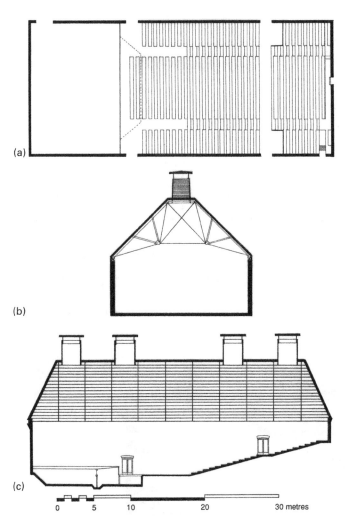

Figure 6.19 (a) Plan and (b), (c) sections of the Maltings Concert Hall, Snape

Figure 6.20 The Maltings Concert Hall, Snape

Figure 6.21 The Maltings Concert Hall, Snape

offer a delightful environment for the high-quality music provided within. The hall remains the major performing space for the Festival.

The auditorium is rectangular in plan, using existing red brick walls which were grit-blasted and sealed. To achieve a suitable internal volume the walls were extended upwards by 1 m. The roof structure was completely new with a 45° gabled section. For the roof, two layers of 25 mm tongued and grooved timber boarding were used, set at 45° relative to one another. The timber roof trusses and steel ties are all exposed on the auditorium interior, yet the construction is stiff enough to ensure little low-frequency absorption. Externally the roof is hung with tiles. The hall contains 824 seats on a single rake. The seats, inspired by those at Bayreuth, are cane on light wood frames. The large stage is cantilevered, offering space underneath the stage front for use as a pit for opera performances. Though the hall is used for various purposes, it was fundamentally constructed as a concert hall but has also proved successful for opera.

In acoustic terms, the exposed roof elements can be expected to contribute to good diffusion. Use of seats without upholstery is unusual in modern auditoria, as it results in large changes of reverberation time with occupancy. The mid-frequency extremes with and without audience were measured in 1967 (*Architects' Journal*, 1967b) as 2.0 seconds and 3.5 seconds. The latter value presents problems of excessive liveness in rehearsal conditions. Since the opening, two variable elements have been introduced in the stage area: an extensive 230 m^2 heavy curtain that can surround the whole stage on three sides and a set of substantial scattering timber screens, which can be placed behind the performers. This combination of absorbent material and reflectors on stage is commendable, since it allows control of hall reverberation without jeopardizing the players' environment.

Subjective characteristics

The acoustics of the Snape Maltings are widely reputed to be good. Forsyth (1987), for instance,

refers to the 'acoustic warmth, presence and clarity – perhaps unsurpassed in modern halls'. Excellence of this calibre does however need to be placed in perspective. It is probably true that the Maltings has the best acoustics of the four halls discussed in this chapter and incidentally the results of the author's subjective survey concurred with that view. But achieving good acoustics for an 800-seat hall is much easier than for a full-size concert hall. One could argue that many other halls of similar size have recognizably non-ideal elements. The Maltings is a modest-size hall, with a suitable reverberation time and a plan form acknowledged for providing good acoustics. One is left with the question of whether other aspects in its design make a special contribution to its success; the gabled cross-section is clearly an unusual feature.

In 1985 a performance by a string quartet plus piano was sampled. The curtain was extended round the stage, but the scattering reflecting screens were placed about half-way back on the stage, well behind the performers. As with the 'best' concert halls discussed in Chapter 5, the Maltings scored well on each of the individual scales on the questionnaire: clarity, reverberance, source broadening, intimacy and loudness. Responses on each of the scales were similar for the two positions tested (in the middle and rear of the front Stalls seating areas), except in the case of intimacy, which was significantly higher at the position closer to the stage. In the case of clarity, the range of response at individual positions was large, indicating that for some listeners it was less than optimum.

The judgements of overall quality were both 'Very Good'. This too should be placed in context. The sampled position in the Queen Elizabeth Hall (section 6.5) which was closest to the stage was similarly judged. The Maltings surpasses the London hall by being uniformly impressive. In more qualitative but personal terms, the hall offers a highly blended, rich spatial experience, with a warm bass sound. The dynamic response is also excellent, soloists can command a magical silence in this auditorium. Both early and late reflected components are audible as they should be. The hall performs well for chamber groups and chamber orchestras. With a full symphony orchestra the sound is, as one would expect, particularly loud; this delights some listeners but would be considered excessive by others. The hall has a reputation for good balance and blend for opera.

Objective characteristics

The fact that the seats are not upholstered in the Maltings creates problems for extensive objective measurements. Measurements were therefore made with compliant audiences of school-children, though unfortunately the occupancy was less than complete. Less confidence should therefore be

placed in these results than with other halls. Benjamin Britten tended to err on the side of longer reverberation times. A measurement undertaken in 1967 (*Architects' Journal*, 1967b) shows an occupied mid-frequency reverberation time of 1.9 seconds. It is interesting to observe that a substantial curtain has now been added around the stage, which when extended allows the reverberation time to be reduced to about 1.6 seconds. This was the condition for the concert mentioned above.

In other respects, behaviour is much as expected on the basis of the hall volume and reverberation time. The objective source broadening is predictably high due to the narrow hall width. But in the

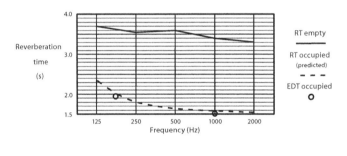

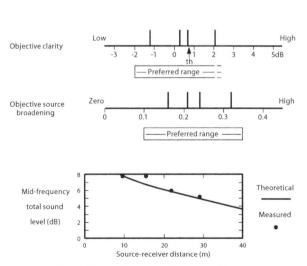

Figure 6.22 The Maltings Concert Hall, Snape: objective characteristics. Empty reverberation time (RT) from Architects' Journal (1967b). Predicted occupied RT with curtain across rear stage wall and five performers on stage

case of the total sound level (Figure 6.22), the close agreement between measured and predicted is surprising. At the time of the objective measurements in the Maltings, the stage curtain was fully extended but no sound scattering screens were present. Both in the Queen Elizabeth Hall and Cambridge Music School Auditorium discussed in this chapter, the total sound is reduced by an absorbent stage curtain. (This behaviour has also been observed in an acoustic model of the Barbican Concert Hall (section 5.10), tested with and without absorbent in the stage area.) There is no obvious explanation for why the fully exposed curtain in Snape Maltings does not appear to influence the total sound level here. Whether the form of the Maltings is significant in this respect is as yet an unanswered question.

Conclusions

The Snape Maltings are widely acknowledged to provide listening conditions among the best in the UK. It is always difficult though to establish reasons for excellence. The small size, the narrow rectangular plan form and appropriate reverberation time all contribute to an acoustic space which would be considered near optimal by most acoustic specialists. A major question remains to be answered: is the cross-section form significant, or would an equal volume space with a flat horizontal ceiling perform as well? A model study could perhaps illuminate this point.

6.7 Faculty of Music Auditorium, Cambridge

Halls in schools, colleges and universities have to house a wide range of activities. In the case of the Cambridge University Music Faculty Auditorium by Leslie Martin, Colen Lumley and Ivor Richards, the uses include teaching, music rehearsal and public performance, for groups ranging from soloists to full orchestra and choir (Martin *et al.*, 1978; Martin, 1983). Such formidable diversity, particularly when coupled with a restricted budget, involves some acoustic

compromises. But these compromises can be acceptable in halls with modest seating numbers.

The Music School Auditorium houses a maximum audience of 496 in a main raked seating area with additional loose seats in 'boxes' in stepped side galleries. To maximize internal volume for acoustic purposes, the roof structure, lighting and much ventilation ducting are contained within the ceiling space of the hall. Visual screening above the audience is provided by an egg-crate structure formed in plywood. Natural lighting was considered desirable for the teaching condition; it penetrates from the rear sides of the hall.

Considerable flexibility is available for the stage, mostly operated by hand. The level of the stage and gangway are identical, which allows the front three rows of loose seats to be covered in to extend the stage (Figure 6.26). Between this area of loose seating and the main stage is a further set of removable panels over a pit for small opera performances. A rectangular curtain track allows a heavy curtain to be used across the rear stage wall or brought forward to act as a stage curtain. Alternatively it can be bunched behind movable triangular vertical towers (*periaktoi*). As a complement to floor-level adjustability, there are rotating panels over the stage. They can act as acoustic reflectors for music performance or, when placed vertically, allow for stage lighting or the hanging of minimal scenery.

The form of the hall is basically rectangular in plan. The cross-section includes a part-gabled soffit. Both features are considered acoustically beneficial. Surfaces, except for the curtain, are hard apart from the gallery sides, which are of timber panelling. The elements of major acoustic significance are the auditorium volume and the nature of the ceiling space, with its ducting and structural elements. The seating is lightly upholstered and less absorbent acoustically than normal. Reverberation time variation can become severe in a hall like this, where the performing area is large relative to the audience but may be occupied by between one and, say, 70 performers. Unfortunately the variation is in the opposite direction to the optimum: a longer reverberation time is appropriate to a symphony orchestra,

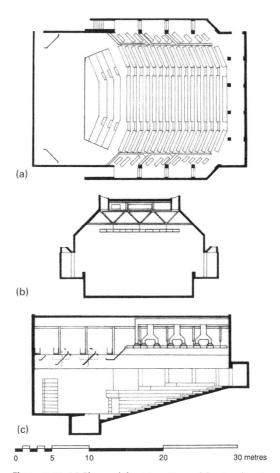

Figure 6.23 (a) Plan and (b), (c) sections of the Faculty of
Music Auditorium, Cambridge

yet the large performer numbers will give this condition the shortest time.

Subjective characteristics

The acoustics in this auditorium have been
sampled at two concerts in 1985: the first with a
performance by a string quartet and the second
by a symphony orchestra. In the case of the string
quartet, the audience was only about one quarter
of full capacity. With lightly upholstered seats the
reverberation time is very sensitive to occupancy.
The front three rows were covered over at this

concert giving a high estimated reverberation
time of 1.95 seconds. Reverberance was predictably judged as high, yet clarity was also universally
judged as adequate. Nuances of tone and dynamics projected well. For some listeners the room
response was deemed as excessive, though overall
judgements were on the boundary of 'Good'. Intimacy, envelopment and loudness were all judged
well. Characteristics were similar at the two test
locations.

For the orchestral concert, the stage with
extension was fully occupied by performers and
the audience was nearly full capacity, giving a

Figure 6.24 The Faculty of Music Auditorium, Cambridge

Figure 6.25 The Faculty of Music Auditorium, Cambridge

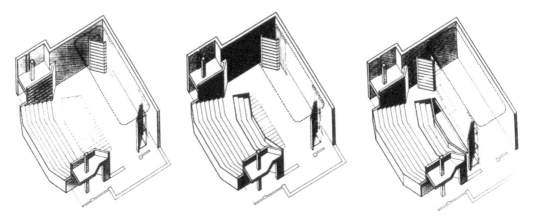

Figure 6.26 Different stage arrangements in the Faculty of Music Auditorium, Cambridge

predicted reverberation time of 1.35 seconds. Reverberance, in this case, was judged lower, but not radically so. Clarity was also surprisingly rated lower than with the quartet, but this may be due to a disappointing performance. In other respects responses were similar, with again a 'Good' assessment overall. Such a response to a symphony concert with such a short reverberation time is remarkable. A reasonable interpretation is that the proximity to the orchestra, the high degree of intimacy and probably most significantly the loud sound all compensate for the otherwise rather dead acoustic.

Objective characteristics

The measured reverberation time without audience is affected by the light upholstery on the seats. The measured value of 2 seconds was taken with curtain extended across the rear stage wall. If the seating was fully upholstered, it is estimated that the reverberation time would be as much as 0.3 seconds shorter. The estimated occupied time with a small chamber group and full audience is 1.5 seconds. Sensitivity to occupancy, both by audience and performers, is clear. The frequency characteristic of the reverberation time drops somewhat in the bass in the empty measured situation, which is probably appropriate for this type of hall.

As regards other measures, early decay time (EDT) is slightly higher and objective clarity slightly lower than expected (Figure 6.27). Low objective clarity was measured in the 'boxes' due to obscured sightlines. Objective source broadening is high, in line with the narrowness of the hall. The effect of the curtain is seen on the total sound level; it produces an average reduction of 1.3 db (with the effect on the early and late parts very similar).

Discussion and conclusions

The results clearly indicate a successful hall for a wide variety of uses. Given that the reverberation time was predicted to vary considerably with occupancy, choice of auditorium volume was important to ensure that the highest and lowest reverberation times encountered are acceptable. That choice of volume appears to have been well made.

Further comments can be made about the design. A curtain offers a valuable variable acoustic element; in this hall it can produce a change in reverberation time of about 0.2 seconds. Yet owing to its location it also controls conditions for the musicians, who in this hall generally seem to prefer to have it obscuring the rear stage wall. Separating the functions of controlling the reverberation time and controlling the stage acoustic environment by having two or more curtains would be preferable. It

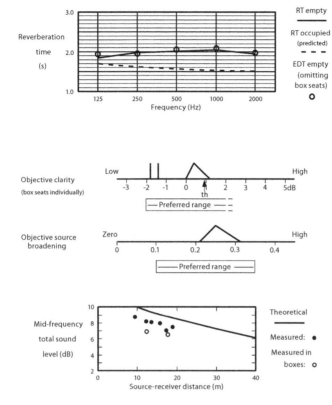

Figure 6.27 The Faculty of Music Auditorium, Cambridge: objective characteristics with curtain across rear stage wall. Predicted occupied RT with small chamber group on stage

is also likely that musicians would prefer more scattering on the surfaces around the platform.

Another question which arises is: why do the acoustics not score as highly as Snape Maltings, when they both have a similar plan form and cross-section? There is no obvious answer to this, though the nature of the elements in the ceiling of the Cambridge design may be significant. The acoustic effect of the egg-crate structure deserves attention. Results of tests on a model version of slats of similar dimensions but only going in one direction are shown in Figure 6.28. Low frequencies bend round such structures with ease but at high frequencies the acoustic transparency was measured as worse than the visual. High frequencies will be reflected by the egg-crate up into the roof space but **backwards**

towards the stage, leaving a reduced level of ceiling reflection to the audience. Objective evidence that this is significant for the audience is only slight in this hall (the higher EDT and lower objective clarity noted above are consistent with these observations). The case is far from proven, but use of such egg-crate type details in larger halls should be treated with caution.

The Cambridge Music Faculty Auditorium has been designed with a maximum reverberation time of about 2 seconds for low-occupancy situations. This was judged as acceptable even for a string quartet. With lightly upholstered seating the variation of reverberation time is large, though the lowest extreme of around 1.4 seconds was also considered satisfactory. If funds had allowed more

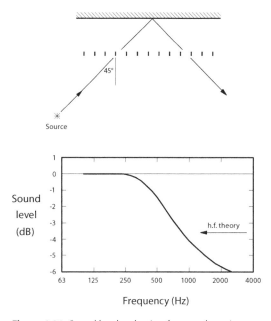

Figure 6.28 Sound level reduction for sound passing at 45° angle through regular slats, 280 mm high spaced 900 mm apart, with reflection off hard ceiling. At high frequencies, behaviour is worse than predicted according to the fraction of open area (high frequency theory). Result derived from model measurement

upholstery, a smaller acoustic variation could have been achieved. Though the design involved this and other compromises, it provides a valuable model for successful acoustic design of a small hall. A very similar design auditorium has been used by the same architects for the Royal Scottish Academy of Music and Drama in Glasgow (Stonehouse, 1988).

6.8 Conclusions

It is almost a platitude among acousticians that designing a small auditorium is easy and much less demanding than a full-size symphony concert hall. The objective reasons for this are that the early reflections are usually plentiful and look after themselves, whereas performers and listeners tolerate a wider range of reverberation times in small halls. Acoustic failures among chamber music halls are rare. Yet this should not engender smugness. There

remains the need to project faithfully the nuances of tone which individual instruments possess. A particular delight of chamber music is the harmonic interplay of individual parts; these should be fully audible. The whole should be an intimate experience involving close rapport between performer and audience.

Good acoustic design involves the inevitable selection of an appropriate reverberation time, which determines the required ceiling height. Room form contributes to the sense of occasion created, as well as the acoustics of the hall. The conservative designer will select a rectangular plan, known to work reliably well. More original forms have advantages for larger audiences, particularly in promoting a shared experience. It is a brave design team which embarks on 'music in the round', owing to the directional nature of many musical instruments.

The value of sound scattering treatment is as much a question of debate for small as it is for large halls. Adequate diffusion remains however a generally accepted goal. Detailed discussion of four chamber halls has shown that the acoustic effects of unusual features should be considered carefully, particularly in the case of suspended elements.

References

General

Beranek, L.L. (1962) *Music, acoustics and architecture*, John Wiley, New York (reprinted 1979 Krieger Publishing Co., Huntingdon, New York)

Forsyth, M. (1985) *Buildings for music*, Cambridge University Press.

Hoffman, I.B., Storch, C.A. and Foulkes, T.J. (2003) *Halls for music performance: another two decades of experience 1982 – 2002*. Acoustical Society of America, Woodbury, New York.

Lord, P. and Templeton, D. (1986) *The architecture of sound*, Architectural Press, London.

Meyer, J. (1990) Zur Dynamik und Schallleistung von Orchesterinstrumenten. *Acustica*, **71**, 277-286.

Meyer, J. (2009) *Acoustics and the performance of music*, 5th edition. Springer, New York.

Talaske, R.H., Wetherill, E.A. and Cavanaugh, W.J. (eds) (1982) *Halls for music performance: two decades of experience 1962-1982*, American Institute of Physics, New York.

Sabine, W.C. (1922) *Collected papers on acoustics*, Harvard University Press (reprinted 1964 by Dover, New York).

References by section
Section 6.1

Bagenal H. and Wood, A. (1931) *Planning for good acoustics*, Methuen, London.

Mee, J.H. (1911) *The oldest music room in Europe*, John Lane, London.

Section 6.2

Cremer L. and Müller, H.A. (translated by T.J. Schultz) (1982) *Principles and applications of room acoustics*, Vol. 1, Applied Science, London.

Hidaka, T. and Nishihara, N. (2004) Objective evaluation of chamber music halls in Europe and Japan. *Journal of the Acoustical Society of America*, **116**, 357-372.

Watson, F.R. (1923) *Acoustics of Buildings*, Wiley, New York.

Section 6.3

Blundell Jones, P. (1988) Scharoun: Kammermusik. *Architectural Review*, March, **183**, 42-51.

Cremer, L. Keidel, L. and Müller, H. (1956) Die akustischen Eigenschaften des grossen und des mittleren Saales der neuen Liederhalle in Stuttgart. *Acustica*, **6**, 466-474.

Fütterer, T. (1988) The acoustics of the new Chamber Music Hall in Berlin (West). *Proceedings of the Institute of Acoustics*, **10**, Pt. 2, 339-353.

Jones, D.K. (1972) Design of a medium-sized music auditorium. *Applied Acoustics*, **5**, 83-90.

Meyer, J. (1998) Acoustical aspects of chamber music for strings. *Proceedings of the 16th International Congress on Acoustics, Seattle, Vol. III*, 1481-1482

Section 6.5

Architects' Journal (1967a) Concert halls on the South Bank, London. 26 April, 999-1018.

Jencks, C. (1968) South Bank Arts Centre. *Architectural Review*, July, **144**, 14-30.

Moro, P. *et al.* (1968) The Queen Elizabeth Hall: an appraisal. *Royal Institute of British Architects Journal*, June, **75**, 251-257.

Section 6.6

Architects' Journal (1967b) Concert hall for the Aldeburgh Festival of Music and the Arts. 13 September, 687-691.

Forsyth, M. (1987) *Auditoria: designing for the performing arts*, Mitchell, London

Section 6.7

Martin, J.L. (1983) *Buildings and ideas, 1933-83; from the studio of Leslie Martin and his associates*, Cambridge University Press.

Martin, J.L. *et al.* (1978) Faculty of Music, University of Cambridge. *Architectural Review*, July, **164**, 18-23.

Stonehouse, R. (1988) Theme and variations. *Architects' Journal*, 11 May, 46-68.

7 Acoustics for speech

It is a common observation that good acoustics for speech and music are generally incompatible. Natural speech communication to the back of a concert hall is difficult, whereas music performed in a theatre sounds lifeless and dead. Thus far these differences can be explained just on the basis of the reverberation time. Optimum values of 1 second for speech and 2 seconds for symphonic music are often quoted. For good music acoustics, several different considerations need to be addressed of which the reverberation time is only one. The situation is simpler with speech. If speech is intelligible and background noise is not intrusive, dissatisfaction with finer points is unlikely. Many spaces with a reverberation time of close to 1 second work well for speech. This situation has led to a certain complacency in acoustic design of theatres. Several recent theatres, at least in Britain, have been built with poor speech intelligibility.

Perhaps a reflection of this complacency is the remarkable absence of literature discussing acoustic design of theatres. In fact, the basic principles are simply stated (Moore, 1964, provides a rare presentation). Concern for reverberation time and the elimination of echoes are standard requirements for any auditorium. Good sightlines are important for acoustic as well as visual reasons. Likewise keeping the audience close to the stage is preferable both acoustically and visually. The major aspect which is specifically acoustical is the concern for adequate early reflections. The need for these is strongly influenced by the type of theatre. In proscenium theatres where the actors and most of the audience face

each other, the direct sound is often strong enough for good speech conditions. But in theatres where audience may be behind the actor, intelligibility can easily become a problem. To maintain the distance range for intelligible speech, surfaces must be oriented to provide early acoustic reflections. Nevertheless in these surround-type theatres the usable distance range is generally smaller than in a frontal proscenium-type theatre.

The problem with theatre design is knowing the limits for satisfactory acoustics. The boundary between good and poor intelligibility proves to be a fine one and prediction simply from drawings is difficult. It is necessary to begin by discussion of the particular characteristics of human speech.

7.1 The nature of speech

Speech in any language consists of vowels and consonants. Voiced sounds, which include vowels, are produced by air from the lungs being pushed through the vibrating vocal chords. This broadband frequency sound is filtered by resonant cavities in the throat, mouth and nose to give it a spectrum that in part characterizes an individual's voice. Movement of the tongue and lips enables different vowel sounds to be produced. This is a relatively efficient way of generating sound, so that vowels are consistently louder (by 12 db on average) than consonants. Consonants are all impulsive in character; they can be either voiced or unvoiced with the latter being particularly quiet. Their duration is shorter than vowels. Because discussion will be turning to the value of reflections on the basis of their delay in ms, actual durations deserve mention: vowels commonly last for 90 ms and consonants 20 ms. Unfortunately the greater energy in vowels makes masking of consonant sounds by vowels a common occurrence in spaces with late reflections or long reverberation times (Figure 2.13).

Speech involves a wide range of frequencies, from below 125 Hz to above 8000 Hz. The male voice obviously contains more low-frequency energy than the female voice. This occurs because of a longer opening in male vocal chords, which produce a lower 'driving' frequency to the voice system. Figure 7.1 shows the spectrum of an actor's voice, which happens to be typical for normal speakers as well. Two curves are included in Figure 7.1, because there are two ways of presenting the same data. The most common way is to place a microphone in front of a speaker in an acoustically dead environment and measure the sound pressure level. This is the spectrum important for normal conversational communication and, of course, for sound system design. But in rooms we generally receive more reflected sound than direct sound so another factor has to be considered: that the human speaker is directional. Human speakers, loudspeakers and most other sound sources are virtually omni-directional at low frequencies and become progressively more directional at higher frequencies. The second spectrum in Figure 7.1 is a sound **power** spectrum, representing the total amount of sound power generated by the speaker in all directions that enters the room.

The high-frequency components are relatively weaker in this case, because they are concentrated in front of the speaker. Consonants contain mainly high-frequency energy.

While discussing the spectrum of the voice, Figure 7.2 shows an interesting comparison of spectra for a single actor speaking at two different voice levels (the sound power levels are 73 and 79 dBA re. 10^{-12} W). The low frequencies have remained the same, whereas the spectrum at 500 Hz and above has been uniformly increased. There is little intelligibility information at low frequencies so the louder voice spectrum is particularly suitable for good projection. The amount of sound power in the voice is minuscule, typically around 0.1 mW (milliwatt); an actor's voice being slightly louder than normal speech level. Voice power varies widely between individuals.

In the open air and in large spaces, such as theatres, the directivity of the human voice becomes important. It is unfortunate that it is the high frequencies which are most directional because these are the most important for intelligibility. The normal method to present directivity is in the form of a polar plot; Figure 7.3 is the polar plot for mean speech frequencies (derived from Moreno and Pfretzschner, 1977). A more approachable method to present the same data is to draw a contour of constant sound pressure (Figure 7.4). This has been drawn with a labelled maximum of 50 units to allow comparison with a famous quotation by Sir Christopher Wren in a discussion on church design:

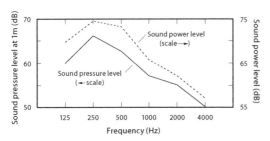

Figure 7.1 Frequency spectrum of an actor's speech presented as a sound pressure level measured on-axis 1 m in front of the speaker (solid line, l.h. scale) and as a sound power spectrum (dashed line, r.h. scale)

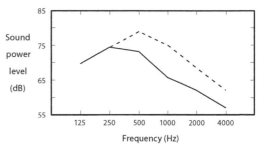

Figure 7.2 The (sound power) frequency spectrum of an actor at two voice levels

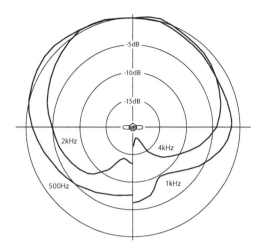

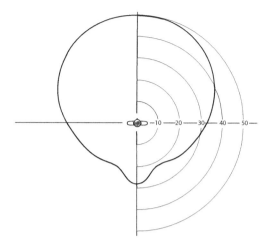

Figure 7.3 Polar diagram for a human speaker at different frequencies (third-octave results after Moreno and Pfretzschner, 1977)

Figure 7.4 Constant sound pressure contour for a speaker, based on results in Figure 7.3 averaged at 500 Hz, 1 kHz and 2 kHz

Concerning the placing of the pulpit, I shall observe – a moderate voice may be heard 50 feet distant before the preacher, 30 feet on each side and 20 behind the pulpit; and not unless the pronunciation be distinct and equal, without losing the voice at the last word of the sentence, which is commonly emphatical, and if obscured spoils the whole sense.

In fact the comparison between the sound pressure level contour, which is relevant in the open air, and the situation in an enclosed church is not entirely correct because the enclosed space can provide useful early reflections to improve matters. Wren was however talking of a general church environment, where early reflections would often not be present. The comparison between Wren's values and those in Figure 7.4 is quite close. The simplicity of this type of experiment has commended itself to several people. Some further examples are mentioned below.

There are two reasons for the directivity of the human speaker: the finite size of the mouth and the location of the mouth in the head. However, the mouth is small relative to the wavelengths of most speech sounds and the shadowing by the head is

found to be the major concern. In Greek theatre, face masks were traditionally used that may have enhanced the voice level in the manner of a megaphone. Part of this enhancement would be due to the mask making the voice more directional.

Several of the characteristics of human speech mentioned here are illustrated in Figure 7.5, which shows sound levels at a low and high speech frequency for a short phrase. Both the levels in front of and behind the speaker are given. At 500 Hz the energy is predominantly due to the vowels, which are of relatively long duration. There is only a small difference between the level in front of and behind the speaker at this frequency. At 4 kHz much of the energy is associated with the brief consonant sounds and levels behind the speaker are substantially reduced. The relative levels at the two frequencies are also correct compared with the base lines in Figure 7.5; the lower frequency contains less intelligibility information but is louder. In design terms, the greater importance of high frequencies does have a compensating advantage that useful reflecting surfaces do not need to be too large (section 2.6.2).

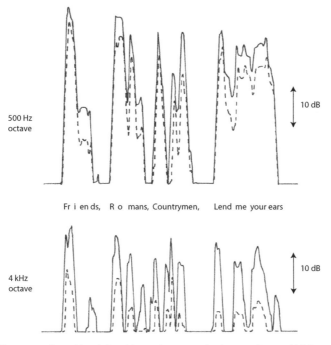

500 Hz
octave

10 dB

Fr i en ds, R o mans, Countrymen,   Lend me your ears

4 kHz
octave

10 dB

Figure 7.5 Sound level time history for a speech phrase at low and high frequencies. Solid line, in front of speaker; dashed line, behind speaker

7.2 Speech in the open air

In the open air, the acoustic situation is dominated by the direct sound. This attenuates rapidly owing to spherical spreading. Speech intelligibility in the open air is a function of the ratio between the speech signal and disturbing noise. Signal-to-noise ratios in excess of 10 db are necessary. But the signal level is enhanced if early reflections also arrive at the listener. Any reflections from surfaces close to the speaker (or listener) are valuable, whether they are off hard ground, or surfaces to the side or behind.

The most famous open-air theatres, not least for their acoustics, are those of classical times. It is appropriate to consider in a little detail why the ancient Greek theatres perform so much better than we might expect. The distance limits for speech transmission in quiet conditions have been measured in the Mohave desert by an eminent acoustician, Knudsen (1932, p. 498). He found that normal speech in this quiet location with no wind could be heard satisfactorily at a distance of 42 m in front of the speaker, 30 m to the side and 17 m behind, with a profile similar to that in Figure 7.4. In the Greek theatre at Epidauros the furthest seat is 70 m from the stage. Could early reflections explain this difference?

A less informed, but more flamboyant, experiment was conducted by Gustave Lyon (Andrade, 1932), who advised on the Salle Pleyel, Paris (section 4.4). Two observers were suspended below small balloons which could be moved independently. By the time they were 11 m apart, the speaking voice was found to be quite inaudible, whereas at night at ground level over perfectly smooth water a normal voice could apparently be heard a mile away. Lyon concluded that sound reflections are indispensable for speech transmission. In fact, Lyon's conclusion is correct but the value of reflections is grossly overstated by this comparison. In the first experiment the distance of transmission is determined by an unspecified background noise level; otherwise he should have obtained a figure similar to Knudsen's 42 m. In the second example propagation must have been enhanced by vertical air temperature gradient effects.

The true effect of early reflections is to act as if the energy of the direct sound were increased (by simple energy addition). In the case of a single early reflection off a hard surface, the sound energy is doubled, giving a 3 db increase. Since sound level is reduced by 6 db for every doubling of distance (the energy is reduced to a quarter), the net effect of a single reflection is to increase the distance limit of satisfactory listening by a factor of 1.4 ($= \sqrt{2}$). Thus in the ancient Greek theatre, the presence of a reflection off the circular 'orchestra' region, the floor area in front of the actors, increases the limiting distance from 42 to 60 m. Additional reflection from behind the actor and perhaps the advantage gained from the face mask make speech transmission over 70 m look realistic. But the additional requirement is that the background noise must be quiet. One reason for the success of the ancient Greek theatres was that disturbing environmental noise was generally quiet when they were in use.

7.3 Speech in rooms

In the case of music in enclosed spaces, a multi-dimensional situation prevails at the subjective level. Auditorium design features have to be assessed regarding their influence on the various subjective characteristics. Questions like 'is an increase in reverberance desirable if it is accompanied by a loss of intimacy?' make concert hall design a delicate balancing act. Is the situation the same with speech? Do we have to concern ourselves with acoustic intimacy, the spatial attributes of the sound and with voice quality?

To elucidate these questions, a brief questionnaire survey was conducted at public performances in three of the theatres discussed in the next chapter (Barron, 1986). The questionnaire was closely related to the one used for music (Figure 3.7), with

additional scales for intelligibility and ease of listening. The seat positions tested included some with poor speech intelligibility. The results of the exercise showed that listeners were responding consistently on three attributes: intelligibility (including ease of listening), intimacy and reverberance. The crucial result was that only intelligibility was rated as significant for judgement of the acoustics overall. In other words, listeners were able to assess acoustic intimacy and reverberance but they were indifferent to the degree of these attributes. But a slight qualification needs to be added: any exercise of this sort is dependent on the range of situations tested. None of the theatres had an extremely short reverberation time (say below 0.5 seconds), which one expects would be interpreted as providing too stark and dry an acoustic.

This exercise suggested that, for drama theatre acoustics, intelligibility must be the prime concern. Latham and Newman (1982) have, for instance, suggested introducing the concept of 'speech quality' but this appears to be a very secondary consideration for theatres. Nor is there any evidence that spatial characteristics of the sound or any preference for early lateral reflections are relevant in speech situations. The following discussion is therefore limited to the requirements for good speech intelligibility.

In an enclosed space, speech transmission is influenced by the signal-to-noise ratio, just as it is in the open air. For instance, a noisy ventilation system can easily make speech difficult to understand. But the reflection characteristics of a room need also to be considered. As well as early reflections which are desirable for speech communication, there will be late reflections which reduce intelligibility. In the extreme situation of a cathedral space with a long reverberation time, the late reflections render speech incomprehensible at all but short distances from the speaker. On the other hand in a small space, such as a domestic room, the reverberation time is short enough (typically 0.5 seconds) for the late reflections to be too weak to undermine intelligibility. In a small room, the density of early reflections is also high enough that even the orientation of the speaker relative to the listener is almost irrelevant. Reverberation time is clearly a determinant of speech intelligibility but is it sufficient?

A common method to measure the intelligibility of speech is for a speaker to read nonsense syllables of the consonant–vowel–consonant form; listeners record what they hear, from which a percentage syllable articulation (PSA) is calculated. To allow for room sound, test syllables are preceded by short phrases. Knudsen used this technique for the experiment mentioned in the previous section. The same technique was used during the 1920s to measure speech intelligibility as a function of reverberation time. This relationship between PSA and reverberation time (Knudsen, 1932, p. 391) is much quoted in textbooks. A scheme was subsequently developed to enable prediction of intelligibility on the basis of reverberation time, voice level and background noise level. It is now difficult to be sure what conditions these curves relate to and in particularly what speaker–listener distances. Measurements in theatres indicate categorically that intelligibility varies both within the auditorium and with changes of source orientation, whereas measurements of reverberation time are generally independent both of position and of the direction of the source. The reverberation times of the theatres described in the next chapter range from 0.7 to 1.2 seconds. A short reverberation time proves to be desirable but is certainly not a guarantee of adequate speech intelligibility. For the prediction of reverberation time in theatres, see section 7.8 below.

The nature of the room effect on speech is illustrated in Figure 7.6 (after Kurtovic, 1975) for the most common case, where a consonant sound is masked or partially masked by a preceding vowel. If a constant source is switched on in a room, the sound pressure builds up approximately exponentially to a steady value. A brief sound causes a build-up in the same way but does not reach the same maximum value. In both cases the sound intensity decays after switch-off, also approximately exponentially. Since vowels are both louder and of longer duration than consonants, a long room decay time (i.e. reverberation time) causes masking

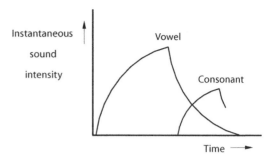

Figure 7.6 The sound intensity versus time history in a room for a vowel followed by a consonant (after Kurtovic, 1975)

of consonants. Because the elements of speech are of much shorter duration than real reverberation times, a finer measure than reverberation time is required. This measure needs to take account of the details of the early reflection sequence. It will involve some averaging since masking of an individual consonant is dependent on the relative levels and delays of the particular consonant and preceding vowel sound. The simplest form of measure is one based on the ratio between early and late sound energy received by the listener. The appropriate duration for the early sound is determined by the duration of speech sounds. Typical speeds for speech are 5 syllables per second, which gives an average of 15 speech sounds per second, assuming a consonant–vowel–consonant situation. The mean duration of a speech sound is thus around 70 ms. Measures of intelligibility have used durations for the early sound of between 50 to 80 ms. Most of the discussion here will revolve around one of the oldest measures suggested: the **early energy fraction** or *Deutlichkeit* of Thiele (1953), measured from the impulse response of the room between the source and receiver. A similar measure has been used for clarity in the case of music (section 3.3). The early energy fraction is simply the fraction of the total energy which arrives within 50 ms of the direct sound. It is measured with a short impulsive sound.

Early energy fraction

$$= \frac{\text{Energy arriving within 50ms of direct sound}}{\text{Total energy arriving}}$$

Measured values of the early energy fraction in excess of 0.50 are considered satisfactory for speech intelligibility. The relationship between the early energy fraction and speech intelligibility has been established by Boré (Kuttruff, 1991, p. 190).

The reason for the choice of the early energy fraction as opposed to other measures is not that it relates most closely to speech intelligibility; some other measures are slightly better. But these other measures are conceptually more complicated. This makes prediction of values of these measures from basic architectural details difficult. The early energy fraction has the advantage that it can be used as a predictive tool. It can also be measured with relative ease.

Two determinants of speech intelligibility have been discussed in this section: the signal-to-noise ratio and the room impulse response. In any practical situation both should be considered, though from an analytic point of view they can be measured independently. A more technical review of other measures suitable for speech intelligibility is given in Appendix D.1.

7.4 Is there an optimum profile for theatre design?

The particular characteristics of the human voice make it tempting to imagine that they might dictate the appropriate form for theatre design. This idea was already exploited by George Saunders in his *Treatise on theatres* of 1790. Under the guise of a scientific approach starting from experiments with the voice, Saunders manages to argue the need for a theatre with a plan conforming to his preference for the circle. His understanding of acoustics is quaint and regarding his explanation of an echo simply wrong. But his discussion of the optimum theatre form makes a valuable starting point (Barron, 1991).

Saunders conducted the speech transmission experiment on 'a calm day ... (over) an open plane'. His key result is reproduced here in Figure 7.7. His maximum distance is 92 ft (28 m); we can suppose again that some background noise influenced the result. His limiting distance in front of the speaker

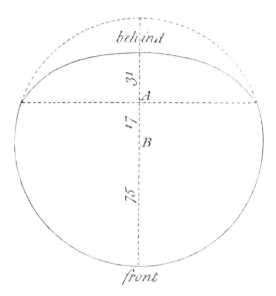

Figure 7.7 Saunders' (1790) measured profile of the limit of intelligible speech

follows a segment of a circle, which subtends 205° at its centre. Modern measurements concur with Saunders' supposition that the contour is circular in front of the speaker; the forward part of the curve in Figure 7.4 is virtually circular comprising a 220° arc. In his treatise, Saunders then argues that a major segment of a circle is also the optimum plan form for a theatre. His prejudice is supported by his enthusiasm for the Grand Théâtre in Bordeaux of 1780: 'All persons acquainted with (this) theatre are unanimous in their decision in its favour. They all agree that the voice of the actor spreads more equally in this than in any other theatre. ... The smallness of its size is much in its favour'. In the Bordeaux theatre, the major segment of the circular plan subtends 260° at its centre (Figure 7.8).

If we are to pursue this line of argument further, it is necessary to distinguish between the **form** of the equal intelligibility curve (or equal sound pressure in the open air, Figure 7.4) and the **absolute distance** values, defined for instance by the maximum distance of transmission. If background noise is audible in the open air, the maximum distance decreases but the form of the curve for acceptable intelligibility remains the same. If the equal sound

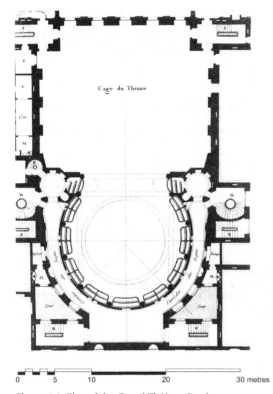

Figure 7.8 Plan of the Grand Théâtre, Bordeaux

pressure curve is relevant to optimum form for theatres, is this form appropriate for all sizes of theatre or should the distance limits in very quiet conditions be used as a limiting profile within which a theatre building should be built? This last approach was used by Izenour (1977) in his extensive review of world theatres. Izenour used a three-dimensional rotation of the profile measured by Knudsen in a quiet open space (section 7.2). This was superimposed on isometric drawings of large auditoria, with the implication that points outside the profile would be inadequate acoustically. But even in the open air, this profile is inappropriate due to reflections. In enclosed spaces, reflections make the absolute distance values of Knudsen's result irrelevant. Wren, for instance, gives a maximum range of only 15 m in a church, nearly a third of the open-air value. An acoustic profile for an enclosed space has to take account of acoustics of the auditorium, including

its reverberation time. But perhaps the form of the equal sound pressure curve is still relevant?

To resolve this question for an enclosed space we need to refer to our acoustic measure relating to speech intelligibility, the early energy fraction. Certain assumptions also need to be made. The advantage of the early energy fraction is that we can subdivide the sound received by the listener into an early component and a late one. An identical approach has been used for music auditoria. A major virtue of this temporal subdivision is that behaviour of the two components is fairly distinct and can be related with some success to architectural features. The early sound consists of the direct sound and early reflections. The direct sound can be predicted if there is good line of sight. It will be suggested below that a reasonable theoretical approach is to assume that the number of early reflections is constant, so that the early sound energy is simply a multiple of the direct sound energy. This multiple will be called the **early reflection ratio**. Direct sound plus an early reflection as strong as the direct sound produces a ratio of 2.0.

The behaviour of the late sound is strongly influenced by the reverberation time: longer reverberation times produce a higher level of late sound and thus undermine speech intelligibility. Traditional theory (section 2.9) claims that the reflected sound level in a space is constant. If we assume that the late sound level is constant in the space (not entirely the same as the reflected sound level but a sensible starting point), then late sound behaves just as if it were background noise. The shape of the limiting intelligibility profile in this case would be identical to the equal sound pressure curve.

Late sound in theatres, and other auditoria, is however found not to be constant. It decreases with distance away from the source (section 3.10.5 and Figure 7.14(b)). The effect of this is to elongate the limiting intelligibility profile along the main axis. Figure 7.9 shows the profile generated for a theatre of volume 5000 m³, a reverberation time of 1 second and an early reflection ratio of 3. (The late sound is calculated according to the scheme in Appendix D.3; with the source on the proscenium line, 28 per cent

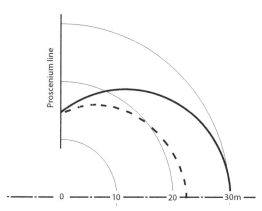

Figure 7.9 'Optimum profile' for a theatre of 5000 m³ with a 1 second reverberation time (solid line). The dashed line shows the profile of constant direct sound pressure level, which would be relevant if the late sound were constant

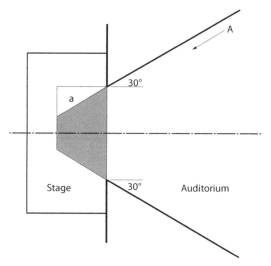

Figure 7.10 The plan view of a stage through a proscenium opening. Spectator A cannot see the area 'a', leaving only the shaded area visible to all. 30° angle of fan is a typical criterion

of the source energy is absorbed in the stagehouse and can therefore not contribute to the late sound.) In this case the maximum distance for speech propagation is predicted as 30 m. The profiles generated by this method tend to be circular with the centre some way in front of the speaker. The effect of the decrease in late sound with distance is to extend a 220° segment (Figure 7.4) into a 260° one in Figure 7.9. The value 260° is identical to the subtended angle in the Bordeaux theatre (Figure 7.8)! Interestingly this exercise has produced a form less elongated than the horseshoe plan which dominated theatre design for so long. But there is a good visual reason for limiting the auditorium at the sides. If the spectator is sitting further from the centre line than the edge of the proscenium, the back corners of the stage are lost from view. In contemporary theatre a maximum view angle of 30° is a common norm (Figure 7.10). For the Baroque theatre with its obsession for forced perspective effects, an extremely deep stage was necessary which required even narrower angles of view.

In the argument above, a revised intelligibility profile was derived for an enclosed theatre. While the result supports George Saunders' suggestions of 200 years ago, it is important to place this conclusion very much in context. Firstly the profile in Figure 7.9 shows the limiting distances beyond which speech will not be intelligible. Any form within this profile is appropriate. By following the profile the maximum plan area is achieved, in theory. But any exercise of this sort will generate concave curved surfaces, which are likely to produce focusing of sound, unacceptable in modern theatres. The line of a specific profile is also dependent on several assumptions about such things as reverberation time and the number of early reflections. The major assumption which undermines an exercise of this sort is that the number of reflections (or more precisely the early reflection ratio) is constant. In reality a theatre design can enhance the number of reflections, and this exercise illustrates the need for this at remote seats.

Another major objection to the intelligibility profile has to be raised: it is relevant to a specific orientation of the speaker. If the speaker rotates 90° and points across the stage, the profile rotates with him, so that the optimum profile for a theatre where this is common is simply a circle about the acting position. This conclusion is worth applying to the case of the thrust-stage theatre, which can become marginal acoustically. Interestingly in the two thrust-stage theatres considered in the next chapter, the theatre conforming with this conclusion is the one which performs better.

7.5 Design for speech in theatres

To provide guidelines on appropriate design for theatres, the details as much as the extent of the design must be considered. In the case of speech a single criterion of intelligibility can be applied, as opposed to the multiple criteria with music. Though this simplifies acoustic design, there is the problem that the criterion is more absolute. The distinction between success and failure can be surprisingly small with speech. In small spaces, producing intelligible speech is no problem; a suitable reverberation time is often sufficient to give good conditions irrespective of theatre form. In larger spaces, shorter reverberation times may be appropriate and the number of useful early reflections can become crucial. The directional nature of the human voice proves to be a major concern, so that different theatre types have different limiting dimensions.

For good speech intelligibility, a high early energy fraction is necessary, requiring high early energy and low late energy. The value of early reflections can be quantified roughly as follows: a single reflection will increase the distance range of intelligible speech by a factor of 1.4, and three strong reflections by a factor of two. If surfaces are directed to reflect sound onto the audience, this tends not only to increase early energy but also to have the effect of reducing the late sound (section 3.5), making this technique doubly useful. The late energy in general becomes lower with a shorter reverberation time. The role of reverberation time has tended to be somewhat minimized for speech. For instance, Cremer and Müller (1982, p. 517) present an experimental result, from which they conclude that reducing reverberation time below 1 second will not increase distinctness.

Traditionally a 1 second reverberation time is recommended which does not rise in the bass. But a short reverberation time will cause a small late

energy. For this reason it can be argued that reverberation times shorter than 1 second may be appropriate in theatres. Values of 0.7 or 0.8 seconds can give good intelligibility in a theatre which would otherwise have marginal conditions. However, very short reverberation times are also undesirable. For values below 0.5 seconds the acoustics would sound too dry for comfort, even though intelligibility is likely to be good.

Crude though it is, the auditorium volume per seat is an instructive guide to reverberation time in auditoria (section 2.8.3). Double the volume per seat and one expects the reverberation time to double approximately. That assumes however that no extra sound-absorbing material has been added. From Table 7.1 one observes that volume per seat is only a poor guide to reverberation time in drama theatres. The reason for this is simply that in theatres with large volumes for their seat capacity, a lot of absorbing material has been added to bring down the reverberation time. It is much better to design with a volume giving around 4 m³/seat than to have to add large quantities of sound absorption. Sound absorbing material will reduce sound levels and may also obscure early reflections, neither of which is desirable.

The effect of the speaker turning away is obviously to reduce the direct sound. Ideally in this case there would be surfaces in front of the actor which catch the direct sound and reflect it to arrive early enough in the audience. This should be possible in a proscenium-type theatre with reflecting surfaces in the region to the side of and immediately in front of the proscenium. This is one reason for the acoustic importance of these surfaces. In more open stage situations such techniques are generally impossible and only reflections from above are feasible, such as from a suspended ceiling. In this last case the early energy fraction will reduce substantially as the speaker turns round; so the energy fraction must be high enough with the actor facing the listener in order to be adequate when the actor faces away.

In the proscenium theatre there tend to be more surfaces available to provide useful reflections, particularly side wall surfaces, so that changes for the

Table 7.1 Volume per seat and reverberation times of British theatres. The volume used here is the auditorium volume without the flytower.

Theatre	Volume/seat (m³)	Reverberation time (s)
Proscenium theatres		
Theatre Royal, Bristol	3.4	0.8
Wyndham's Theatre, London	3.4	0.7
Royal Shakespeare Theatre, Stratford	4.3	1.0
Arts Theatre, Cambridge	2.4	0.7
Lyttelton Theatre, National Theatre, London	4.8	1.1
Towngate Theatre, Poole	4.2	0.9
Thrust-stage theatres		
Festival Theatre, Chichester	4.7	1.0
Crucible Theatre, Sheffield	7.3	0.8
Open-stage theatres		
Olivier Theatre, National Theatre, London	11.6	1.0
Barbican Theatre, Barbican Centre, London	3.6	0.8
Theatres in the round		
Roundhouse, Chalk Farm, London	6.9	1.2
Royal Exchange Theatre, Manchester	4.3	0.8

listener due to the actor turning on stage are not normally large. The principal lesson from study of actual theatres is that in larger theatres early reflections are crucial to acoustic success. The ceiling, or more likely suspended ceiling, is a surface that can be particularly valuable to provide acoustic reflections. Their importance for open-stage theatres has just been mentioned. Even in proscenium theatres the ceiling design can have acoustic consequences (see section 8.4.5). There is however a potential conflict in drama theatres between the use of the ceiling for concealing stage lights and for providing acoustic reflections, as shown in Figure 7.11. To ensure that the ceiling is contributing acoustically, the acoustic consultant will have to fight hard, aware that he is at a disadvantage in the design team since his own requirements are less precise and certainly less visual than those of the lighting designer.

In any theatre the particularities of the early reflection situation are substantially a matter of its detailed design. This is pursued in the next chapter by reference to individual examples.

The behaviour of the late reflected sound is less influenced by the particular orientations and positions of acoustically reflecting surfaces. The late sound is affected by the presence of a proscenium opening; an additional effect of the proscenium on the reverberation time is discussed below. A proportion of the energy coming from the speaker will enter the proscenium opening. The fate of this sound energy is dependent on the amount of acoustic absorption in the stagehouse. Frequently the quantity of drapes etc. is such that nearly all of this energy is absorbed and is therefore unavailable for contributing to the late auditorium sound (Figure 7.12). The important measure here is the solid angle subtended at the actor by the proscenium opening. With the picture-frame proscenium scheme, where all action takes place behind the proscenium, values for the fraction of voice energy actually entering the auditorium may be only 40 per cent. The proscenium opening also affects the early sound as well, but in this case the loss is small because most reflections in theatres come from surfaces on the auditorium side of the actor. In a theatre, surfaces can seldom be suspended behind the speaker to reinforce his speech, as is frequently done in lecture halls.

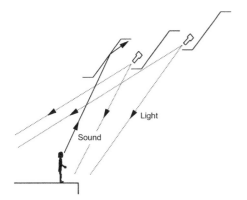

Figure 7.11 Long-section view through a theatre auditorium showing how a ceiling line designed just to conceal stage lighting can result in no useful overhead sound reflections for audience

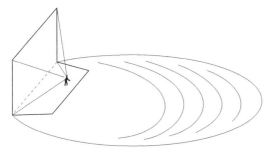

Figure 7.12 Solid angle subtended by the proscenium opening at a speaker. Most sound entering the stagehouse does not return to the auditorium

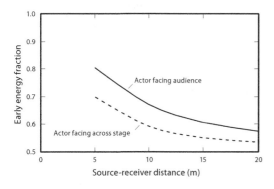

Figure 7.13 Typical early energy fraction behaviour in a theatre at different distances, for an actor facing the audience (solid line) and an actor facing across stage (dashed line). Energy fractions below 0.5 tend to imply poor intelligibility

Typical behaviour of the early energy fraction (and therefore intelligibility) for different distances from the source is sketched in Figure 7.13. The main curve relates to the situation with the speaker facing forwards. Intelligibility will inevitably be good close to the source due to the dominance of the direct sound. At the most remote seats, the energy fraction is generally still decreasing for increasing distance (whereas in concert halls the rate of decrease at remote seats tends to be small). Therefore to restrict the distance to the furthest seat in a theatre represents good acoustic design. The effect of the speaker changing his orientation is to decrease both the direct sound and the level of early reflections from overhead surfaces. This will reduce the early energy fraction, as illustrated. In individual theatres the number and to a lesser extent the direction of early reflections influences actual behaviour.

The energy fraction can rise at the rear of the theatre owing to reflections from the rear wall and from the rear wall/ceiling cornice. These reflections are obviously valuable in a large theatre and should not be obscured. The presence of a gangway round the back of seating will reduce these useful reflections and such gangways are therefore undesirable in a theatre in which intelligibility could be marginal.

The behaviour of sound under a balcony is no different in a theatre from a concert hall (section 3.7). The early sound level is generally little affected but the late sound decreases substantially as one moves beyond the balcony front. This is usually undesirable for music. For speech it will simply guarantee a high early energy fraction! Deep balcony overhangs are thus acceptable in theatres on acoustical grounds as long as good sightlines are preserved. They were used extensively in nineteenth-century proscenium theatre designs, no doubt in the light of experience that speech intelligibility remained good in these positions. A risk associated with deep overhangs does exist though: that the total speech level can become too quiet. This brings us to the other concern for speech intelligibility: the signal-to-noise ratio.

When one comes to measure the acoustics of theatres, the signal-to-noise aspect is measured by feeding signals of known amplitude into a loud-speaker speech source and measuring the received levels at different seats in the auditorium. This enables calculation of how much the theatre is amplifying the voice of the actor. Small theatres will usually provide more amplification than large ones. However, rather than refer to this procedure as amplification, it has been called here a measurement of **speech sound level**. We could present this as a simple sound pressure level, but numbers like 44 db are rather clumsy without a simple point of reference. It is easier to normalize the values (as was done with music). The reference level used in figures in the next chapter is the mean (over direction) of the direct sound level at 10 m. It happens that with this reference level the minimum criterion for speech level emerges as 0 dB. The derivation of this is included in Appendix D.2.

Little in fact needs to be said about speech sound level behaviour in theatres. Speech level is likely to be well maintained in designs with strong early reflections. The level inevitably decreases as one moves away from the stage. It is only in theatres which are at the limit of maximum size for their particular theatre form that speech level is likely to be a problem by itself. From the experience of the twelve theatres discussed in the next chapter, low speech levels only arose in theatres with inadequate early reflections, an absence which usually affected early energy fractions as well.

The design factors influencing reverberation time in theatre auditoria should also be discussed. To achieve an appropriate reverberation time in a theatre does not require the elevated ceiling necessary for a music auditorium. Indeed in theatres without a flytower the room volume can become too large. This is a major risk in the arena-type auditorium, where the ceiling height is determined by the seating arrangement and in particular its rake which tends to leave large volumes over the central stage area (see section 8.5). The presence of a flytower introduces extra absorption at the proscenium opening. Nearly all sound which enters a flytower will be absorbed within it by drapes and scenery. Other things being equal, this allows a larger

auditorium volume in a theatre with a flytower than one without for the same reverberation time. By 'other things being equal' we understand that no additional acoustic absorbent is introduced into the auditorium. In most theatres there appears little reason for additional absorption to be used, for the simple reason that absorption is always detrimental to the resulting speech level. However, in the case of the theatre discussed in section 8.5.2, absorption on the auditorium ceiling has been used, with good reason and success.

Little has been mentioned here about the acoustic conditions for the actor. This is partly because actors are remarkably flexible; many appear willing to tolerate echoes and worse, which would frustrate the more timid speaker. Unfortunately virtually no research has been done in this area. There is no doubt that all performers are happier with some feedback from the auditorium. Without this feedback, one tends to strain the voice, which is a problem for all open-air performance. A good design should provide for the actor some early reflected sound followed by some reverberation from the auditorium. Reflections back to the stage, which arrive so late they are heard as echoes, from such surfaces as rear walls or soffits, should clearly be suppressed. In smaller theatres the situation usually looks after itself. In larger theatres it is wise to consider possible reflection paths back to the stage. Surfaces which scatter the sound are likely to be useful (such as convex profiled balcony fronts).

7.6 The early reflection ratio

There are advantages in a conceptual subdivision of the sound received by the listener into an early and late component. The measurements made in theatres provide a sample which indicates typical behaviour of each component. Figure 7.14 shows behaviour of the early reflected and late sound in a particular theatre. The example of the Crucible Theatre, Sheffield (section 8.5.2) illustrates characteristic behaviour, in this case for a lateral source pointing across the stage away from the listeners (Figure 7.17). It would be normal to display the early

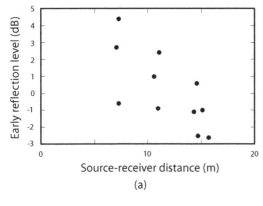

(a)

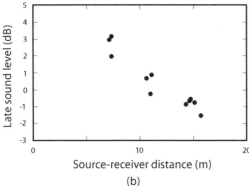

(b)

Figure 7.14 (a) Early reflected and (b) late sound levels measured in the Crucible Theatre, Sheffield, for a lateral source

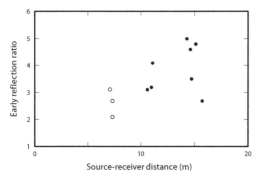

Figure 7.15 The early reflection ratio measured in the Crucible Theatre, Sheffield, for a lateral source. Points close to the source (hollow circles) are omitted when deriving a mean value for the theatre

sound level, but the early sound is highly dependent on speaker orientation. We can remove this orientation effect to some extent by subtracting the direct sound component, which can be calculated from theory. Figure 7.14(a) therefore shows the remaining early energy: the early reflected sound level. Results are generally quite scattered and not amenable to simple prediction techniques. The behaviour of the late sound is, however, reasonably linear with distance. The late sound proves to be predictable with reasonable accuracy on the basis of the reverberation time, auditorium volume and source–receiver distance. One characteristic not illustrated in Figure 7.14(b) is the behaviour of late sound below a balcony; the Crucible Theatre does not have one. In most theatres the late sound becomes weaker at seats overhung by a balcony.

The concept of the early reflection ratio was introduced above in section 7.4. The ratio is given by the early sound energy (up to 50 ms) divided by the (theoretical) direct sound energy. So with direct sound alone the ratio equals one. With direct sound and a strong early reflection, the ratio takes a value of two. The ratio concept has the advantage that it can be used directly with the intelligibility measure, the early energy fraction, but it also has a simple physical interpretation. The number of early reflections is approximately equal to the reflection ratio minus one. If a reflection arrives very early and is as strong as the direct sound, its contribution to the reflection ratio is unity. Later reflections will be less intense than the direct sound and will generally make a smaller contribution to the reflection ratio. However, in a theatre where the listener is to the side or behind the speaker, a reflection can be more intense than the direct sound if it originated frontally from the speaker. This last point is a detail which is rarely obtrusive. The important aspect of the early reflection ratio is that it offers a crude measure of the number of early reflections.

Early reflection ratio values in the Crucible Theatre (for a lateral source) are plotted in Figure 7.15. In all theatres, positions closest to the stage have lower values of the ratio. This is explained physically because reflecting surfaces are usually

Table 7.2 Mean values of the early reflection ratio in British theatres

Theatre	Mean early reflection ratio
Proscenium theatres	
Theatre Royal, Bristol	5.3
Wyndham's Theatre, London	3.0
Royal Shakespeare Theatre, Stratford	4.3
Arts Theatre, Cambridge	5.1
Lyttelton Theatre, National Theatre, London	3.0
Towngate Theatre, Poole	5.8
Thrust-stage theatres	
Festival Theatre, Chichester	2.8
Crucible Theatre, Sheffield	4.0
Open-stage theatres	
Olivier Theatre, National Theatre, London	2.4
Barbican Theatre, Barbican Centre, London	2.3
Theatres in the round	
Roundhouse, Chalk Farm, London	2.1
Royal Exchange Theatre, Manchester	3.3

not close enough to the actor to provide strong early reflections to audience nearby. From a design point of view, these seats are of little interest, since owing to their position they have adequate speech intelligibility. When taking mean values these positions are best ignored. For the remaining positions, the early reflection ratio proves to be reasonably consistent throughout the theatre and this consistency is a characteristic of most of the theatres measured. In addition, the mean early reflection ratio is largely unaffected by rotation of the source.

Table 7.2 lists mean values of early reflection ratios in the theatres discussed in the next chapter. Values of the ratio in proscenium theatres are on average higher than in open-stage type theatres. Within a particular theatre type, theatres with the higher mean early reflection ratios perform better acoustically. The other factor affecting speech

intelligibility is the level of the late sound. This is of course quite independent of the early reflection ratio. Shorter reverberation times and the presence of a proscenium opening reduce the late sound level and thus promote speech intelligibility.

It is instructive to use the early reflection ratio concept to derive limiting distances in theatres. In theatre design we have two criteria to fulfil: an adequate early energy fraction and an adequate speech level. The limiting early energy fraction has been given as 0.50, based on subjective observation. The limiting speech level was mentioned in the previous section; it is derived in Appendix D.2. The total speech level including contributions from reflections and reverberation should exceed the mean level (over direction) of the direct sound at 10 m from the speaker. These two criteria lead directly to a requirement that the early sound must be not less than −3 db relative to the mean direct sound at 10 m. This leads to a series of maximum distances at different directions from the speaker, depending on the value of the early reflection ratio. These limiting distances are listed in Table 7.3. The table shows how valuable is the contribution of early reflections and the detrimental effects of the directivity of the human voice. For example, two strong early reflections are needed for theatre in the round with a maximum distance between actor and audience of 15 m. Four strong early reflections are required to give a range of 20 m at the worst direction relative to the speaker, a range which is achieved by the direct sound alone in front of the speaker.

Limiting distance values in Table 7.3 are derived from consideration of the early sound alone. They assume that the late sound level is optimal, which of course it will not normally be. While the early energy fraction criterion cannot be relaxed, the speech level criterion might be relaxed by a few decibels. This means that the actual values in Table 7.3 can probably be used where the late sound is known to be weak.

It is the hope for most numerical analyses of problems that they will lead to a prediction technique for the future. The scheme discussed here can help indicate the 'ball-park' one is in. It would

Table 7.3 Limiting distances for intelligible speech in theatres, based on the requirement of adequate early sound energy. An angle of 150° to the speaker produces the minimum range

Early reflection ratio	Limiting distances in metres		
	Angle of listener to straight ahead of speaker		
	0°	*90°*	*150°*
1	20.0	13.0	8.9
2	28.3	18.4	12.8
3	34.6	22.6	15.5
4	40.0	26.1	17.9
5	44.7	29.2	20.0

use as parameters the source–receiver distance, listener orientation relative to the speaker, the early reflection ratio, auditorium volume, reverberation time and the solid angle of the proscenium subtended at the source. All these can be derived from drawings with the exception of the early reflection ratio, which has to be estimated with the help of experience (Table 7.2). Such a prediction depends on the estimated number of reflections actually reaching the sensitive areas of seating in a theatre. The appropriate equations for prediction are to be found in Appendix D.3.

More detailed analysis of a theatre design is appropriate in large theatres. Acoustic scale models tested with a directional source are a valuable design aid, which also allow checks to be made in the variety of conditions under which the theatre will be used (see sections 8.6.1 and 11.7). Computer modelling is becoming both more reliable and popular as a design aid. It should work well for theatres when prediction is made with a directional source.

7.7 Acoustic measurement of theatres

The next chapter contains discussion of the results of measurements in twelve theatres. Apart from measurement of the reverberation time, measurements are almost worthless in theatres unless the sound source has the correct directivity. To reproduce the directivity of a human speaker, human head and torso simulators are commercially available. An alternative is to mount a small loudspeaker in an enclosure approximating a human head and torso. Figure 7.16 shows the loudspeaker speech source used for the measurements described here. With a directional source it is clearly important to orientate the source appropriately.

Figure 7.16 The loudspeaker speech source used for theatre measurements

Two orientations were chosen, which represent roughly the best and worst case (Figure 7.17). The source was located at a typical acting distance from the stage front. For what will be called the central source position, the source was facing along the axis of symmetry of the theatre. The lateral source position represented the typical situation for an actor facing across stage. To make measurements of the worst case, the microphone positions were in the half of the auditorium behind this lateral source. A minimum source–receiver distance for the central source of 7 m was used. All measurements reported in the previous section and in the next chapter (e.g. Figures 8.11 and 8.12) are for the frequency range 500–2000 Hz, averaged over the three octaves.

Four main quantities should be measured in theatres:

1. an intelligibility measure (early energy fraction in this case);
2. reverberation time;
3. speech sound level for a standard source;
4. background noise level.

The measurement procedure was similar to that used for music auditoria. Short-duration sounds were fed to loudspeakers to obtain impulse responses at about 12 positions throughout the auditorium. The impulse response can be used to calculate any of the intelligibility measures discussed above. The impulse response by itself is also valuable to indicate arrival times of reflections. Speech sound level was measured with continuous noise signals fed to the directional source and sampled with a standard sound-level meter. The background noise level is measured with ventilation plant running. In practice, audience breathing noise will also contribute to the background (Kleiner, 1980; section 2.7.3).

7.8 Prediction of reverberation time in theatres

The procedure for reverberation time prediction in auditoria has been explained in section 2.8. In a proscenium theatre however, the problem arises that we have two spaces coupled by the proscenium opening: the auditorium and the stagehouse. Technically these are known acoustically as 'coupled spaces'. In practice, it is the auditorium reverberation time which is the major concern. The auditorium reverberant decay may be influenced by the reverberation time of the stagehouse and produce a double slope decay (section 2.8.1 and Figure 2.31); a double slope decay is likely when the reverberation time in the stagehouse is significantly longer than that of the auditorium.

For calculation of the auditorium reverberation time, it is usual to assume that the stagehouse is acoustically dead because generally it contains many sound-absorbing drapes. The Sabine formula is used with the volume of the auditorium alone inserted in the equation. The proscenium opening is treated as an absorbing surface; as absorption coefficients for the opening, coefficients for seated audience have proved to be reasonably suitable.

References

Andrade, E.N. da C. (1932) The Salle Pleyel, Paris, and architectural acoustics. *Nature* **130**, 332–333.

Barron, M. (1986) Speech in theatres: what are the important considerations? *Proceedings of the Institute of Acoustics* **8**, Part 3, 371–378.

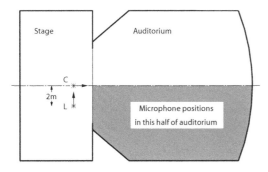

Figure 7.17 Source and microphone positions used for measurements in theatres. C, central source; L, lateral source

Barron, M. (1991) George Saunders'Treatise on theatres' of 1790 and optimum acoustic profiles. *Acoustics Bulletin* **16**, No. 2, 14–17.

Cremer, L. and Müller, H.A. (translated by T.J. Schultz) (1982) *Principles and applications of room acoustics,* Vol.1, Applied Science, London.

Izenour, G.C. (1977) *Theater design*, McGraw Hill, New York.

Kleiner, M. (1980) On the audience induced background noise level in auditoria. *Acustica* **46**, 82–88.

Knudsen, V.O. (1932) *Architectural acoustics*, John Wiley, New York.

Kurtovic, H. (1975) The influence of reflected sounds upon speech intelligibility. *Acustica* **33**, 32–39.

Kuttruff, H. (1991) *Room acoustics*. 3rd edn, Elsevier Applied Science, London.

Latham, H.G. and Newman, P.I. (1982) Subjective preference design criteria for evaluating acoustic quality. *Proceedings of the Institute of Acoustics, Auditorium acoustics and electro-acoustics meeting*, Edinburgh, September.

Moore, J.E. (1964) Theatre acoustics. *Architect and Building News*, 12 August, 309–311.

Moreno, A. and Pfretzschner, J. (1977) Human head directivity in speech emission: a new approach. *Acoustic Letters*, **1**, 78–84.

Saunders, G. (1790) *A treatise on theatres*, London (facsimile by Benjamin Blom, New York, 1968).

Thiele, R. (1953) Richtungsverteilung und Zeitfolge der Schallrückwürfe in Räumen. *Acustica* **3**, 291–302.

8 Theatre acoustics

The history of the theatre and the development of its architectural form are well documented (see general references at the end of the chapter). Yet, as already mentioned, the acoustic design of theatres is by contrast meagrely charted territory. This would be of little consequence if all theatres provided intelligible speech. This chapter traces the historical development of theatre form, seen from an acoustic standpoint. The argument is illustrated by detailed discussion of 12 British theatres of various stage forms. It will become apparent that the conditions for speech are deficient in some of the more recent theatres.

8.1 Classical Greek and Roman theatre

Few acoustical situations are so enveloped in myth as the antique Greek theatre. For some, the Greeks are credited with an understanding of acoustics which still baffles modern science. In spite of our otherwise extensive knowledge of Greek culture, no contemporary accounts survive and much has to be gleaned from archaeological evidence. The earliest documentary discussion is by the Roman Vitruvius (1960) from the first century BC. Interestingly his overriding concern is for acoustics, rather than vision, and this even extends to the rules he gives for seating design. Vitruvius presents us with an elegant theory for the manner in which theatre sites can be unsuitable, namely if they exhibit acoustic dissonance, circumsonance or resonance. The propitious site is consonant:

in which (the voice) is supported from below, increases as it goes up and reaches the ears in words which are distinct and clear in tone. Hence, if there has been careful attention in the selection of the site, the effect of the voice will, through this precaution, be perfectly suited to the purposes of a theatre.

This suggests an appreciation of the value of a floor reflection, but as design advice it is too enigmatic. And it is probably rash to accept Vitruvius as an authority on Greek thinking 300 years before he himself was writing.

Greek theatre developed from festivals of singing and dancing in honour of Dionysus, the Bacchus of the Greeks. A circular platform with an altar at its centre provided the focus, which could be witnessed by a large crowd if they were placed against a hillside. Greek drama evolved with a chorus countered by first one, then three actors. Theatre design, realized first in timber and later in stone, retained the circular platform, or 'orchestra', which was used to set out the concentric seating scheme. The stagehouse behind the orchestra was apparently introduced at the time when the importance of the actors increased at the expense of that of the chorus.

The development of the acoustics of the Greek theatre, which spanned more than two centuries, can be seen as an equally logical empirical development. A similar trial-and-error process was responsible for the optimization of the proscenium theatre design in more recent times. The resulting acoustic situation can be clearly stated. Members

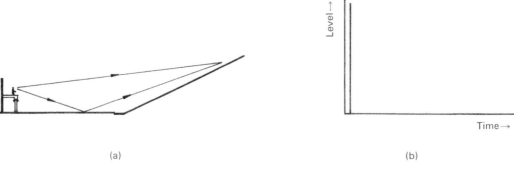

Figure 8.1 Sound propagation in the Greek theatre: (a) sound ray paths, (b) the received impulse response for the listener

of the audience receive the direct sound. This is shortly followed by a reflection from the front of the horizontal orchestra; the reflection occurs both for the chorus on the rear half of the orchestra and for actors on the raised stage. Some further reflections will come from the stagehouse but since this is behind the speakers their reflection energy will be small. Figure 8.1 illustrates the resulting reflection sequence; the effect of the orchestra reflection is to make speech carry 40 per cent further than without it. A distinctive characteristic in the Greek theatre of acoustic significance is the steep seating rake (typically 20°–34°). Though modern information on sound propagation over audience is still incomplete (section 2.6.1), there is no doubt that higher angles of incidence to the seating plane are beneficial. (The situation could however be further optimized: more uniform conditions are achieved with an increasingly steep seating rake as one moves away from the stage.)

For unassisted speech to be audible depends no less on quiet conditions. Many of these theatres are to be found on quiet sites, while for those in urban locations there would have been little residual noise when the majority of the population attended. The advent of motorized transport has disturbed these urban environments, but regular performances can be witnessed at the isolated site of Epidauros. The success of the drama, now just as 2000 years ago, also requires a very quiet audience. To have built theatres able to accommodate 14 000 within such

constraints remains a remarkable achievement, particularly when compared with our contemporary theatres which hold only a tenth of that number.

The theatre at **Epidauros** (Figure 8.2) in the Peloponnese, dating from around 350 BC, was considered by Pausanias in the second century AD as the most perfect of the Greek theatres and fortunately it has remained in a good state of preservation. The furthest seat lies at 70 m from the front of the stage (or the rear of the orchestra), much too far to allow facial expressions to be observed. Greek drama employed masks, which limits the visual problem for the audience to little more than recognition. These masks also performed an acoustic function by providing a short megaphone in front of the actor's mouth.

Much of the individual appeal of the theatre at Epidauros lies in its considerable symmetry. Such a fan shape in plan with an angle of around 210° implies poor conditions both visually and acoustically in the extreme seating areas to the sides of the orchestra. The Greek solution was to reserve these seats for foreigners, latecomers and women. Regarding this problem it is interesting to observe the development in some later theatres, where the rear seating extended much less far to the sides. Figure 8.3 compares the plan of the theatre at **Priene** (near Izmir, Turkey) with the profile for equal sound pressure for an actor on stage facing forward. In this theatre the width limitation also obviated the need for some substantial foundations.

It has already been noted that the stagehouse developed in response to a change in dramatic

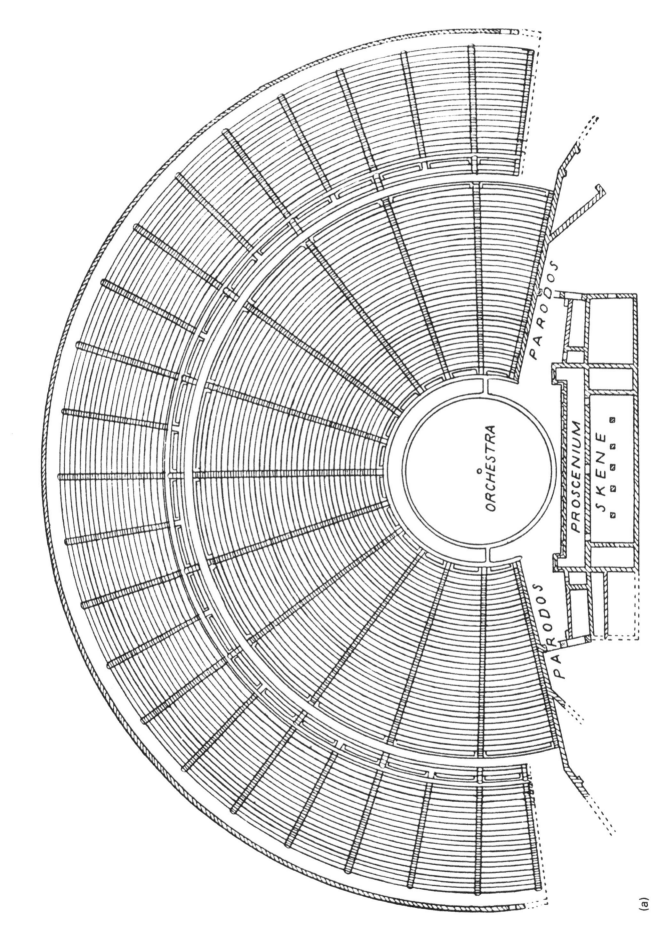

ORCHESTRA

PARODOS

PARODOS

PROSCENIUM

SKENE

(a)

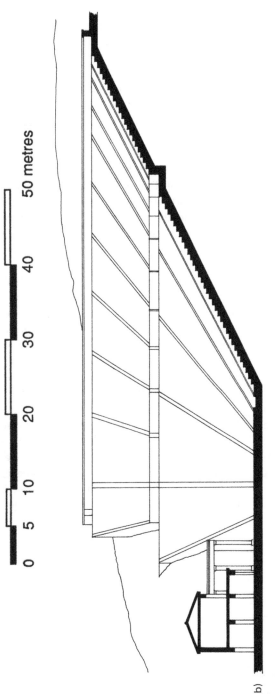

(b)

Figure 8.2 (a) Plan and (b) section of the theatre at Epidauros, Greece

0   5   10   20   30   40   50 metres

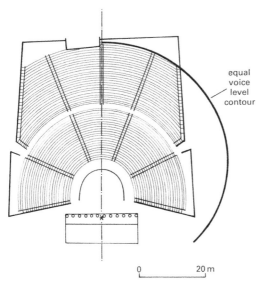

equal
voice
level
contour

0   20 m

* – source position

Figure 8.3 Plan of theatre at Priene, with (superimposed) the equal voice level contour for an actor at the front of the stage, facing forwards. The contour is plotted through the furthest seat on axis

style. Early theatres (such as the Theatre of Dionysus, Athens) had a stage height of only 1–1.2 m, but this was increased to 3–3.6 m in the cases of Priene and Epidauros. The acoustic implication of this change is an increased angle of incidence, α, relative to the seating plane for the direct sound (Figure 8.4). Cremer's measurements (1975) confirm that the direct sound in Epidauros behaves as expected according to the inverse square law, discounting theories

that wind or temperature effects enhance sound in some consistent way. Of almost equal importance is the reflection off the orchestra floor, but Canac (1967) has pointed out that raising the stage is disadvantageous for the reflected wave since its angle to the audience plane, β, is reduced (Figure 8.4). Canac found evidence that this angle b was always greater than 5°. Canac also noted that to preserve the orchestra reflection, it is essential to have only a shallow stage, which indeed is a characteristic of Greek theatres.

Shankland (1973) has measured word articulation in several classical theatres including Epidauros. (Word articulation is always found to be a little higher than the syllable articulation mentioned in section 7.3.) For a speaker at the centre of the orchestra of the Epidauros theatre, he measured a 72 per cent word articulation at the back of the auditorium on the axis of symmetry. Inevitably though, this measurement was made in the empty theatre, when reflections off the stone seating may help. No significant reflection would occur off an audience. Shankland also records that in a Roman theatre the word articulation dropped from 80 per cent to 40 per cent 'when a moderate breeze began to blow'.

The Roman theatre differs in several distinctive ways from the Greek. With knowledge of vaulting techniques, later facilitated by the use of concrete, theatres could be constructed as independent structures on level sites. The Romans developed a highly effective access and escape system beneath the upper levels of seating, with vomitoria feeding into the auditorium. The plan of the theatre included

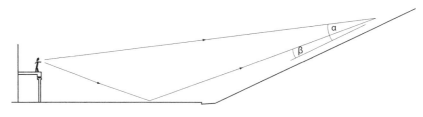

I+

Figure 8.4 Long section of a Greek theatre showing angles to the seating plane for the direct sound (α) and the orchestra reflection (β). As the source is raised, the image position I is lowered, reducing the angle β

a high stagehouse structurally linked to the semicircular auditorium. The orchestra itself was also made semicircular, but since this was normally occupied by senators it could no longer perform a useful acoustic role as a reflector. This in turn required a lower stage level (generally 1.5 m) to maintain visual conditions for those seated on the orchestra. Both the high angle of seating rake (usually 30°–34°) and the smaller size than the large Greek examples explain why the orchestra reflection could be dispensed with in the Roman theatre. The best preserved example is that at **Aspendos** near the southern Mediterranean coast of Turkey (Figure 8.5). This held an audience of 7000 with a distance between the stage front and the most distant seat of 53 m. In this and other Roman theatres there is evidence that a velarium (or canvas cloth) could be drawn over the theatre to shield the audience from the sun. Canvas is in fact mildly reflective acoustically, but little reverberation would occur.

It is not often appreciated that small enclosed theatres also existed in Roman times; for instance, remains survive of one at Pompeii (Izenour, 1992). Structural limitations at the time prevented the Romans from enclosing large-scale theatres and thus avoided definite acoustic problems. An enclosed theatre of the scale of Aspendos would have a reverberation time much too long for speech owing to the excessive enclosed volume. Izenour also discusses another interesting enclosed auditorium: the Odeum of Agrippa in Athens (12 BC). In this case the plan was rectangular but the estimated height of 40 m would give a reverberation time for a fully enclosed space in excess of 5 seconds. But in this building, natural lighting was provided by a substantial opening in the south-facing wall behind the audience; that opening would substantially reduce the reverberation time.

Intelligibility of speech is determined by two acoustic concerns (section 7.3): the reflection sequence and the ratio of speech sound to background noise, or signal-to-noise ratio. In the antique theatre the absence of reverberation means that it is solely a question of signal-to-noise ratio, even if the signal consists of a direct sound enhanced by a few strong early reflections. By enclosing the theatre space, the reverberant sound becomes the disturbing 'noise' and only much smaller audiences can be accommodated. In enclosed space our attention turns to the details of the reflection sequence.

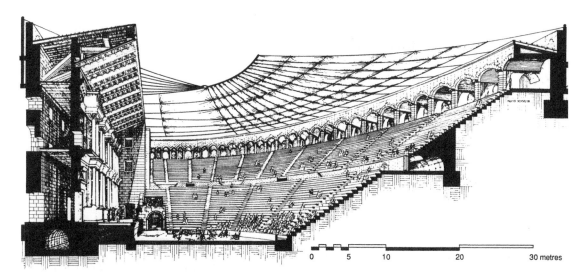

Figure 8.5 Longitudinal perspective section of the Roman theatre at Aspendos (reproduced by permission of Pennsylvania State University Libraries)

8.2 Elizabethan theatre in London

Following the demise of the Roman empire, drama performance went into serious decline and was mainly limited to temporary outdoor structures. The revival of permanent theatre building grew out of temporary theatre structures in Renaissance Italy of the late fifteenth century. Before discussing the development of the proscenium theatre, it is appropriate to turn to the uniquely English design in the Elizabethan theatre. In form these theatres represent an offshoot in the historical development of the theatre, but from a dramatic point of view their importance goes beyond the crucial association with Shakespeare. In the seventeenth and eighteenth centuries theatre development became gradually more obsessed with splendid scenic effects. The earlier Elizabethan theatre played more directly on the imagination. With its lesser reliance on visual realism, it allowed for a particularly intimate relationship between actor and audience.

The open-air, essentially circular Elizabethan theatre lasted in London for 70 years following the first permanent theatre built in 1576. Unlike many European theatres of the time, they were public and fully commercial. Theatre-going was a remarkably popular activity; around 1600 it is estimated that 10 per cent of Londoners saw a play each week (Day, 1996, p. 6). The form of the theatres almost certainly derived from that of contemporary bull- and bear-baiting yards. Several theatres existed in this period but the essential details were consistent. The stage occupied a substantial proportion of the yard, the remaining ground space being occupied by standing audience. A large stagehouse formed the back of the stage and was joined to two or three balcony levels for audience arranged around the perimeter of the building. The most famous of these was the Globe of 1599 on the south bank of the River Thames, rebuilt again in 1613 following a fire. William Shakespeare was a joint shareholder in the Globe and many of his greatest plays were written for the first of the Globe theatres.

The details of these theatres, of the stagehouse, of entrances and exits and inner stages have over the years filled many learned books (such as Hodges, 1968). Several sources of information exist including building contract notes for two theatres. Unearthing the foundations of the Globe and the nearby Rose theatres in 1989 provided incontrovertible evidence, all used for the modern reconstruction of the **Globe Theatre** (Figure 8.6). This project was the dream of the American, Sam Wanamaker, who inspired and seduced people to help him realize the scheme over a 35-year period, only to die three years before the final completion in 1996. The architects were Theo Crosby and Pentagram Design.

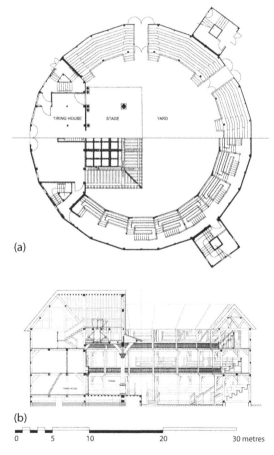

Figure 8.6 (a) Plans and (b) long section of the Globe Theatre, London.

The site of the new Globe is a few hundred metres from the original, whose foundations now lie mostly under a nineteenth-century building and a road. Excavation of what is accessible showed that the original was 20-sided with an outside diameter of 99 ft (30.2 m). The conclusion of various scholars is that the plan followed an *ad quadratum* design (Day, 1996, p. 94), with the front of the stage lying on a diameter of the house. Thus the stage front passes through the centre point of the polygonal plan.

All details of the original have been reproduced as faithfully as scholarship and craftsmanship can make it. The bricks used for the foundation walls were hand-made to Elizabethan measurements, the frame is of green oak held together with wooden pegs, infilling was with traditional lath and plaster. The new Globe is the first building in London to use thatch since the Great Fire of 1666. In two major respects linked to safety, changes were necessary. The new theatre has more exits and the present capacity of 1500 is about half that of the original; Elizabethan audiences were crammed in in ways we would not find acceptable today.

Acoustically the situation in the Globe, either new or old, is fairly straightforward. The absence of a roof means there is little reverberation. Compared with the antique theatres, some reflections will occur within the building, though the open balustrades to the balconies will be mainly sound absorbing when the theatre is occupied. With the furthest seat no more than 25 m from the back of the stage, the direct sound will be strong throughout. Perhaps modern-day audiences are in general quieter than in Elizabethan times when playhouses were noted for their rowdy crowds.

The experience of attending a play at the new Globe certainly feels like stepping back in time, even if some of the rougher edges are now missing. The text is generally easy to hear and surprisingly the noise of road traffic is not obvious. Only the noise of civil aircraft flying low prior to landing provides a serious reminder of the age we inhabit. In the moments in between, this theatre offers a theatrical experience essential for any student of drama.

To return to the seventeenth century: in 1609 the King's Company associated with Shakespeare was sufficiently successful to perform in two theatres. Their new premises at Blackfriars, London, were converted into a theatre for their use. This was an enclosed space with a rectangular plan, which a contemporary performer records did not require the company 'to break our lungs' as did the Globe (Leacroft, 1973). This proves to be the great acoustic gain of enclosure, that the actors receive reflections of their own voices and do not feel tempted to strain their voices. (The other major advantage of enclosure is isolation from external noise.) But as we shall see, the acoustics of enclosed spaces place severe limits on theatre size. Accommodating the 14 000 from the Greek theatre in an enclosed space is unrealistic with unassisted speech.

For the Shakespearean theatre, the victory of the Puritans in the English Civil War brought an end to all public drama and the theatre buildings were demolished in 1644. By the time of the Restoration, theatre design had taken a different line of development. It is only in recent years that a descendant of the Elizabethan theatre has emerged in the thrust-stage theatre. This theatre form provides demanding acoustic challenges (section 8.5).

8.3 The development of the proscenium theatre

Both the antique and English Elizabethan theatres shared an elaborate rear stage wall, known as a *scenae frons*. In the Roman case, this contained entrance ways, interspersed by columns and niches. The Elizabethan stagehouse was of timber but probably became equally ornate with in addition balcony openings above. The oldest surviving Renaissance theatre presents an interesting compromise between the classical *scenae frons* and the new perspective stage. Palladio's **Teatro Olimpico, Vicenza,** of 1584 is a development of the classical idiom in an enclosed space (Figure 8.7). The elaborate *scenae frons* has permanent openings in place of doors, behind which Scamozzi constructed three-dimensional perspective street scenes.

Development elsewhere in Italy had already abandoned the classical models, but the interpretation of the classical design form in an enclosed space exemplified by Palladio's design remained a powerful point of reference. The English Restoration and Georgian theatres retained an extensive forestage with a moderate opening to the rear perspective stage.

The origin of the perspective scene was a logical development from the discovery of perspective which so excited Renaissance artists. It rapidly led to theatres with very deep stages, but only the front of these stages could be used for acting because actors were out of scale with the diminished perspective at the rear. (Acoustically this is just as well since their voices would not have projected well from a distance through a proscenium opening.) The ideal auditorium plan was a matter of much experiment in the first half of the seventeenth century. The semicircle, the U-shape, the horseshoe, the bell-shape, the straight-sided bell and the truncated oval or ellipse were all tried as plan forms. Of

these, the horseshoe became the most common with multiple rows of boxes stacked above each other. Thus the Italian Baroque theatre was born and it proved particularly suitable for opera (section 9.4). The form rapidly spread across Europe, with Inigo Jones as its ambassador in England. It was to dominate theatre design for 200 years, in spite of some obvious shortcomings such as poor sightlines from the side boxes.

The Italian Baroque theatre form subsequently developed in Europe in little more than size and quality, reaching such climaxes as La Scala, Milan of 1778 (section 9.4). Its French contemporary, the opulent **Grand Théâtre of Bordeaux** of 1780 was designed by Victor Louis. The plan of the auditorium here is a segment of a circle (Figure 7.8), a form championed enthusiastically by the English writer on theatre design, George Saunders (Saunders, 1790). This theatre is of considerable interest to the theatre historian, as its timber auditorium, raked stage and stage machinery are virtually unchanged since the original construction. The 1200-seat

Figure 8.7 Teatro Olimpico, Vicenza

theatre is currently used for opera, concerts and ballet (Xu, 1992).

In the latter years of the eighteenth century, social forces were beginning to challenge the subdivision into boxes; the Italian Milizia claimed in 1771 for example, that they were bad for seeing and hearing as well as being immoral. More egalitarian seating arrangements were appropriate to the principles of the French Revolution, which made designers return to classical models for inspiration. Ledoux's theatre at Besançon of 1778 is a rare realization of some of these ideals, with open galleries stepping back instead of vertically stacked boxes. But 100 years later Garnier was still using the Baroque form in the Paris Opéra of 1875. That is not to suggest that architects were not considering alternatives, but that clients would not accept them. Only with the single-minded ambition of Wagner was the mould finally broken in his Bayreuth Festspielhaus of 1876 (section 9.4).

Theatre in England followed a rather independent course, owing to the more commercial basis on which theatres were run. As a guide to theatre development over this period, one has the prime example of the **Theatre Royal** in **Drury Lane, London**, whose auditorium went through no less than seven reconstructions between 1661 and 1922 (see Leacroft, 1973; Mackintosh, 1983). Sir Christopher Wren's theatre of 1674 was slightly fanshaped in plan with two levels of boxes at the sides (Figure 8.8). It contained the extensive forestage already mentioned as a characteristic of Restoration theatre. Wren's theatre also included a feature which was England's single original contribution to the development of theatre architecture prior to 1800: the large undivided balcony. Not only was this a feature of all subsequent Drury Lane auditoria, but it is characteristic of nearly all nineteenth-century British designs.

Robert Adam revised the auditorium in 1775 by adding an extra tier of boxes and replacing Wren's heavy pilasters by much narrower ones to improve sightlines. The design of the Bristol Theatre Royal (section 8.4.1) is thought to be influenced by Wren's design but tellingly this also omits the wide pilasters

in the auditorium. Returning to Adam's Drury Lane, it contained another feature common to many nineteenth-century British theatres: the upper gallery did not fully overhang the lower gallery so the gallery fronts stepped back. (An earlier example of this detail was in Vanbrugh's Queen's Theatre, Haymarket, London of 1705.)

By the late eighteenth century, the Drury Lane theatre had become too small and was totally demolished and rebuilt in 1794 by Holland on a much grander scale. Seating capacity was increased to 3600 compared with the previous 2000. A horseshoe plan was used with now five levels of boxes at the sides and with galleries opposite the stage at the two upper levels. In acoustic and visual terms the theatre stretched the realms of the acceptable. The most remote seat was now 30 m from the actors. A contemporary recorded that:

> the stages of Drury Lane and Covent Garden have been so enlarged in their dimensions as to be henceforward theatres for spectators rather than playhouses for hearers. ... The distant audience might chance to catch the text, but would not see the comment, that was wont so exquisitely to elucidate the poet's meaning, and impress it on the hearer's heart.

The new theatre demanded a profound change in acting style, necessitating a declamatory voice style in order to reach the galleries. Holland's theatre was to burn down in 1809, to be replaced by Benjamin Wyatt's, whose design was strongly influenced by the work of George Saunders (section 7.4). Wyatt followed Saunders' recommended plan form of a truncated circle with the maximum distance from the stage front to the furthest seat now reduced to about 24 m. In spite of its 'scientific' progeny, Wyatt's design had both sightline and acoustic problems. The latter come as no surprise; the acoustic difficulties were probably mainly due to excessive internal volume, leading to excessive reverberation. These deficiencies were apparently remedied in Beazley's reconstructed auditorium of 1823 (Figure 8.9). The form of Beazley's auditorium was a horseshoe in plan, like Holland's. Plan form in this case appeared to have little to do with acoustic success. In general the success of a large proscenium theatre is dependent on a 'tight' design rather than on whichever geometrical form is used. Beazley used a longer, narrower form than Wyatt with smaller balcony overhangs. In addition, Beazley's main ceiling was below the highest seats in the upper gallery. This would restrict the auditorium volume, which we now know as desirable in theatres to avoid excessive reverberation. Moving to the present, the current Theatre Royal, Drury Lane, was built within Wyatt's foyer spaces and dates from 1922. Its capacity is now 2283 but the theatre is no longer used for unassisted voice.

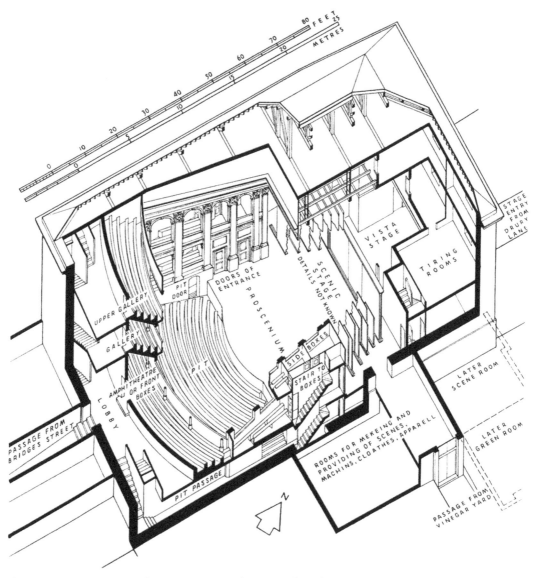

Figure 8.8 Axonometric view of Wren's Drury Lane Theatre, London, of 1674

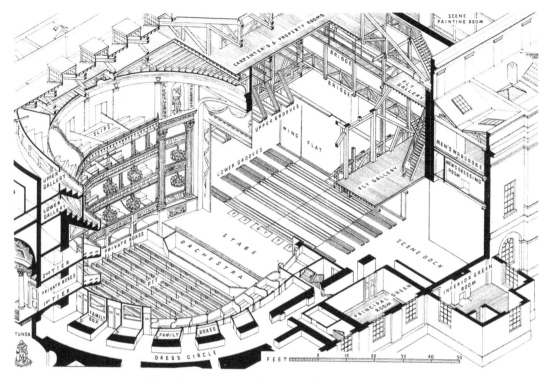

Figure 8.9 Axonometric view of Beazley's Drury Lane Theatre, London, of 1823

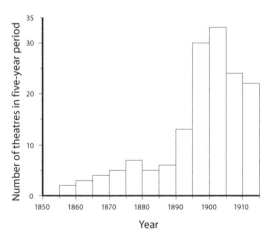

Figure 8.10 Numbers of theatres built in Britain in five-year periods from 1850 to 1915. (Source: Mackintosh and Sell, 1982)

The numerous reconstructions of the Drury Lane theatre were due to it being a Patent theatre, when theatre numbers were heavily restricted by law. Theatre building took off after 1850 and by the first years of the last century almost six new British theatres were opening each year (Figure 8.10). By 1914, 1000 theatres flourished in Britain, only to be eclipsed by the 'bioscope' and 'talking picture'. Most of these theatres operated as music halls or for variety and occasional spectaculars rather than serious drama. The interior decoration was coloured by escapism with eclectic mixtures of Baroque, oriental, Rococo and classical details. The charm of these theatres is unmistakable and given their variety it is remarkable how few architects were responsible. Phipps, Matcham, and later Sprague dominated the scene and amply satisfied their commercial clients. These theatres fully exploited the opportunities offered by steel to cantilever balconies without needing obstructive columns as in the past. In most instances these architects showed considerable concern for sightlines and produced theatres with good acoustic conditions. Their experiences and thoughts on acoustic design were unfortunately never recorded.

Recent history of the proscenium theatre is less flamboyant but nevertheless interesting acoustically. For our detailed examination, we return to the eighteenth century to discuss the oldest British theatre still in regular use.

8.4 The acoustics of six proscenium theatres

The emphasis now shifts from a historical survey to a detailed analysis of twelve British theatres, for each theatre scaled plans and sections and two photos are included. Firstly six conventional proscenium theatres are considered; their basic details of size, capacity etc. are listed in Table 8.1. The table includes the reverberation time, which has an influence on speech intelligibility, though the influence is greater in larger auditoria. The volume per seat is also listed; in auditoria without substantial areas of added absorbing material, this parameter tends to be well related to reverberation time. It provides a simple rule-of-thumb measure; a value of about 4 m^3/seat is recommended for theatres (section 2.8.3).

The measured objective acoustic characteristics in the six proscenium theatres are summarized in Figures 8.11 and 8.12. In the previous chapter, it was concluded that for theatre, two acoustic objective measures should be considered: the early energy fraction and the total speech level. These were measured in each theatre at between ten and thirteen positions throughout the auditorium.

As explained in section 7.1, the orientation of the speaker/actor is very important for acoustic behaviour, so in each theatre two orientations were used. The arrangement of the central and lateral orientations are illustrated in Figure 7.17; for the central orientation the sound source with the directivity of a human speaker is facing directly into the auditorium, for the lateral orientation it faces across the stage. The acoustic conditions are inferior for the lateral source orientation (except in theatre in the round where source orientation relative to the auditorium is irrelevant). The central source location was selected as a typical acting position relative to the stage front. For the central orientation, a minimum source–receiver distance of about 7 m was used. Measurements were made with stage sets in place.

Figure 8.11 shows the range and mean values of the early energy fraction for both source positions. The criterion value for adequate speech intelligibility is greater than 0.50. Figure 8.12 shows

Table 8.1 Basic details of the six British proscenium theatres. Volume is of auditorium only, excluding the flytower. Reverberation time is at mid-frequencies (500–2000 Hz) in the unoccupied theatre.

Theatre	Date	Seats	Volume	Proscenium		Reverb. time (s)	Volume/ seat (m^3/seat)	Acoustic consultant
				Width (m)	Height (m)			
Theatre Royal, Bristol	1766	638	2170	7.4	6.0	0.8	3.4	–
Wyndham's Theatre, London	1899	724	2490	8.4	7.6	0.7	3.4	–
Royal Shakespeare Theatre, Stratford	1932–2007	1459	6310	9.0	7.6	1.0	4.3	H. Bagenal
Arts Theatre, Cambridge	1936	655	1576	8.0	5.0	0.7	2.4	–
Lyttelton Theatre, London	1976	890	4292	14.4	8.8	1.1	4.8	H.R. Humphreys
Towngate Theatre, Poole	1978	584	2433	9.9	5.5	0.9	4.2	–

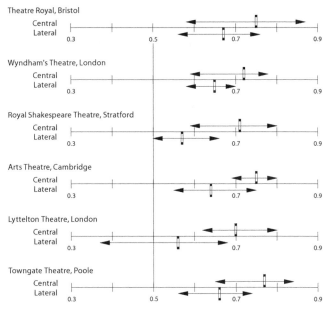

Figure 8.11 Measured values of the early energy fraction (500–2000 Hz) for a speech directional source in six proscenium theatres. The range and mean of values is given for both central and lateral source orientations. The criterion is not less than 0.50 for satisfactory speech intelligibility

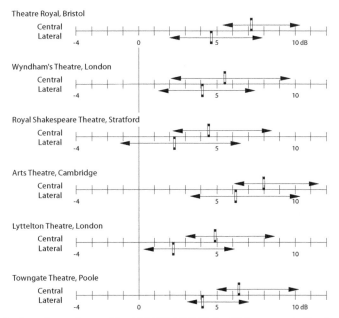

Figure 8.12 Measured values of speech sound level (at 500–2000 Hz in dB) for a speech directional source in six proscenium theatres. The range and mean of values is given for both central and lateral source orientations. The criterion is not less than 0 dB

corresponding values of the total speech sound level; in this case the criterion is for values greater than 0 dB. The measurements were taken during the period 1982–3; most comments refer to the conditions at that time.

8.4.1 Theatre Royal, Bristol

This theatre was designed by James Paty in 1766 at a time when the Puritans were still a force to be reckoned with. For this reason the original theatre contained no formal frontage on the road, but was approached through the ground floor of a house. The date of the theatre makes it contemporary with

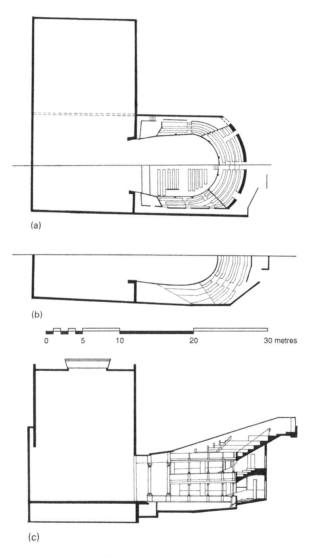

Figure 8.14 (a), (b) Plans and (c) long section of the Theatre Royal, Bristol

Figure 8.13 The Theatre Royal, Bristol

Figure 8.15 The Theatre Royal, Bristol

Wren's Drury lane. The architect was certainly in possession of details of the London theatre but the design is unlikely to be a faithful copy. The two theatres were however comparable in basic dimensions.

The plan form contains a splay at the stage end and the original theatre had the stage extending to the end of this splay (Figure 8.14). Beyond, the balconies are almost parallel, terminating in a semicircle opposite the stage. This is possibly the first occasion that the continental horseshoe form was used in Britain. The upper gallery is thought to date from 1800. The gilded filigree surface ornamentation is also nineteenth century, so the character of the theatre is not fully true to its Georgian origins. The stagehouse was completely rebuilt in 1972 (*Architectural Review*, 1973). This intimate auditorium contains 638 seats.

From an acoustic standpoint, this theatre amply demonstrates the virtues of the small auditorium. The reverberation time itself is short (0.8 seconds). The frequency characteristic is flat, which together with the short reverberation time is likely to guarantee good intelligibility in a space of this size. Objective measures of intelligibility do show high values, which are not much affected by movement of the source. The worst measured position was at the side of the second balcony; this occurs because of the poor sightlines (and 'sound lines') to the upper side seats which occur with the horseshoe form. Poor sightlines were a characteristic often quoted by critics of the horseshoe plan theatre but at nearly all other seats this theatre form has acoustic benefits. Detailed analysis shows a rich early reflection sequence in this theatre which can be associated with the modest width of the auditorium as well as the proliferation of reflective balcony fronts. In the horseshoe plan form no seat is far from a useful reflecting surface. In the Bristol Theatre Royal the low ceiling further guarantees good conditions in the gallery.

Most theatres of the eighteenth century had concave surfaces for the balcony fronts opposite the stage, often with concave rear walls as well. These are anathema for modern acoustic design owing to the focusing they produce. Measurements in the Theatre Royal clearly indicate that this focusing occurs here, for instance in the rear Stalls. Subjectively the effect may be perceived by audience as false localization onto sound from behind. In buildings with such a rich history, such defects always seem to be acceptable.

8.4.2 Wyndham's Theatre, London

This example comes from the golden age of British theatre building before the First World War and is a jewel among a distinguished cast. It was designed in 1899 by Sprague, one of the major theatre architects of the period, who was responsible for nearly all London's new playhouses at the turn of the century. The style is Louis XVI handled with typical eclecticism. Painted panels adorn both the balcony fronts and the glorious circular ceiling (in the manner of Boucher). The proscenium design is at an ultimate stage of development. The forestage, so prominent in Georgian times, had gradually shrunk in size. This theatre illustrates the final point of that trend, where all the stage sits behind a picture-frame, which is expressed architecturally as such. Typical for the time, the design exploits the possibilities offered by cantilevered balconies. The gross design used in Wyndham's Theatre with deep overhangs is characteristic of many larger theatres of the time, but this theatre has only a modest capacity of 724.

In several respects, the acoustic behaviour of Wyndham's Theatre is comparable with that of the Theatre Royal, Bristol. They are of equivalent size with very similar reverberation times. Values for the objective intelligibility measures are similar and are not much influenced by the orientation of the source. At the more remote seats, however, there is a difference between these two theatres: the sound level is significantly quieter in Wyndham's Theatre in the seats furthest from the stage. These seats are in the Balcony, the highest seating level in Wyndham's Theatre (always intended as cheap seats), and to a lesser extent towards the rear of the Stalls. In this theatre early reflections are somewhat compromised by the fan shape in plan. The elaborate box arrangements in front of the proscenium will

Figure 8.16 Wyndham's Theatre, London

Figure 8.17 Wyndham's Theatre, London

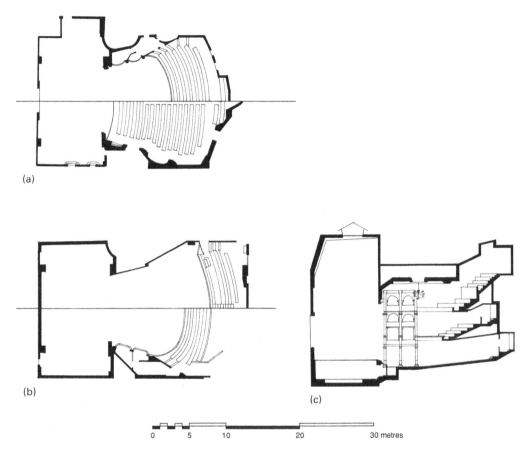

(a)

(b)

(c)

| 0 | 5 | 10 | 20 | 30 metres |

Figure 8.18 (a), (b) Plans and (c) long section of Wyndham's Theatre, London

render this acoustically crucial region poorly reflecting. However, the intelligibility does not suffer in either the rear Stalls or in the Balcony because both the early and late sound decrease together. The later reverberant sound drops off in the Stalls owing to the deep overhang by the Circle, and in the Balcony it likewise drops off due to the small opening between the Balcony region and the main auditorium.

In a theatre of this modest size, the sound level at remote seats is unlikely to be too quiet, and in Wyndham's Theatre it is indeed loud enough. The crucial concern is the balance between early and late energy, which ensures good intelligibility. The acoustic development of the nineteenth-century

proscenium theatre was an unrecorded exercise in trial and error, with results at least in the surviving examples that are seldom disappointing. Wyndham's Theatre is an acoustical success as well as a visual delight.

8.4.3 Royal Shakespeare Theatre, Stratford-upon-Avon

The first Shakespeare Memorial Theatre (of 700 seats) was built in a charitable act of faith that the power of Shakespeare's plays could sustain such a local enterprise. It opened in 1879 amid much scepticism from the London press. By 1926 when it burnt down, the annual Festival at Shakespeare's

Figure 8.19 The Royal Shakespeare Theatre, Stratford

Figure 8.20 The Royal Shakespeare Theatre, Stratford

birthplace had become an essential element of English theatrical life. The very success of the Festival had exposed many inadequacies in the first theatre; George Bernard Shaw disliked the building to such an extent that on hearing of the fire he sent a telegram of congratulations to the Chairman of the Governors.

The 1932 theatre considered here was designed by Elizabeth Scott following a competition. It shocked the local people and the traditionalists who expected a repeat of the mock-Tudor of its predecessor. It shocked many who had simply never seen a free-standing theatre before and had not realized the huge size of a modern flytower. The building was an expression of its age, when modernism dictated a clear expression of function. If latterly it seemed less than beautiful, it proved to be highly adaptable into a good theatre.

The auditorium was fan shaped and clearly in the tradition of Bayreuth (section 9.4). With a balcony, this can and does provide excellent sightlines but at the expense of a sense of intimacy. In an act of some deference to Elizabethan theatre, the Shakespeare Memorial Theatre at its opening in 1932 contained a deep forestage. This was rarely used by players and created a no-man's-land between actor and audience, a gap that was further enhanced for those in the Circle by large areas of blank side walls. Major modifications began in 1951 when a narrow tier was taken down both side walls from the Circle, itself now extended by an extra front row (Mackintosh, 1977). The proscenium area was totally remodelled ten years later with an apron forestage. In 1972 a second tier from the Balcony level was run along the side walls above that from the Circle. Then in 1976 a final 'complete' transformation was achieved when two levels of gallery seating were run round the back of the acting area behind the proscenium, tying in with the side tiers. This scheme almost completely concealed the proscenium arch; a cinema-like auditorium had been converted into an Elizabethan stage surrounded by playgoers. These changes, all introduced to improve the actor–audience relationship, had at each stage increased the audience capacity, from an original 1000 to over

1500. In 1961 the theatre had also acquired a new name: the Royal Shakespeare Theatre.

By 1982 when these measurements were taken, the audience galleries on stage had gone; the proscenium was again visible but a raked forestage, the width of the proscenium arch, jutted out prominently into the Stalls. A vast canopy sat over the forestage and two additional sound-reflecting panels were suspended from the auditorium ceiling.

Hope Bagenal, who has been mentioned at length under British concert halls, was the acoustic consultant for the 1932 theatre (*Architect and Building News*, April 1932, p. 108). In the crucial area around the forestage, convex splays reflected sound from the side and a canopy reflected from above. The extended forestage was also likely to provide acoustically valuable reflections. These three elements were to recur in Bagenal's later concert hall designs: the convex splays in plan in Watford Town Hall (section 5.4) and the canopy and forestage reflector in the Royal Festival Hall (section 5.5). Though in each case their acoustic value is a matter of debate for concert hall use, they are certainly likely to be valuable for theatre. In his account of the acoustic design, Bagenal also mentions that a lot of absorption had to be added to limit the reverberation time to below 1.5 seconds. He calculated the final value as 1.4 seconds by the Sabine formula, but in line with other buildings of his at the time this is probably an overestimate. Projected uses in 1932 included opera for which a longer reverberation time is desirable, but for speech 1.4 seconds is likely to be too long. Plenty of wood panelling was provided 'to brighten musical tone'; we would now welcome such treatment to limit low-frequency reverberation for speech.

The fact that the acoustic consultant found it necessary to introduce large areas of absorption would normally imply that the volume of the theatre was too large. In view of the subsequent large increases in seat capacity, this excessive volume was a fortunate error, so that the measured reverberation time in 1982 was 1.0 seconds, the optimum value for speech. Nearly all the original absorbent treatment had by then been removed. The dated timber work in front of the proscenium was an early casualty

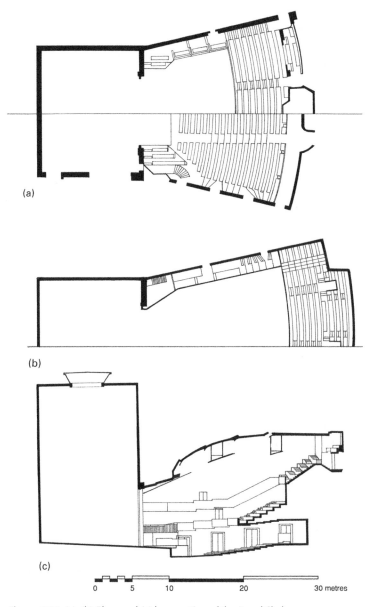

(a)

(b)

(c)

0   5   10   20   30 metres

Figure 8.21 (a), (b) Plans and (c) long section of the Royal Shakespeare
Theatre, Stratford

in modifications to the theatre, to be replaced by
a substantially larger reflecting canopy above the
stage. The 1982 stage arrangement happened to
benefit from a large forestage capable of providing
a strong floor reflection.

Although this is the largest capacity theatre
considered among the twelve here, it behaved well
acoustically, but only just. The changes with source
orientation were greater in the Royal Shakespeare
Theatre than in the two theatres considered above.

With the lateral source, the minimum values of the objective intelligibility measures were on the criterion value of 0.5. A full analysis of good and bad seats and reasons for their behaviour would be tedious but the extremes deserve brief discussion. As frequently happens, the situation at the remote Balcony was quite good with many reflections providing a loud sound. A particularly quiet sound occurred deep under the overhang in the Stalls, but since both the early and late sound were affected, intelligibility remained good there. One of the worst seats for intelligibility was in the front row of the Circle. This seat did benefit from a reflection off the overstage canopy but it missed, one might say by chance, good reflections off the side walls. These side reflections would have been present in the 1932 theatre but, by adding side tiers, some seats benefited acoustically and some suffered.

The Royal Shakespeare Theatre was closed in 2007 prior to gutting the interior of the auditorium. During its 75 years, it had undergone a remarkable organic development; such trial-and-error processes can be appropriate for theatres. Whereas the original auditorium had too large an internal volume, this proved fortunate for subsequent developments. The theatre's latter arrangement demonstrated that satisfactory acoustics can be achieved for a 1500 audience in a proscenium-type theatre. Specific acoustic elements in the auditorium were limited to a large reflecting canopy over the forestage plus other reflectors suspended from the ceiling.

The demise of the much-altered 1932 auditorium was principally caused by the intractable separation between actors and audience. During the current building programme, a temporary courtyard theatre has been built, with a thrust stage plan and two galleries. The new auditorium will follow this courtyard form with a reduced capacity of 1030 but the furthest seat will now be only 15 m from the stage as opposed to 29 m in the older theatre. The new theatre is described as 'the largest tiered thrust stage auditorium in the world, with over 1000 seats'; the similarity with Elizabethan theatres is striking.

8.4.4 Arts Theatre, Cambridge

A major claim to fame for the Arts Theatre is that it was founded and financed by John Maynard Keynes, the father of modern economics. Keynes was born in Cambridge and became a Fellow of King's College. Around 1930, *avant-garde* experiments were being conducted in Cambridge by Terence Gray at the Festival Theatre of ca. 1816, located outside an earlier city boundary. By the mid-1930s its fortunes were declining and a theatre in the centre of the city was proposed. The Arts Theatre opened in 1936 and has provided a distinguished and varied fare to the people of Cambridge as well as providing many British actors and theatre directors with their first experience on a professional stage.

The architect, G. Kennedy, had to accommodate the theatre at the centre of a block of shops and college residences. The small site of yards and gardens of King's College lodging houses demanded a compact design. As another university college could not be persuaded to part with a small piece

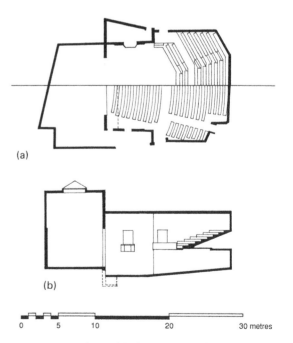

(a)

(b)

| 0 | 5 | 10 | 20 | 30 metres |

Figure 8.22 (a) Plan and (b) long section of the Arts Theatre, Cambridge

Figure 8.23 The Arts Theatre, Cambridge

Figure 8.24 The Arts Theatre, Cambridge

of land at one side of the site, the stage is small and has a back wall built at an angle. The auditorium seats 655 on two levels. The interior decor is austere 1930s. The walls are mainly timber panelling except those opposite the stage, which are covered in carpet, presumably to suppress acoustic reflection back to the stage.

This is a simple proscenium theatre with ample wall and ceiling surfaces to provide acoustic reflections. In a modest theatre such as this, the number of stage lights is too small to create any conflict with acoustic requirements. The result is a loud sound, except in Stalls seats furthest from the stage under the balcony overhang. The reverberation time is short (0.7 seconds); this is caused by the relatively low ceiling and is no disadvantage in a theatre space. The resulting intelligibility is good and remarkably uniform throughout the theatre. Comparison between the acoustic behaviour of the Arts Theatre and the Lyttelton Theatre shows that simplicity can be a virtue.

8.4.5 Lyttelton Theatre, Royal National Theatre, London

The Royal National Theatre contains three theatre auditoria. The Cottesloe is a small studio theatre, too small to be acoustically critical. The Lyttelton Theatre is the intermediate-size theatre with a conventional proscenium organization, while the Olivier Theatre is an open-stage design which is discussed in section 8.6.1. The bizarre machinations prior to Britain having its first purpose-built national theatre will also be considered there.

The Lyttelton Theatre seats 890 and has continental-type seating on two levels in a frontal, almost cinema-like arrangement. The proscenium arch is highly adjustable in both width, height and on plan it can move 1.6 m from above the stage front, back into the stagehouse. The proscenium itself is dark and barely defined other than by lighting. In line with the architect's enthusiasm for concrete, the auditorium itself is visually delineated by a series of vertical concrete walls parallel to the main axis of symmetry. Theatre critics are less than enthusiastic

about the light spill these walls inevitably catch; in acoustic terms these walls and the ceiling are the principal surfaces capable of providing early sound reflections.

The three National Theatre auditoria contain elaborate lighting schemes, capable of accommodating up to 750 luminaires each. This number can be compared with between about 100 and 200 in a commercial London theatre, depending upon the type of show involved. The rational behind this 'saturation rig' is that the theatres are expected to be used in repertory, with the possibility of a matinée play closely followed by a different play for the evening performance. Alternatively rehearsal on stage and public performances can follow each other with minimum delay. In these circumstances there is no time to reorientate, refocus and change filters on luminaires so each production has its own set of independent lights. From a lighting point of view there is compelling logic to this, but it poses immediate acoustic problems. This plethora of lights has to be housed somewhere and sound entering a traditional lighting slot is not available for acoustic reflection into the auditorium (Figure 7.11). In addition, most stage lighting is located close to the proscenium opening in the area which is most valuable for providing early sound reflections. The implications for the Lyttelton Theatre (Figure 8.28) are that only sound from a fairly narrow angle in plan actually enters the auditorium.

The large number of lamps considerably complicates the design of the ceiling as well. The eventual suspended ceiling design in the Lyttelton Theatre has few sections capable of providing sound reflections to seating areas other than to the rear of the Circle (Figure 8.29). The resulting sound was labelled by Victor Glasstone (*Architects' Journal*, 1977) as 'remote and hollow; from under the balcony one strains, but acclimatises; but from the balcony one strains, and does not'. If this criticism is severe overall, it is at least accurate in pointing a finger at the weakest seating area. The measured results indicate good conditions for the performer facing the auditorium but deficient intelligibility with an actor facing away across stage. Before discussing the

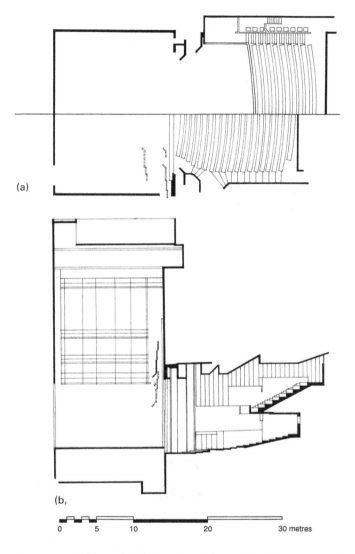

Figure 8.25 (a) Plan and (b) long section of the Lyttelton Theatre, National Theatre, London

area of weakness at the front of the Circle, certain general comments deserve to be made. The reverberation time is 1.1 seconds, caused by an overgenerous auditorium volume. (In intelligibility terms, reducing the reverberation time to 0.8 seconds would increase the early energy fraction values by about 0.10). The speech level in the theatre is quiet, which is attributable to both the detailed design in front of the proscenium arch as well as to some early reflection deficiencies. The measured levels are generally at least 3 db quieter than in the Arts Theatre, Cambridge, at comparable positions; the seating capacity of the Lyttelton is of course 40 per cent greater but that would only account for half the 3 db difference. The major deficiency proves to be in the early reflections when the source points

Figure 8.26 The Lyttelton Theatre, London

Figure 8.27 The Lyttelton Theatre, London

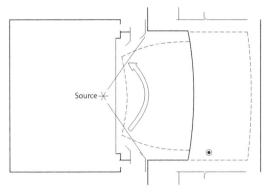

Figure 8.28 Plan of the Lyttelton Theatre showing the angle of the auditorium subtended at the source: 110° in plan out of a maximum 360°. ☉, the Circle seat position with the lowest measured intelligibility for a source pointing away across stage

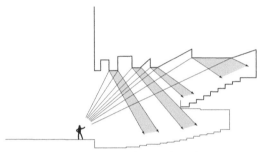

Figure 8.29 Long section of the Lyttelton Theatre showing areas which receive reflections from the ceiling

across stage, which leaves marginal and poor conditions in the front half of the Circle.

In detail the worst measured seat was to the side in the front Circle, behind the source (Figure 8.28). This, of course, is a seat with a weak direct sound because of its orientation but this could be compensated by early reflections. Some early reflections arrive, such as off the side wall opposite the source, but the reflection possibilities turn out to be too few. A good ceiling reflection would probably remedy the situation. Again from Table 7.2 in comparison with the Arts Theatre, Cambridge, which also has a single gallery, the number of early reflections throughout the Lyttelton auditorium is almost half that of the Arts. The Lyttelton front Circle is the only area where this deficiency is crucial.

The Lyttelton Theatre is of a modest size for a proscenium design, which should mean reliable acoustic conditions. All theatres have their weakest seats but in most the intelligibility is always just large enough. The Lyttelton Theatre acoustics are compromised by an excessive auditorium volume and by the detailed design of the proscenium area and ceiling. This detailed design has been complicated by the need to accommodate an immense number of stage lights. The proscenium theatre is the most reliable stage form acoustically but its success does depend on certain surfaces being available to provide acoustic reflections. The example of the Lyttelton Theatre confirms that theatre acoustics cannot be taken for granted, particularly when elaborate lighting schemes are envisaged.

8.4.6 Towngate Theatre, Poole

The Towngate Theatre was the second auditorium in the Poole Arts Centre, adjacent to the concert hall originally called the Wessex Hall (section 5.9). The Arts Centre has now been rebranded the Lighthouse and the Towngate Theatre is now the Lighthouse Theatre. It is a proscenium theatre with well-raked seating on a single level, with a capacity of 600. The front seating row may be removed to provide a small orchestra pit. Since the theatre opened in 1978, its activities and those of the concert hall have been strongly related to the community they serve.

The acoustic characteristics of the Towngate Theatre were admirable in several respects. The reverberation time is short (0.9 seconds) and though it rises in the bass, the theatre is small enough for this not to be serious. The measured intelligibility was high and not much affected by source orientation. Particularly high values for intelligibility measures were found at the furthest seats from the stage. Examination of the theatre long section shows an exemplary design for the suspended ceiling, with angled sections capable of enhancing the number of overhead reflections to the rear-most seats.

However, the theatre had two serious acoustic faults. If an acoustician's eye passed over the long section, it seems to have missed the auditorium

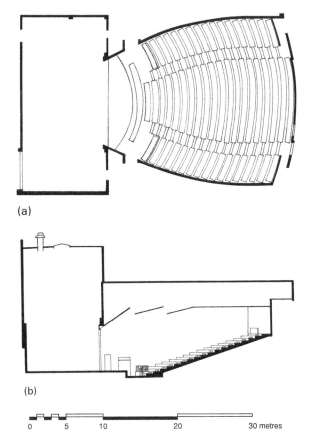

(a)

(b)

0   5   10   20   30 metres

Figure 8.30 (a) Plan and (b) long section of the Towngate
Theatre, Poole

plan. The three concave wall surfaces have a certain
architectural logic about them but they were potentially disastrous acoustically. In many auditoria
there is concern to suppress the sound reflection off
the wall opposite the stage, since this can be perceived as an echo by actors on stage. If this surface
is concave as well as being reflective, the echo level
is enhanced, especially if the centre of curvature is
close to the acting position, as it was here. In the
Towngate Theatre, the path length from the stage
to the rear wall and back is about 50 m, with a travel
time of 150 ms. The stage impulse response (Figure
8.33) clearly shows reflections arriving around this
time. For actors on stage this was a very obvious
echo; it frustrated their performances but they

somehow managed to adapt to it. Placing acoustic
absorbent on this surface was an obvious remedial treatment. Acoustic tiles were installed back in
1978, but the next day a painter rendered them the
same colour as the adjacent walls. Paint on acoustic tiles blocks the very pores which make them
acoustically absorbent, so the echo in this theatre
persisted. In fact the degree of focusing appeared
to be such that a single layer of (unpainted!) acoustic tile would in any case not be adequate. A more
elaborate solution was required with more efficient absorbing treatment than acoustic tile and/or
remoulding of the surface itself.

The curved side walls also produce focusing and
particularly impressive echo effects were produced

Figure 8.31 The Towngate Theatre, Poole

Figure 8.32 The Towngate Theatre, Poole

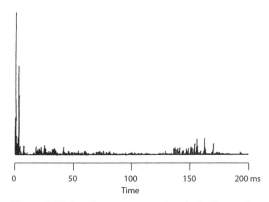

0   50   100   150   200 ms
Time

Figure 8.33 Impulse response on stage in the Towngate Theatre. The reflections, which arrive between 140 and 170 ms after the direct sound, are heard as an echo

with hand claps in the rear of the auditorium. However this echo situation would not have been one normally excited by performers. Single focused reflections are produced by the concave side walls on their own for a sound source on stage. Curtains had been placed on front portions of the side walls to suppress focusing, but their value was limited. In practice, this focusing was probably less serious than the echo returning to the stage but it was disturbing for those at the rear of the auditorium.

The second acoustic fault in this theatre was the background noise level of NR35, which is at least 10 db higher than the normal criterion. Given the relatively loud speech level in this theatre, such a noise level should not in fact impair intelligibility, but it compromises theatrical performances. A dramatic hush should not these days be inhibited by the roar of the ventilation system.

The Towngate Theatre operated and still operates as a successful theatre. Since the theatre was measured, the curved rear auditorium wall has been straightened. Acoustic absorbent has been applied both to this rear wall and the rear of the side walls. In addition the upper section of the front side walls has been cut away to reveal the plane walls that were behind with the extra space now used for extra lighting positions. These various modifications have most likely resolved the quirky focusing effects.

8.5 The thrust-stage theatre

Tyrone Guthrie summarized his view of the theatrical experience by saying that it 'makes its effect not by means of illusion, but by ritual' (Guthrie, 1960). This view led to his invention of a new stage known as the thrust-stage, in which the audience surrounds the stage on three sides. His experiment at the Edinburgh Festival in 1948 spawned a series of theatres in North America and Britain, many of which were designed under his guidance. The two large British examples will be discussed in detail below.

Guthrie's dissatisfaction with the proscenium theatre sprang from his observation that these theatres were trying to be more and more natural by employing progressively more artifice. His argument runs that at a time when cinema and television can provide more effective naturalistic expression, the role of theatre must surely move from creation of spectacle to concentration on the text. This can be achieved by dispensing with much of the scenic trappings inherent in the proscenium system and by providing more intimate contact between actor and audience. Simultaneously, the thrust stage offers a three-dimensional experience in which the actor can be at the focal point of an enclosing audience rather than the two-dimensional arrangement of spectators looking at a screen. The thrust stage is not appropriate to all forms of drama, but it is particularly suitable for the playing of Shakespeare and other pre-Restoration plays. The Elizabethan theatre made no pretence at naturalistic scenery and the rapid changes of location indicated in Shakespearen texts make a mockery of attempts at full naturalism. In compensation, the texts frequently include descriptions to set the scene, whose power is only diminished by literal scenic interpretations.

The first thrust-stage presentation in 1948 was not in fact of Shakespeare but of a play by Sir David Lindsay. Written in thick vernacular Scots in about 1540, *Ane Satire in the Thrie Estaites* was then virtually unknown. In order to stage the play for the Edinburgh Festival, Guthrie surveyed all manner of halls in Edinburgh and selected the Assembly Hall of the Church of Scotland. The hall is rectangular with a raised platform jutting out from the middle of one side, surrounded by seating on three sides. The production was so successful it was repeated at two subsequent festivals. The next experiment was conducted in 1953 in Ontario, Canada, when the modest town of Stratford decided to mount a Shakespeare festival. On this occasion a special tent housing 1500 was constructed around a purpose-built thrust stage. Success also attended this enterprise so that a permanent auditorium was built in 1957 to replace the tent. There followed the Chichester Festival Theatre of 1962 and the Tyrone Guthrie Theatre, Minnesota of 1963, all designed with Guthrie's involvement.

Critics of the thrust-stage arrangement raise the problem that with 180° or more of encirclement, the actor cannot relate to the audience in its entirety. Particularly for spectators to the side of the stage there will be frequent occasions when actors are

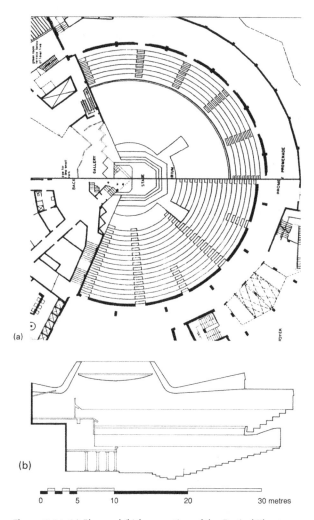

Figure 8.34 (a) Plan and (b) long section of the Festival Theatre, Stratford, Ontario

facing away. Enthusiasts, however, consider that with suitable acting techniques, this is a small price to pay for the flexibility and pace that this stage arrangement allows. But from an acoustic point of view the critics' point has a direct parallel: the human voice is directional so that hearing behind the actor becomes difficult. True, the same audience size can be accommodated closer to the stage in the fan-shape plan of a thrust-stage theatre than in the more linear proscenium situation. But the fan-shape plan is notoriously problematic acoustically, for music in particular, and one reason for this is the limited number of surfaces available to provide early sound reflections. There is a further minor acoustic complication introduced by omitting an opening into a flytower. In a proscenium design some of the voice energy disappears into the flytower and is not then available to contribute to late (disturbing) reflected energy in the auditorium. Without a proscenium opening, all the speech energy enters the auditorium and the design must ensure that most of it arrives at the listeners early enough. Good acoustics in a thrust-stage theatre design is not achieved lightly.

The **Festival Theatre, Stratford, Ontario** (by Rounthwaite and Fairfield), contains seating on two levels in a 230° fan (Figure 8.34). Though this theatre accommodates a remarkable 2258 audience, it contains only 16 rows in the Stalls with no seat at either level further than 20 m from the stage front. The suspended ceiling of the theatre is stepped to provide acoustically useful near-horizontal surfaces separated by 'risers' containing stage lights (Tanner, 1960; Bradley, 1986). Predictably this ceiling has become progressively more perforated to accommodate even more lighting, no doubt with detrimental effects on the acoustics.

The auditorium at Stratford is criticized in theatrical terms for an excessive angle of fan. The Greeks too were aware of the complications for those at the extremities of the fan; in both the ancient and modern fan-shaped theatre there are visual and acoustic problems, making a distinctly non-intimate situation for these seats. A poor sense of intimacy is also claimed for the balcony seating in Stratford, separated as it is from the stage by empty space. Nevertheless this first thrust-stage design functions better acoustically than its immediate successor. The better ceiling design at Stratford is the probable cause of this.

For the two British thrust-stage theatres as well as the four open-stage theatres, measured acoustic results are given in Figures 8.35 and 8.36. Basic theatre details are listed in Table 8.2.

Table 8.2 Basic details of two thrust-stage theatres, two open-stage theatres and two theatres in the round. Volume is of auditorium only, excluding the flytower. Reverberation time is at mid-frequencies (500–2000 Hz) in the unoccupied theatre.

Theatre	Date	Seats	Volume	Proscenium		Reverb. time (s)	Volume/ seat (m³/seat)	Acoustic consultant
				width (m)	height (m)			
Festival Theatre, Chichester	1962	1395	6585	–	–	1.0	4.7	J. McLaren
Crucible Theatre, Sheffield	1971	982	7120	–	–	0.8	7.3	H. Creighton
Olivier Theatre, London	1976	1160	13,500	17.6	12.6	1.0	11.6	H.R. Humphreys
Barbican Theatre, London	1982	1166	4155	22.0	12.6	0.8	3.6	H. Creighton
Roundhouse, London	1979 scheme	703	4828	–	–	1.2	6.9	D.K. Jones
Royal Exchange Theatre, Manchester	1976	684	2917	–	–	0.8	4.3	D.K. Jones

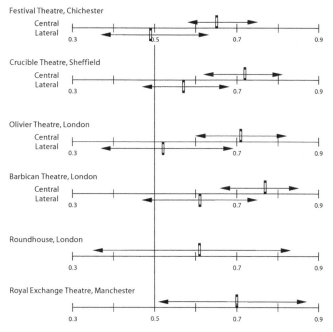

Figure 8.35 Measured values of the early energy fraction (500–2000 Hz) for a speech directional source in two thrust-stage theatres, two open-stage theatres and two theatres in the round. The range and mean of values is given for both central and lateral source orientations (except for theatres in the round). The criterion is not less than 0.50 for satisfactory speech intelligibility

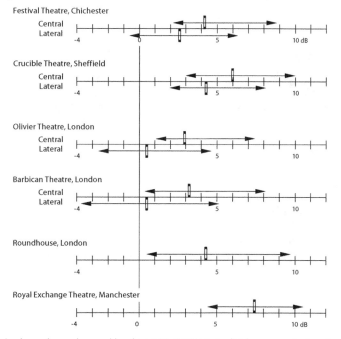

Figure 8.36 Measured values of speech sound level (at 500–2000 Hz in dB) for a speech directional source in two thrust-stage theatres, two open-stage theatres and two theatres in the round. The range and mean of values is given for both central and lateral source orientations (except for theatre in the round). The criterion is not less than 0 dB

8.5.1 Festival Theatre, Chichester

The biography of the genesis of the Festival Theatre, called *The impossible theatre*, was written by the theatre's founder, Leslie Evershed-Martin (1971). Inspired by seeing a television programme about the Canadian Stratford theatre, Evershed-Martin achieved the remarkable feat of mobilizing a cathedral town of only modest size to privately finance and build a new theatre with a form then new to Europe. The obvious need to keep costs to a minimum has resulted in limitations which were not appreciated at the time it was built.

The construction of the theatre is of some interest. The architects, Powell and Moya, selected a hexagonal building plan with the amphitheatre raised on stilts. This allows access to the seating area from vomitoria, much like the Roman theatre system. The lightweight roof was claimed to be the first completely suspended roof in the world: cables link the points at opposite corners of the hexagon allowing the roof simply to sit on the external walls. The roof of woodwool slabs also serves as the ceiling for the auditorium. In comparison with the Stratford theatre, the angle of fan of the walls adjacent to the

stage is only 120°. This is a feature in Chichester's favour, whereas the larger number of rows from the stage, 20 in the Chichester theatre, is likely to be disadvantageous acoustically. The side balconies in the Chichester theatre form a neat solution; given the need to maximize the seat capacity it does so without making those seated in the balcony feel remote from the stage. The total capacity is 1400 seats.

One problem with the amphitheatre form is that it tends to produce a high room volume leading to excessive reverberation time. This is minimized in the Chichester theatre by a shallow seating rake of 13–19°, but obviously at the expense to sightlines. The reverberation time has been further reduced by using absorbent unscreeded woodwool for certain areas of the ceiling. The central lantern and the outer 4 m around the perimeter of the roof are absorbent, leaving the main roof area reflective for simple reflections down onto the audience. The resultant reverberation time was measured unoccupied as a respectable 1.0 second.

But it is the theatre form which compromises the acoustic behaviour. For an actor facing forwards along the centre line, no problems arise. The

Figure 8.37 The Festival Theatre, Chichester

Figure 8.38 The Festival Theatre, Chichester

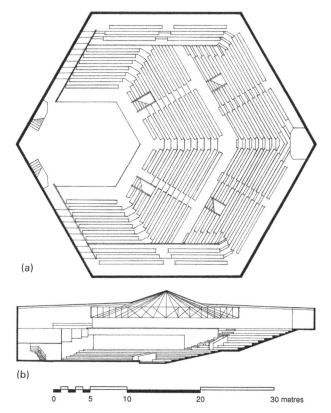

(a)

(b)

| 0 | 5 | 10 | 20 | 30 metres |

Figure 8.39 (a) Plan and (b) long section of the Festival Theatre, Chichester

lowest measured value of the early energy fraction is 0.58. On average the early reflection ratio is 2.7, meaning that 2.7 times the direct sound energy arrives within the useful 50 ms (section 7.3). The seats with the weakest direct sound are naturally those furthest round to the side, but these locations are helped by a reflection off the wall adjacent to the stage. However, with the actor facing across stage, the early reflection ratio is still 2.7 but this is inadequate to provide satisfactory intelligibility. The lowest energy fraction value was measured as 0.37 in the Gallery, indicating obvious difficulties. Figure 8.40 shows the seating areas with poor intelligibility in the Festival Theatre for an actor pointing across stage. Again the reflection off the wall adjacent to the stage is beneficial to neighbouring seating; these seats behave better acoustically than their equivalents at similar distances from the stage. In the remaining seats more than a few metres from the stage there are simply not enough early reflections for adequate intelligibility. This is the case in spite of the fact that the late reflected sound (detrimental to intelligibility) is lower than might otherwise occur. This late sound behaviour is characteristic of the wide-fan plan (see section 4.5). At two measurement positions the measured total level is also just below the criterion of acceptability, indicating possible signal-to-noise problems for these seats as well. Minor echoes also tend to be audible in theatres with insufficient early reflections and they do occur in this theatre.

The thrust-stage design is a potentially marginal situation acoustically. In the case of Chichester, the seating extends rather further from the stage than in other designs. The ceiling arrangement is not good at providing strong reflections; not only is the ceiling rather remote, it is also only visible from the stage through a host of lights and girders. The ceiling has rightly been made absorbent in patches in order to restrict the reverberation time, but this prevents second-order reflections off the perimeter cornice, which could be useful for the remote seats. Even the gangway round the perimeter of the theatre is an element best left outside rather than inside the auditorium. The acoustic problems arise

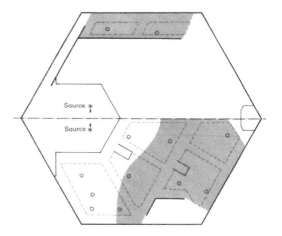

Figure 8.40 Plan of the Festival Theatre, Chichester, showing areas of seating with poor intelligibility (shaded) for a source facing away across stage. Upper half is balcony level, lower half is stalls level. O, measurement positions

when actors face away across stage. In proscenium theatres this orientation is common enough, though theoretically possible to avoid. In the thrust-stage theatre it is essential for actors to perform in this direction and all possible early reflections must be provided to make it work.

An obvious partial remedy to the acoustic problems in the Chichester theatre would be a suspended acoustic reflector at the underside of the ceiling gantry. Whether this is possible in practice is not known, but problems have already been encountered owing to the limited extra loading which the structure can sustain.

8.5.2 Crucible Theatre, Sheffield

Under the guidance of Tanya Moiseiwitsch, a longtime associate of Guthrie's, the designers of the Crucible Theatre (Renton Howard Wood Associates) aimed to incorporate all the lessons learnt from earlier thrust-stage designs. In particular they hoped to avoid the disappointments of the Chichester theatre. Instead of the hexagonal plan, they chose an octagon, which allowed a 180° (or semicircular) fan. The seating capacity was set at

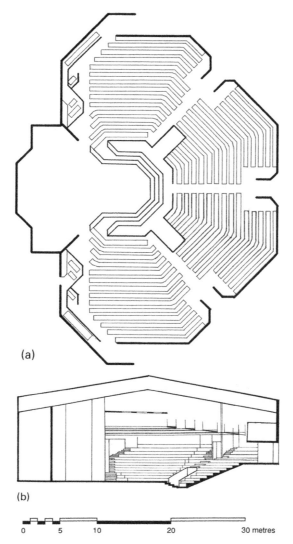

(a)

(b)

0   5   10   20   30 metres

Figure 8.41 (a) Plan and (b) long section of the Crucible
Theatre, Sheffield

1000 instead of 1400, while the seating rake was
increased to 21–28°. The stage was made narrower but with the same depth as Chichester. The
centre of the octagon is at the optimum position
of command for an actor on stage. In these various
ways the focus of the house onto the stage is dramatically increased. Problems with flexibility in the
use of the stage were minimized by making the

whole stage demountable. An opening behind the
stage allows a high degree of freedom in this difficult transition area between stage and auditorium.

The Sheffield theatre was claimed in 1972 to
be 'undoubtedly the best theatre to evolve from
the original (Edinburgh) Assembly Hall experiment' (*Architectural Review*, February 1972, p. 91).
In practice it achieves high annual attendances,

Figure 8.42 The Crucible Theatre, Sheffield

Figure 8.43 The Crucible Theatre, Sheffield

with performances of drama, dance, music and, best known, snooker. The success involves good acoustics, but the means employed to achieve this are different from those in earlier thrust-stage theatres.

The steep seating rake produces a large volume because ceilings are generally not designed to fall from the edge to the centre of an amphitheatre. The predicted reverberation time with a simple shell would be in excess of 2 seconds, much too long for speech use. The ceiling, the only substantial surface apart from the seating area, has therefore to be made absorbent (Creighton, 1978). In the Crucible Theatre the ceiling is of acoustically absorbent unscreeded woodwool. In addition, partially absorbent cork slabs are mounted on the upper perimeter walls to reduce reflected sound energy into the ceiling level of the auditorium. The acoustics of an enclosed space thus becomes changed in the direction of the acoustics of an open amphitheatre. The transition is far from complete though, since the resulting reverberation time is 0.8 seconds. This is long enough that we still have to be concerned with the influence of the late reflected sound on intelligibility. The measured reverberation time frequency characteristic does rise substantially in the bass in this theatre, which is not recommended practice in speech environments. Woodwool is a typical porous absorber and thus little material in the auditorium as a whole is available to absorb at the low frequencies.

With a wide fan-shape plan the ceiling becomes the most valuable reflecting surface: hence the need for a reflecting suspended ceiling in this design. Above the stage is a lattice with a high proportion of open area for lighting purposes. To some extent it will provide sound reflections back to actors. Visible in the long section in Figure 8.41(b) are the four suspended reflectors in front of the stage. In plan these reflectors each follow the profile of the stage front and are effective in giving good reflections for all members of the audience. In simple terms though, this might seem inadequate to give good intelligibility. It was noted for the Chichester theatre that the early energy within 50 ms (or early reflection

ratio) was 2.7 times that of the direct sound, a value which proved inadequate. A single reflection off a suspended ceiling would not make this ratio greater than 2. In reality more reflections do arrive, giving a mean early reflection ratio of 3.5 in Sheffield (Figure 7.15). But it is the physical size of the auditorium as much as the detailed design which produces good speech conditions.

A simple maxim for good auditorium design is that no seat should be too far from a reflecting surface. In the Crucible Theatre this criterion is fulfilled by having only 14 rows of seating. At the rear of the audience local reflections usually occur to make conditions satisfactory. An exception to this generality occurs in the Chichester theatre, where the remote seats are separated from the wall by a gangway and the ceiling immediately above them is acoustically absorbent. The wall design at the Sheffield theatre is responsible for good conditions for the rear three rows at least. For audience close to the stage, the acoustics in any theatre are good simply due to the short distances to the performers. This leaves an intermediate zone in the auditorium where problems can occur. With a short distance between the stage and the most distant listener, the size of the intermediate zone is kept small. The detailed situation in this intermediate zone in the Crucible Theatre has been analysed for a seat in the eighth row. Here in addition to the reflection from the suspended ceiling there appear to be both a second-order reflection path involving the stage floor and the suspended ceiling and also scattered reflections off the rear wall.

The measured results for objective intelligibility are, with one minor excursion, all above the criterion values. This is a considerable achievement which designers of future thrust-stage theatres might like to study. The Crucible Theatre is interesting for a second reason in that the volume is excessive from an acoustic standpoint yet the reverberation time has been brought down by making the ceiling absorbent. This has been compensated by installing suspended reflectors. Even further reflectors could have been placed round the perimeter also directing reflections onto the audience.

8.6   The open stage

The open stage is not a widely accepted category as a theatre type but the use of the term by Ham (1981) for theatres with non-existent proscenium openings is useful. The open stage can be seen as a hybrid between the proscenium and thrust-stage theatres. The forestage produces an intimate relationship between actor and audience whilst flying facilities over the main stage avoid the rather stark character of thrust-stage productions. The two examples discussed here were the result of many years experience by the two leading subsidized theatre companies in Britain. They were both designed as large theatres suitable for the classical repertoire.

In the circumstances there is little point attempting to draw a line through historical predecessors of these designs. One might refer to Reinhardt's Grosses Schauspielhaus of 1920 in Berlin which though vast allowed a communion of actors and audience in a single space. (It also had predictable acoustic problems with an audience of 3500 and excessive auditorium volume.) But the Grosses Schauspielhaus proved to be only a two-year experiment and no single open-stage design has yet become seminal, at least in international terms. The potential dramatic rewards of the open-stage form are considerable, allowing for a high degree of flexibility in staging. In acoustic terms the situation can be expected to be intermediate between the proscenium and thrust-stage theatres.

8.6.1 Olivier Theatre, Royal National Theatre, London

The path to the definitive decision to build a National Theatre in Britain is the stuff of dramatic comedy in itself. Four different sites were acquired or allocated for the theatre and then released for other projects. The idea of a National Theatre was first voiced in the mid-nineteenth century. But the first tangible scheme did not emerge until 1904 under the auspices of the critic, William Archer, and Harley Granville-Barker, actor, director and author. By 1913 a site

had been purchased and fund-raising was gathering some momentum, only to be dissipated by the onset of war. The same cycle repeated itself with the Second World War with a different site and a design by Lutyens. By 1951 the site had moved to the South Bank of the Thames adjacent to the new Royal Festival Hall. A foundation stone was considered appropriate in the heady days of the Festival of Britain, so the then Queen (mother of the present Queen) obliged in 1951 by laying one 'at a temporary location'. A year later the stone was moved to the next site on the South Bank but this was not to be the final location. The eventual theatre was built out of concrete, no foundation stone supports the building and the initiation ceremony was performed with shovels but no royalty! (The stone now sits in the theatre foyer.) The final theatre site is still on the South Bank but a little downstream from the Royal Festival Hall, with commanding views of the city and river.

In 1963 a National Theatre company was formed with the famous actor Laurence Olivier as its first director. Their base was the Old Vic Theatre in south London. Denys Lasdun was then appointed as architect for the new theatre. In the words of Olivier: 'We didn't want anyone who has designed a theatre before, we wanted to find a new soul'. Three auditoria were finally built within the same complex but it was the largest theatre, now called the Olivier Theatre, which was the subject of protracted development between a group of advisers and the architects. Laurence Olivier had recently been directing at the Chichester Festival Theatre. The thrust stage has its attractions but also limitations in that actors cannot relate to the whole audience at once. During the discussions Peter Hall and John Bury developed the concept of the 'point of command': that there is a point about 2½ m back from the front of the stage from which an actor can command the attention of a whole audience within an arc of 135° without having to move his head. This led to the plan of the theatre as a square with the stage at one corner.

> We searched for a single room embodying stage and auditorium ... (in) an open relationship which looked back to the Greeks and Elizabethans and at the same time, looked forward

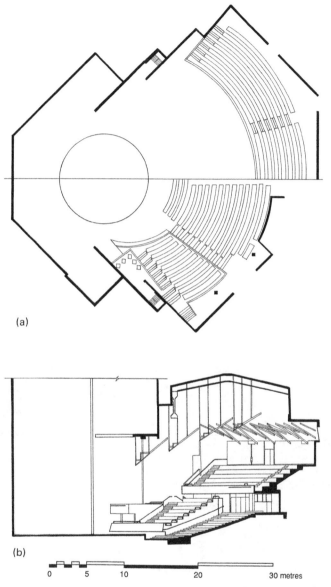

(a)

(b)

0   5   10   20   30 metres

Figure 8.44 (a) Plan and (b) long section of the Olivier Theatre, National Theatre, London

to a contemporary view of society in which all could have a fair chance to see, hear and share the collective experience of human truths.

(Denys Lasdun, *Architectural Review*, 1977)

The selected plan was a 90° fan, but since the stage projects into the space the angle subtended by the audience at the front of the stage is much larger. There is no proscenium opening in the normal sense though vertical and horizontal surfaces define an opening, albeit too large to be expressed visually. Visual limits are defined by the set for each production. A well-publicized photograph underlines the debt the seating arrangement owes to classical Greek theatre; it shows the architect and Peter Hall (National Theatre director 1973–1988) at Epidauros. In the Olivier Theatre the seating is wisely elevated to different levels. However, in the name of equality an overhang has been avoided making a substantial floor area for an unremarkable seating capacity. The arena form has been enclosed in the grand manner with a vast internal volume to

the auditorium; the true ceiling is screened from view by an array of suspended panels.

Suspended ceiling design in a theatre has to satisfy visual, stage lighting and acoustic considerations. With the saturation lighting scheme of about 750 luminaires, similar to that used in the Lyttelton Theatre, there is little scope for accommodating acoustic requirements in the forward sections of the ceiling. In the event three of the four forward (black) suspended ceiling panels, which each span most of the width of the theatre, are oriented in such a way that sound from actors is reflected off them into the upper ceiling space. The panel closest to the flytower was found to be involved in a disturbing echo for actors on the forestage; the echo path also involved reflection and focusing by the curved balcony front. This suspended ceiling panel was therefore made mainly absorbent soon after the opening, which was unfortunate from the point of view of the audience as the panel would have provided them with a valuable additional reflection. (A few years after the measurements reported here, absorbent material

Figure 8.45 The Olivier Theatre, London

Figure 8.46 The Olivier Theatre, London

was placed on the curved balcony front, allowing the suspended panel to be made reflective again. Other minor changes at the same time have now resolved outstanding speech problems.)

At the time of the opening of the Olivier Theatre, it was clear that there were intelligibility problems. Complaint letters were being sent indicating precise seating areas. By that time an acoustic 1:8 scale model existed at Cambridge, which had been built for research purposes, but was completed too late to influence the design. Model tests (conducted by the author and Raf Orlowski) soon isolated the cause of the problems. Echoes had been suspected but these were more a symptom than the cause of poor intelligibility. A lack of early reflections was the reason for the difficulties and this lack tends to make echoes more audible. Reorientation of elements of the suspended ceiling would have helped but was deemed too expensive. Substantial areas of absorbent material were placed above the suspended ceiling, which by lowering the reverberation time effected some improvement in speech conditions. The major acoustic problem however remained.

Measurements in the Olivier Theatre showed a respectable final reverberation time of 1.0 second. Intelligibility for an actor facing into the auditorium is good. But, as also occurs in the Chichester theatre, serious problems arise with an actor facing across stage. The seating area with speech problems was, somewhat surprisingly, the front few rows of the Circle. Simple analysis showed the reason for this. The multitude of small reflecting panels at the rear of the auditorium are there for acoustic reasons and they produce good conditions at the remote seats. But they do not reflect sound to the front of the Circle and these seats have virtually no other source for early reflections. The fan shape in plan does not encourage useful reflections to the centre of the audience. Figure 8.47 shows the reflection situation off the walls; surface A looks useful but analysis shows that the reflections it provides arrive too late to be helpful in many conditions and, for the source positions tested, it reflects more disturbing late energy into the auditorium than useful early sound. For this reason, this surface, which is black in the theatre, was made predominantly absorbent acoustically. A lack of early reflections also exists in the Stalls, but proximity to the source maintains adequate intelligibility for these seats.

The poor intelligibility due to inadequate early reflections into the front rows of the Circle was the most severe problem in this theatre. But a subsidiary problem also exists regarding the sound levels in the audience for an actor on stage. Measured values are below criterion with the source facing across stage. This too is caused by the paucity of early reflections. Since little useful sound is reflected down onto the audience, it eventually arrives as disturbing late sound, which has had to be absorbed as far as possible above the suspended ceiling (and by surface A mentioned above).

The lessons of the Olivier Theatre are clear. The open stage design requires careful acoustic design, particularly of the suspended ceiling. In addition, the immense volume of this auditorium (13 500 m³) is inappropriate for a theatre. The value of 11.6 m³/seat in the Olivier Theatre is considerably above the recommended 4 m³/seat and values in other British theatres (Table 7.1). With excessive volume, extra absorbing material is necessary in order to bring down the reverberation time to give acceptable intelligibility. This material however results in low speech sound levels. A tighter design is more likely to succeed, as illustrated by the next example.

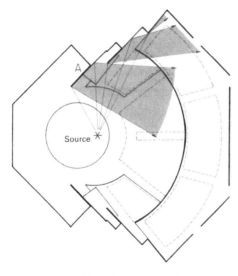

Figure 8.47 Plan of the Olivier Theatre showing areas which receive reflections from the side wall. The reflection from wall surface A arrives too late to enhance intelligibility

8.6.2 Barbican Theatre, Barbican Arts Centre, London

Buried in the huge volume of concrete of the Barbican Arts Centre, by architects Chamberlain, Powell and Bon, is this drama theatre, which for many years was the London home of the Royal Shakespeare Company. The Barbican Theatre holds exactly the same size audience as the Olivier Theatre, but in an auditorium of only one third of the volume. This compactness has been promoted by having no aisles within the theatre. Circulation is kept outside and each row has an entrance door which before the performance automatically closes while the house lights are being dimmed. The effect has been described as a real *coup de théâtre*. The auditorium has another novel feature: the balconies step forward rather than back as they go up. In this obvious manner the distance from the stage does not increase at the higher levels. A sense of intimacy is preserved for all without the 'irrelevant polemic about egalitarianism and Greek amphitheatres' (Glasstone, 1982).

Curiously this balcony scheme was originated for a smaller theatre with a totally different client, the Guildhall School of Music and Drama, which is a neighbour of the Arts Centre but now has separate facilities. Richard Southern was the consultant responsible at that stage in the late 1950s. The Royal Shakespeare Company (RSC) became involved in 1964 and RSC members, especially Peter Hall and John Bury, worked out the design with the architects. The design in plan was again inspired by the 'point-of-command' theory, mentioned under the Olivier Theatre above. The angle of the walls to the centre line is less here than in the Olivier so that few people are at extreme angles to the stage. The result is an exciting working theatre, though its very compactness makes it difficult to photograph satisfactorily.

The reverberation time in this theatre is a short 0.8 seconds. The values of the objective intelligibility measure, the early energy fraction, are all above criterion, with one minor exception. However, though the acoustics of the theatre work, this proves to be

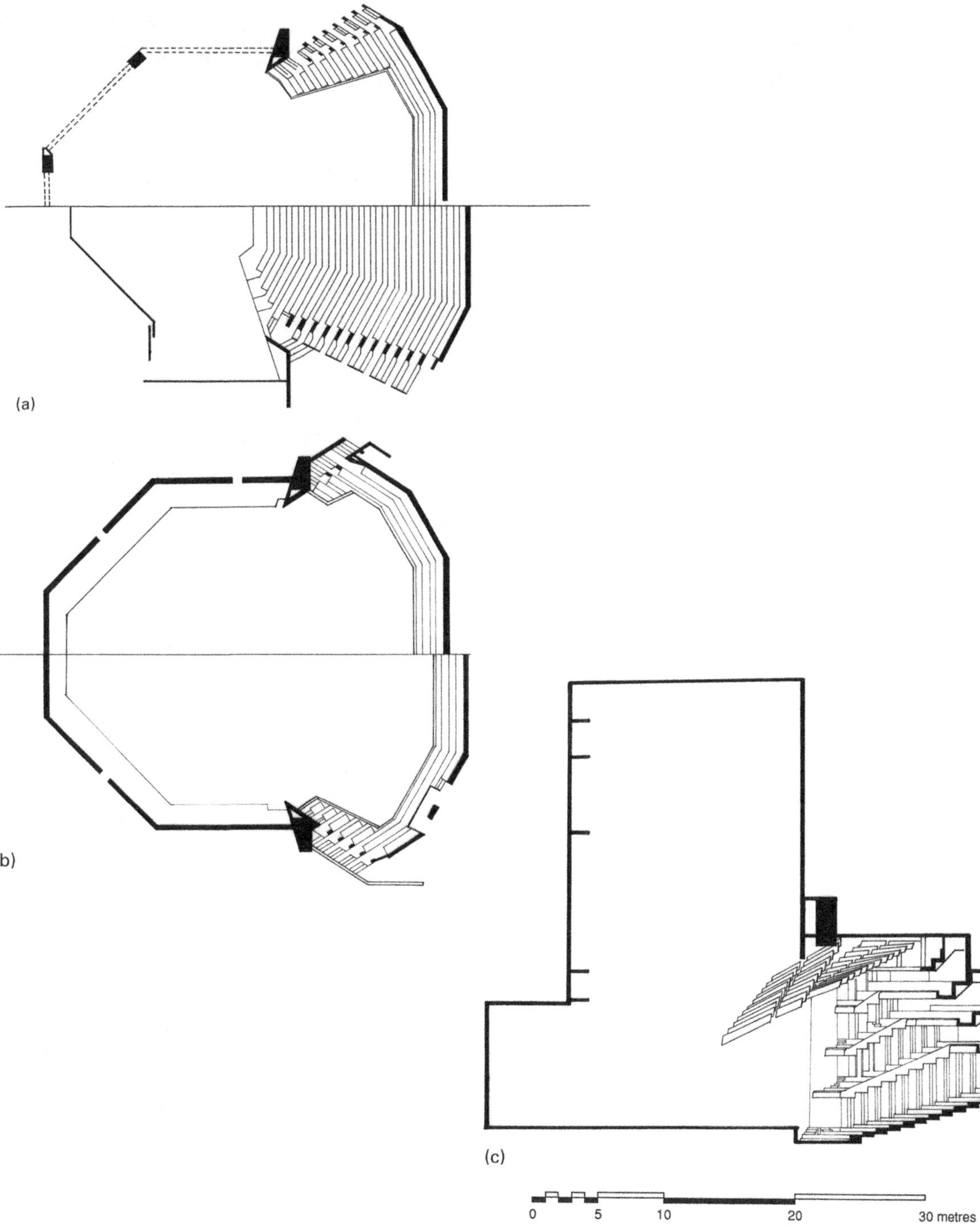

(a)

(b)

(c)

0   5   10   20   30 metres

Figure 8.48 (a), (b) Plans and (c) long section of the Barbican Theatre, Barbican Centre, London

Figure 8.49 The Barbican Theatre, Barbican Centre, London

due to a balancing act which is more delicate than one might imagine. Many references have been made in this chapter to the need for good early reflections to assist speech intelligibility. It is in fact in this area that the Barbican Theatre does not perform well (Table 7.2). The simple measure of this, the early reflection ratio, takes a mean value of 2.3 in the Barbican Theatre. If we restrict comparison to the other theatre of similar form, the ratio is identical in the Olivier Theatre. Yet the intelligibility is distinctly better in the Barbican Theatre. The reason can be summarized by saying that this theatre performs well because it is a tight design with overhung seating.

Good speech intelligibility depends on a high ratio of early to late reflected energy. Keeping all the audience close to the stage is automatically beneficial. In the Barbican Theatre design all the remote seats are close to walls and other reflecting surfaces, so that early reflections are present where they are most needed. The other consideration is the need for a weak late sound energy. In this theatre a large opening into the stagehouse contributes by 'absorbing' a lot of the sound energy which might otherwise arrive late for the audience. (This explains incidentally why an early reflection ratio of 2.3 is acceptable in the open-stage theatre, whereas a value of 2.7 is inadequate in the Chichester thrust-stage design.) The balcony scheme in the Barbican Theatre makes all balcony seats overhung, except for a few seats at the side. Late sound at overhung seats is attenuated giving satisfactory ratios of early to late sound. The penalty with these influences is some low sound levels, with an especially low value at the rear wall corner of the first balcony. This theatre has the potential at just a few seats of suffering from signal-to-noise difficulties.

It remains to discuss the reasons for limited early reflections in the Barbican Theatre. The gross shape of a fan in plan is not good for reflections, but as

Figure 8.50 The Barbican Theatre, Barbican Centre, London

with the thrust stage this can be compensated by good suspended ceiling design. But the potential for providing good ceiling reflections is limited in a design with balconies stepping forward as they go up. The visual screen in the form of slats suspended below the ceiling, as used in this theatre, probably has the effect of inhibiting rather than enhancing ceiling reflections.

In conclusion, acoustic design of an open-stage theatre with balconies stepping forward is not easy. The very compactness of the design would make insertion of specific acoustic reflecting surfaces difficult. But it is this compactness which as much as anything guarantees acoustic success in the Barbican Theatre. For instance, the seating areas with intelligibility problems in the Olivier Theatre are at the front of the balcony, 16 m from the stage front. In the Barbican Theatre, the same distance from the stage front brings one into the last three rows of the Stalls and these seats benefit acoustically from the proximity of nearby surfaces. Similar benefits accrue in the balconies with no seats further than 18 m from the stage front.

8.7 Theatre in the round

Theatre in the round is one of the most primitive forms of drama. The street performer will draw around him a circle of spectators and the sophistication of his act can be judged by his ability to engage the attention of all his audience without favour. Circular earth mounds in Cornwall, of which some survive, are thought to have been used as theatres (Southern, 1975). Many temporary circular arrangements were probably common for theatre in medieval times and it can be argued that the form slowly evolved right up to Shakespeare's day. It is only more recently that the particular intimacy of theatre in the round has been rediscovered. Several examples exist in America, from the first professional centre stage in Dallas, Texas in 1947 (by Margo Jones) to the 2000-seat Casa Mañana, Fort Worth, Texas. In Britain this theatre form was championed by Stephen Joseph, who founded small theatres in the round in Scarborough and Stoke-on-Trent (Joseph, 1967).

Theatre in the round is an extreme stage form in that the stage is totally enclosed by audience.

The acoustic problems with the thrust stage occur when audience is behind the actor. With a centre stage there is always some audience behind the actor. There is also no wall at the back of the stage to provide useful reflections. Only the wall surfaces surrounding the seating are available for acoustic reflections, with only the ceiling as an additional reflecting surface, though this will be somewhat obscured by lighting. If the theatre is circular in plan, steps also have to be taken to avoid focusing by the walls. Several designs were proposed by Norman Bel Geddes in the 1920s, which would have failed seriously for this reason. Not only were his plans circular, but they were enclosed by enormous concave domes. (Fortunately perhaps none progressed beyond the drawing board.)

Acoustic limitations with theatre in the round mean that substantially fewer people can be accommodated for natural speech use than with the thrust stage. Successful acoustic design depends on optimizing every surface in the space to a precise acoustic function, without producing any compromising focusing. The benefits for speech intelligibility which are available from balcony overhangs (which reduce the late reflected sound) have also to be exploited. Two British examples are now discussed which illustrate the probable borderline in terms of audience numbers for this type of theatre. By chance, both theatres are built inside existing Victorian buildings. In both cases acoustic advice was given by D.K. Jones.

8.7.1 The Roundhouse, Chalk Farm, London

Built in a converted circular railway engine shed (originally constructed in 1846), the Roundhouse Theatre always had a makeshift air about it. Funds were always scarce. But in its heyday it was a popular venue, acclaimed for its *avant garde* productions. The playwright Arnold Wesker originally established Centre 42 there in 1964. Its golden era was the late 1960s and early 1970s, when the Royal Shakespeare Company among others played there in a thrust-stage arrangement. Contemporary

classical music, jazz and popular music were also performed; it was even used for some contemporary music concerts as part of the BBC Promenade concert series. Conditions for speech were always a problem and when the theatre was converted into the round in 1978/9, the acoustics were seriously considered. The shed itself has of course a huge volume but without funds to seal off the auditorium in more solid materials, thick drapes round the perimeter and a stretched canopy over the central area contained the theatre space. In spite of the impermanent nature of this theatre its acoustics warrant brief discussion.

There was only one obvious acoustically reflective surface in this theatre: the floor. The canvas canopy reflected a little, but its transparency to sound waves was an obvious consequence of its low mass. In these circumstances it is perhaps surprising that the theatre performed satisfactorily, with only seats in the balcony showing inadequate conditions when the actor faced away. The acoustics of this theatre approached those of an outside space. This is not good for early sound, though the mean measured early reflection ratio was 2.0. The late reflected sound is fortunately also very low, showing that the thick drapes are effectively absorbing sound. The risk with 'open-air' acoustics is that the sound level may become too low. With the furthest seat only 15 m from the centre of the stage the criterion level was in fact achieved here. However, background noise was a problem in this theatre, sandwiched as it was between a railway and a busy road. Signal-to-noise problems did occur.

The measured reverberation time, which at 1.2 seconds would normally be considered excessive, was not of much significance in this theatre, since reflections only arrived late and attenuated by the drapes. The behaviour of sound here as it affects intelligibility is more typical of a shorter reverberation time. Only at low frequencies does the reverberation intrude. Thick drapes are effective as porous absorbers, but do not function efficiently at low frequencies. Thus the effective volume for low-frequency reverberation was the whole building volume rather than just the auditorium volume.

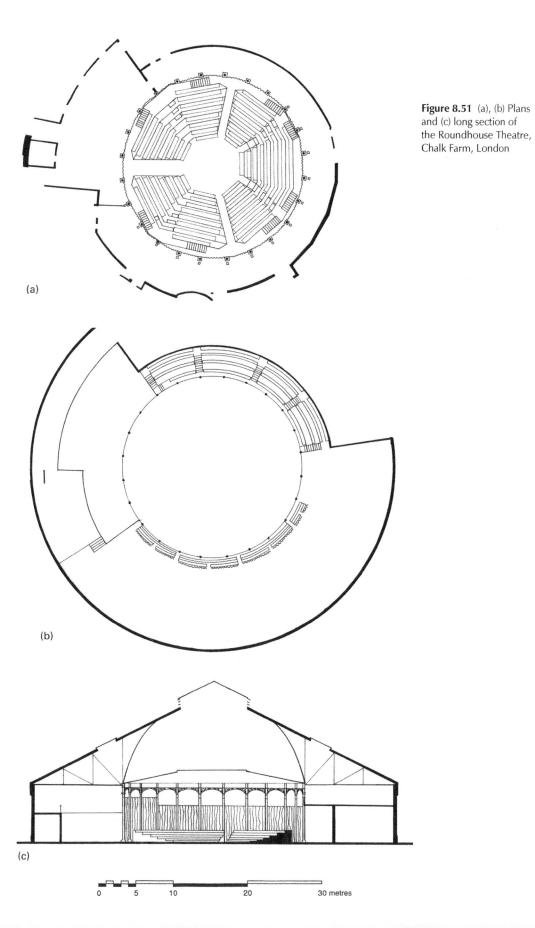

Figure 8.51 (a), (b) Plans and (c) long section of the Roundhouse Theatre, Chalk Farm, London

(a)

(b)

(c)

0   5   10   20   30 metres

Figure 8.52 The Roundhouse Theatre, Chalk Farm, London

This led to an excessive low-frequency reverberation time, somewhat detrimental to intelligibility.

In retrospect, conversion from a thrust stage to theatre in the round may have been a rash move in this building. The earlier thrust stage had an extensive timber canopy over it which must have been acoustically useful. The acoustics were never optimized in that form but the acoustic demands for theatre in the round are more severe. The Roundhouse theatre in the form discussed here closed in 1983 owing to inadequate turnover.

After long periods standing empty, the Roundhouse underwent a major refurbishment, reopening in 2006 as a multi-purpose space for theatre, music of all varieties, dance etc.; stage arrangements in-the-round, thrust and end stage are available. A significant element of the acoustic treatment (designed by Paul Gillieron Acoustic Design) was adding an isolated roof weighing 220 tonnes, resting on 400 springs to match the sound insulation of the 750 mm brick walls. Large areas of fixed absorption have been added to the walls and

soffit to bring down the empty reverberation time without seats to 3 seconds. Occupied reverberation times ranging between 2 and 1 second (at 500 Hz) are now available using curtains as the variable element.

8.7.2 Royal Exchange Theatre, Manchester

From the mid-eighteenth century the Royal Exchange building figured as a major element in Manchester's growing prosperity. The present building, completed in 1874, was used for 94 years as the exchange for the cotton trade. This grand example of Victorian architecture forms a most unlikely home for a fascinating theatre in the round. The theatre 'module' is suspended from the four central piers of the building, which support the domed roof structure of the Exchange (Figure 8.54). It is surrounded by 0.3 ha (3/4 acre) of polished golden parquet floor. The intention throughout was that the auditorium should present a lightness

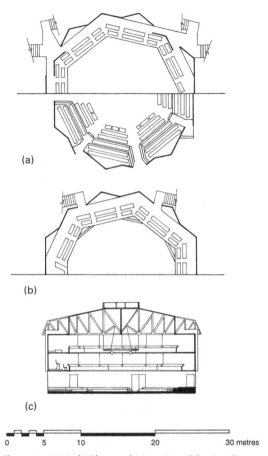

(a)

(b)

(c)

| 0 | 5 | 10 | 20 | 30 metres |

Figure 8.53 (a), (b) Plans and (c) section of the Royal
Exchange Theatre, Manchester

and character precisely opposite to that of the building around it. The theatre envelope is built of steel decking and glass, supported by circular steel trusses. All ancillary accommodation for the theatre is in other enclosed parts of the Victorian building.

Initial experiments were conducted in the Royal Exchange in 1973 with a temporary structure providing theatre in the round with a tented roof. Their success led to the present design which was painstakingly developed between the theatre company, the architects (Levitt Bernstein Associates) and consultants. A compact design seating 700 was required. While the enclosing building provides environmental protection, its structural capabilities

were less than ideal. The maximum floor load was for 450 people, so the upper levels are independently supported from the central piers. Fire requirements dictated substantial areas of glass to allow the audience to see if the surrounding exchange building caught fire. The final design has an acting area totally surrounded by audience. The two shallow galleries contain specially raised seats in their second rows to preserve good sightlines. The result is a highly intimate space in which all the audience is within 11 m of an actor at the centre of the stage.

Several aspects about this theatre provide unusual acoustic challenges: the acoustics of the

Figure 8.54 The exterior of the Royal Exchange Theatre, Manchester

Figure 8.55 The Royal Exchange Theatre, Manchester

surrounding building, the modest acoustic isolation provided by the theatre envelope (due to weight limitations) and the basically circular plan (Jones, 1978). The measured reverberation time in the Royal Exchange building is 4.5 seconds. Rather than hope to be completely isolated from this acoustic environment, the clients wanted to exploit the late decay beyond the theatre envelope, particularly for chamber music. The top 2 m of the theatre walls were therefore made with opening louvres to allow variable coupling. These were left partly open for music performance, giving a classical double slope reverberant decay. For theatre, the acoustic separation (with the louvres closed) is claimed to be 32 dB. This allowed exciting sound effects from the huge reverberant Exchange space to be heard inside. But it was realized that rather greater separation was appropriate to avoid the audible 'ringing' (i.e. reverberation) when an actor shouts.

Elaborate steps were taken to avoid focusing by the theatre walls, or the colouration due to inter-reflection between parallel surfaces which can occur in a building with an even number of sides. A seven-sided plan form was chosen, which is rotated by half-a-side at each gallery level. The compact nature of the design produced a short reverberation time of 0.8 seconds. Though no specific low-frequency absorbent was included in the theatre, the incidental absorption by the glazing and steel decking of the roof structure limited the low-frequency reverberation time to less than 1 second. Speech intelligibility in the theatre was good. Four aspects are responsible for this: (a) short distances between actor and audience; (b) the proximity of all audience to wall/ceiling surfaces capable of providing early reflections; (c) the short reverberation time; and (d) the balcony overhangs which obscure late sound.

A refurbishment of the theatre in 1998 involved some minor acoustic changes by Arup Acoustics (Faria, 1999). To accommodate a more elaborate flying system, the theatre roof was raised by about 600 mm. This had the effect of increasing the theatre reverberation time, which was welcomed as the acoustics are now less dead for music use.

The louvre system which offered variable acoustic coupling between the theatre and surrounding space was replaced by opening roof doors, which are better sealed in their closed position than their predecessors. The Royal Exchange Theatre is not only one of the most exciting theatre buildings in Britain but it is also an acoustical *tour de force*. It illustrates successful resolution of many of the acoustic concerns voiced in this review, which in the extreme case of theatre in the round are essential for good acoustics.

8.8 Conclusions

The need to provide good sightlines for a whole audience introduces severe design constraints in auditoria. In theatre the options become even further reduced with the introduction of scenery. The distance to the furthest spectator is also limited for purely visual reasons (to about 20 m), if facial expression and gesture are to be registered. Many aspects in theatre have progressed immeasurably since classical times, yet the basic elements of the human voice and hearing system and the fundamental nature of human communication have obviously remained constant. A certain circularity can be observed in the development of basic theatre form. During much of the twentieth century the nature of the theatrical experience was brought into question. But surprisingly theatre designers have found themselves frequently turning back to earlier models for inspiration. Acoustics has often intruded into these developments as an awkward necessity. It is obvious that any theatre form will work with a small enough audience. Commercial pressures to maximize audience sizes have ensured that the limits of each theatre form have been established on a trial-and-error basis. It is the hope and expectation that modern acoustic techniques can avoid acoustic failures in theatres.

Remarkably little has been published in the literature about acoustics for theatre, presumably because acoustics for speech are much simpler to handle than for music. Whereas with music several acoustic factors have to be juggled simultaneously, with speech the overriding concern is with intelligibility. For speech other concerns rarely intrude to the point where compromise is required. When one proceeds to look at the objective acoustic requirements for good speech intelligibility, two aspects are found to deserve attention: the proportion of energy which arrives early and the signal-to-noise ratio. Of these two, the proportion of early energy is the major concern. From experience of six proscenium theatres, the signal level is not likely to be deficient in this theatre form. In the more open stage arrangements, signal level problems tended to occur in the theatres with inadequate fractions of early energy, owing to deficient early reflections in each case. Excessive background noise levels can occur in any theatre of course, but noise is usually intrusive by itself by the time it begins to interfere with speech intelligibility. A noise level criterion of close to NR20 for larger theatres is recommended.

This review of 12 theatres has clearly shown that the type of theatre is a significant determinant of the acoustic performance. A major influence is the degree of enclosure of the stage by audience, which varies from small angles for proscenium theatre to 360° for theatre in the round. Because of the directional nature of the human speaker, the wider the degree of enclosure the shorter must be the distance to the furthest seat in the house and the more critical the acoustics become. The proscenium theatre, carefully designed, is likely to be able to accommodate the largest capacity audience.

Detailed analysis here has shown consistent characteristics in the theatres with good acoustics. Reverberation time values of 1.0 second are generally recommended for speech use. This survey has shown that shorter values are common and there seems no reason not to recommend shorter values of 0.8 seconds in larger theatres. In the case of the Olivier Theatre, for example, there would be a potential gain in the early energy fraction of 0.10 due to a modest reduction in reverberation time from 1.0 to 0.8 seconds. This is a typical result and the intelligibility gains from such a reverberation time change are not to be ignored in marginal designs. A short reverberation time is achieved either by having a

small auditorium volume or by installing an absorbent ceiling with careful suspended ceiling design. A second basic comment on theatre form is appropriate to all auditoria: the avoidance of concave surfaces.

In the case of proscenium theatres, small designs automatically behave well. The Baroque theatre that dominated design for so long is usually reliable because no member of the audience is far from a potentially useful reflecting surface. A clear lesson of analysis of these theatres is that for speech intelligibility deep overhangs often behave well, because the overhang obscures much of the late reflected sound detrimental to intelligibility. Yet problems can arise in the proscenium theatre with excessive auditorium volume and inadequate surfaces to provide early reflections.

In theatres with greater degrees of enclosure of the stage by the audience, the acoustic situation rapidly becomes marginal due to the directional nature of the human voice. In the examples studied here, problems were principally caused by inadequate early reflections. It is a simple matter to give good conditions at the most remote seats since these can be surrounded by reflecting surfaces. Seats close to the stage are unlikely to present problems. It is in the intermediate zone that acoustic complications are most likely. One might call this the 'bare midriff phenomenon'. Theatres in which this zone is small (i.e. with short distances to the furthest seats) are minimizing potential acoustic difficulties. In larger open-stage designs, good early reflections for this intermediate zone are essential. Acoustically reflecting suspended ceilings are generally necessary, which must not be compromised by the demands of lighting designers.

Compact design in theatres is often beneficial for the acoustics. It gives a short reverberation time and no audience member is too far from an acoustically reflecting surface. The opposite extreme with a high ceiling takes one back to the classical Greek open-air theatre when the stage has a high degree of enclosure. By making the theatre ceiling absorbent, the reverberation time can be made sufficiently short, but this has to be compensated by extremely careful suspended ceiling design.

The borderline between satisfactory and inadequate intelligibility is a surprisingly fine one. Many smaller theatre designs can be guaranteed acoustically simply on the basis of precedent. But for larger theatres, especially for the thrust stage, open stage and theatre in the round, the prediction of success or failure is difficult. Modelling with either acoustic scale models or computer models should be able to help avoid that 'midriff phenomenon'.

References

General

Glasstone, V. (1975) *Victorian and Edwardian theatres*, London: Thames and Hudson.

Izenour, G.C. (1977) *Theater design*, New York: McGraw-Hill.

Joseph, S. (1968) *New theatre forms*, London: Isaac Pitman.

Leacroft, R. (1973) *The development of the English playhouse*, London: Eyre Methuen.

Leacroft, R. and Leacroft, H. (1984) *Theatre and playhouse*, London: Methuen.

Mackintosh, I. (1993) *Architecture, actor and audience*, London: Routledge.

Mackintosh, I. and Sell, M. (eds) (1982) *Curtains or a new life for old theatres*, Eastbourne: John Offord (Publications).

Pevsner, N. (1976) *A history of building types*, (Chapter 6), London: Thames and Hudson.

Vitruvius (translated by M.H. Morgan) (1960) *The ten books on architecture*, New York: Dover Publications.

References by section

Section 8.1

Canac, F. (1967) *L'acoustique des théâtres antiques. Ses enseignements*, Paris: Centre National de la Recherche Scientifique.

Cremer, L. (1975) The different distributions of the audience. *Applied Acoustics* **8**, 173–191.

Izenour, G.C. (1992) *Roofed theatres of classical antiquity*, New Haven: Yale University Press.

Shankland, R.S. (1973) Acoustics of Greek theatres. *Physics Today*, **26**, 30–35.

Vassilantonopoulos, S.L. and Mourjopoulos, J.N. (2003) A study of ancient Greek and Roman theater acoustics. *Acta Acustica with Acustica*, **89**, 123–136.

Section 8.2

Day, B. (1996) *This wooden 'O', Shakespeare's Globe reborn*, London: Oberon Books.

Gurr, E. (1998) *Shakespeare's Globe*, Reading: Spinney Publications.

Hodges, C.W. (1968) *The Globe restored, a study of the Elizabethan theatre*, Oxford: Oxford University Press.

Section 8.3

Mackintosh, I. (1983) The rise and fall of the Georgian playhouse 1714–1830. *AA Files*, **4**, July, 16–23.

Saunders, G. (1790) *A treatise on theatres*, London (facsimile by Benjamin Blom, New York, 1968).

Xu, A.Y. (1992) Invisible acoustics in the restoration of the Grand Théâtre de Bordeaux. *Proceedings of the Institute of Acoustics*, **14**, Part 2, 137–145.

Section 8.4.1

Architectural Review (1973) **153**, February, 123–131.

Section 8.4.2

Mander, R. and Mitchinson, J. (1975) *The theatres of London*, London: New English Library.

Section 8.4.3

Architect and Building News (1932) 22 April, 90–117.

Architectural Review (1932) **71**, April, 219–231 and 241–256.

Beauman, S. (1982) *The Royal Shakespeare Company, a history of ten decades*, Oxford: Oxford University Press.

Ellis, R. (1948) *The Shakespeare Memorial Theatre*, London: Winchester Publications.

Mackintosh, I. (1977) Scene individable or poem unlimited. *Architectural Review*, **159**, January, 59–64.

Pringle, M.J. (1994) *The theatres of Stratford-upon-Avon 1875–1992*, Stratford: Stratford-upon-Avon Society.

Section 8.4.5
Architects' Journal (1977) 12 January, 59–89.

Section 8.5
Bradley, J.S. (1986) Acoustical comparison of three theaters. *Journal of the Acoustical Society of America*, **79**, 1827–1832.
Guthrie, T. (1960) *A life in the theatre*, London: Hamish Hamilton.
Tanner, R.H. (1960) Acoustical design and performance of the Stratford (Ontario) Festival Theatre. *Journal of the Acoustical Society of America* **32**, 232–234.

Section 8.5.1
Architects' Journal (1962) 4 July, 25–40.
Evershed-Martin, L. (1971) *The impossible theatre*, London and Chichester: Phillimore.

Section 8.5.2
Architectural Review (1972) **151**, February, 82–94.
Creighton, H. (1978) The Crucible Theatre, Sheffield. *Proceedings of the Institute of Acoustics* **4**, 14-8-1 to 14-8-2.

Section 8.6
Ham, R. (1981) Buildings update, Leisure: Theatres and performance spaces. *Architects' Journal*, 12 August, 309–323 and 19 August, 355–368.

Section 8.6.1
Architects' Journal (1977) 12 January, 59–89.
Architectural Review (1977) **159**, January, 2–70.
Elsom, J. and Tomalin, N. (1978) *The history of the National Theatre*, London: Jonathan Cape.
(1977) *The complete guide to Britain's National Theatre*, Tadworth, Surrey: Heinemann.

Section 8.6.2
Glasstone, V. (1982) A triumph of theatre. *Architects' Journal*, 18 August, 31–44.

Section 8.7
Joseph, S. (1967) *Theatre in the round*, London: Barrie and Rockcliff.
Southern, R. (1975) *The medieval theatre in the round*, revised edn, London: Faber and Faber.

Section 8.7.2
Architectural Review (1976) **160**, December, 356–362.
Faria, J. (1999) The acoustic design of the refurbished Royal Exchange Theatre, Manchester. *Proceedings of the Institute of Acoustics*, **21**, Part 6, 123–128.
Jones, D.K. (1978) The Manchester Royal Exchange Theatre. *Proceedings of the Institute of Acoustics* **4**, 14-9-1 to 14-9-2.

9 Acoustics for opera

9.1 Introduction

Opera has always been an extravagant art form. There is not only the expense of an orchestra and the need for full stage facilities, but certainly since the early eighteenth century opera singers have commanded high salaries. The expense of opera led to early performances being limited to court surroundings, yet when first presented for public consumption, opera started a theatre-building boom which can be likened to that for variety and music hall entertainment in the second half of the nineteenth century. In the twentieth century, opera spawned the musical, exploiting all the tricks electronic sound has to offer. One might have imagined that in a scientific age Dr Johnson's 'exotic and irrational entertainment' would disappear as a cultural irrelevance. In reality the opposite has happened; television, radio and recordings have made opera accessible as never before. If opera remains élite, it is simply because the expense of its production limits the number of venues where it is performed. The pursuit of opera is lavish in Germany and of course in its original home of Italy. As a monument to the bicentenary of the French Revolution in 1989, Paris built a completely new opera house at the symbolic location of the Place de la Bastille. In Britain, public subsidy for new opera houses is considered frivolous; no new public opera house was built in the twentieth century, only the privately funded Glyndebourne Opera House was completed in 1993. Cardiff gained a new 1700-seat opera house in 2004 (Newton, 2006), which forms the main space in the Wales Millennium Centre.

Opera, as an art form in which music was more than incidental, developed in Italian court circles in the late fifteenth and early sixteenth centuries. Monteverdi was one of its earliest masters. Many of the early productions were in temporary performance spaces, but often spectacular in their staging. Public opera dates from 1637 in Venice when the first purpose-built opera house, Teatro San Cassiano, was opened. This house was also distinctive in that for the first time the orchestra was placed immediately in front of the stage, instead of musicians performing on the stage itself. Following its public launch, Venice became obsessed with the new art form; at least 16 new theatres for opera were built in the city between 1637 and 1700.

Because they were run on a commercial basis, public opera was inevitably less lavish than most court entertainments. But elaborate stage decor still remained a crucial element of the performance, with cloud machines and magical illusions. An early design survives of the Teatro SS Giovanni e Paolo in Venice, originally built in 1639, then remodelled specifically for opera in 1654 with five tiers of boxes (Figure 9.1). This illustrates how early the traditional baroque theatre form became established for opera, which then dominated design for so long. In this case the plan is U-shaped. The deep stage able to create dramatic perspective effects was another hallmark of its time.

In traditional baroque designs, the distinction between drama theatres and opera houses has generally been small. There should be a subtle difference in sightline design: for seating other than the stalls

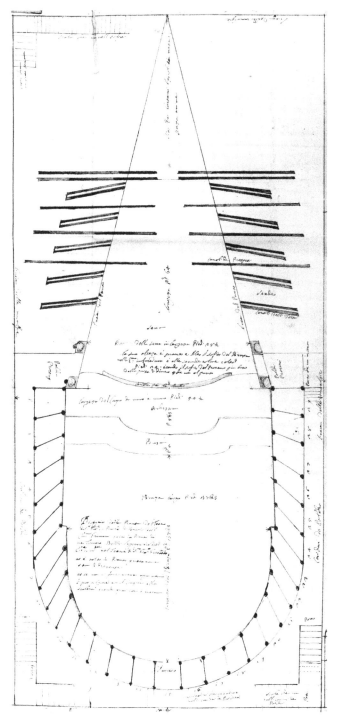

Figure 9.1 Plan of SS Giovanni e Paolo theatre, Venice, 1654 (by courtesy of the Trustees of Sir John Soane's Museum)

the orchestra pit should be visible for opera. Opera houses have tended to have boxes, which suited the élite clientele. The advantage of the box is not acoustic. For all except those at the very front of the box, the appreciation of the room sound is diminished by the partial enclosure provided by the box itself. Yet it was this design form with boxes which was used for the first large-scale houses: Teatro San Carlo, Naples, of 1737 and Teatro alla Scala, Milan, of 1778 (discussed further in section 9.4). Opera houses prior to 1900 deserve our attention for at least two reasons. Firstly they were built principally for performance of contemporary operas. And secondly many operas, among them those of Mozart, were composed with a specific house in mind for their premières. Many of these early houses were more intimate than is common nowadays. It is necessary to understand the acoustic nature of these original opera houses, if we wish to reproduce the intentions of the composer and his colleagues.

Of all auditorium forms the opera house is the most constrained in terms of design. The need for singers and orchestra to view the conductor makes the proscenium stage inevitable, with the orchestra in a pit between stage and stalls seating. Even if alternative positions for the orchestra might be considered viable with a closed-circuit television link to enable singers to see the conductor, the need for international artists to be able to perform standard repertory with minimum rehearsal makes this an unsuitable proposition for grand opera. The design of the auditorium is also highly constrained. Sightline considerations determine a maximum angle of view in plan relative to the stage of typically 30° (Figure 9.2). The maximum audience distance from the stage is limited both for sightlines and acoustics; a figure of 30 m is generally applied for opera. Given these constraints it is no wonder that the baroque theatre form has been so dominant and is still used for some new opera houses today.

The development of the baroque theatre has already been discussed in section 8.3; Forsyth's (1985) *Buildings for Music* contains a more complete treatment specifically for opera houses. Both for opera and drama, U-shaped, horseshoe or elliptical plan forms were used. The Galli-Bibiena family were the designers for many eighteenth-century opera houses, including such jewels as the Markgräfliches Opernhaus, Bayreuth, of 1747. This and several other of their houses had a bell-shaped form in plan. The significance of these various plan forms is more for sightlines than acoustics, especially in the smaller houses. The biggest acoustic risk with these curved forms is of focusing, which would be most severe with an elliptical form with a flat stalls floor. Steep raking of the floor is likely to remove this risk by making the area of focus occur above the heads of any audience.

The major challenge to traditional design was due to Wagner with his Bayreuth Festspielhaus of 1876, which returned to classical Greek or Roman forms. From the 1950s, hybrids of the baroque and Wagner solutions were tried; the Opera Bastille of 1989 might be seen as a major product of this approach. Since the 1990s the trend has been to return to the intimacy of the baroque theatre form, while avoiding some if its traditional acoustic shortcomings.

In common with all auditoria, the size or audience capacity strongly influences the ease or difficulty of design. A small house with less than 800

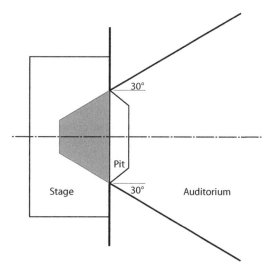

Figure 9.2 The effect of angle of view on visibility of the stage

seats is relatively uncritical in terms of acoustic design, though this size is hardly economic. The medium size, from 800 to 1500 seats, can be considered the optimum in which careful design is required but few serious compromises are likely to be involved. At the large scale of over 1500 seats, compromises have to made, with a larger than ideal distance to the furthest seats and probably deeper balcony overhangs.

Before discussing in more detail the various design forms that have emerged during 370 years of development, it is necessary to begin an acoustic analysis with discussion of the character of the sound sources in opera.

9.2 The elements of opera house design

A major puzzle of opera performance is how do individual singers make themselves heard over an orchestra? It is clearly impossible for a singer to generate the same sound power as an orchestra, which for grand opera may have as many as 100 players. Clues to the solution of this riddle have only been suggested in recent years. Of major importance is an understanding of the characteristics of the trained singing voice.

9.2.1 The singing voice

The human voice is a fascinating organ. The vocal chords determine the frequency of the (fundamental) note being sung. Singers from soprano to bass each have a compass of roughly two octaves (Figure 2.5). But as with musical instruments, though the fundamental represents the lowest note present, there are many multiples or harmonics of that fundamental frequency, which are produced simultaneously. The sound character of different instruments is due to the relative intensities of these harmonics. For the singer the intensity of the various harmonics characterizes the individual voice and is strongly influenced by the acoustic behaviour of the vocal tract between the vocal chords and the lips. The

vocal tract acts as a resonator which enhances certain frequencies and suppresses others. In this respect it is no different from the resonator effects associated with the 'tubes' in all wind instruments. The distinctive characteristic of the human voice is the ability to modify the resonant or **formant** frequencies by changing the shape of the vocal tract (section 2.2). The different vowel sounds in speech are produced in this way, principally by movement of the jaw, as well as the body and tip of the tongue. Manipulation of the formant structure of the voice is the main element at the disposal of singers to enhance their vocal output.

In major discussions of the acoustics of the singing voice, Sundberg (1977, 1995) mentions that adult males can vary the frequency of their **lowest formant** from about 250 to 700 Hertz; that is over one and a half octaves. The basic formant frequencies in adult males and females are similar, so this presents a problem for sopranos singing towards the top of their range (at around 1000 Hz). Sundberg found that by raising the lowest formant frequency to near the note of the frequency being sung, the soprano can considerably increase her vocal output with no extra voice effort. The singer does this by opening her jaw wider. The penalty the singer pays for this manipulation is to compromise the purity of vowel sounds.

The most intriguing aspect of Sundberg's discussions concerns the notion of a 'singing formant', which he claims is typical of voiced sounds sung by professional male singers and altos. The frequency range of this formant lies between about 2500 and 3000 Hz. Its value can be seen in Figure 9.3, which shows how the spectrum of the orchestral sound is decreasing rapidly above 1000 Hz. The singing formant is present regardless of the pitch, the particular vowel and the dynamic level. Since its frequency occurs where orchestral sound is relatively weak, it allows the singer to be heard above a louder orchestral sound even though the fundamental singing sound may be masked. It is thought that the extra formant is produced by lowering the larynx. In architectural terms, the presence of a singing formant implies that the design should

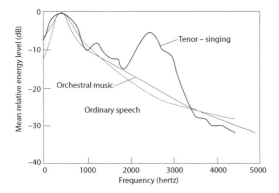

Figure 9.3 Spectra of orchestral sound and the singing voice, showing the 'singing formant' around 3 kHz (after Sundberg, 1977)

enhance projection from the singers at frequencies around 3 kHz.

The other aspect of singers which should be considered is the directional pattern of their sound output. This has been measured by Marshall and Meyer (1985) for a baritone. (More recent studies of singer directivity and power by Cabrera and Davis, 2004, and Katz and d'Alessandro, 2007, are also of interest.) The radiated directions measured by Marshall and Meyer are illustrated in Figure 9.4 for the horizontal and vertical planes. The figure shows the inevitable increase in directivity with higher frequency. For instance, for sound energy in

the 'singing formant' which would mainly fall in the 4 kHz octave, the majority of the energy is within only ±35° horizontally of straight ahead. The voice is however more directional (that is, confined to a smaller angle) vertically than it is to the side. The behaviour at 2 kHz also deserves mention since the maximum sound output is not straight ahead, but is directed downwards at 20°. However, limited measurements on female voices suggested that this directional peak may not extend to them. Marshall and Meyer conclude with the design implications of their results:

1. Noticeable differences in timbre will occur for singers turning more than 40°, which will become detrimental beyond 80°. Reflections, especially just beyond the side of the proscenium, should be used to minimize these effects.
2. A reflection off the stage in front of the singer is especially valuable and should be exploited where possible.
3. Side reflections are likely to be more valuable than overhead reflections and should, where possible, be within 60° of the singer's direction of view.
4. Choirs should be arranged on steeply raked steps, with a slope of at least 1:1 (= 45°).

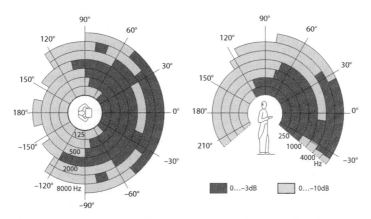

Figure 9.4 Directivity pattern for a baritone singer (after Marshall and Meyer, 1985)

9.2.2 The orchestra in the pit

The reasons for placing the orchestra on a raised stage in a concert hall are to optimize both sight-lines and acoustic transmission. By placing the orchestra in a pit there is no longer a direct sound path between the players and the stalls audience. Sound of course bends round corners but the amount of bending is frequency-dependent (Figure 9.5), with high frequencies bending less than low frequencies. It results in a characteristic lack of brilliance for orchestral sound in the stalls of opera houses. This can readily be appreciated during opera overtures but becomes less noticeable when singers are performing. The attenuation caused by 'bending', or diffraction to use the correct term, is a function of the extra path length involved in clearing the barrier, in this case the pit rail. This extra path length is reduced by a good stalls seating rake. For sound from the pit, high-frequency reflections to the stalls can be usefully enhanced by some light diffusing treatment on surfaces adjacent to the pit

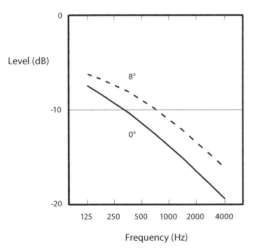

Figure 9.5 Reduction of sound level due to diffraction round the pit rail, for two stalls floor rakes

and from side balcony fronts (if these are present). Pit design is discussed in section 9.3.

9.2.3 Subjective aims

Opera house design has to contend with several conflicting requirements. For instance in simple acoustic terms the optimum reverberation time for speech intelligibility is near 1 second and for orchestral music nearly 2 seconds. In a free design it is possible to have a long reverberation time with strong enough early reflections to provide intelligible speech. But in most opera houses, the additional design constraints and the pressure to accommodate the maximum audience tend to make such a scheme impossible to achieve. To what extent are compromises acceptable?

Starting with the question of intelligibility and the sense of reverberation, it can be asked whether speech needs to be intelligible for opera. An admittedly rather *ad hoc* survey suggested that during true lyrical singing only about 10 to 15 out of every 100 words are clearly heard by an unprepared listener (Alexander, 1982). It is not however acceptable to conclude that this allows for inferior projection of speech. An opera house will always be used at times for unaided speech (such as in operetta) and this should be intelligible. Whether a singing line against an orchestra will also be intelligible involves more delicate balances associated with the relative loudness of voice and orchestra. Canny composers have used repetition to convey meaning where the demands on the singer, such as singing at the top of their range, make true enunciation difficult.

As well as intelligibility for the singer, the orchestral sound must be sufficiently clear in order to convey musical detail. With a reverberation time shorter than the optimum for music, this is rarely a problem in opera houses. An audible room sound from the orchestra is also necessary, to avoid the sense of just looking at the performance.

In terms of loudness, the balance between singer and orchestra is the crucial aspect. An operatic performance is a failure if the singers cannot be heard above the orchestra. This is of course an aspect within the control of the conductor, but room acoustics also has an important role to play in enhancing the level of the singers' sound as much as possible. Surfaces just in front of the proscenium are valuable for selectively enhancing the singers' sound. The absolute level of sound, which was found important in concert music, is no less important for opera. Again it probably contributes to the impression of acoustic intimacy.

Finally there is the question of whether source broadening (section 3.2) is also important for opera. This effect produced by lateral reflections in concert halls appears to be characteristic of the best concert acoustics. No studies have been conducted to show its relative importance for opera, though it would seem to be a subsidiary aspect for opera. Enhancing the received level of the singer's voice is likely to take priority over the direction from which it arrives. In other words overhead reflections have their place in opera houses. A sense of source broadening is probably only really necessary for the orchestral sound. With the orchestra in the pit, the degree of source broadening will automatically be high in the stalls. Measurements in four conventional British opera houses showed reasonable values for objective source broadening in each.

Thus in subjective terms the aims in opera house design are that speech should be intelligible, and that the orchestral sound should have adequate clarity but also provide a suitable sense of reverberation. The balance of voice and orchestra sound must favour the former, with both components loud enough at all seats. A sense of source broadening is subsidiary to the requirements of balance. The realization of these aims in opera house design depends on correct location and orientation of room surfaces and selection of the appropriate auditorium reverberation time.

For the subjective assessment of British opera houses reported in section 9.6, a questionnaire very similar to that for concert halls was used (Figure 3.7). The significant difference was that in place of a single 'Clarity' scale, two scales were used: 'Voice clarity' and 'Orchestral clarity'. The balance scale for 'Singers re. orchestra' becomes very important for opera.

9.2.4 Design in plan

Theatre designers have one area which they consider crucial in an opera house: the area where the regions of the performers and the audience overlap (Figure 9.6). The situation in many concert halls is different, with audience fully surrounding the performers. In the opera house, this crucial area used to contain a forestage, but very few forestages have in fact survived (Mackintosh, 1992, 2002). Mackintosh has campaigned for their reintroduction, both to bring back real involvement between singer and audience, as well as offering acoustic advantages. The forestage brings singers further towards the audience and offers a floor reflection when they are further up stage.

Plan design in an opera house has to originate from the proscenium opening. Proscenium width for grand opera is typically 14 m, but larger sizes up to 18 m are also found. Smaller widths of 10 m or below are common in more intimate houses. Proscenium height up to 9 m is common for grand opera.

In discussing the orientation of wall surfaces for opera, there are two surfaces above all which deserve careful design and detailing. The primary aim in opera house design is to enhance the singers' sound relative to the orchestra. The surfaces immediately in front of the proscenium are closest to the performers and are able to perform this role most effectively. By appropriate orientation, useful reflections for the singers only can be directed off these proscenium splays into the auditorium (Figure 9.7). The optimum orientation will usually be less splayed than would be used to 'join' the proscenium to the side walls. Compromise on this point should

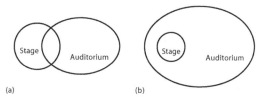

Figure 9.6 The performer/audience relationship in (a) an opera house and (b) a large concert hall

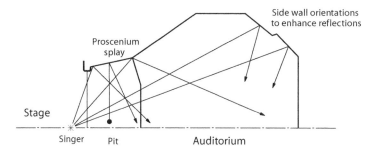

Figure 9.7 Reflections associated with opera house design form in plan

be strongly resisted, as should severe reduction of the size of the surface in order to accommodate stage lighting.

In practice, plane proscenium splays are unlikely to be the most appropriate. Some diffusion at lower levels will scatter singers' sound both back into the stagehouse and down into the pit. At higher levels one can argue for quite large horizontal surfaces fixed to the splays to reflect stage sound down into the auditorium. Small scattering elements on the proscenium splay can also be justified to reflect a little high-frequency energy from the orchestra into the stalls seating area. With such complex treatment of a highly visible surface, it is no surprise to find in the Deutsche Oper, Berlin (Figure 9.21), for instance, that the proscenium splay is placed behind a visual, but acoustically transparent, barrier.

Nearly all opera houses have some form of splay in the main side walls in plan. With a small angle of splay, it may be in order to omit any special orientations. In a fan-shaped plan, the weakest seating area tends to be the centre stalls about three-quarters of the way back from the stage. The side wall at lower level can be oriented in 'concertina' fashion to direct lateral reflections to this seating area (Figure 9.7). In the case of rear walls, these are often curved in traditional opera houses. There is no doubt that this causes disturbing focusing for particular areas of seating in the stalls. In modern designs rear walls should consist of plane surfaces. The use of curved balcony fronts can be acceptable if the front is either inclined or made sufficiently diffusing to avoid focusing where audience are seated.

9.2.5 Ceiling design

The surface immediately in front of the proscenium above the orchestra pit is also potentially very valuable to enhance singers' voices. The design principles are similar to those for the proscenium splay. If the ceiling slopes up from the proscenium to a flat ceiling (Figure 9.8), the singer's voice is reflected to the balconies, which already receive a good reflection off the flat ceiling, rather than into the stalls where extra reflections are required. Fasold and Winkler (1976) suggest a series of horizontal surfaces above the pit (Figure 9.9), which have the dual virtues of providing a reflection down into the stalls for the singer's voice and of directing orchestral sound back into the pit. (Over-stalls reflectors can however be excessive; Halmrast, 2002, has provided a well-documented case study.) A problem arises though with such schemes in larger opera houses

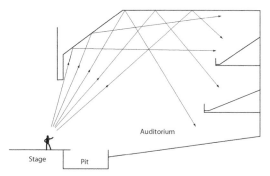

Figure 9.8 A sloping ceiling in front of the proscenium tends to direct reflections to the upper balconies, which are already well served by the horizontal ceiling

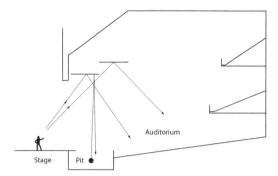

Figure 9.9 Reflections from horizontal surfaces above the orchestra pit (after Fasold and Winkler, 1976)

where the reflection delays easily become too large. In circumstances where the surface above the pit cannot be placed low enough, it and the remainder of the front part of the ceiling should be made acoustically diffusing, as was done in the new Metropolitan Opera House, New York (Figure 9.22). Late ceiling reflections are less apparent as echoes if they are proceeded by other reflections, such as scattered sound from the balcony fronts in traditional designs.

A comprehensive acoustic ceiling design was proposed by Cremer, Nutsch and Zemke (1962; also discussed in Cremer and Müller, 1982) for the Deutsche Oper, Berlin (Figure 9.10). By using a reverse-tilt ceiling scheme all seating areas received a ceiling reflection, with some desirable duplication in the stalls. Reflection delays were everywhere less than 50 ms with the exception of into the pit. Figure 9.21 shows the ceiling design used in the final opera house. Cremer's scheme in Figure 9.10 suggests yet a further possibility. It is tempting to consider whether suspended acoustically reflective ceilings might have a place in opera houses. Many of the reasons which make them inappropriate in concert halls no longer apply. For the opera singer the extra reinforcement is welcome, while for the orchestra the shrill tone quality associated with strong overhead reflections would hardly occur, at least in the stalls where the direct sound is weak. In larger opera houses, the conflict between the need for early reflections and for a large auditorium volume to provide a relatively long reverberation time is no less acute than in concert halls. After the disaster of the Philharmonic Hall, New York, it is a dangerous suggestion to offer, but suspended reflective surfaces in front of the proscenium are a possible solution to these conflicting requirements.

The ceiling is the major surface available for acoustic reflection and it should be treated as such. Domed ceilings, which became popular in the nineteenth century, had the acoustic effect of concentrating sound in inappropriate directions,

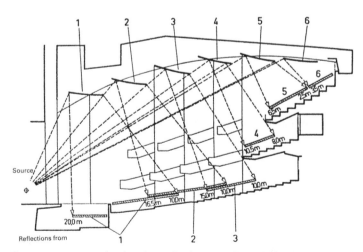

Figure 9.10 Proposed ceiling design for the Deutsche Oper by Cremer, Nutsch and Zemke (1962)

often depriving those in the highest seating level of a ceiling reflection. It is amusing to note that some theatres used *trompe d'oeil* domes painted onto flat ceilings; Adam's Drury Lane theatre of 1775 and Wagner's Bayreuth Festspielhaus of 1876 are two examples. A major acoustic fault of many theatres of the nineteenth century is the focusing produced by concave domes (Xu, 1999).

In modern designs an intrusion more difficult to counter in the ceiling area is stage lighting. Lighting slots create traps for sound, which in the numbers frequently found in contemporary theatres can easily leave little ceiling available for acoustic purposes (Figure 7.11). Suspended lighting gantries can be made acoustically transparent but are unpopular because of light spill. In the design of an opera house this conflict of interests is unlikely to be resolved without heated discussion!

9.2.6 Balcony and box design

The major acoustic effect of being under a balcony overhang is reduction of late reflected sound (section 3.7). For speech this is often not serious unless the sound level becomes too quiet, because it increases the proportion of early sound and thereby raises intelligibility. For concert performance the reduction in room sound is perceived after only a few rows of overhang, making listeners feel detached from the performance. With opera being halfway between the world of speech and music, the criterion for balcony design is likely to be intermediate between the concert hall and the drama theatre.

The degree to which sound is reflected from the main auditorium into the overhung space is also an influence. In fact, considering this leads to the conclusion that the late sound situation below overhangs is worse in opera houses than in concert halls due to the presence of the proscenium opening (Figure 9.12). In the main, stage sets are acoustically more sound absorbing than reflective, so that in general little reflected sound reaches audience members from this direction. This can leave a particularly small solid angle from which late sound can

be received below an overhang. In opera houses however a compensating factor of considerable subjective significance occurs: at seats below overhangs the balance is shifted in favour of the singers.

Two geometrical criteria have been proposed for balcony overhangs: Beranek (1962) considered the ratio between the depth of balconies and the opening height (Figure 9.11), while Barron (1995) found in concert halls some evidence that the vertical angle of view was more appropriate. Beranek's criterion for opera is a ratio depth to height of not more that two. Examination of an example with a ratio of slightly over two is instructive. Listening tests at a rehearsal in the London Coliseum (section 9.6.2) gave an overall acoustic assessment at a well-overhung seat of 'Reasonable'. However in this theatre the relatively long reverberation time by opera standards of 1.5 seconds will increase the perceived sense of reverberation, not only in the main volume of the theatre but also in overhung seats. The improved balance may be a mitigating factor from an acoustic point-of-view, but visually such a degree of overhang is unfortunate. As suggested for balcony overhangs for concert halls (section 3.7), the vertical angle of view (θ in Figure 9.11) may be a better parameter; for opera a recommended minimum angle of 30° is probably suitable.

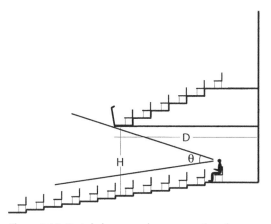

Figure 9.11 Basic balcony overhang proportions for an opera house. The vertical angle of view, θ, should be greater than 30°

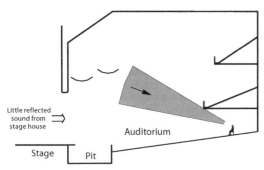

Figure 9.12 Long section of an opera house, showing the small angle from which late sound can arrive at seats beneath an overhang

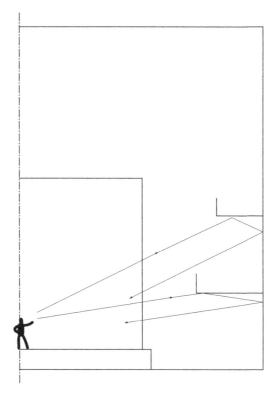

Figure 9.13 Half cross-section of opera house showing possible reflections off balcony cornices

In order to enhance the sound level underneath a balcony overhang, balcony soffit or rear-wall reflections can be used. As already discussed in section 3.7, these schemes do succeed in increasing the sound level in overhung seats but they do little to counter the main problem of inadequate late reflected sound. Nevertheless in opera, where overhead reflections are probably more acceptable, such extra reflections may well be desirable. The most valuable technique is to maintain a large opening for an overhang, but this of course runs contrary to accommodating a large audience.

In traditional baroque theatres, balconies run roughly horizontally along the side walls. This has distinct advantages for those in the stalls, who can receive scattered reflections off soffits and side walls (Figure 9.13). There are however visual and acoustic disadvantages for audience in horizontal side balconies, particularly those high up and close to the stage. The normal remedy, which is certainly effective for sightlines, is to rake balconies up from the stage end, but this provides a loss of acoustic reflections both to the main stalls and for those at the front of balconies facing the stage. The only logical resolution of this dilemma seems to be to use a series of balcony sections which step down the side walls.

Boxes are a common feature of eighteenth- and nineteenth-century opera houses. Considering the acoustics in a box in terms of angles of view, both vertical and horizontal, it is clear that any sense

of reverberation decreases rapidly as the listener moves away from the front of the box. Boxes do not constitute good acoustic design.

9.2.7 Long section design

Seats overhung by balconies are inevitable in all but the smallest of opera house designs. This leads to the question of how one best disposes seats in a large auditorium. From the visual point of view, proximity to the stage is a distinct advantage, as is a view of most of the acting area behind the proscenium. Deep overhangs are unwelcome visually but, as Reichardt (1979) has suggested, the sense of isolation is less severe under an overhang at stalls level. In acoustic terms, stalls seating suffers poor balance but the visual conditions there guarantee satisfied spectators paying high ticket prices!

The best seats acoustically in many houses are to be found in 'the gods', where there is usually good balance and a loud sound due to reflection off the ceiling. This argument can be used to justify having a deep upper gallery. There appears to be no virtue in deep intermediate sections of overhung seating, and seats in sections of balcony attached to the side walls are further disadvantaged for sightline reasons.

9.2.8 Reverberation time design

The conflicting acoustic requirements for speech intelligibility and music performance obviously bear on the question of the appropriate reverberation time for opera houses. A plot of auditorium volumes versus mid-frequency reverberation time for ten opera houses in Figure 9.14 indicates a range between 0.9 and 1.8 seconds. Short reverberation times are common in opera houses predominantly designed for drama use, as occurs in the case of Buxton Opera House. But a reverberation time of only 1 second is certainly shorter than ideal for music. Conversely, a reverberation time of 1.8 seconds is ideal for orchestral music but is unlikely to leave speech intelligible. If for instance it were just

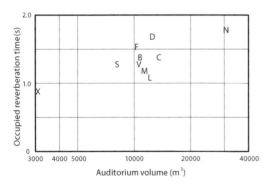

Figure 9.14 Volumes and mid-frequency occupied reverberation times of ten opera houses: B, Deutsche Oper, Berlin; C, London Coliseum; D, Semper Oper, Dresden; F, Festspielhaus, Bayreuth; L, Covent Garden, London; M, La Scala, Milan; N, Metropolitan Opera House, New York; S, Sydney Opera House, Opera Theatre; V, Staatsoper, Vienna; X, Buxton Opera House

the operas of Mozart that one wished to perform, the situation would be fairly simple. (The reverberation times of houses used for the premières of Mozart operas were in fact close to 1 second: Meyer, 1986.) But for most late Romantic operas longer times are required, particularly for Wagner. Unfortunately variable acoustics are unlikely to be possible to accommodate these varying requirements. For instance, with sound level at a premium, movable absorbent drapes to reduce the reverberation time are unlikely to be acceptable. One is thus left with a compromise requirement. Reverberation times for opera between 1.3 and 1.8 seconds appear appropriate, with the point on this range chosen according to taste.

At the time of writing, a current trend in new European opera houses is to design with a long reverberation time of around 1.7 seconds. This was inspired first by the rebuilt Semper Oper in Dresden (1300 seats, 1985) and the subsequent success of the Gothenburg Opera House (1390 seats, 1994), which have reverberation times of 1.8 and 1.7 seconds. Both these houses are further discussed in section 9.4.

To predict the reverberation time of an opera house, the usual Sabine equation is reasonably reliable. To represent average stage absorption, the proscenium opening can be treated as having an absorption coefficient of one (Hidaka and Beranek, 2000). For a simple assessment, working with the volume per seat is very convenient, but in concert halls it tends to be unreliable (section 2.8.3). It turns out that estimating reverberation time according to the volume per seat works better for opera houses (though less accurate than volume divided by audience area); Figure 9.15 demonstrates a reasonable empirical relationship. On this basis, for a 1.7 second mid-frequency reverberation time, a volume of almost 10 m³/seat is needed.

One solution to the conflicting requirements of speech and music, which appears worth considering, is to enhance early reflections to the point where speech is intelligible in spite of a long reverberation time. This happens automatically for seats below balcony overhangs. But enhancing early

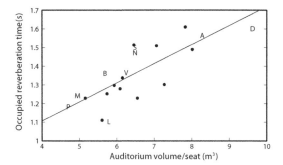

Figure 9.15 Relationship between occupied reverberation time and auditorium volume per seat in 21 opera houses. Equation of the best-fit line is RT = 0.1 × vol/seat + 0.7. (Data from Hidaka and Beranek, 2000.) [A – Teatro Colon, Argentina; B – Deutsche Oper, Berlin; D – Semperoper, Dresden; L – Covent Garden, London; M – La Scala, Milan; N – Metropolitan Opera, New York; P – Opera Garnier, Paris; S – Festspielhaus, Salzburg; V – Staatsoper, Vienna.]

reflections in most opera house designs is a highly constrained exercise, removing much of the flexibility one would like to exercise. Two designs are interesting in this respect: Orange County Performing Arts Centre (section 10.6), which is reputed to work well for opera, and the opera house of the New National Theatre, Tokyo (Beranek, Hidaka and Masuda, 2000).

The appropriate frequency characteristic for the reverberation time in opera houses is also intermediate between requirements for speech and music. A longer bass reverberation time is undesirable for speech. Neither is it desirable for the orchestral sound in an opera house in the stalls, since high-frequency sounds are suppressed due to absence of line-of-sight into the pit. A maximum rise of 20 per cent at 125 Hz relative to the mid-frequency reverberation time is likely to be acceptable. At the other extreme of the frequency range, the presence of the 'singing formant' suggests that reverberation times at 2 and 4 kHz should be maintained as long as possible. The scope for this is limited since the audience is by far the major porous absorber. Excess porous absorber in the form of curtains, drapes etc. should therefore be avoided.

In most spaces, to provide the correct reverberation time involves appropriate selection of the ceiling height. But for opera there are other claims on the ceiling location, both for stage lighting and to produce acoustic reflections into the stalls which are early enough. In the traditional baroque opera house design with many balconies surrounding a central volume (as in La Scala, Milan), the main element of flexibility is the size in plan of the central auditorium volume. If the ceiling is already high enough, by extending this central volume the reverberation time is increased. An alternative scheme was used as long ago as 1878 by Semper in Dresden, whose opera house included extra reverberant spaces around the upper gallery level (Reichardt, 1985); these spaces have also been incorporated in the rebuilt house. The need to limit the rise in low-frequency reverberation time will generally imply additional low-frequency absorption being applied to the auditorium surfaces. Panel absorbers are an obvious possibility. But with most surfaces already performing some specific role or other, ingenuity may be necessary. Fasold and Winkler (1976) describe using slots around the dome of the Staatsoper, Berlin, which feed into concealed enclosed volumes so that they act as Helmholtz absorbers.

9.3 Stage and pit design

For opera performance, the acoustic conditions on stage are also of importance, particularly for the singers. The only surfaces behind the proscenium opening with the potential for supplying early reflections to singers are elements of stage sets. There is much in principle to be said for a stage set which is highly reflective, built out of solid timber rather than canvas. In most houses the expense of this measure rules it out. Where solid stage sets are used, such as in the Vienna Staatsoper, the effect contributes to the acoustic character of the house. But of course not all opera scenes use scenery. Indeed open-air scenes often sound reverberant due to an absence of scenery in the stagehouse, when acoustically they should be dead! Excessively reverberant stagehouses should in any case be avoided by installation of surface absorption in the stagehouse and by storing sufficient drapes, sets etc. above the stage.

The acoustic requirements for singers on stage have to date been the subject of few studies; the most valuable is by Marshall and Meyer (1985). On the basis of experiments with synthetic sound fields, they found that singers were most responsive to reverberation, as opposed to specific early reflections. If early reflections were going to help, they had to arrive with a delay of less than 35 ms, which implies a maximum distance for a reflective surface of only 6 m from the singer. In terms of implications for opera house design, some reverberation will always be present in the auditorium. For the singers to hear it requires surfaces in the auditorium visible from stage by which reverberant sound can be reflected. If the evidence of these experiments with synthetic sound fields is correct, few surfaces can be close enough to help the singer with early reflections. Reflections from solid stage sets are a possibility, as are scattered reflections from surfaces just beyond the proscenium.

Providing support for singers can always be assisted by electronic foldback. Such a simple remedy cannot however provide a full solution to the problems of opera house pits. The conflicting requirements of orchestra pits are among the most difficult to reconcile of any acoustic enclosure. Reports over recent years have begun to shed light on the options in pit acoustics. But before discussing the pit itself, one significant finding from a questionnaire survey of 5000 opera-goers has to be mentioned. Invariably, the opera-goers considered the orchestra too loud relative to the singers (Mackenzie, 1985). Improving the balance can be achieved by use of the partially covered pit (Figure 9.16) but this has to be weighed against the musicians' reluctance to play in them. The partially covered pit is also demanded for economic reasons, since pits projecting into stalls seating space reduce revenue from what would otherwise be prestige seats.

The cause of musicians' grievances over covered pits are clear enough (Naylor, 1985). There are often

several ergonomic problems, principally lack of space. The need to maintain line-of-sight to the conductor frequently results in a raked pit with additional raised platforms or boxes to provide further changes in floor level. This substantially reduces the available space for the musicians. For these reasons a more generous floor space allocation than for concert halls is necessary. Suitable visual contact between players is in addition often difficult to achieve in raked pits.

Turning to the acoustic conditions, excessive sound level in covered pits constitutes not only a disturbing environment but also a potentially dangerous one. Naylor reports measured values in the woodwind section for complete acts of an opera of 86–92 dBA (equivalent sound level, L_{eq}, which can be compared with the current European limit of 85 dBA for the eight-hour working day). Peak levels, occurring perhaps two to three times per opera, ranged from 105 to 110 dBA, generally due to percussion noise. On the question of the impact of a partially overhung pit, Gade and Mortensen (1998) record that sound levels in the old Copenhagen opera house pit increased about 3 db with the overhang. It comes as no surprise to find that some musicians playing in pits do suffer permanent hearing loss.

Such high sound levels make ensemble playing very difficult. They are not in fact exclusive to covered pits; levels in open pits are in some conditions quite similar. Another problematic issue is communication from the stage. The ability of an instrumental player to play in time with a singer is obviously important, yet most performers commented that they could not hear the singers as well as they would like to. For the performer himself some sense of the sound he is producing returning from the auditorium is likewise a characteristic of most performing environments. The amount of sound returning to the back of the pit is generally very small. In acoustic terms the overhung section of a pit behaves like a coupled space to the auditorium, with double slope reverberant decays (see section 2.8.1). The initial slope of the decay is determined by the local acoustic conditions in the pit

0   1   2   3 m

Figure 9.16 Typical layout for the partially covered orchestra pit (courtesy of G.M. Naylor)

with decay rates corresponding to reverberation times of only 0.35–0.7 seconds. Thereafter the decay behaves as in the auditorium. The net result is that the performer in the overhung part of the pit is only likely to hear an auditorium sound when nobody nearby is playing.

A solution to these various problems is not easy to achieve in most cases. The principal non-acoustic variables are the physical dimensions of the pit, the degree of cover and the height of the opening. Surfaces of the pit can be treated by absorption or diffusion but while, for instance, absorption will lower

sound levels it will also reduce communication. To aid communication, there is of course the possibility of electronic assistance, but would that be acceptable to the musicians?

In long section, typical dimensions are given in Figure 9.17. The top of the orchestra rail is generally in line with the stage. The degree of overhang extends in some houses considerably further than 2 m mentioned in the figure. For example, it is 4.7 m in the opera theatre of the Sydney Opera House. Values of 3 m overhang or no more than 40 per cent of the total pit area should probably not be

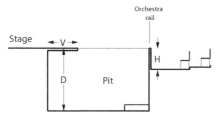

9.17 Typical opera house pit dimensions: stage overhang V, 1–2 m; pit depth D, 2.5 – 3.5 m; orchestra rail height H, 1 m

exceeded in large-scale houses. The height of the opening should preferably be not less than 2.5 m; it should be higher for pits with larger overhangs to allow all musicians to maintain some visual contact with the auditorium. The ideal solution to the question of pit floor levels is to have the complete floor area on lifts. If costs do not permit this, then the design must allow for plenty of flexibility to permit changes to be made on the basis of experience.

In plan the size of the pit will normally be calculated on the basis of the number of musicians to be accommodated. Naylor and colleagues recommend an allocation of 1.5 m² per player in pits (significantly higher than the value of 1.1 m² per player for concert hall platforms). This measure by itself introduces a significant element of flexibility. However recent evidence from another survey suggests that even this criterion may be too limiting (Gade *et al.*, 2001). The Royal Opera in Stockholm sent questionnaires out to other houses worldwide and received 46 responses. The average pit area per musician was 1.7 m², with a maximum of 2.1 m². Two-thirds of pits were criticized for lack of space and excessive sound levels. Only the pits with floor areas of 2.1 m² per musician were judged as having acceptable sound levels and ease of ensemble. Another interesting statistic was that 75 per cent of houses had an adjustable depth of pit and, of these, 85 per cent used the feature.

The proportions of the pit should also not be excessive; in general, the width should be less than four times the length (measured along the long axis of the house). Since in the proscenium area some degree of splay in plan is almost inevitable, a cranked or curved orchestra rail is often used.

An on-going discussion concerns the appropriate acoustic treatment of the pit surfaces. In logical terms some absorbent in a partially covered pit looks desirable. Absorption would achieve two goals: it would reduce the sound level in the pit itself and reduce the orchestral sound level in the auditorium thereby improving balance. From the survey quoted above, balance was generally felt to be at the expense of the singers. Though the orchestral level can in theory be controlled by the conductor, there is a limit to how quiet individual instruments can be played, especially woodwind. The problem with absorption in the vicinity of the musicians is that it reduces even further the acoustic support they receive from the performing space, which is in any case already lacking in covered pits. Mackenzie (1985) reports on model experiments on one particular pit where they achieved a reduction of 3 db in the auditorium level by making the pit ceiling absorbent. A similar reduction would occur for the reflected sound level in the pit. While these changes by themselves are desirable, such treatment reduces the ability of performers to hear themselves and their colleagues. Without sufficient feedback, players are prone to play too loud, defeating the intention of the absorbing treatment.

There may however be a case for frequency-selective absorption: Meyer (1998) discusses vertical standing waves in overhung sections of pits which may prove very distracting to players of instruments at lower frequencies. Tuned absorbers on the soffit of the overhang may help with this problem.

Less controversial is the question of scattering on surfaces in pits. Since scattering is appreciated for platform performing locations, there seems good reason for it being used more extensively in orchestra pits. A radical and apparently successful experiment was conducted by Harkness (1984) in the pit of the opera theatre of the Sydney Opera House. Not only did he temporarily install patches of absorbent and scattering around the surfaces of this substantially overhung pit, but he also introduced vertical screens between different instrumental groups. The rationale behind this was that some instruments are weaker than others and only by separating the weak from the loud instruments can the former hear themselves. He reports that 'the response to the removal of these (dividing) walls was very strongly negative'. One should add though that the size of this pit in plan is generous, allowing ample space for dividing walls and diffusing boxes.

No mention has as yet been made about Wagner's orchestra pit in Bayreuth with its hood extending from the orchestra rail blocking the view of all the audience into the pit. This pit design has proved to be uniquely suitable for its location but there are no copies of his scheme. This has occurred because no other opera house exclusively performs Wagner's operas and the mystical sound of the Wagner pit proves unsuitable for the works of any other composer.

In an illuminating discussion of pit design, Blair (1998) recommends maximizing the height of overhung pits and having a sloping ceiling. He cites two opera houses as being particularly liked by conductors and pit orchestras: the Opéra Garnier in Paris and the Teatro Colón of Buenos Aires. These houses both have good late reflections back to the pit from surfaces above the pit, from tier fronts and domes (!). However neither house has a particularly generous size pit, and this seems the most important parameter to emerge from recent studies. But while large pits may favour the musicians, they tend to leave a big gulf between the audience and the stage. Reconciling these conflicts is a major challenge in new opera houses.

9.4 Opera house form and acoustics

Though opera house design is highly constrained, certain different forms have evolved. The acoustic implications of these various solutions can be illustrated by example. In traditional opera houses, there were two basic forms of accommodation in the balconies: in boxes or in open galleries. Extreme versions with boxes throughout are found in Italy, whereas in England open galleries became the norm. Between these two extremes we find, for example, the Paris Opéra Garnier (1875) and the

Teatro Colón, Buenos Aires (1908), which both have boxes more open than the famous Italian examples. The Staatsoper, Vienna (1869) on the other hand, has a mixture of boxes and open galleries.

The centre of Italian opera and Italy's most famous house is the **Teatro alla Scala** in **Milan** (1778, rebuilt 1946). Its plan is a classical horseshoe with boxes in six tiers; the auditorium volume is 11 250 m³. Figure 9.18 shows the house in its current state; originally this and other houses of the eighteenth century had significant forestages, which have now virtually all been removed. Its seating capacity of around 2300 was considered enormous for its time. This number initially appears impressive given that the furthest audience is only 31 m from the stage front. But the house suffers from the fault of traditional baroque theatres that sightlines at the sides and particularly at upper side levels are very poor. Acoustically the situation in La Scala (and its close Italian relatives) is special; its reputation is certainly among the best in the world. The particular acoustic character of these houses derives from a main auditorium volume surrounded by bands which are alternately reflective (balcony fronts) and acoustically absorbent (box openings). The boxes are deep and well upholstered so that virtually no sound entering them will re-emerge into the main volume. The openings in La Scala are small, only 45 per cent of the height from one layer of boxes to the next. This allows a rich reverberant sound field to build up in the main volume. The small distance between opposing balcony fronts also produces a good sense of acoustic intimacy, with a sound which is 'clear, warm and brilliant' (Beranek, 1962, 2004).

Access to this intimacy and rich reverberant sound is however strongly limited. Only those in the stalls, the front row of seats in the boxes or in the more open upper galleries can fully appreciate this luxurious sound. For those at the rear of the boxes, the experience is one of being a spectator but not a participant. Izenour (1977) also notes a further minor problem that echoes can be detected in the stalls in La Scala, though the ceiling is in fact plane. For opera houses of this size, the travel time for the

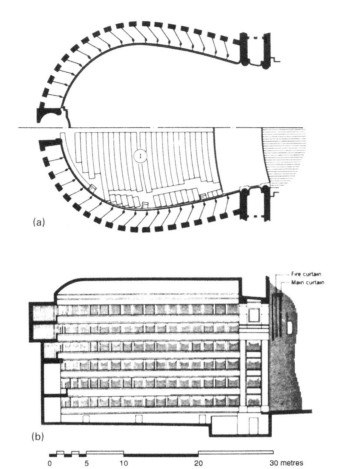

(a)

(b)

Fire curtain
Main curtain

| 0 | 5 | 10 | 20 | 30 metres |

Figure 9.18 (a) Plan and (b) long section of the Teatro alla Scala, Milan

ceiling reflection to the stalls is simply too long. Lower reflecting surfaces immediately in front of and above the proscenium opening could replace these long ceiling reflection paths by shorter ones (at the expense of some audience positions, which anyway have poor sightlines). In spite of the high ceiling, the reverberation time in La Scala is only a modest 1.2 seconds.

The open gallery, often stepping back at higher levels, is England's particular contribution to theatre design. London's premier opera house, the **Royal Opera House, Covent Garden** (1858), has three levels of open gallery with fronts arranged vertically above one another (Figure 9.19). The upper gallery extends back well beyond the others. The acoustics of Covent Garden are more democratic than the Italian houses, though fairly marked differences are audible between seating areas. The orchestral sound in the stalls is muted since there is no line-of-sight to the pit. As one moves back behind a balcony overhang, the degradation of sound is clearly audible, with a diminution of the sense of reverberant sound (section 9.2.6). The best acoustics in houses like this are often to be found in the upper gallery, which benefits from an unobstructed early ceiling reflection. Even in the gallery seats furthest

Figure 9.19 The auditorium of the Royal Opera House, Covent Garden, London

from the stage, the sound is better than one might expect from the limited size of the opening to the main auditorium volume. In Covent Garden there is a concave dome, but while odd quirky echoes are audible, they constitute only a minor irritant. The acoustic success of houses like Covent Garden is due mainly to the intimate character of their sound. This probably has much to do with the presence of balcony fronts which, because of the limited width of the main open volume of the house, are capable of providing early acoustic reflections to most seats in the main body of the house, whereas for seats in the upper gallery the ceiling performs the same function. With a reverberation time of only 1.1 seconds, Covent Garden is biased in favour of the libretto at the expense of orchestral tone. The acoustics of this house are treated in more detail below.

Traditional architecture with elaborate surface decoration extended to the First World War. A late example, much loved by those that know it, is the **Teatro Colón** in **Buenos Aires** of 1908, designed by V. Meano. It shares with La Scala a horseshoe plan

and five levels of balcony. The seating capacity is however larger at 2487 with a substantial auditorium volume of 20 570 m³. This results in a long reverberation time estimated as 1.8 seconds (Beranek, 1962, 2004). Another feature which distinguishes it from La Scala is that the upper two balcony levels are more widely spaced vertically than those lower down. The house is particularly popular with performers; reflections from surfaces in front of the proscenium and from balcony fronts and soffits provide valuable contributions.

The baroque theatre, generally based on the horseshoe plan form, dominated theatre design for 200 years, to the virtual exclusion of all else. Its success as a form had as much to do with social custom as any theatrical virtues: seeing and being seen was a primary concern in these houses. This may explain why so little was done to implement simple remedies to the baroque theatre's greatest failing, that of poor sightlines from the side boxes. Acoustically the baroque theatre and its descendants have proved remarkably appropriate for opera, even at the largest scale discussed below. Little distinction was drawn in design between theatre and opera use, except for the obvious inclusion of an orchestra pit. Virtually all Europe's legacy of famous opera houses predates scientific acoustics and in them we find reverberation times a little shorter than is now considered optimum for opera.

The great break with traditional design occurred with Wagner's **Festspielhaus** in **Bayreuth** of 1876 (architect: O. Bruckwald). The seeds of Wagner's scheme evolved over many years. In the preface to the text of his *Ring des Nibelungen* in 1862, Wagner urges the construction of a Festspielhaus with a concealed orchestra and amphitheatrical seating. With characteristic single-mindedness, Wagner finally incorporated these revolutionary elements in Bayreuth and produced a house uniquely adapted to the performance of his operas. With an auditorium volume of 10 300 m³, the seating for 1800 is in a continuous well-raked fan (Figure 9.20). Piers topped by columns delineate the fan and conceal entrances to the seating rows, which are in the arrangement we now refer to as 'continental seating'. Though Wagner's seating arrangement was designed to eliminate social distinctions, his principal sponsor, King Ludwig II, requested two levels of gallery at the rear for high society. This proved to be a fortunate detail, since it suppresses the echo back to the stage, which would otherwise occur off the ceiling and curved rear wall.

Concealment of the orchestra fulfilled both a visual and acoustic function. To Wagner no distraction should interrupt the view of the stage. The gap between the auditorium and the stage produced by the concealed pit is continued by a double proscenium arrangement designed to create a *mystischer Abgrund* as part of the dramatic effect (*Abgrund* is variously translated as 'abyss' or 'precipice' according to taste). Acoustically the location of the orchestra behind a curved screen suppresses the level of orchestral sound, so that singers can compete successfully with the 130-piece orchestra. In Bayreuth a mysterious sound emerges from an invisible source but with a strange tone, absent in higher harmonics and brilliance. The Wagner pit has provoked much controversy, but it is used nowhere else, and

seems only appropriate to its designer's music. To an enthusiast, 'a vital artistic effect is given by the reverberation and the subdued orchestra combined, namely a remoteness and unity' (Bagenal and Wood, 1931).

The reverberation time is the other aspect which distinguishes the Festspielhaus from other houses. With a value of 1.55 seconds it is longer than most other famous opera houses before 1900. Again this value is highly appropriate to Wagner's music but too long, certainly, for classical Italian operas. Coming a generation before Sabine, the resulting reverberation time value must be seen as fortunate chance. Fortune also limited the low-frequency reverberation time (1.75 seconds at 125 Hz). This occurred because financial problems dictated a light-weight construction for the auditorium as opposed to the projected marble palace. A longer low-frequency reverberation time would have further upset the tonal balance, already compromised by the overhung pit.

The Bayreuth Festspielhaus proved highly influential to drama theatre design, less so to opera house design, until perhaps the second half of the twentieth century. But there is one exception in the example of the Prinzregententheater in Munich of 1901. Wagner's Festspielhaus had originally been destined for Munich, but King Ludwig's money never materialized to finance it. Only after Wagner's death did the city of Munich realise that there were financial advantages in emulating the success of the Bayreuth Festival. Littmann's copy of the Bayreuth house was named after the Prince Regent, who effectively ruled after King Ludwig. But though the seating arrangement matches Bayreuth, two details were changed which illustrate acoustic characteristics of the original. The Munich house has a continuous rear wall which produces severe echoes back to the stage in consequence of the second-order reflection off the ceiling and curved rear wall (Cremer and Müller, 1982, Vol. 1, p. 76). Littmann also rationalized the plan: instead of the piers and columns in Bayreuth which convert a rectangular plan into a fan shape, the Prinzregententheater has walls following the splay of the

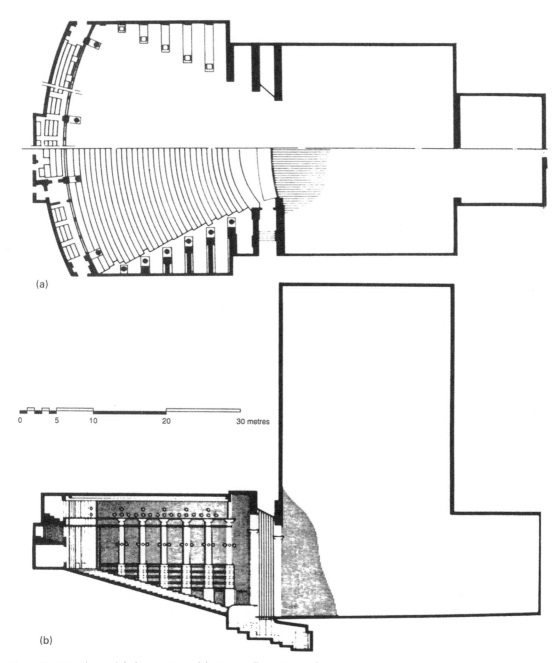

(a)

0   5   10   20   30 metres

(b)

Figure 9.20 (a) Plan and (b) long section of the Festspielhaus, Bayreuth

audience. Acoustically this proves less satisfactory than the highly sound scattering scheme at Bayreuth. Fan-shaped plans with angles much wider than the Prinzregententheater are not favoured for opera, any more than for concert halls.

The most intense building programme of opera houses during the twentieth century was undertaken in Germany in the 1950s and 1960s to rectify war damage. The majority of these were modern houses rather than reconstructions of the baroque theatres which preceded them (Schubert, 1971). The new auditorium forms can be loosely grouped into those which are descendants of the baroque form and those deriving their main inspiration from the Bayreuth solution. Most of the houses have open galleries opposite the stage. The manner in which these galleries extend down the sides of the auditorium provides a further opportunity for subdivision. Either continuous gallery sides are used or elements are segmented along the walls, to the extent in some houses of having small seating sections cascading from the rear towards the stage. Unfortunately no survey of the acoustics of these German opera houses has been made.

A significant example from these German opera houses is the **Deutsche Oper** in western **Berlin** by the architect F. Bornemann, completed in 1961. Its success extends to reasonable/good acoustics. The Deutsche Oper is clearly inspired by the Bayreuth Festspielhaus with a well-raked stalls floor (Figure 9.21). Two gallery levels extend the full width at the rear; the lower one is continued in steps down the side walls, while above are individual 'sledges' relating to the upper gallery level. The acoustic design, principally by Cremer, is well documented (Cremer, Nutsch and Zemke, 1962). The acoustician was anxious to use the side walls to direct reflections to the centre of the stalls seating, the traditional problem area in opera houses. But with structural decisions already made, this proved impossible to accommodate. The 'boxes' on the side walls are bounded by surfaces radial to the stage, which is logical for sightlines but of little value acoustically. Only in the proscenium frame area was appropriate orientation of vertical surfaces possible to provide

reflections to the main floor. Raised horizontal panels are also attached on this surface to reflect sound between stage and pit. There remained some freedom in the design of the ceiling. Cremer's proposal has already been discussed in the previous section (Figure 9.10). The eventual scheme in the Deutsche Oper consists of narrower elements (in the long section), which acoustically are less directed but more scattering. Such detailed acoustic design of the ceiling can easily be destroyed by lighting bridges. Crucially the acoustic consultant was able to limit their number in the ceiling area to only one. To anyone with the experience of designing such auditoria, this was a remarkable achievement! With an auditorium volume of 10 800 m³, the reverberation time is 1.35 seconds at mid-frequencies. Extensive timber panelling on the walls restricts the low-frequency reverberation time at 125 Hz to 1.65 seconds (Cremer and Müller, 1982).

The pit in the Deutsche Oper is partly covered by the stage, which offers several advantages. For the singers, it allows them to come forward further on the stage and thus be able to 'see the auditorium ceiling' and exploit acoustic reflections off it. For performers further back on stage, the extension of the stage increases opportunities of a useful floor reflection. In use only the strings, woodwind and harps perform in the exposed pit areas, which contributes to the good balance between singer and orchestra in this house.

The Deutsche Oper holds an audience of 1900, large by opera house standards. Its acoustic success probably stems from a multitude of features. Among these is a favourable gross form with a modest maximum width in the auditorium. Advice was given on the orientation of nearly all interior surfaces, to optimize early reflection conditions. Unfortunately this advice could not be followed for the side walls; these do not promote lateral reflections, which is an audible defect. The balcony overhangs are modest here and overall the open arrangement of the house must promote a diffuse reverberant sound.

Being the largest opera house world-wide in terms of seating capacity is a dubious virtue. The

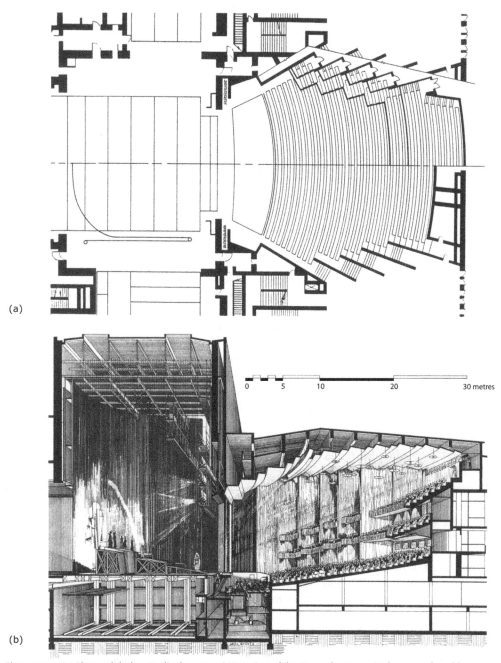

(a)

(b)

Figure 9.21 (a) Plan and (b) longitudinal perspective section of the Deutsche Oper, Berlin (reproduced by permission of Pennsylvania State University Libraries)

Metropolitan Opera House (1966) in the Lincoln Center, **New York**, holds an audience of 3800 in a 30 500 m³ auditorium (Figure 9.22). Accommodating this number involves severe compromises. A significant proportion of the audience has a poor view of the stage and the furthest seat in the auditorium is a staggering 53 m from the stage front. In the circumstances it was fortunate that the architect, W.K. Harrison, heeded the advice of his acoustic consultants, C.M. Harris and V.L. Jordan. A disaster equal in scale to that of the Philharmonic Hall next door could easily have occurred.

The 'Met' follows the conventional baroque form with four gallery levels opposite the stage and five rows of boxes along the side walls. The plan form is a gentle splay with a maximum auditorium width between balcony fronts of 29 m. Balcony fronts are segmented to prevent focused reflections back to the stage; the underside of the balcony fronts also directs sound down to audience below. Cremer and Müller (1982) comment on the detailed planning of the side boxes in this house. By keeping the rows of boxes almost horizontal and spacing them further apart at higher levels, lateral reflections off the soffits and side walls down into the stalls are still possible. The disadvantage of this scheme is a visual one: sightlines from horizontal boxes are poor. In his account of the design, Jordan (1980) stresses the importance he attached to the design of the proscenium zone. Extensive surfaces to the sides and above provide an acoustic connection between the stage and audience areas. In these various ways sufficient early reflections are achieved throughout the main auditorium, providing good intelligibility for the audience.

The best seats acoustically in opera houses tend to be in the upper gallery. Only in these seats do audience receive strong reflections from the majority of the ceiling. In a large house this justifies extending the upper gallery way beyond the galleries below. The ceiling in the Metropolitan Opera House is carefully segmented to scatter sound, providing reflections into the main seating areas. Only two major lighting galleries interrupt the ceiling profile. Izenour (1977) describes the acoustic results

of this design as 'two entirely different sounds: intimate, balanced and opera-house-like in the orchestra (stalls) and ring balconies; big and more concert-hall-like in the upper stadium balcony'. The reverberation time is around 1.7 seconds (Beranek, 2004), a surprisingly high value given its favourable reputation for intelligibility.

The solution adopted in New York was forced on the designers by the financial requirement for a high seat number. Opera is now virtually always a subsidised art form and designing for 2000 good seats is surely better than 1500 good and 1500 bad. With 2000 seats both visual and acoustic requirements have a hope of being satisfied. It is instructive that acceptable acoustics can be achieved for an audience of more than 3500. This modern opera house convincingly demonstrates that with minor modifications, such as removal of focusing situations and limitation of balcony overhangs, the baroque theatre form is a justified survivor for opera.

The **Opéra Bastille** in **Paris** with 2700 seats has a plan, at least at stalls level, which is similar to that of the New York 'Met' (Suner, 1990). Taking its inspiration more from Wagner than the baroque form, after the initial splay in plan from the proscenium the side walls are parallel. This is another large house; it was built (at great expense) to celebrate the bicentenary of the French revolution in 1989 (architect: C. Ott). The acoustic design was by Müller-BBM in collaboration with the CSTB of Grenoble (Müller and Vian, 1989). There are two levels of balcony, with balcony fronts that are closer to the stage at the side than in the centre. A reverberation time of 1.55 seconds was achieved with an auditorium volume of 21 000 m³. But opera house design at this scale inevitably involves compromises, one of which is the 46 m distance of the furthest seat from the stage front. Responses to the acoustics have been mixed (Beranek, 2004). A further example whose design owes a little debt to the 'Met' is the opera house of the New National Theatre in Tokyo, which opened in 1997. A novel feature of this 1810-seat house is what the acoustic consultants call an 'acoustic trumpet' in front of the proscenium to

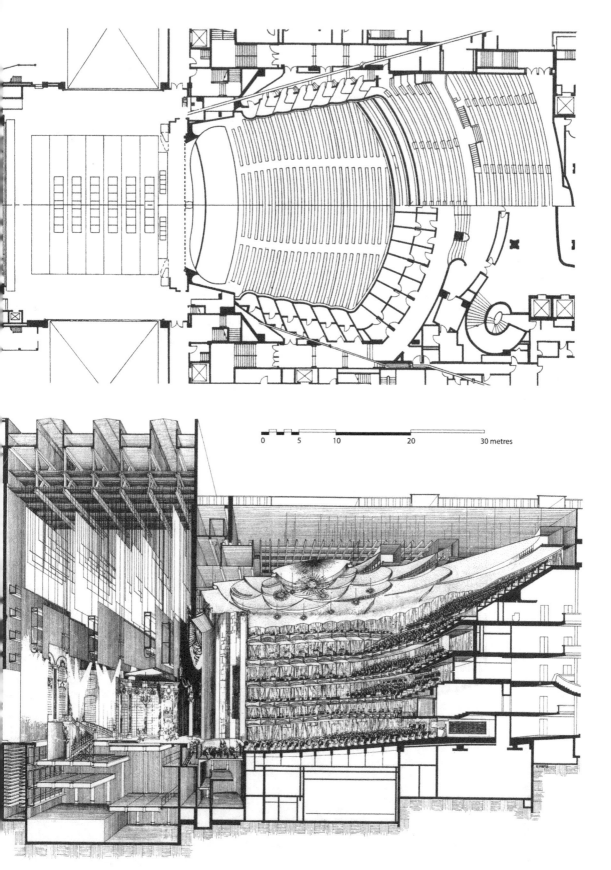

Figure 9.22 (a) Plan and (b) longitudinal perspective section of the Metropolitan Opera House, New York (reproduced by permission of Pennsylvania State University Libraries)

enhance early reflections from the singers (Beranek, Hidaka and Masuda, 2000).

While the baroque theatre was closely associated with class distinctions, Wagner's scheme in Bayreuth was considered more egalitarian. Yet can a distant seat and one near the stage really be considered equal? And is equality such a virtue that it justifies the impersonal nature of many 'modern' houses derived from Wagner's model? In the latter years of the twentieth century, several designers have returned to the baroque form with its horseshoe plan and stacked balconies because it offers a more involved and intimate experience for both performers and audience. The new Glyndebourne opera house in England is a key example of the new trend, discussed in detail in section 9.6.4.

As well as reconsidering the baroque theatre, another recent trend among some designers of modest-size houses has been to seek longer reverberation times. A reference design has been an opera house designed twice by Gottfried Semper, a friend of Wagner, originally in 1841 but rebuilt and slightly modified following a fire in 1878. The **Dresden Staatsoper** is conventional in design with

a horseshoe plan and four levels of balcony; room surfaces are highly decorated providing scattered sound reflections. Reichardt (1985) carefully documents the evidence that Semper understood the acoustic significance of many features of his design. Extra volume was introduced at high level that would enhance reverberation. Semper's opera house disappeared in the destruction of Dresden in 1945 but it was scrupulously rebuilt by the then east German government in 1985 with Reichardt as acoustic consultant. Though the visual details were reproduced, some modifications were made, such as reducing the number of rows in the highest balcony from ten to four. Now frequently known as the Semperoper, it seats 1309 with an auditorium volume of 12 500 m³ and reverberation time of 1.8 seconds.

The lush sound of the Semperoper inspired the acoustic design by Akustikon of the low-cost **Gothenburg Opera**, which opened in 1994 (Figure 9.23). To the untutored eye, these two houses look totally different, one lavish in its decoration with barely a plane surface to be seen, the other with black walls, plain surfaces and rather minimalist in character. However, given the modest dimensions

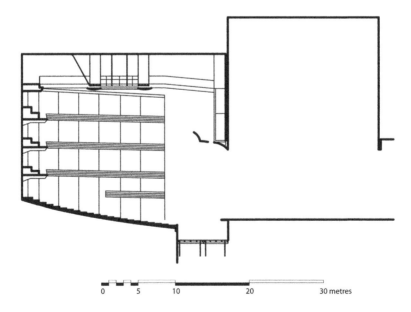

0   5   10   20   30 metres

Figure 9.23 Long section of the Gothenburg Opera House

of most surfaces in the Gothenburg opera house, diffraction from finite-size surfaces in that house is equivalent acoustically to reflection off the obviously scattering surfaces in Dresden (section 2.3).

The plan form selected for the Gothenburg Opera was an elongated hexagon, which combines the inevitable fan-shape in front of the proscenium with a reverse splay at the rear valuable for early lateral reflections (Gustafsson, 1995). There are three horizontal balconies spaced sufficiently far apart vertically to allow worthwhile reflections from side walls and soffits. With 1310 seats the furthest seat is only 27 m from the stage. Predictably the auditorium ceiling is high to create sufficient volume (of around 12 000 m³) for the long reverberation time of 1.7 seconds.

One of the characteristics of the Semperoper that the designers wished to recreate in Gothenburg was the relatively low clarity for sound from the pit as opposed to high clarity for the singers' sound. To achieve this, there was very careful design and orientation of surfaces suitable for acoustic reflections, particularly surfaces close to the proscenium. This is a house which deserves careful study. The acoustic designer has also produced an interesting report on the variation of the acoustics of this house with different stage sets, a topic often ignored (Gustafsson and Natsiopolos, 2001).

9.5 Objective assessment of opera houses

As with other auditorium types, this chapter concludes with a detailed discussion of four opera houses in Britain. The assessment will include consideration of objective measurements made in the individual houses. A discussion of objective assessment for opera is thus necessary at this point. For measurements in concert halls, one can now refer to an international standard, but the standard does not as yet address opera houses. The choice of measures will vary between authors; a modest list seems appropriate.

A key difference between opera and other forms of auditorium is that two separate sound sources have to be considered: the singers on stage and the orchestra in the pit. The requirements for each source are different. The following selection of measures encompasses clarity, both for singers and orchestra, the sense of room sound, both due to early lateral reflections and reverberation, the loudness and balance between singers and orchestra:

- reverberation time
- early decay time
- objective voice clarity
- voice sound level
- objective orchestral clarity
- orchestral source broadening
- singer/orchestra balance.

The measures used are identical to those used in concert halls and theatres, for the orchestral and voice sound respectively. The relevance of each measure, appropriate values and the manner in which they are presented now follows. For each of the opera houses, figures containing the objective characteristics are given that are similar to those used for concert halls in Chapters 5 and 6.

Reverberation times for opera have been discussed above in section 9.2.8. Optimum values of 1.3 to 1.8 seconds are recommended, with a slight rise at low frequencies. The choice of time depends on the expected repertoire and the taste of the client. A short time is appropriate for operas such as Mozart's in which the text is rapid and precision is considered paramount; a longer time would be preferred by many for more romantic operas.

As with concert halls, measured values of reverberation time are presented at five octaves both for the unoccupied and occupied condition. The unoccupied measurements were made with an omni-directional source on stage (curtain open). A characteristic of the British houses is to have short reverberation times for reasons which will be discussed below.

As with concert halls, the early decay time (EDT) is considered to be a measure of the sense of reverberation. It has been measured in the unoccupied opera houses with the same source conditions as reverberation time. In the figures, the mean EDT is given omitting results from overhung seats. As usual, low EDT values are found underneath overhangs; each of the four houses has an overhung position with EDT in the region 0.6 – 0.8 seconds. The criterion for EDT is essentially as for reverberation time. In particular if the reverberation time is already short, one does not want an even shorter early decay time.

Sound from the singers is assessed in terms of its clarity and level. For these measurements the 'speech source' with the directivity of a human speaker (Figure 7.16) was used. But whereas in theatres two source orientations were used, only one was used in opera houses: the less critical one with the source facing directly into the auditorium. Source locations of 2 – 3.2 m from the stage front were chosen.

For voice clarity, the speech intelligibility measure employed for theatres in Chapter 8, the 50 ms early energy fraction, D_{50}, is again used. As proposed in section 9.2.3, there is a good argument for speech to be intelligible in opera houses; the criterion is thus again for values greater than 0.5. In the figures, the mean and range of the energy fraction are presented; values are the mean for the octaves 500, 1000 and 2000 Hz. Results are for all measurement positions; measurements were made at between 12 and 15 positions in the four houses.

Voice level is measured and presented as in drama theatres, and the same criterion of greater than 0 db is suggested as appropriate.

For measurements relating to the orchestral sound, an omni-directional source was used placed roughly centrally in the open area of the pit. Identical measurements were made to those used for concert spaces. Thus for orchestral clarity, the 80 ms early-to-late sound index, C_{80}, is used, averaged over the octave frequencies 500, 1000 and 2000 Hz. Because reverberation times for opera are shorter than for symphonic music, a slightly higher optimum range of –1 to +3 db has been used here.

Orchestral source broadening is measured as for concert spaces, using the four octaves between 125 and 1000 Hz. Higher values are to be expected in

the stalls because the direct sound is blocked by the orchestra rail (Barron, 2000).

The orchestral sound level (averaged over 500, 1000 and 2000 Hz) was also measured for the omnidirectional source in the pit, but for economy the data has not been presented explicitly in the figures with objective characteristics. For this quantity a criterion of greater than –2 db may be appropriate, less stringent than the 0 db suggested for concert spaces. The data is however used to quantify a crucial concern for opera: the balance between sound from the singers and orchestra. To obtain the balance, the difference is taken between the singers' sound level and the orchestral level in dB.

The significance of this level balance requires brief discussion. The balance figure assumes that the two sound sources have equal power, whereas in practice singers are seldom as loud as their accompanying orchestra. To obtain a true balance, the actual sound power difference would have to be taken into account. The method used here has the virtue of simplicity and gives convenient decibel values. (Note that measured values here are for singers represented by a directional source facing into the auditorium; different values will result if singers are represented by an omni-directional source.)

No criterion for this balance measure is available so existing data must be used to provide a preliminary recommendation. What is striking from measurements in British houses is how small the range of balance values is: on average about 4 db in each house. Designers and house managements have at their disposal the possibility to change pit arrangements to alter the balance between singer and orchestra. Perhaps these measurements indicate the results of trial-and-error development of pit design in each case. Areas with poor balance prove to be highly localized and specific to the particular house. In the figures of objective data, a tentative preferred range between +1 and +4 db is included; higher values correspond to better audibility of singers. Recent research by Prodi and Velecka (2005) provides basic support for this approach.

Table 9.1 Basic details of British opera houses. Volume is of auditorium only, excluding the flytower. Reverberation time is at mid-frequencies in the occupied house (estimated from unoccupied values for the Coliseum and Buxton house). Data for Covent Garden is for the house prior to the 1999 renovations.

Opera house	Date	Seats	Volume (m³)	Proscenium		Reverberation time (s)
				width (m)	height (m)	
Royal Opera House, Covent Garden, London	1858	2120	12,250	13.5	12.0	1.1
London Coliseum	1904	2354	13,600	15.3	9.0	1.4
Buxton Opera House	1903	946	3100	9.1	6.1	0.9
Glyndebourne Opera House	1993	1209	8300	12.0	9.4	1.25

9.6 British opera houses

Until 2004, Britain could only claim two purpose-built opera houses: Covent Garden and Glyndebourne. Most opera is performed in theatres from the era before the First World War. In the case of auditoria which are homes to opera companies, suitable refurbishment has been undertaken. In many cases the result is houses with short reverberation times, such as is found in Buxton Opera House. The name, Buxton Opera House, in fact derives from what was considered socially acceptable; it was originally built as a variety theatre. The London Coliseum is an exception to the trend regarding reverberation time. Subjective tests (described at the end of section 9.2.3) were conducted during the 1980s in the two London houses in rehearsal conditions. Basic details of the opera houses are given in Table 9.1. For each British opera house, there are figures containing plans, long sections and two photographs.

9.6.1 Royal Opera House, Covent Garden

From the start, the Covent Garden theatre led a life of privilege. John Rich had acquired the second of the 'letters patent' for dramatic productions, originally issued in 1660 by Charles II. His previous theatre had staged John Gay's *The Beggar's Opera*, a parody of *opera seria* which 'made Gay rich and Rich gay'. The proceeds were no doubt valuable for Rich's new theatre at Covent Garden, which opened in

1732. For the next hundred years, the Theatre Royal, Covent Garden, enjoyed the 'duopoly' situation shared with Drury Lane (section 8.3). As with Drury Lane, rebuilding of the theatre to optimize capacity and performance was frequent (Leacroft, 1973). The original small theatre was fan-shaped in plan and was used for mixed dramatic purposes. The horseshoe form was first used for the auditorium remodelling of 1794 by Holland. The same plan form has been consistently used since (a fire in 1808 necessitated complete reconstruction). But it was not until 1846 that the house became principally associated with opera. Before then, Her Majesty's Theatre in the Haymarket had been the home of Italian opera. Dissatisfaction with the management in the Haymarket prompted the conductor and most of the best singers to switch their loyalties to Covent Garden. The interior was again remodelled and the name changed to the Royal Italian Opera House for its first opera season in 1847. But the new interior did not survive long; a fire, started during a *bal masqué*, completely destroyed the theatre again in 1856.

The present theatre of 1858 was designed by E.M. Barry at right angles to its predecessors. Motivated mainly by the major use for opera, Barry selected the horseshoe form again with four levels of boxes/galleries. Barry (1860) justified his choice of plan form as follows:

> The requirements of a London theatre for the Italian Opera are very peculiar, and differ in many respects from those of an ordinary playhouse ... Thus the Royal box being on the grand tier, the latter is the resort of fashion, ..., it is obviously a

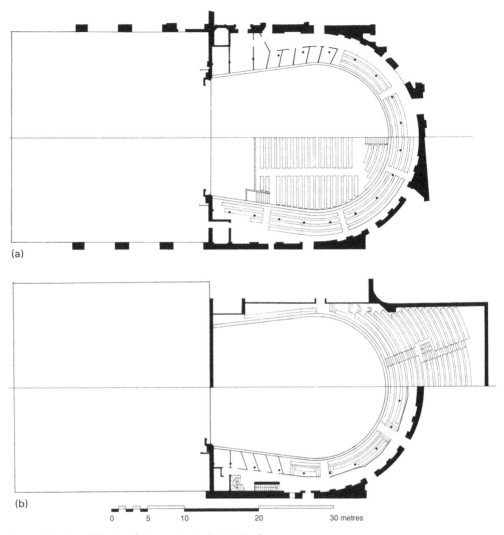

(a)

(b)

0   5   10   20   30 metres

Figure 9.24 Plans of the Royal Opera House, Covent Garden

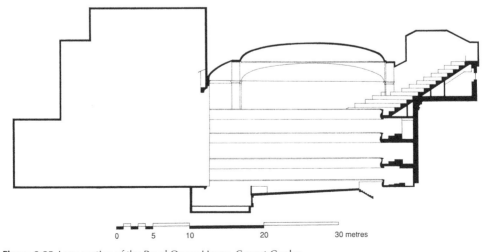

0   5   10   20   30 metres

Figure 9.25 Long section of the Royal Opera House, Covent Garden

Figure 9.26 The Royal Opera House, Covent Garden

great desideratum to obtain as many boxes as possible on this level, where it is found that even the side boxes (in spite of the drawback of position) are always eagerly sought for ... The alleged acoustical advantages of a horseshoe plan were likewise taken into account.

Opera was not the only use, however. A playhouse arrangement was more appropriate for the winter pantomimes, and for this reason the box sides and backs were removable. Many of the original boxes have now been permanently replaced by conventional seating. True to the English tradition, the upper gallery has always been open, extending well beyond the rear wall below. Also characteristic for its time is the domed ceiling, though this introduces the potential for disturbing focusing of sound. Plans and sections are given in Figures 9.24 and 9.25; Figures 9.19 and 9.26 show interior views.

Since becoming London's principal opera house, Covent Garden has had oscillating fortunes. In its golden eras, it attracted the world's greatest singers such as Melba and Caruso. In its lean periods, financial problems have never been far away. In spite of its status, until recently there was a continual lack of official support for the theatre, the royal association being little more than symbolic.

In its earlier years, Italian was the ubiquitous language for Covent Garden opera (performances of Wagner's *Lohengrin*, *Tannhäuser* and *Die Meistersinger* included!). The Italian obsession had waned sufficiently by 1892 for the 'Italian' to be dropped from the name. Since then there have been periods in which only English versions were sung. Now this rôle has been assigned to the English National Opera company at the Coliseum, while Covent Garden presents works in their original languages. The present opera seasons continue to attract international artists to performances of world class. Though the house now absorbs the largest slice of the English Arts Council grant, this is modest by the standards of other European countries. The house is also used by the Royal Ballet.

In the late 1990s Covent Garden was extensively redeveloped by architects Dixon Jones BDP, reopening in December 1999 (K. Powell *et al.*, 1999). The flytower was replaced, side stages were added, a studio theatre built and rehearsal studios for the ballet and opera constructed. Changes to the main auditorium were designed to improve visual and acoustic conditions but might well be missed by a casual visitor. They included increasing the stalls rake, removing the carpet, cutting back box dividers, enlarging the pit and replacing the seating and the ventilation system to give a new background noise level of PNC15. The opening to the highest section of seating, the Amphitheatre, was enlarged and extra volume was created such that the reverberation time rose slightly at the same time as the seating capacity was increased from 2101 to 2157, with 65 standing positions increased to 100.

The measurements and subjective assessment recorded here were made prior to the recent modifications, but the character of the house has changed only slightly.

Subjective characteristics

The acoustic situation in baroque-type opera houses varies considerably between seating areas. Seats in the stalls often suffer from the lack of line of sight to the pit musicians, upper side seats have to contend with poor viewing (and hearing) from the stage, and overhung seats can be at a disadvantage. The highest gallery frequently has the best acoustics, but is this undermined by the large distance from the stage? Beranek (1962) cited two conductors who considered that the orchestral sound lacked brilliance in this house. He also quotes further comments about the difficulties of hearing the violins in the stalls. In this author's survey, the Royal Opera House was tested at four locations but this is far short of including all the acoustic conditions to be experienced in this house. The locations were all facing the stage, with one in the stalls, one position in the Grand Tier (first balcony level) plus positions at the front and rear of the Amphitheatre.

Concerning the main subjective attributes, the clarity of the orchestral sound was judged consistently high but this was at the expense of a good sense of reverberation. Loudness and envelopment were reasonable. Only regarding intimacy and clarity of the voice do the responses vary significantly, decreasing in both cases as the distance from the stage increases. Several individual comments were also made, such as of the 'muddy' character to the orchestral sound in the stalls. At the front of the Amphitheatre some listeners noted odd directional effects; these appear to be caused by focusing from the ceiling. The balance at the rear Amphitheatre is very good, but being 35 m from the singers, there is an inevitable reduction in the impact of the sound here.

But what is remarkable about Covent Garden is that the overall sound is so uniformly judged,

in spite of the fairly large differences in character which can be perceived. All four positions were judged as 'Good' (though in the case of the most distant Amphitheatre seat the assessment sits on the border with 'Reasonable'). The balance judgements on the four positions were similar and of acceptable balance (objectively the values were between 1.7 and 2.6 dB, in the lower half of the suggested preferred range).

Objective characteristics

Measurements in Covent Garden were made in 1980 and '82 before the major renovations (Figure 9.27). The reverberation time is short for opera; the occupied value measured by Parkin *et al.* (1952) was 1.1 seconds. The recent renovations have probably increased the value by 0.1 seconds (Newton, 2001). At mid-frequencies the unoccupied early decay time (EDT) is significantly shorter than the unoccupied reverberation time. It is thus not surprising that the subjective judgement was of lack of reverberance. As for the physical reason for the short EDT, this is not obvious; a lack of reflecting surfaces in front of the proscenium might perhaps be a contributory factor?

Speech intelligibility is above criterion in most seats but below in some positions in the Amphitheatre. One notes that the values are worse than measured in proscenium drama theatres (Figure 8.11) but the seat capacity in Covent Garden is much larger and as a consequence the furthest seat is a distant 36 m from the stage front. For the same reasons, there are some positions below criterion for voice level.

The range of measured balance is the largest of the four opera houses. For a large opera house designed before any useful science of acoustics existed, this is perhaps not surprising. Now one would adjust heights between balconies and attend to orientation of useful surfaces.

While some criticisms can be levelled at the acoustics for sound from singers, measured values for sound from the orchestra are within preferred ranges. The high clarity will be influenced by the short reverberation time.

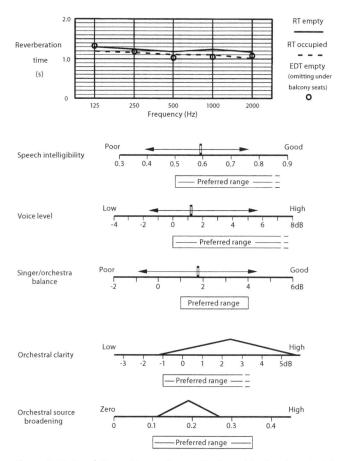

Figure 9.27 Royal Opera House, Covent Garden: objective characteristics

Conclusions

Covent Garden is the flagship opera house in Britain and its age is certainly part of its charm. The consequence of its age for the acoustics is a short reverberation time (and shorter EDT), which is perceived as a rather dry acoustic. In other respects, there are variations to be detected between different seating areas, but often there appear to be virtues to compensate for local deficiencies. The acoustics of the Royal Opera House, Covent Garden, may not sit right at the top in the world league but the experience of attending an opera there is one to be savoured.

9.6.2 The London Coliseum

While the rebuilding in 1858 of the present Covent Garden precedes the spectacular growth of public entertainment in theatres, the London Coliseum of 1904 comes from the height of the theatre-building boom (Figure 8.10). Its instigator was the highly successful self-made theatre entrepreneur, Oswald Stoll. His architect was the most prolific theatre architect of all time, Frank Matcham. Matcham designed over 80 new theatres, and reconstructed at least 50 more, in an idiosyncratic style, criticized by some for its eclecticism. But his ability to build inexpensive theatres often on constrained awkward

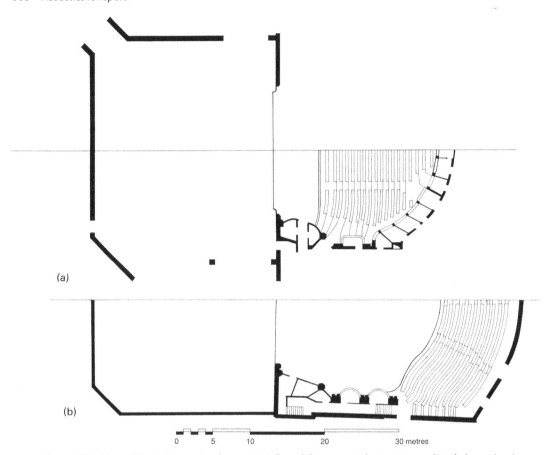

(a)

(b)

Figure 9.28 Plans of the Coliseum, London, at (a) stalls and (b) upper circle (i.e. intermediate balcony) level

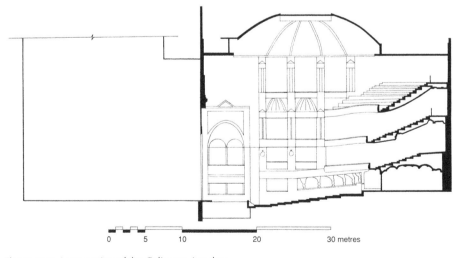

Figure 9.29 Long section of the Coliseum, London

Figure 9.30 The Coliseum, London

Figure 9.31 The Coliseum, London

sites, satisfying sightline and acoustic requirements, the whole decorated with an element of fantasy, proved irresistible to his many clients (Walker, 1980). The Coliseum was to be the largest theatre in the country, with all its major dimensions larger than the Drury Lane theatre of the time (the present Drury Lane auditorium dates from 1922). The stage too was built on a grand scale with a 22.8 m revolve in three independent concentric sections. The theatre was said to

> combine the social advantages of the refined and elegant surroundings of a Club; the comfort and attractiveness of a Café, besides being the Theatre de Luxe of London and the pleasantest family resort imaginable.

The Coliseum advertises its exterior with a large tower, prominently visible from Trafalgar Square. For the auditorium, Matcham's solution to the problem of size without serious loss of the sense of intimacy was to create a compact central volume surmounted by a dome with three deep balconies (Glasstone, 1980). The auditorium width steps out from that of the Stalls to 31 m in the balconies. To optimize sightlines with this substantial width, the balcony fronts droop towards the auditorium sides. In fact, Matcham took out a patent on the slender cantilever system he developed to bridge the spans in the Coliseum. The whole is probably Matcham's most important work.

The Coliseum was built as a variety theatre. A famous spectacle in its early years was re-enactment of the 'Derby' with live horses and jockeys racing 'against' the stage revolve. Many famous artists have performed there (Barker, 1957), among them Diaghilev's ballet company after the First World War. In 1968 a full restoration was undertaken including installation of a large orchestra pit, in preparation for the company now known as English National Opera. The house is now used for opera performed with English libretti, as well as short ballet seasons. The current capacity of 2354 is the largest of any working theatre in London. At the time of writing, it is undergoing another refurbishment.

The most obvious potential problem with the design of this theatre is the deep balcony overhangs. The acoustic performance of this problem seating area was one of the interests of a subjective survey undertaken in this house. Two other aspects of interest were the response to the relatively long reverberation time by UK standards of 1.4 seconds and the effect of the dome on the sound in the upper balcony.

Subjective characteristics

The acoustics of the Coliseum were sampled at four positions, three of them exposed to the main volume of the house and one in the Dress Circle substantially overhung by the balcony above. Judgements at this overhung seat are in many respects different from those at the exposed seats. Since the overhangs are particularly deep in the Coliseum, this is to be expected. With regard to many of the subjective dimensions one can speak of the response in the exposed seats of the house as a unified group.

The voice was judged as slightly less clear than in Covent Garden, though clarity of orchestral sound was judged equally in both houses. In line with the reverberation time, the Coliseum was judged as more live than Covent Garden, except at the overhung seat. Envelopment and loudness were both judged adequate in the Coliseum but intimacy varied substantially with distance from the stage. Balance between singer and orchestra for the exposed seats was judged as uniform. All exposed seats were judged overall as 'Good' regarding their acoustics.

Relative to the exposed seats, the overhung seat was assessed as unreverberant, less enveloping, the least intimate and the quietest in sound level. These are all negative attributes predictable for underbalcony behaviour (section 9.2.6). However, in view of the relatively long reverberation time in the Coliseum, the degree of reverberance at the overhung seat is greater than it would otherwise be; in fact it was judged as comparable to the reverberance at all seats in Covent Garden. In one respect though the overhung seat is at an advantage, which was perceived subjectively: the balance of sound at this position favours the singer at the expense of the orchestra. The overall judgement at this seat was 'Reasonable', which is surprisingly high in the circumstances. The Coliseum was built for entertainment which frequently involved music and its acoustic success suggests that Matcham had a clear understanding of the risks he was taking.

So far the comments have related to average subjective response. In the exposed seats the following differences were apparent to the author. In the Stalls the voice was strong but, without a view into the pit, the orchestra sound was highly diffused. At the front of the Dress Circle the balance was more natural though voices lacked a certain incisive clarity. In the Balcony (the highest level) the sound seemed quieter than expected, perhaps because the dome obscures strong ceiling reflections.

The exercise in the Coliseum provided another most valuable lesson: that different listeners are seeking different characteristics in the sound they hear. Whatever the differences in sound character which listeners perceived, they judged the relative virtues of the four positions differently. For instance, out of seven listeners, two judged the Balcony seat the best whereas two judged it the worst acoustically. In the case of concert hall acoustics, listeners appear to fall into perhaps three different groups regarding preference. A similar variety of taste exists

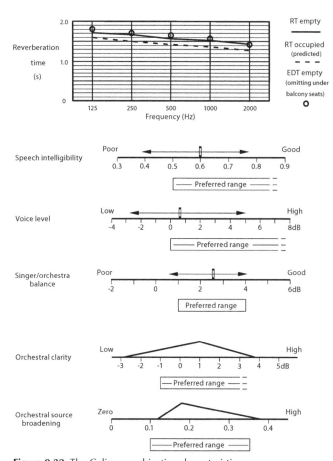

Figure 9.32 The Coliseum: objective characteristics

with opera and because of the extra complication of two sound sources, the number of different groups may even be more for opera.

Objective characteristics

Acoustic measurements were made at 12 positions in the Coliseum in 1982 with an opera set on stage (Figure 9.32). The unoccupied reverberation time is just over 1.5 seconds at mid-frequencies, estimated as about 1.4 seconds with audience. In terms of perceived reverberation, it is good to observe in this auditorium that the early decay time is slightly more than the corresponding (unoccupied) reverberation time.

It is interesting to compare the Coliseum with Covent Garden regarding reverberation time. The volume per seat of both houses is identical at 5.8 m^3/person, yet the reverberation time of the Coliseum is 0.3 seconds longer than the 1.1 seconds in Covent Garden. The likely reason for this is that in the Coliseum a significant proportion of the audience absorption is shielded by the deep overhangs and does not contribute absorption that depresses the reverberation time.

Speech intelligibility in the Coliseum covers a wide range, just as in Covent Garden. Again the seat capacity of the auditorium influences this situation, with the worst positions being towards the front of the highest balcony (the dome may well compromise a good ceiling reflection). Speech level is again easily accounted for: the low mean value reflects the audience capacity and the lowest values are found underneath balcony overhangs.

While conditions for voice are similar in Covent Garden and the Coliseum, the balance between singer and orchestra is much better behaved in the Coliseum, indeed the range of the measured balance is remarkably small. There is a good match between objective and subjective balance, with the highest values in each case at the overhung seat.

For the orchestral sound, the range of clarity is wide with the highest values at overhung seats. Measured source broadening is reasonable.

Conclusions

The acoustics of the Coliseum are impressive. Although designed as a variety theatre, it functions well for opera with a satisfactory reverberation time and sufficient clarity in most locations. The deep balcony overhangs give seats which are judged reasonable acoustically, even though visually they are too remote. The dome does however appear to create disappointing conditions in the front half of the upper balcony. Overall the Coliseum is a testament to what must have been a deep understanding of acoustics developed by its architect, Frank Matcham, a knowledge which was sadly never recorded.

9.6.3 Buxton Opera House

Buxton, in the Peak District, is England's highest market town. Its warm mineral waters were known and used in Roman times. After years of neglect, substantial improvements to the town were made in the late 1700s by the Duke of Devonshire. In spite of its small population of only about 20 000, its popularity as a Victorian spa town was sufficient to support the construction of an 'opera house' in 1903. The architect was again Frank Matcham. Though it was being built simultaneously with the London Coliseum, it is known that Matcham was also closely involved in construction work in Buxton (Walker, 1980). With an audience capacity of only 946, it represents a masterpiece of intimate theatre design. The detailing is characteristically Edwardian, both broader and bolder than theatres of the previous era. The opera house was scrupulously renovated by Arup Associates in 1979, with the only modification being the enlargement of the pit (Sugden, 1979). Apart from changes in seating, the house survives virtually as it was in 1903. Indeed it still includes a working gas 'sunburner' at the centre of the auditorium dome for house lighting and ventilation purposes.

The Buxton Opera House has two balcony levels with an extension of the Upper Circle behind a parapet. This Gallery seating area is clearly separated

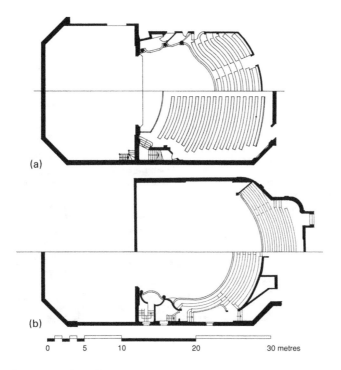

Figure 9.33 Plans of Buxton Opera House

Figure 9.34 Long section of Buxton Opera House

Figure 9.35 Buxton Opera House

Figure 9.36 Buxton Opera House

from the main house volume, in line with social distinctions of the time. The main auditorium is surmounted by a sumptuous oval saucer-domed ceiling decorated with six painted panels. The name Opera House is in fact somewhat a misnomer, designed to confer respectability at a time when theatres were considered suspect. While basically used for drama, the house is the central venue for the annual Buxton Festival, which includes opera performances. The acoustics have not been subjectively sampled in this house but are known to be good.

Objective characteristics and conclusions

As one expects for an auditorium basically intended for drama, the reverberation time is short at 0.9 seconds (Figure 9.37), in line with a small volume per seat of 3.3 m³/person. The consequences of the short reverberation time are predictable: the clarity for orchestral sound and intelligibility of speech are both high. A similar situation is found in several British concert halls discussed in Chapter 5 which also have inadequate volume; a small auditorium volume results in a short reverberation time and provides surfaces for strong early reflections. High clarity is at the expense of a good sense of reverberation. As usual, the lowest sense of reverberation is found under the balcony overhangs.

Because the audience capacity is small for opera, being less than 1000, the measured sound levels are high. Yet the data presented in Figure 9.37 conceals a local variation: sound levels for both sound

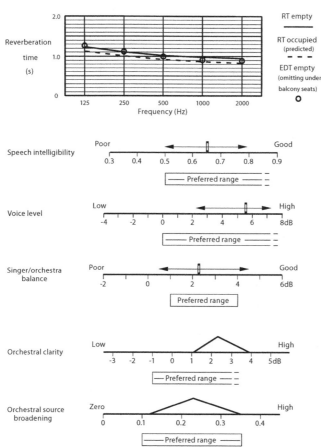

Figure 9.37 Buxton Opera House: objective characteristics

sources are noticeably less in the Gallery, which must surely be a consequence of the reduced opening size into this highest seating area. However the balance between singers and orchestra remains good throughout and there is reasonable source broadening.

It is easy to criticize the Buxton Opera House for its short reverberation time but this is simply a consequence of its being principally designed for drama, not opera. However for many audience members, the intimacy and visual beauty of this auditorium are likely to provide ample compensation for this acoustic shortcoming.

9.6.4 Glyndebourne Opera House

John Christie's idea in 1934 of adding a wing to his country house containing a 300-seat auditorium and inviting the public to a short festival of opera struck many people as crazy at the time (Binney and Runciman, 1994). Yet from the start both audiences and performers were seduced by the intimacy and quality of the performances plus the opportunity to enjoy the delightful gardens and setting in the chalk Sussex Downs. The summer opera seasons soon sold out and the first of many modifications were made to the modest opera house. Ultimately the seat capacity reached 830 but the auditorium was the product of piece-meal renovations with a low reverberation time of 0.8 seconds.

Ticket demand continued to grow. In the late 1980s, George Christie, son of John, took the brave decision to demolish what had never been a distinguished auditorium and seek an architect to build a larger capacity, fully equipped new opera house. A limited competition involving nine leading British architects was held, which was won by Hopkins Architects. To limit the closure period, a fast-track building programme was used with maximum use of off-site construction. The old house closed on 24 July 1992 and the new house was completed by December 1993, ready for the 1994 festival starting in May. Remarkably this was at the time only the third purpose-built public opera house in Britain, the second was the old Glyndebourne (Davies *et al.*, 1994).

The brief was specific that an intimate atmosphere took precedence over technology. Hopkins' solution was to build a free-standing compact building containing auditorium, stagehouse and administrative offices. Great efforts were made to minimize the apparent bulk of the opera house, to avoid it over-powering Glyndebourne House and to maintain the rather contradictory mix of formality and informality which is a hallmark of the Glyndebourne experience. Hopkins ultimately produced a rigorously honest building with no applied decor, using a minimum number of materials constructed with great care and precision.

The dominant material is red load-bearing handmade brick, bonded with traditional lime putty mortar to avoid the need for expansion joints. The same minimalist approach extends to the auditorium which uses just three materials: exposed brick walls, high-quality concrete ceilings and timber flooring, balcony fronts and internal 'furniture'. Most of the timber in the auditorium is reclaimed pitch pine, originally shipped from America during the nineteenth century and used in industrial buildings of that time. It is a strong material with a delightful mature orange glow about it.

The auditorium plan started as fan-shaped. But with the engagement of Theatre Projects Consultants, Iain Mackintosh suggested a return to the baroque form of the horseshoe (Mackintosh, 1995). This traditional form dates from the earliest days of opera but it had at times become unacceptable, associated with wealth and privilege particularly around the time of the French Revolution. The fan-shape plan, though apparently egalitarian, fails to create a shared experience for performers and audience. In the Baroque theatre the sharing extends to the audience itself, in many cases they can see each other; feeling part of the whole contributes to the experience.

The genesis of the plan form can be seen in a very basic diagram, Figure 9.38 (Mackintosh, 1993). Two circles, one the domain of the actor, the other the audience, overlap in a fish-shaped area, known as the *vesica piscis*. Remarkably the two circles can be seen in the plan of the completed building

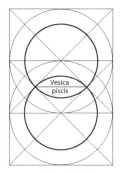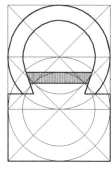

Figure 9.38 Geometric origins of the Glyndebourne Opera House plan according to ad quadratum principles (after Mackintosh, 1993)

(Figure 9.39), the circle of the actor generating the curved surface on either side of the proscenium opening, the circle of the audience generating the line of the balcony fronts, the rear auditorium wall and the auditorium roof. The region of intersection is shared between the crucial elements for opera: the conductor, orchestra and downstage soloists.

The closest historical precedent to the plan form of Glyndebourne is to be found in the Grand Théâtre of Bordeaux of 1780 (Figure 7.8), which incidentally has a similar seat capacity. But Mackintosh was clearly also excited by the 'sacred geometry', which he considers offers a circulation of energy between the actor and audience. Yet as an auditorium form the circle presents an acoustical dilemma: how to avoid the focusing created by so many concave surfaces? This problem demanded considerable ingenuity from the acoustic consultants, Arup Acoustics (Harris, 1995).

For the architect, the requirement was to increase the audience capacity by 50 per cent without losing the sense of intimacy (the final total is 1209 seats, with 42 standing and 12 wheelchair positions). By having two balconies and a larger maximum width, the most distant seat is 29 m from the stage front. This is in fact 3 m closer than it was in the old auditorium. A frequent issue in opera auditoria is the conflict between acoustic and stage lighting design (Figure 7.11). In this auditorium, conflicts are minimized by suspending stage lighting principally at three locations: on the advanced lighting bridge in front of the proscenium, in front of balcony fronts and in the circular roof area.

A major decision for opera is the choice of reverberation time, should one place greater emphasis on the words or the music? Historically there have been national differences, the French for instance with their literary tradition have favoured the words, whereas the Italians have been more concerned with melody, harmony and timbre (Brzeski, Sugden *et al.*, 1994). Glyndebourne opera had a tradition for performing Mozart and smaller scale operas for which a shorter reverberation time is considered suitable. Times of 1.5 – 1.6 seconds as found in the London Coliseum and the Bayreuth Festspielhaus were considered too long. The goal for Glyndebourne was 1.4 seconds, which with strong early reflections provided from the side to maintain clarity was felt to be the most appropriate. The selected auditorium volume was 8300 m³, giving a volume per seat of 6.9 m³/person.

The balcony fronts, which are a maximum of 17–18 m apart, are a major element for provision of side reflections. The potential focusing due to the concave balcony profile in plan was overcome by making the balcony fronts convex in vertical section. Indeed the profile shape varies along the length of the individual fronts to optimize the reflection situation.

Dealing with concave wall surfaces is less easy; two methods have been used in this auditorium. The preferred solution is to place convex surfaces to obscure the concave surface behind. At the highest

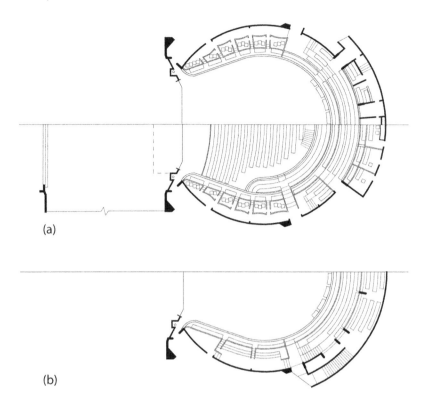

(a)

(b)

Figure 9.39 Plans of Glyndebourne Opera House

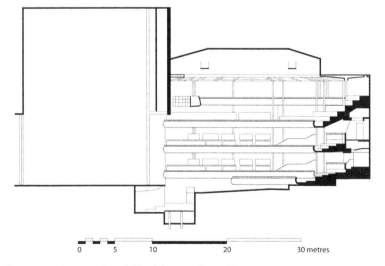

Figure 9.40 Long section of Glyndebourne Opera House

Figure 9.41 Glyndebourne Opera House

Figure 9.42 Glyndebourne Opera House

level in the concrete ceiling, this results in a 'pie-dish' profile. Behind seating, timber convex elements have been placed in front of the concave brick wall (Figure 9.43). However in positions opposite the stage, because of insufficient available depth and in some cases for visual reasons, absorbing panels have been used, with slotted timber over porous absorbent. This successfully deals with the focusing problem but introduces unwanted absorption that reduces the reverberation time.

To investigate the early reflection situation and check the effectiveness of the various techniques for preventing focusing, the acoustic consultants tested a 1:50 scale acoustic model. Computer modelling would not have been appropriate given the difficulties that current programs have with curved surfaces. Reflection paths were also traced optically in the scale model using a laser.

Some surfaces of acoustic significance are missing from this auditorium, namely those located adjacent to the proscenium. The original proposal by Mackintosh (1993) contained fixed boxes which bridged between the balconies and just in front of the proscenium, while adjacent to the proscenium movable boxes in towers were included to give flexibility. In the hands of the architect, the boxes were all removed, the balconies were extended towards the proscenium with a gap left between the proscenium and the auditorium. Surfaces adjacent to the proscenium can perform a valuable function reflecting sound into the auditorium from the singers but not from the orchestra. The acoustic consequence of their absence here may be detrimental to balance between singers and orchestra.

Another contentious issue in opera houses is pit design and in particular the degree of overhang. In the Glyndebourne house, because singers are often young with voices yet to develop their full power, it was decided to use a larger overhang than would otherwise be considered. The maximum dimension between the orchestra rail and stage front is 4 m, the overhung depth is 3.7 m, with a total pit area of 112 m^2.

Objective and subjective characteristics

The measured mid-frequency reverberation time for the occupied auditorium is 1.25 seconds (Figure 9.44), a little shorter than the goal of 1.4 seconds. The reasons for the lower value are likely to be a slight reduction in auditorium volume made late in the design process as well as the absorbing treatment to control focusing. The lower reverberation time will favour clarity at the expense of a sense

Figure 9.43 View of rear wall of Glyndebourne Opera House showing convex and sound absorbing panels

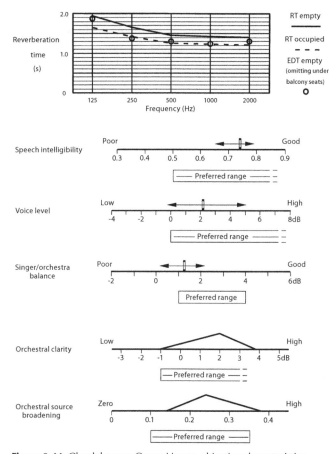

Figure 9.44 Glyndebourne Opera House: objective characteristics

of reverberation, words being favoured relative to the feeling of being surrounded by the music.

The measured quantity relating to intelligibility for a source on stage shows remarkably uniform high values, which bears testament to the care with which early reflections were preserved in the design. The spread of values for other measures is more typical, influenced in several cases by characteristic behaviour under balcony overhangs.

The voice level for listeners is also satisfactory relative to the criterion of 0 dB. Compared with other British houses, the Glyndebourne value is less than that measured in Buxton but more than in both the London houses. This is as one expects since a major influence on sound level is the amount of sound-absorbing material in the auditorium, which is of course substantially influenced by the seating capacity. A more detailed analysis shows however that the measured speech level in Glyndebourne is a little quieter than expected for its seat capacity. The absence of reflecting surfaces around the proscenium may contribute to this.

Proscenium area design may also be significant when one comes to consider the balance between the singers and the orchestra. The average balance in Glyndebourne is lower than in the other three British opera houses measured. The source position in the pit was in the open part. Because of the deep pit overhang in this house, the measurement for the pit source may be louder than it would be if averaged over many positions throughout the pit. Apparently subjective balance is not a problem here in practice.

For the two measures relating specifically to the orchestral sound, behaviour is entirely satisfactory, suggesting a clear, spatially rich sound.

The acoustics of the new Glyndebourne Opera House have been uniformly praised. A systematic subjective survey has not been undertaken. A single listener's views are of limited value, but at a central position in the upper seating level the sound was judged as clear both for singers and orchestra, as well as being loud, intimate and spatially surrounding. Overall good to very good, the only criticisms related to less reverberance than one would have liked and a balance slightly to the detriment of the singers. These subjective judgements all tie in well with the objective measurements.

Conclusions

The new Glyndebourne Opera House is the only surviving purpose-built house in Britain from the twentieth century. Its auditorium follows the traditional baroque plan form with close attention to a circular geometry. Internal finishes are simple but executed with great care and attention to detail. This combination of geometry and simple but attractive finishes creates an intimacy appreciated by nearly all who have experienced it.

The circular or at least curved geometry is visually very seductive. The soprano Alison Hagley has commented (*Time Magazine*, 13 July 1994): 'It's really more a circle than a horseshoe, and on stage I am part of that circle. The audience is my friend and I am theirs'. But this geometry presents an acoustical dilemma: concave surfaces focus sound. In the case of balcony fronts, the concave form in plan can and has been compensated by a convex vertical profile. Treating concave walls is more difficult; both convex and absorbing panels have been used in Glyndebourne and have succeeded in eliminating audible focusing. But the by-product of this treatment has been a reverberation time shorter than intended.

Another significant feature for the acoustics is the absence of sound-reflecting surfaces adjacent to the proscenium opening; this may affect the balance between the singers and orchestral sound a little. But overall the house offers as intimate an acoustic experience as one can expect. In visual terms the design cleverly reconciles the traditional horseshoe plan with modern architectural sensibilities. As a medium-size opera house design the new Glyndebourne Opera House is likely to remain a landmark for many years to come.

9.7 Conclusion

Of all auditorium types, the architectural form of opera houses has changed the least during its existence as a public space of entertainment. With a proscenium opening and a pit separating the stage from stalls audience, opera houses have considerable design constraints. The horseshoe plan form, which was already established for opera in the seventeenth century, has proved admirably suited to social opera-going. So successful has it been that in several countries the visual shortcomings have been ignored for the sake of tradition. It was only in the second half of the last century, particularly in Germany, that a radical reassessment of opera house design was undertaken. Many of these designs avoided deep balcony overhangs, which had become a major fault of large house arrangements. Unfortunately many of the rebuilt German houses predated the discovery of the importance of early acoustic reflections. In plan, design for sightlines was often good, but this tended to generate surfaces radial to the stage, which were not suitable for providing acoustic reflections. Providing good sightlines and useful acoustic reflecting surfaces became the great challenge in opera house design. Since around 1990 however the trend has been to return to the baroque theatre form, the German Wagner-inspired forms being criticized as lacking in visual intimacy. More recently still, some opera houses with longer reverberation times have and are being built. Providing good acoustics throughout the audience seating area is a major challenge, particularly if specifically acoustic elements need to be concealed.

The subjective aspect which emerges as probably the most important for opera is the balance between singer and orchestra. A British survey of opera-goers found that the majority wanted the balance shifted more in favour of the singer. The historical situation is of considerable interest here. Meyer (1986) has investigated how balance has changed between Mozart's time and our own. During this period opera orchestras have become larger and opera houses have expanded in size and seat capacity, which leaves the question of whether the two developments have compensated for each other. What analysis reveals is that to achieve sound levels equivalent to 20 string players in the historical Viennese theatres requires in modern large houses string sections of up to 70, as specified by Verdi. The number of woodwind players has however remained constant, so their current contribution is relatively quieter than in Mozart's day. Turning to the balance between singer and orchestra, the sound output of singers has presumably remained constant. For the orchestra, as well as changes in numbers of players, the lowered orchestra pit since the nineteenth century has reduced sound levels, while development of instruments has made them louder. The great secret of opera singers appears to be the singer's formant (section 9.2.1). Looking at the current situation, Meyer concludes:

> With regard to the balance between the singer and the orchestra, the sound power of the singer in *forte* thus corresponds approximately to that of the historical string group and by comparison, the wind group's *forte* approximates the peak values of the singer's sound power. As the aforementioned numerical values demonstrate, at 3000 Hz the singer's formant can distinctly rise above the sound power level of the orchestra, and without this formant, the singer would have difficulty making himself heard above the orchestra. At these frequencies, the violins and oboes can indeed radiate a considerable number of sound components, but the direct radiation of these is weakened by the orchestra's low-lying position in the orchestra pit. In contrast, the directivity of the singer's voice is to his fullest advantage.

If the above explains how a single singer can hope to compete with 70-odd instrumental musicians, it still behoves the consultant to encourage that balance as far as is possible. As Shawe-Taylor (1948) has said:

> Opera is a composite art, and one of its principal attractions has always been the thrill of a great voice greatly used. This is a truth which no opera house in the world can safely neglect.

The crucial surfaces at the consultant's disposal are those adjacent to the proscenium. Owing to the proximity to the two sound sources, the singers and orchestra, these surfaces offer the greatest potential for selective reinforcement of the singer. The other major variable affecting balance is orchestra pit design. It is inevitable during the lifetime of a pit that changes will be made. Designers should plan for this by offering substantial flexibility, with generous floor space for the expected number of musicians, generous height allowance in the covered section of the pit and if possible a floor on lifts.

Opera house design, which seemed static for so long, entered a new phase with post-war reconstruction. Paradoxically perhaps, in several new houses designs have involved revisiting the baroque form. But it is only now that it can be informed as well in acoustic as it is in visual and performance terms. There is the potential for modern opera houses to perform better acoustically than their predecessors of 30 or more years ago.

References

General

Beranek, L.L. (1962) *Music, acoustics and architecture*, New York: John Wiley.

Beranek, L.L. (2004) *Concert and opera houses: Music, acoustics and architecture*, 2nd edn, Springer, New York.

Beranek, L.L., Hidaka, T. and Masuda, S. (2000) Acoustical design of the opera house of the New National Theatre, Tokyo, Japan. *Journal of the Acoustical Society of America* **107**, 355–367.

Cremer, L. and Müller, H.A. (trans. Schultz, T.J.) (1982) *Principles and applications of room acoustics, Vol. 1*, London: Applied Science.

Forsyth, M. (1985) *Buildings for music*, Cambridge University Press and MIT Press.

Hidaka, T. and Beranek, L.L. (2000) Objective and subjective evaluations of twenty-three opera houses in Europe, Japan, and the Americas. *Journal of the Acoustical Society of America* **107**, 368–383.

Izenour, G.C. (1977) *Theater design*, New York: McGraw-Hill.

Jordan, V.L. (1980) *Acoustical design of concert halls and theatres*, London: Applied Science.

Mackintosh, I. (1993) *Architecture, actor and audience (Chapter 11)*, London: Routledge.

Marshall, A.H. and Meyer, J. (1985) The directivity and auditory impressions of singers. *Acustica* **58**, 130–140.

Meyer, J. (1986) Some problems of opera house acoustics. *Proceedings of the Vancouver Symposium on Acoustics and Theatre Planning for the Performing Arts*, August, pp. 13–18.

Orrey, L. (1972) *A concise history of opera*, London: Thames and Hudson.

Sugden, D. (1994) The opera house: complexities and contradictions. *Proceedings of the Institute of Acoustics*, **16**, 461–469.

Vermeil, J. (1989) *Opéras d'Europe*, Paris: Edition Plume.

References by section

Section 9.1

Newton, J. (2006) The acoustic design of the Donald Gordon Theatre, Cardiff and Operaen, Copenhagen. *Proceedings of the Institute of Acoustics*, **28**, Part 2, 36–43.

Section 9.2

Alexander, A. (1982) *About the house* (the magazine of the Friends of Covent Garden), Spring, 1982.

Cabrera, D. and Davis, P. (2004) Vocal directivity measurements of eight opera singers. *Proceedings of the 18th International Congress on Acoustics, Kyoto*, Paper Mo5.C1.5, Vol. I, 505–506.

Cremer, L., Nutsch, J. and Zemke, H.J. (1962) Die akustischen Massnahmen beim Wiederaufbau der Deutschen Oper Berlin. *Acustica* **12**, 428–433.

Fasold, W. and Winkler, H. (1976) *Bauphysikalische Entwurfslehre, Band 5: Raumakustik*, Berlin: VEB Verlag.

Halmrast, T. (2002) The influence of a large reflector over the orchestra pit in an opera house. *Proceedings of the Institute of Acoustics*, **24**, Part 4.

Katz, B. and d'Alessandro, C. (2007) Directivity measurements of the singing voice. *Proceedings of the 19th International Congress on Acoustics, Madrid*, Paper MUS-06-004.

Mackintosh, I. (1992) The theatrical and acoustic significance of the forestage in opera house design and practice from the 18th to the 21st centuries. *Proceedings of the Institute of Acoustics*, **14**, Part 2, 103–113.

Mackintosh, I. (2002) The mystery of the disappearing forestage in the 18th and 19th century opera house and its relevance to both modern staging and the audibility of the singer over the orchestra today. *Proceedings of the Institute of Acoustics*, **24**, Part 4.

Reichardt, W. (1979) *Gute Akustik – aber wie?*, Berlin: VEB Verlag.

Reichardt, W. (1985) Die akustische Projektierung der Semper-Oper in Dresden. *Acustica* **58**, 253–267.

Sundberg, J. (1977) The acoustics of the singing voice. *Scientific American* **236**, 82–91.

Sundberg, J. (1995) Acoustics of the singing voice. *Proceedings of the 15th International Congress on Acoustics, Trondheim*, Vol. III, 39–44.

Section 9.3

Blair, C.N. (1998) Listening in the pit. *Proceedings of the 16th International Congress on Acoustics, Seattle*, Vol. I, 339–340.

Gade, A.C. and Mortensen, B. (1998) Compromises in orchestra pit design; a ten year trench war in the Royal Theatre, Copenhagen. *Proceedings of the 16th International Congress on Acoustics, Seattle*, Vol. I, 343–344.

Gade, A.C., Kapenekas, J., Gustafsson, J.I. and Andersson, B.T. (2001) Acoustical problems in orchestra pits; causes and possible solutions. *Proceedings of the 17th International Congress on Acoustics, Rome*, Vol. III, 10–11.

Harkness, E.L. (1984) Performer tuning of stage acoustics. *Applied Acoustics* **17**, 85–97.

Mackenzie, R.K. (1985) The acoustic design of partially enclosed orchestra pits. *Proceedings of the Institute of Acoustics, Reproduced Sound*, Autumn Conference 1985, pp. 237–243.

Meyer, J. (1998) Sound fields in orchestra pits. *Proceedings of the 16th International Congress on Acoustics, Seattle*, Vol. I, 337–338.

Naylor, G. (1985) Problems and priorities in orchestra pit design. *Proceedings of the Institute of Acoustics* **7**, Part 1, pp. 65–71.

Section 9.4

Bagenal, H. and Wood, A. (1931) *Planning for good acoustics*, London: Methuen.

Barron, M. (1995) Balcony overhangs in concert auditoria. *Journal of the Acoustical Society of America* **98**, 2580–2589.

Cremer, L., Nutsch, J. and Zemke, H.J. (1962) Die akustischen Massnahmen beim Wiederaufbau der Deutschen Oper Berlin. *Acustica* **12**, 428–433.

Gustafsson, J.-I. (1995) The acoustics of the new Gothenburg Opera. *Proceedings of 15th International Congress on Acoustics, Trondheim*, Vol. II, pp. 485–488.

Gustafsson, J.-I. and Natsiopolos, G. (2001) Stage set and acoustical balance in an auditorium of an opera house. *Proceedings of 17th International Congress on Acoustics, Rome*, Vol. III, pp. 36–37.

Müller, H.A. and Vian, J.-P. (1989) The acoustics of the new 'Opéra de la Bastille' in Paris. *Proceedings of the 13th International Congress on Acoustics, Belgrade*, Vol. 2, 191–194.

Reichardt, W. (1985) Die akustische Projektierung der Semper-Oper in Dresden. *Acustica* **58**, 253–267.

Schubert, H. (1971) *The modern theatre – architecture, stage design, lighting*, London: Pall Mall Press.

Suner, B. (1990) Ott à l'Opéra Bastille. *L'Architecture d'aujourd'hui*, Avril 1990, 126–135.

Xu, A.Y. (1999) Domes and echoes in classical opera houses. *Proceedings of the Institute of Acoustics*, **21**, Part 6, 249–255.

Section 9.5

Barron, M. (2000) Measured early lateral energy fractions in concert halls and opera houses. *Journal of Sound and Vibration* **232**, 79–100.

Prodi, N. and Velecka, S. (2005) A scale value for the balance inside a historical opera house. *Journal of the Acoustical Society of America* **117**, 771–779.

Section 9.6.1

Barry, E.M. (1860) On the construction and rebuilding of the Royal Italian Opera House, Covent Garden. *Papers read at the RIBA, Session 1859–60*, pp. 5364.

Leacroft, R. (1973) *The development of the English playhouse*, London: Eyre Methuen.

Newton, J.P. (2001) Room acoustics measurements at the Royal Opera House, London. *Proceedings of 17th International Congress on Acoustics, Rome*, Vol. III, pp. 34–35.

Parkin, P.H., Scholes, W.E. and Derbyshire, A.G. (1952) The reverberation times of ten British concert halls. *Acustica* **2**, 97–100.

Powell, K. *et al.* (1999) House style. *Architects' Journal*, 2/9 December, 29–44.

Shawe-Taylor, D. (1948) *Covent Garden*, London: Max Parrish.

Sheppard, F.H.W. (1970) *The Theatre Royal, Drury Lane, and the Royal Opera House, Covent Garden*, London: Survey of London XXXV.

Section 9.6.2

Barker, F. (1957) *The house that Stoll built*, London: Frederick Muller.

Glasstone, V. (1980) *The London Coliseum* ('Theatre in Focus' series) Cambridge: Chadwyck-Healey.

Walker, B.M. (ed) (1980) *Frank Matcham, theatre architect*, Belfast: Blackstaff Press.

Section 9.6.3

Sugden, D. (1979) A restoration and the birth of a festival at Buxton. *Cue Technical Theatre Review*, Nov/Dec. 1979, pp. 18–19.

Walker, B.M. (ed) (1980) *Frank Matcham, theatre architect*, Belfast: Blackstaff Press.

Section 9.6.4

Binney, M. and Runciman, R. (1994) *Glyndebourne: Building a vision*. Thames & Hudson, London.

Brzeski, S., Sugden, D., Thornton, J. and Turzynski, J. (1994) Engineering an opera house: the new Glyndebourne. *The Arup Journal* 3/1994, 3–9.

Davies, C. *et al.* (1994) Glyndebourne. *Architectural Review*, June 1994, 35–66.

Harris, R. (1995) The acoustic design of the Glyndebourne Opera House. *Proceedings of the Institute of Acoustics*, **17**, Part 1, 107–114.

Mackintosh, I. (1993) *Architecture, actor and audience*, Routledge, London, pp. 144–147.

Mackintosh, I. (1995) The origin and antecedents of the Glyndebourne auditorium. *Proceedings of the Institute of Acoustics*, **17**, Part 1, 99–106.

Minors, A. (1995) To hear, to see and to be seen? *Proceedings of the Institute of Acoustics*, **17**, Part 1, 115–123.

Swenarton, M. (1994) Arcadian overtures: Hopkins at Glyndebourne. *Architecture Today*, No. 48, May 1994, 30–39.

Section 9.7

Shawe-Taylor, D. (1948) *Covent Garden*, London: Max Parrish.

10 Acoustics for multi-purpose use

10.1 Introduction

The use of auditoria for more than one purpose is far from new; indeed we have the prime example in the Redoutensaal, Vienna, of the classical rectangular concert hall which developed out of the ballroom. Flat floor arrangements for balls were also found in theatres, such as in the Opéra de Versailles, France, of 1770 and Smirke's 1809 Covent Garden theatre. (This proved to be the undoing of Covent Garden: it caught fire and was destroyed in 1856 during a *bal masqué*.) But the conscious design of auditoria to accommodate more than one acoustic type of performance is relatively recent. It has become increasingly apparent that for economic reasons auditoria dedicated to just one single use are often unrealistic in all but large cities. A degree of flexibility in use is now becoming the norm. The stage extension which converts into an orchestra pit or the stage shell which allows for orchestral performance in a proscenium theatre are both common features now taken for granted. But neither of these involves a major change in the acoustics. It is the possibility of variable acoustics which will be the main interest in this discussion.

One discovery in recent years for those working in acoustics research has been that for physical changes to be perceived subjectively they must be substantial. Small patches of acoustic absorbent or changes in orientation of small surfaces generally have no significant effects. Acoustic character is more a question of gross shape than small detail. This means that, for variable acoustics to be meaningful, major changes are required. This is certainly the case with variable reverberation time, which often proves to be the most valuable acoustic change one can achieve.

Optimum reverberation time is principally related to programme but different uses often require different seating arrangements. In terms of auditorium form, one has at one extreme the central performing area with arena-type seating and at the other extreme all the audience facing in a single direction looking through an opening. Each use has a maximum seat capacity associated with it, owing to visual and acoustic constraints. Table 10.1 lists approximate limiting seat capacities, distance limits from the stage (for visual purposes) and optimum reverberation times. In small auditoria the reverberation time criteria become less stringent. For a hall containing several hundred seats, a short time can be acceptable for small-scale music performance (section 6.2). However in larger halls, a longer reverberation time becomes necessary for music but this creates risks of speech being no longer intelligible.

To accommodate two or more acoustic uses generally means that a reverberation time change is desirable. Introducing acoustic absorbent is the simplest approach, but this has the additional effect of reducing sound level, which may be unacceptable for the lower reverberation time configuration. Popular music on the other hand is performed quite happily at outdoor locations, so that for popular music in enclosed spaces a highly absorbent acoustic space is preferred. But it is the auditorium which has to house dramatic theatre as well as orchestral concerts which poses the most serious problems. It

Table 10.1 Basic limitations on auditorium size and reverberation time as a function of programme

Use	Maximum seat capacity	Maximum audience distance from stage (m)	Optimum reverberation time (s)
Popular music	–	–	<1.0
Drama theatre	1300	20	0.7–1.0
Opera and ballet	2300	30	1.3–1.8
Chamber music	1200	30	1.4–1.8
Orchestral music	3000	40	1.8–2.2

is appropriate first to discuss the options in creating variable acoustics.

10.2 Variable acoustic elements

Variable acoustics depends either on electroacoustic systems or physical adjustable elements. The possibilities among physical variable acoustic elements are considered first.

From the Sabine equation (section 2.8.2), the two variables influencing the reverberation time are the internal volume and the amount of acoustic absorbing material. Of these, changes in volume are in principle more attractive but in practice often difficult to incorporate within a functioning auditorium. Varying the volume has the advantage over varying the acoustic absorption that there is little penalty in terms of sound level.

10.2.1 Variable auditorium volume

There are basically two methods to provide variable acoustic volume: by a movable panel/partition or by a movable shutter system. The movable panel option is not unusual in the case of vertical partitions. These however are generally used to vary the floor area and with it the seating capacity. The effect on the reverberation time is often small. Non-vertical movable surfaces offer the chance to open up extra volumes with little absorbent in them, thus introducing a reverberation time change. Movable ceilings are the most common solution for this type of variable-volume hall.

Early examples of movable ceilings were tried in the 1960s and '70s in North America, mostly using

rotating ceiling sections to close off higher seating areas (Izenour, 1977). The rather mixed success of these variable auditoria appears to have prevented many new halls following their example. Now at the turn of the millennium, new halls with variable volume are becoming more common. Two examples illustrate interesting possibilities for vertically movable elements.

The creation in 1999 of a concert space **Sala São Paolo** with a seat capacity of 1509 in a railway station in the city of **São Paolo, Brazil**, is an intriguing story (Nepomuceno *et al.*, 2002a, 2002b; Beranek, 2004). The architects for the conversion were Dupré Arquitetura with the acoustics handled by Acústica & Sônica of São Paolo and Artec Consultants. An internal rectangular-plan space open to the sky proved to have proportions similar to those of classical concert halls. This courtyard was covered over but, for present purposes, the interest of this example is the ceiling made of 15 independently movable panels, which can alter the height between 12.2 and 23.7 m. This provides a volume change from 12 000 to 28 000 m³ and occupied reverberation times between 1.9 and 3.0 seconds. The movable panels are loose-fitting so with panels at different heights, the upper space is acoustically linked to the lower. For conferences etc. a reverberation time of 1.5 seconds can be achieved when absorbing banners are placed in the void above the panels. This system of movable panels clearly offers a highly variable acoustic space, though such variability is not cheap.

In the second example, a single large ceiling element moves through a vertical distance of 10 m to provide variable acoustics (Orlowski, 1999; Allen *et al.*, 1999). The **Milton Keynes Theatre** in England (architects Blonski Heard Architects, acoustic consultants Arup Acoustics) has to accommodate drama, musicals, opera, ballet, light entertainment and orchestral concerts, with a bias towards drama use. The auditorium form is as for a theatre with two balconies and a full height flytower (Figure 10.1). The movable ceiling at its highest position provides additional volume to provide a longer reverberation time (RT) for orchestral music. The intermediate ceiling height provides acoustics suitable for drama to an audience of 1250, while at the lowest level the upper balcony is closed off to give drama conditions for 850. Drapes over sections of side walls contribute to reduce RTs for drama, while an orchestral enclosure is used to surround musicians on stage. The reverberation times range from 1.1 to 1.5 seconds, with good speech intelligibility for drama. This theatre, which opened in 1999, has provided the modest-size town of Milton Keynes with a highly flexible multi-purpose auditorium at a modest cost.

The alternative shutter system which allows a suspended ceiling to be opened or closed would appear to offer acoustic variability at very little cost. Many auditoria have a void above the ceiling, usually containing structure and ventilation ducting. Adding this volume to the acoustic volume of the auditorium can provide useful reverberation time gains. As a strategy, it does however present two pitfalls to the unwary: is the coupling area with the void adequate and is the nature of the void such that it enhances reverberation? Experience suggests that in the open condition, an open area of greater than 40 per cent is needed. The void space must behave acoustically as a reverberant volume; if there are significant acoustically absorbent or scattering surfaces in the void, the extra volume may not make a worthwhile contribution to reverberation time. The obvious economical solution, in which the void contains roof structure and ventilation ducting, may not work in practice for these reasons. Any scheme of this sort should be tested in an acoustic model with the elements in the void scrupulously reproduced.

An impressive low-cost auditorium, which was designed with panels that link a ceiling void to the auditorium for acoustic variability, is to be found in **Auckland, New Zealand** (Valentine and Day, 1998). The **Bruce Mason Theatre**, designed by Avery Jasmax Architects with acoustics by Marshall Day Acoustics, was opened in 1996. The theatre has

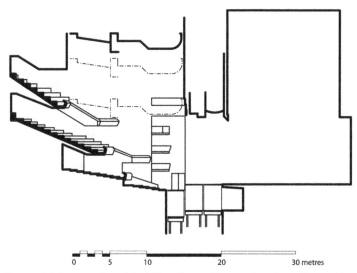

| 0 | 5 | 10 | 20 | 30 metres |

Figure 10.1 Long section of the Milton Keynes Theatre, showing ceiling positions

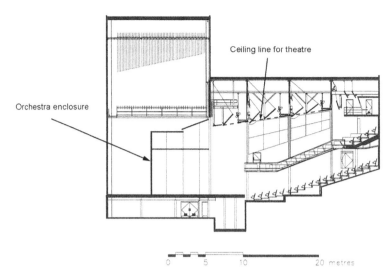

Figure 10.2 Long section of the Bruce Mason Theatre, Auckland, New Zealand, in the symphony mode with the location of ceiling panels for theatre indicated

900 seats with a conventional flytower. The ceiling consists of panels hinged at their bottom edge, the majority of which can be opened to enlarge the acoustic volume (Figure 10.2). In theatre mode with the panels lowered, the ceiling provides strong overhead reflections appropriate for speech. In symphony mode the panels are raised with only about 31 per cent of the ceiling providing overhead reflections, a proportion that is appropriate for music use. A 1:25 scale model was tested, among other things to establish the optimum orientation for the panels in symphony mode. Rather than the expected vertical orientation, the optimum angle for the panels to maximize reverberation time was found to be 30–40° to the vertical, parallel to sound paths from a typical source position. In other words, the optimum corresponds to getting the maximum sound energy **into** the ceiling void. An orchestral shell is used on stage for musical events. In the theatre mode, some ceiling panels have to be opened for stage lighting purposes; acoustic absorbing material has to be introduced into the ceiling void in this condition to avoid double slope reverberant decays. A small amount of absorbing material is also deployed for theatre on the front

side walls to assist in suppressing the reverberation time for that use. In practice, the auditorium has well satisfied the client's ambitions, offering a mid-frequency early decay time change from 1.1 to 1.6 seconds (unoccupied measurement, the occupied reverberation time change is likely to be the same or slightly larger).

In summary, varying the volume of auditoria has been tried for many years but several examples have offered disappointing degrees of variability. With care however, this option can be successful and further ingenious solutions can be expected in the future. One successful variable-volume solution, which has been tried several times, is further discussed in section 10.4 below. In a theatre space with a flytower, it is possible to seal off the flytower at ceiling level and use the drama stage as the orchestra platform. By these means, an auditorium volume change can be sufficient to give appropriate reverberation times for both drama and music.

10.2.2 Reverberation chambers

Reverberation chambers also involve an increase in effective auditorium volume but differ from the schemes discussed in the previous section regarding the size of coupling area. With a smaller coupling area, the auditorium volume and the chamber volume will act as coupled spaces. When measured in the auditorium the decay should have a double slope (Figure 2.31(b)); the first slope is determined by the auditorium and the second slope by the reverberation time of the chambers themselves. For correct operation, it is again crucial to have adequate coupling area and sufficiently reverberant chambers. In simple terms, enough acoustic energy has to enter the chambers and then bleed gradually back into the auditorium.

For the listener, variable volume should influence both running and terminal reverberation (section 2.8.1). Reverberation chambers tend only to influence terminal reverberation; listeners are only aware of the chambers during pauses in musical performance, when the full sound decay is audible. Listeners at these moments may get the impression of a space larger than the one they can appreciate with their eyes.

Reverberation chambers involve substantial volume spaces being built around the auditorium with 'doors' providing the variable coupling with the main auditorium. Chambers have been included since 1989 in concert spaces consistently, but also virtually exclusively, by Artec Consultants (New York). They were first used in the 1989 Eugene McDermott Concert Hall, Dallas by architects Pei Cobb Freed & Partners (Beranek, 2004) and Birmingham Symphony Hall (section 5.14). Subsequently both the chambers and the openings into them have become larger, as in the 1999 Lucerne Concert Hall by Architectures Jean Nouvel (Beranek, 2004). The value of reverberation chambers is questioned by several acoustic consultants, particularly in view of their expense. Their potential role is probably limited to spaces for music of different types, so that one might set the acoustics as less reverberant for Mozart and maximum reverberance for Mahler.

10.2.3 Variable acoustic absorption

Variable absorption is the commonest variable feature, particularly so in North America. In small halls a retractable curtain is enough to change acoustic character in perceptible ways. Curtains placed in the stage area can be extended to mute loud instruments positioned close to walls. In some cases curtains can obscure later reflections occurring with particular source placements. But in order to influence the reverberation time, the area of adjustable absorption must be very large, in fact comparable in size to that of the audience area. This makes variable absorption a progressively more extravagant option the larger is the hall in question. Before discussing where absorbing material can be placed, the physical realization of such absorption should be discussed.

In smaller rooms, hinged panels have been used with one side acoustically hard and the other acoustically absorbent. If the absorbent side of the panel faces another absorbent surface when it is closed, a transition from a fully reflective to a fully absorbent surface is theoretically possible. The degree of change is dependent on the quality of seal in the reflective condition. An interesting variant of this arrangement was used by Bickerdike Allen Partners in an orchestral recording and rehearsal hall in the Hong Kong Academy for Performing Arts (Lord and Templeton, 1986, p. 172). Motorized flaps located over modular absorbers can either screen the absorber or expose it. An intermediate arrangement with alternate flaps lowered provides lateral reflections (Figure 10.3). A reverberation time change from 2.3 to 1.9 seconds at 500 Hz in the unoccupied hall was achieved. These techniques are however only really appropriate to smaller halls.

The most common technique in modern halls for variable absorption is to use acoustic banners. Banners are usually free hanging and can be retracted from the auditorium through slots or folded into well-sealed boxes when not required. For maximum absorption both the weight and porosity of the banner must be optimized. Figure 10.4 shows the typical absorption achieved, which is characteristic

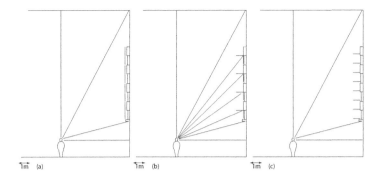

Figure 10.3 Variable absorption in the Hong Kong Academy for Performing Arts. The three arrangements of flaps allow three degrees of absorbent to be exposed, with a high number of lateral reflections produced in the intermediate scheme. (a) Flaps – all closed, few early reflections, long reverberation time; (b) flaps – half open, many early reflections, medium reverberation time; (c) flaps – all open, few early reflections, short reverberation time

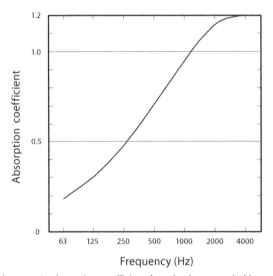

Figure 10.4 Typical achievable acoustic absorption coefficient for a freely suspended banner, based on the area of one side of the banner

of a porous material in that it is effective at mid-frequencies but not at low frequencies. If balconies are constructed with a gap between the back of the balcony and the auditorium walls so that banners can be lowered behind them, substantial areas of banners can be used and large reverberation time changes are possible.

The major problems with variable absorption are firstly the effect on sound level and secondly

possible suppression of early reflections. Neither problem is of any significance with amplified sounds; indeed extra absorption is then usually desirable for those very reasons. But for unassisted speech or music performance, a serious compromise can arise. The effect on sound level arises directly from the traditional (or revised) formulae, which state that the total sound level depends on the total absorption (section 2.9). Introducing banners increases that total absorption and makes the (reflected) sound quieter. In a large hall it may be undesirable to squander acoustic energy in this way.

In the case of the second problem, it is difficult to locate absorbent material so that it will only influence late reverberant sound and not early reflections. Wall surfaces are generally required to provide early reflections, yet these can be the most convenient locations for banners. Substantial reverberation time changes produced by introducing absorption carry this penalty, which will often be unacceptable for unassisted sound sources.

10.2.4 Movable reflectors

In the history of concert hall acoustics, suspended horizontal reflectors have not been a success. The story of the New York Philharmonic Hall (section 4.8) is still vivid in many memories. The less extreme example of orchestral reflectors which were used in British halls during the 1950s, for instance, went out of favour owing in part to the shrill sound quality they impart to string tone. The American consultants, Bolt, Beranek and Newman, continued to use overstage reflecting saucers which could be adjusted in height with audible effect, it was claimed (e.g. Roy Thomson Hall, Toronto, section 4.11). Artec Consultants currently use large continuous overstage canopies which serve the performers and the front stalls (sections 4.11, 5.12 and 5.14). However, these more recent American examples do not provide reflections for the majority of the audience.

An interesting example of a movable reflector is to be found in the Queen Elizabeth Hall of 1967 in London (section 6.5). This is for music use only, but confronted at the time with an uncertain verdict on overstage reflectors in earlier British halls, a movable surface was installed. This reflector in its lowered position provides an overhead reflection to the seating; in its raised position it does not. Subjective tests proved less convincing than had been anticipated. For orchestral forces the majority view was that an overhead reflection was undesirable, whereas for piano solo it was preferred.

Yet for speech use, the arrival direction of reflections at the audience is immaterial and overhead reflectors can be highly valuable, if not essential in some types of drama theatre (section 8.5). There is thus the possibility of movable reflecting panels which for speech are lowered into positions where they introduce additional early reflections. The reflecting panels can in fact each be relatively small because only higher frequencies are required for speech. However it also needs to be borne in mind that more than a single added reflection is likely to be required for a significant acoustic transition.

If the reflectors are placed in front of and above the performing area, a further advantage can be gained: not only would they provide extra reflections but they would also reduce the sound energy reaching the upper volume of the room which will later fall on audience as reverberant sound. The overstage reflector in the Royal Albert Hall, London, behaves in this dual manner (section 5.1). The use of movable overhead reflectors to enhance speech intelligibility does appear to offer considerable potential, though it has been little used.

While overhead reflectors are the more obvious option, movable lateral side reflectors are also a possibility. One can envisage a reflector hinged at its base which is only lowered say for unassisted speech or music. There appears to be interesting scope for further experimentation with movable reflectors for variable acoustics.

10.2.5 Variable scattering

Changing a surface from being plane to being acoustically scattering requires some ingenuity, but is certainly possible. For instance, a horizontal ceiling surface could become transformed into a coffered surface with slats lowered through slots from above. Free-standing scattering elements could also be lowered to introduce scattering. But the acoustic value of such schemes is generally unlikely to justify the expense. A change in degree of scattering of a surface carries less impact than change in absorption or orientation. Again, large areas would have to be changed for audible effect. Adjustable stage enclosures offer the only likely location for surfaces with changeable scattering, though in this case a straight exchange of surfaces might be used.

10.3 The totally variable acoustic space

The ultimate in acoustically variable space must be the Paris **Espace de Projection** at **IRCAM** (Institut de Recherche et Coordination Acoustique/ Musique) (Figures 10.5 and 10.6). It was completed in 1977, with Piano and Rogers as architects and Peutz et Associés as acoustic consultants. Though the Espace is also used for public concerts, its *raison d'être* is as a research facility for the institute, which is primarily devoted to contemporary music composition and performance. The auditorium is rectangular in plan and section with a total volume of 6800 m³. The ceiling is made of three sections which can be lowered or raised independently, offering a volume range of 4:1. The whole ceiling and all but the lowest section of the four walls consist of panels, each panel containing three rotatable prisms (Figure 10.7). The orientation of the prisms within the panels alternates from horizontal to vertical

on the walls; a similar direction change is found in the ceiling. Each prism has one face absorbent, one specular reflecting and one diffuse reflecting surface. The state of the prisms in each panel can be altered independently by remote control. This allows either plane reflecting surfaces, highly scattering surfaces or absorbing surfaces, as well as intermediate combinations to be arranged. On the absorbing face of the prism, half are predominantly low-frequency absorbing and the other half are mainly high-frequency absorbing.

Regarding its acoustic performance, the reverberation time with 400 audience and 50 performers is a maximum of about 2 seconds. With a combination of a lowered ceiling and absorbing prisms, this can be reduced to 0.5 seconds (Peutz and Bernfeld, 1980). This hall provides a fascinating resource for research into the effects of geometry and surface character on acoustics for the listener as well as the performer. It has of course enabled IRCAM to optimize the acoustics for their public concerts. But it offers no model for normal auditoria; the cost of its acoustic variability is high and indeed by music auditorium standards 400 seats is small. In the real world, priorities have to be stated and adhered to.

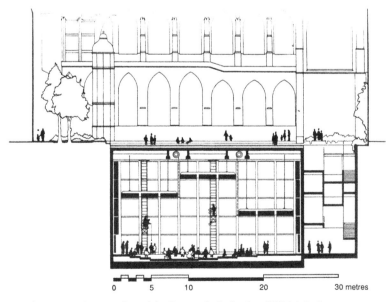

Figure 10.5 Long section of the Espace de Projection, IRCAM, Paris

Figure 10.6 The Espace de Projection, IRCAM

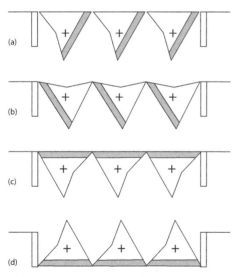

Figure 10.7 Possible arrangements in section of the rotatable prisms used on the walls and ceiling of the Espace de Projection, IRCAM: (a) specularly reflective; (b) scattering; (c) absorbing; and (d) highly scattering

10.4 Meeting the requirements of both speech and music

The greatest challenge confronting the designer of a multi-purpose hall is to have to accommodate both unamplified music and unassisted speech within the same space. By addressing this problem, the solutions to less extreme problems associated with other mixes of use will in the main be covered. The major concern in any multi-purpose situation is generally to provide appropriate reverberation times.

Before turning to reverberation time change, it is worth mentioning one aspect which can be considered common to requirements for both speech and music. Intelligible speech depends on a strong early sound relative to the late reverberant sound. For music the energy balance should be shifted towards the later sound, and this is a major objective consequence of having a longer reverberation time. Enhanced early reflections with music increase the sense of clarity and if the reflections arrive from the side they also provide a desirable sense of envelopment. Strong overhead reflections are considered undesirable for music, but reflection direction is of no concern with speech. Additional lateral reflections therefore, which arrive early enough to enhance speech intelligibility, can be viewed as beneficial for both uses. The further promising possibility of an overhead reflector being lowered specifically to assist speech has already been mentioned in section 10.2.4. But there may only be scope for successful implementation with reflectors in smaller theatres. For more than 1000 seats, one is forced to design a theatre space with balcony overhangs and low ceilings in order to meet the criterion of intelligible speech.

Each acoustic use of a hall has an optimum reverberation time associated with it. In the case of concert and drama use, reverberation times of 2 seconds and 1 second are ideal, though the optimum time for music decreases for smaller halls (section 6.2). But to achieve such a reverberation time change by natural means is decidedly difficult. The two variables to hand are the auditorium volume and the total acoustic absorption. It is simplest to illustrate the argument with the basic Sabine reverberation time equation:

$$T = \frac{0.16V}{(S_A \alpha_A + \Delta A)}$$

The reverberation time T is a function of the auditorium volume divided by the total acoustic absorption. The total absorption has here been stated in terms of the audience absorption, given by the product of the audience area, S_A, and its absorption coefficient, α_A, plus additional absorption ΔA from other surfaces in the auditorium. This formula indicates that the reverberation time can be varied by changing the volume, the audience seating area or the additional absorption. In practice, the scope for changing any of these is often circumscribed, bearing in mind that the minimum change worth considering is at least 10 per cent.

Changing the auditorium volume by a substantial amount is often the most rational approach on acoustic grounds, though it is usually a costly exercise. The options have already been aired in section 10.2.1. Variable volume is not commonplace, but some solutions have been successful. More modest volume changes have in some buildings been incorporated by using reverberant chambers round the perimeter, which are coupled acoustically to the main volume. This appears to be valuable in some North American halls, such as the Centre in the Square, Kitchener, Ontario, Canada (section 4.11 and Forsyth, 1987, p. 102). The feature is also found in the Semper Oper, Dresden (Reichardt, 1985). More elaborate chamber systems have been discussed in section 10.2.2.

Even less common are attempts to control reverberation time by changing the audience area. If the seating is upholstered, then in principle by removing seating the reverberation time will be extended. This is of course a fully realizable option with bleacher seating, but both architecturally and financially it makes little sense on the scale required for an adequate acoustic change. To put it simply, hall managers will fill seats if they can.

The third variable component is additional acoustic absorption not associated with seating.

In a small auditorium this can be handled with curtains, which can either cover hard surfaces or be stowed away. In North America, the use in large auditoria of acoustic banners has become commonplace. For radical reverberation time changes, the areas of banner required are considerable and comparable to the total area of seating. In section 10.2.3 the drawbacks associated with introducing extra absorption were mentioned. Total sound level is a function of absorption alone (section 2.9) and is nominally independent of room volume. Additional absorption produces a quiet sound and, when large areas are involved, it may also be unavoidable to have absorbent in positions which obscure valuable early reflections as well. Such reservations no longer apply when the sound source is amplified.

10.5 Variable acoustics through electronics

After all the limitations associated with achieving acoustic changes by physical means, the chance of affecting alterations at the press of a button is highly attractive. But the systems we are referring to are complex and expensive, though often cheaper than their physical equivalents. The goal for an electronic system is for it not to be perceptible as such. Not surprisingly the audibility of electronic artefacts increases with the degree of change introduced. For anyone contemplating purchasing one of these systems, there is no substitute for actually experiencing an installed system.

At its simplest, electroacoustics is found in virtually every auditorium in a public address system. In this case the microphone is placed close to the speaker, while the loudspeakers are directional pointing at the audience. In a reverberant space like a concert hall, directing the loudspeaker sound exclusively at the (absorbent) audience becomes crucial, since if the loudspeaker sound excites the reverberation it will undermine the speech intelligibility provided. But while a public address system provides variable acoustics of a sort, it carries a severe penalty for performing arts in that the sound is no longer lifelike. A major element in the unnaturalness of the simplest system arises because the listener localizes on the loudspeaker nearest to him. This can be overcome (within the limitations set by the precedence or Haas effect, section 3.2) by placing a delay in the signal from the microphone. If the first sound the listener receives comes from the speaker or performer, then the directional illusion can be maintained. The listener localizes on the true speaker and this can occur even when more energy actually arrives from the loudspeakers. This is the basis of the more sophisticated sound reinforcement systems used in churches, as well as the Delta Stereophony System. With radio microphones, these systems are admirable for musicals and where the audience size is larger than can be reached by natural acoustic means. But for more intimate drama, there remains the problem of a bland sound which does not vary when the actor turns around. There are many in the theatre who fear the day might arrive when electronic assistance becomes accepted for drama.

The **Delta Stereophony System** (DSS) not only offers a public address system without losing directional information but also enhances the acoustics of the space. The name is informative since a Greek 'delta' is the mathematical symbol used for a delay (Steinke, 1985). The system is an elaboration of the standard public address system with delays to maintain the directional illusion. As already mentioned this depends on the precedence or Haas effect: if the first sound arrives from the same direction as the source but is followed by other reflections which may be as much as 10 db louder, then the listener perceives the sound as coming from the source. The microphones in the DSS are placed close to the performers, as you would for popular music. The signals are fed typically to 10 loudspeakers with delays arranged to provide not only additional level reinforcement but also, if required, a sense of spatial impression (section 3.2) from lateral reflections. Listeners receive perhaps eight reflections from different loudspeakers. The result is a sound simulation substantially more sophisticated and lifelike than simple public address. It was initially developed for the 'Palast der Republik' in former East Berlin, which

with 5000 seats could not provide suitable natural acoustics for any use. It has since been installed in concert and multi-purpose halls, as well as open-air venues.

Whenever a microphone, amplifier and loudspeaker are in a room, there is the risk of feedback or howl-round, which everyone has at some time experienced. Two of the variable acoustic systems described below exploit feedback, whereas the other systems have some mechanism for suppressing feedback. It is because of feedback that the microphone needs to be near the speaker for a public address system.

The main concern of an electronic variable acoustics system is to enhance reverberation. For instance, if one wants to accommodate speech and orchestral music, in principle one can build a low-ceiling hall with a 1 second reverberation time suitable for speech and extend the time electronically for music. This saves the expense of providing the large auditorium volume required for music. The available systems divide into two groups: those that stimulate the room's reverberation and those that generate the additional reverberation electronically. The discussion starts with three examples of the former.

While reverberation is one of the key acoustic aspects in rooms, the early reflections are also important; some enhancement systems also generate early reflections that can enhance those of the existing hall. In a more general sense, another concern in halls is sound level; enhancement systems will always increase sound level. In large halls, a noticeable level increase is often desirable, while in small halls electronically enhanced music can become too loud.

As already mentioned, feedback can be used constructively in rooms; if the gain of a microphone–amplifier–loudspeaker chain is turned down slightly to avoid feedback, it is found that the reverberation at the howl frequency has been extended. Both Assisted Resonance and Multiple Channel Reverberation are multi-channel systems which use this effect in a controlled manner to enhance reverberation over a broad frequency range (Krokstad, 1988).

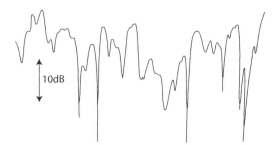

Figure 10.8 Frequency response between a fixed loud-speaker and microphone in a room over the frequency range 1800–2200 Hz

The basic difference between the two systems is that with Assisted Resonance the channels operate at different frequencies whereas with MCR they are all broadband. Figure 10.8 shows a typical room frequency response between two points; it contains random peaks and troughs. It will be different for any other two points in the room. By arranging channels with microphones and loudspeakers at many different positions in the room, some considerable degree of randomization is achieved.

Assisted Resonance was invented by Parkin to solve the low-frequency reverberation time problems in the Royal Festival Hall, London (Parkin and Morgan, 1965). Its success led to its installation in several multi-purpose halls to provide variable acoustics. The filtering to restrict the frequency of each channel is primarily done by placing the microphone in a tuned resonator (Figure 10.9). For stability the microphone location is selected so that it occurs at a frequency peak. In an Assisted Resonance system for variable acoustics, typical channel frequency spacing is 3 per cent. The electronic gain of the channels determines the reverberation time; it must of course be tightly controlled, since excessive gain leads to audible sound colouration. Figure 10.10 illustrates the reverberation times with and without Assisted Resonance in a system (by AIRO Ltd) using 72 channels, with the highest at 922 Hz. The change is clearly larger at low frequencies. In part this is due to the design aim of a smooth frequency characteristic, but extending

Figure 10.9 Resonators containing microphones for the assisted resonance system at Central Hall, York University

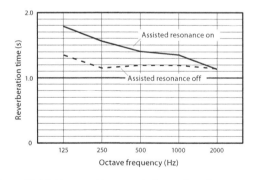

Figure 10.10 Measured (unoccupied) reverberation times at Central Hall, York University. Solid line, with assisted resonance system; dashed line, natural reverberation time

Figure 10.11 (a) Microphones and (b) ceiling loudspeakers in the multiple channel reverberation (Philips) system in the Austria Center, Vienna

Assisted Resonance to much higher frequencies has not proved possible. The higher frequency limitation occurs because the system depends on the peaks in the room response (Figure 10.8) being constant in space, whereas with temperature changes etc. they will move. The movement of peaks is more serious at higher frequencies. The problem of stability limits the contribution that Assisted Resonance can provide and the system is no longer marketed. The system in the Royal Festival Hall was finally switched off in 1998.

While Assisted Resonance uses many narrowband channels, **MCR** (Multiple Channel Reverberation) uses many broadband channels. Each channel will extend reverberation time at the peak frequencies in its response, but the channel gain must be kept low for colourless amplification. The system was originally proposed by Franssen (1968), but the first reported full installation was by Dahlstedt (1974); it was marketed by the Philips company for many years (Figure 10.11). The MCR system increases both reverberation time and sound level simultaneously. To produce a sound level increase of 3 db in a room with the MCR system, about 100 broadband channels are required. To optimize the system it is however necessary to introduce electronic filters in individual channels to equalize the transducer and room responses. Typically four filters per channel might be used. Selection of the filters is one reason why installation of a Multiple Channel Reverberation system is an elaborate exercise.

A much more recent system from France is based on electro-acoustic reflectors, which are capable of reflecting more energy than they receive. In this way such reflectors have the effect of reducing the effective amount of acoustic absorption in a room and thus increasing the reverberation time. An electro-acoustic reflector involves careful design to avoid feedback in spite of the microphone and loudspeaker being very close to each other. The system, named **CARMEN** (*Contrôle Actif de la Réverbération par Mur virtuel à Effet Naturel*), has been successfully installed in several auditoria, typically using 20–30 electro-acoustic reflectors (Vian, 2004).

The remaining enhancement systems rely on electronically generated reverberation to supplement the natural room reverberation. While public address requires microphones close to the source and the systems discussed above have remote microphones, the following electronic solutions have medium-distance microphones. To allow this without creating feedback, each has an ingenious solution, which in several cases is not publicized. Time-variant processing is one solution used. Four current systems will be mentioned.

The **ACS** (Acoustical Control System) comes from Delft, Holland, as described by de Vries and Berkhout (1988). The procedure is linked to wavefield synthesis. Typically 16 microphones and 100 loudspeakers are used. The ACS system links most microphones to most loudspeakers, simulating a reflection pattern which may be selected to a desired pattern based, for instance, on a hall with preferred acoustics. Reverberation time increases of over 100 per cent can be achieved. A major claim of the ACS is that different acoustic elements such as reverberation time and the sound level of the reverberant component can be controlled independently.

The Lexicon company in New England has developed **LARES** (Lexicon Acoustic Reinforcement and Enhancement System); see Griesinger (1992). The system can be constructed by purchasing the hardware and connecting up the components, as opposed to most other systems which require installation by the manufacturer. Usually 2–4 microphones at medium distance are used, which are connected to banks of time-varying reverberators and many (typically 50) distributed loudspeakers. The system is effective and in its simpler manifestations economical as well.

SIAP (System for Improved Acoustic Performance) also comes from Holland, due to Prinssen and Kok (1994). The system claims to be able to add missing reflections and reverberation. Not only can the system extend the reverberation time etc. appropriate for music, but also on a different setting, add reflections to improve intelligibility for speech. Microphones, typically 4–8 in number, are mounted about 10–12 m above the stage front edge; 150 loudspeakers are generally used. Digital processing is used for 'decorrelation in the input and output stages'; this is the patented secret behind SIAP, which provides major reductions in feedback.

Finally, Poletti from New Zealand has developed **VRAS** (Variable Room Acoustics System). In his description of the system, Poletti (1994) distinguishes between in-line electronic systems, such as standard public address, and non-in-line systems such as Assisted Resonance etc. VRAS is a non-in-line system but it is claimed to also offer most of the advantages of in-line systems. The reverberation enhancement is described as equivalent to a passively coupled room. A typical system in a large auditorium might have 24 microphones in three locations connected to 64 loudspeakers. Reverberation time increases of over 100 per cent are possible.

From the above it is clear that much ingenuity has been expended in the search for electronic systems which provide uncoloured acoustic enhancement. A verbal description can only explain the principles behind systems; their success must be judged by listening to installed systems. Experience has shown that there are definite limits to changes that are considered subjectively acceptable (Gade, 1995). These systems are now used in quite a few auditoria, which use electronic variability to allow them to accept a wide variety of types of performance. Electronic systems are however not a simple panacea for all acoustic ills; they work best in halls which are well designed acoustically. Electronic systems also require regular maintenance, as well as control by informed operators.

Many musicians are unhappy about playing in spaces with electronic enhancement systems, so these systems are unlikely to be found in the main auditoria. Even when a little assistance from an electronic system might overcome an acoustic fault beyond remedy by physical acoustic measures, managers and performers often take the view that they prefer to live with the acoustic limitations. This is the approach taken for the 2007 refurbishment of the Royal Festival Hall in London.

10.6 The multi-purpose auditorium with flytower

10.6.1 The 'proscenium concert hall'

This type of multi-purpose auditorium is common in North America but rare in Europe. The label 'proscenium concert hall' describes the basic elements: a concert hall-size auditorium with a proscenium and full flying facilities. An orchestral shell is used to provide a stage enclosure for concert use (the acoustic value of these stage enclosures has been studied by Bradley, 1996). Orchestras can also be placed in an optional pit in front of the stage. The principal uses of these halls are for orchestral concerts, opera, ballet and musicals. In acoustic terms the degree of variation is generally small, since speech is handled by a sound reinforcement system. The change from concert to proscenium use involves a small change in reverberation time due to exposure of the (absorbent) proscenium opening. Since a shorter reverberation time is required for opera and other stage events than for concerts, this inevitable change is welcome. Two examples will be discussed here: the first is characteristic of the North American type, while the second constitutes an intriguing evolution.

The **Uihlein Hall** of 1969 is the major auditorium in the **Milwaukee Center for the Performing Arts, Wisconsin,** with 2327 seats and an auditorium volume of around 25 000 m³ (Figures 10.12 and 10.13). Designed by H. Weese, the acoustic consultants were Bolt, Beranek and Newman (Talaske, Wetherill and Cavanaugh, 1982; Izenour, 1977). The 22-ton demountable stage shell intrudes well into the stagehouse, enlarging the auditorium volume for concert use. Indeed the shell is so large with a maximum height of 13.7 m, that convex plastic panels are suspended above the orchestra at 7.6 m to provide both support for the musicians and reflected sound to the stalls and mezzanine seating levels.

The reverberation times achieved in the Uihlein Hall of 1.9 seconds for concerts and 1.65 seconds for proscenium use, particularly opera, are close to

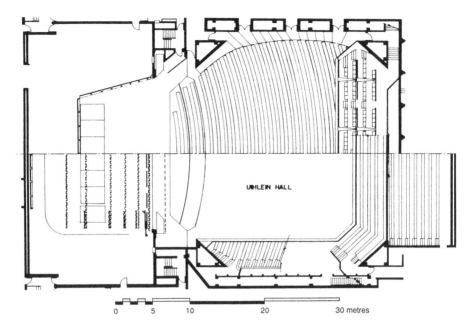

Figure 10.12 Plan of the Uihlein Hall, Milwaukee Center for the Performing Arts (reproduced by permission of Pennsylvania State University Libraries)

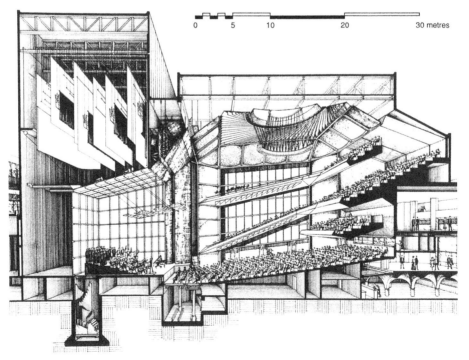

Figure 10.13 Longitudinal perspective section of the Uihlein Hall, Milwaukee Center for the Performing Arts, in its orchestral layout (reproduced by permission of Pennsylvania State University Libraries)

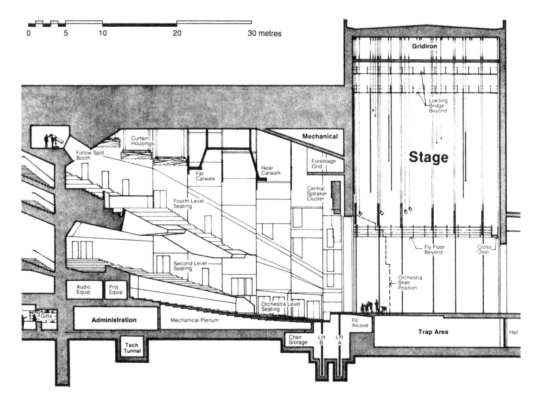

Figure 10.14 Long section through Segerstrom Hall, Orange County Performing Arts Center

optimum. Accommodating this audience number all facing the stage is difficult. This hall employs the rare feature of flying balconies. The virtue of these is primarily acoustic, since they allow sound to reach the audience from behind as well as from in front. The flying balcony mitigates against the rather one-dimensional acoustics experienced underneath deep overhangs. The gross dimensions of the auditorium are large, even by concert hall standards with the furthest seat 43 m from the stage front and an auditorium width of 36 m. Writing in 1977, the theatre consultant of the Center, Izenour (1977), rated the Uihlein Hall as 'one of the best concert hall–opera houses on the North American continent'.

The **Segerstrom Hall** of the **Orange County Performing Arts Center, California** of 1986 offers the same range of performance uses but for an even larger audience size of 2906 (Hyde and von Szeliski,

1986; Hyde, 1988). Confronted by the acoustic requirement of a large auditorium volume and the visual requirement of a suitable angle of view through the proscenium opening, the acoustic consultants, Marshall, Hyde and Paoletti, realized a basic conflict. Visually the only way to house this audience number within the constraints of distance, view through the proscenium and maximum seating rake is to use a fan-shaped envelope. The fan shape is now notoriously unpopular on acoustic grounds for concert use (section 4.5). Its major problem is the lack of useful reflecting surfaces close enough to audience in the centre of the seating, particularly surfaces capable of providing reflections from the side. Marshall proposed the radical solution of an oblique Stalls level with an asymmetrical arrangement for the remaining seating (Figures 10.14 and 10.15). The auditorium form has been most aptly described by the principal architect, Charles

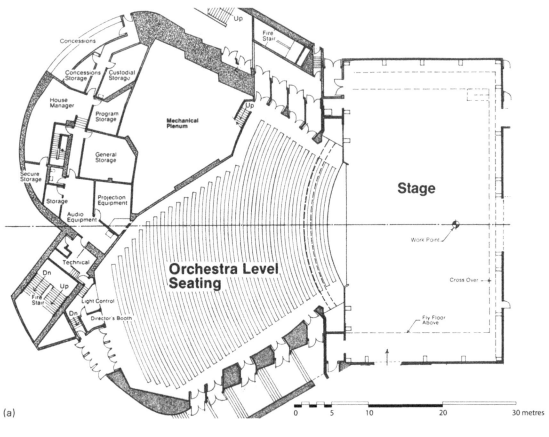

(a)

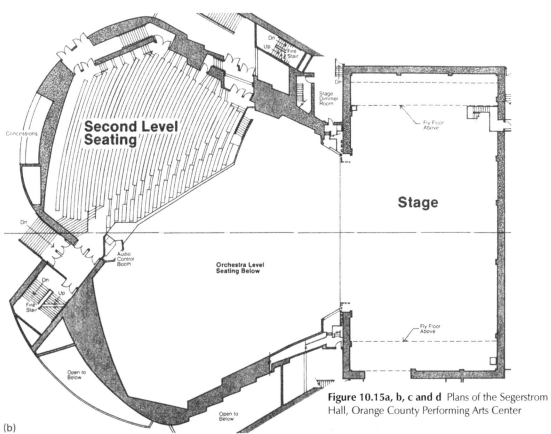

(b)

Figure 10.15a, b, c and d Plans of the Segerstrom Hall, Orange County Performing Arts Center

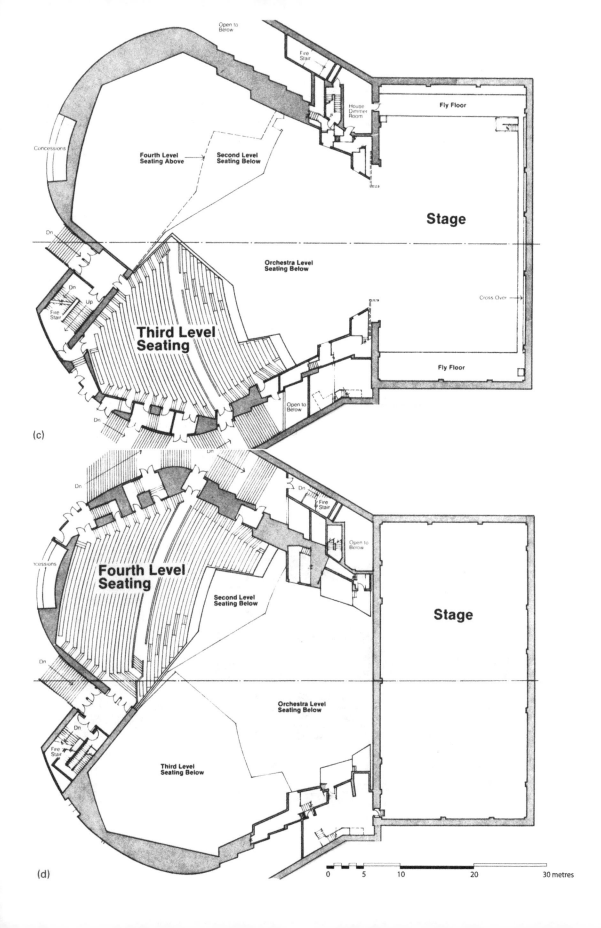

(c)

Open to
Below

Fire
Stair

House
Dimmer
Room

Fly Floor

Concessions

**Fourth Level
Seating Above**

**Second Level
Seating Below**

Stage

Orchestra Level
Seating Below

Dn

Dn

Up

Fire
Stair

**Third Level
Seating**

Cross Over

Fly Floor

Dn

Open to
Below

Dn

(d)

Dn

Dn

Dn

Fire
Stair

Open to
Below

icessions

**Fourth Level
Seating**

**Second Level
Seating Below**

Stage

Dn

Orchestra Level
Seating Below

Fire
Stair

Dn

**Third Level
Seating Below**

0   5   10   20   30 metres

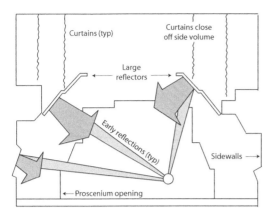

Figure 10.16 Cross-section sketch of Segerstrom Hall illustrating reflection paths. Variable absorption may be added to the upper volume

Lawrence of CRS Sirrine: 'If you had a typical, symmetrical (fan-shaped) hall with two seating levels, an orchestra and one balcony, and an earthquake spilt it down the middle so that one half sank 10 feet (3 m), then you'd have four seating levels. The split would develop vertical surfaces, smaller wall surfaces between floors that were once a contiguous level. Those walls can reflect sound into the next lower tier.' Thus in an auditorium with a volume of 27 800 m³ and a maximum width of 49 m, the effective maximum width for any seating level is only 28 m. In addition, with only two levels at any position in plan, the height of the overhangs could be made unusually high at 6 and 7 m. This feature proves to have interesting acoustic consequences.

Considerable attention was paid to optimize the reflection situation for all seats. The cross-section (Figure 10.16) contains an elaborate series of inclined reflecting surfaces to provide lateral reflections from above. In order to minimize sound level drop-off at remote seats, there are reverse splay walls in plan at the rear of each of the four seating levels. The stage itself is relatively conventional with a massive shell for music performance; it contains ensemble reflectors at heights between 7 and 10 m. The interior of the hall is shown in Figures 10.17 and 10.18

Objectively the Segerstrom Hall behaves as a good concert hall (Appendix C) with a reverberation time of 2.0 seconds occupied, a high proportion of early lateral sound and sound levels virtually all above criterion level. These are particularly impressive achievements for an audience size of

Figure 10.17 Segerstrom Hall, Orange County Performing Arts Center

10.18 Segerstrom Hall, Orange County Performing Arts Center

nearly 3000. Measurements at seats below over-hangs show none of the usual deficiency of sound level (mainly from deficient late energy) that is nor-mally found. Subjectively the hall has been well received, with good uniformity and a particularly surprising suitability for solo, chamber music and opera performances. Compared with the classical rectangular halls, both have an enveloping sound character, but Segerstrom Hall is more immediate with less of the sensation of being surrounded by reverberating sound. This hall constituted a brave and exciting development in auditorium design, so far not yet copied. It demonstrates the rewards that can result from reconsidering from first principles the implications of the performing requirements of an auditorium.

10.6.2 'Enclosing' a drama stage

Some of the most successful schemes for variable acoustics involve making a change in a way which makes architectural as well as acoustical sense. Use of both a volume change and total absorp-tion change to alter the reverberation time also offer the chance of less extreme degrees of each. For the substantial acoustic change from concert hall to drama, a particularly interesting solu-tion is available. In a theatre with flying facilities, the stagehouse volume is comparable and often larger than the volume of the auditorium. If some of this volume can be used as part of the audito-rium volume for the concert configuration, a sig-nificant reverberation time increase is possible. For this approach to work for concerts, the absorbent drapes in the flytower which are used for drama have to be sealed off with acoustically reflective panels. On the other hand for drama the stage-house needs to be fairly absorbent.

An example of this approach is discussed in section 11.8. The technique has also been used in several halls in the Netherlands by Akoestisch Adviesbureau Peutz (Peutz and Klomp, 1986; Talaske and Bonner, 1986). In the **'de Maaspoort' hall** in **Venlo** (Figure 10.19), the proscenium opening is also enlarged for the concert condi-tion, using the fire curtain to change the height. For the drama condition, absorbent roller curtains are lowered from the ceiling to provide additional

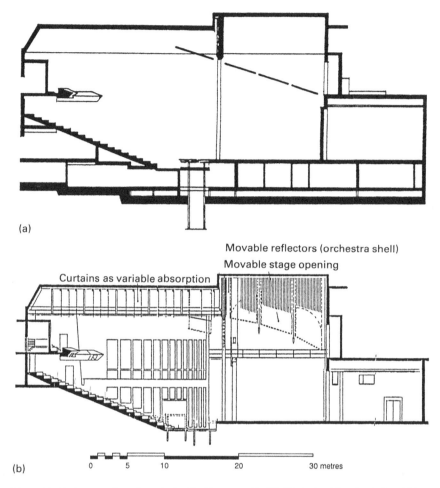

Figure 10.19 Long sections through the 'de Maaspoort' hall in Venlo for (a) orchestral and (b) drama conditions

absorption. The hall seats 750 and the reverberation time changes from 1 second for drama to 1.8 seconds for concerts (both figures with no audience). Adviesbureau Peutz's conclusion about these halls is that, in order to effectively act as a single acoustic space, the proscenium opening which couples the orchestral volume to the hall needs to be huge! Adviesbureau Peutz's latest multi-purpose space, the **Theatre 'De Spiegel'** in **Zwolle** of 2006, uses a drama theatre as its starting point rather than a concert hall (Luykx *et al.*, 2007); it has a capacity of 850 for drama and 1000 for opera and concerts.

Interestingly this approach might be considered as the preferred option regarding halls reviewed in Chapter 11.

A slight variant of the de Maaspoort approach has been used in the **Concertgebouw** in **Brugge/ Bruges** in Belgium (Butcher, 2002). While a wide range of events had to be catered for, the prime concern was good acoustics for concert and opera use. The hall has full flying facilities but the upper flytower can be closed off for concert use. The auditorium is basically designed as an opera house but for concert use there is an orchestral shell and the empty backstage volume is coupled to that of the main auditorium to extend the reverberation time. The hall seats 1200–1400; variable absorption is also available.

It has taken a while to appreciate the sheer scale of physical changes necessary in larger auditoria in order to effect worthwhile reverberation time changes. Substantial changes of volume, large areas of sound-absorbing material or reorganization of the auditorium are needed. The use of movable wall elements on air castors in the Derngate Centre, Northampton (section 11.8) offers further ingenious solutions. On the other hand, the alternative in a modest-size hall is to choose a compromise reverberation time adequate for both speech and music. Fasold and Winkler (1976) recommended values of 1.3 – 1.4 seconds for halls of 800–900 seats, though designing for adequate speech intelligibility for this seat capacity and reverberation time is demanding. It must be stressed that compromising the reverberation time is much more acceptable in a smaller space, as speech intelligibility is better in a small hall and shorter reverberation times are more acceptable for music in smaller spaces (Figure 6.2).

10.7 Conclusions

Like Tolstoy's unhappy families, each multi-purpose hall is multi-purpose after its own fashion. But it is only in recent times that the differing requirements for different uses have been seriously addressed. The first concern is to establish the priorities of use. This should be left to and indeed demanded of the client. It is all too easy to provide facilities which are either inadequate or too elaborate for the performers and the community they are serving. The success of a performing arts centre relies on its establishing a good reputation and this is only possible if seats are filled and performers enjoy performing there.

How the acoustics should be tailored to fit various uses has been the main concern of this chapter. The architectural forms of drama theatres, opera houses and concert halls have evolved and established optima both in shape and size, which are distinct for each use. In particular, certain maximum seating

numbers apply to each type of auditorium. Design limitations are principally due to the combination of visual and acoustic requirements. For example, if a hall is to work for unassisted speech, then audience sizes over 1000 demand good acoustic design and in excess of 1200 it becomes difficult to meet speech intelligibility criteria. However, if the same hall is to be used for music, these seating numbers are modest by concert hall standards and it may be that not much more than reverberation time need be considered specifically for the music condition. For rare events, the compromise of using fewer seats for an unusual configuration can be appropriate. But it is halls at the top limit of seating capacity for a major use where problems can arise.

To meet the acoustic requirements, considerable ingenuity is required for more than one performance type when electronic assistance is ruled out. If one is fortunate, a single acoustic condition is acceptable; this is particularly appropriate for small halls. In larger halls, reverberation time design is generally the primary concern. Orchestral concert conditions demand the longest reverberation time; opera, chamber music and amplified speech require moderate times; while unassisted speech needs the shortest reverberation time. The normal recommended values are around 2 seconds for orchestral music and 1 second for speech. Achieving this degree of variation is difficult by natural means, but in medium- and large-size halls the possibility of providing variable acoustics without electronics has to be seriously considered.

Physical variable acoustic elements have the distinct advantages that they require no specialist control or maintenance and that they are unlikely to create an unnatural sound quality. However the scope for variable acoustics is not large. Variable auditorium volume is potentially the most valuable but also difficult to accommodate. Variable absorption is often easier to include but suffers from the risk that sound levels can become too quiet. However, reduced sound level is definitely preferred for use with amplified music, but the quantity of extra absorbing material required to convert a concert hall into a popular music venue is unlikely to be realizable. Movable reflectors have to be sufficient in number to produce audible effects. None of these possibilities has become widely accepted but, with multi-purpose design becoming more sophisticated with time, realistic solutions are emerging. In practice, changes within auditoria for acoustic purposes are frequently operated incorrectly. The 'best' solutions combine an acoustic change with an operational one, which guarantees it is used correctly!

Electronically variable acoustics is becoming more common and there are now many systems to choose from. They offer acoustic variability at costs likely to be less than physical options. Many performers however remain unhappy with the thought of electronic assistance.

The development of the sophisticated multi-purpose hall has been a question of supply and demand. The realization that halls in smaller towns and cities need to address multi-purpose use has evolved with economic prosperity. The ability to satisfy those demands has depended on growth in acoustic understanding. It is natural that the growth of knowledge should have occurred first for single-purpose venues. Solutions to the acoustic problems of multi-purpose use are likely to remain one of the most interesting areas of progress in coming years.

References

Allen, I. *et al.* (1999) A star is born. *Architects' Journal*, 28 October 1999, 34–41.

Beranek, L.L. (2004) *Concert halls and opera houses: Music, acoustics and architecture*, 2nd edn, Springer, New York.

Bradley, J.S. (1996) Some effects of orchestra shells. *Journal of the Acoustical Society of America*, **100**, 889–898.

Butcher, H. (2002) Acoustic design of the Concertgebouw Brugge. *Proceedings of the Institute of Acoustics*, **24**, Part 4.

Dahlstedt, S. (1974) Electronic reverberation equipment in the Stockholm Concert Hall. *Journal of the Audio Engineering Society*, **22**, 627–631.

Fasold, W. and Winkler, H. (1976) *Bauphysikalische Entwurfslehre, Band 5: Raumakustik*. VEB Verlag fur Bauwesen, Berlin.

Forsyth, M. (1987) *Auditoria – designing for the performing arts*. Mitchell Publishing, London.

Franssen, N.V. (1968) Sur l'amplification des champs acoustiques. *Acustica*, **20**, 315–323.

Gade, A.C. (1995) Possibilities and limitations in the use of reverberation enhancement systems for small multi-purpose halls. *Proceedings of 15th International Congress on Acoustics, Trondheim*, Vol. II, 465-8.

Griesinger, D. (1992) Uncolored acoustic enhancement through multiple time-variant processors. *Proceedings of the Institute of Acoustics*, **14**, Part 2, 179–186.

Hyde, J.R. (1988) Segerstrom Hall in Orange County – design, measurements and results after a year of operation. *Proceedings of the Institute of Acoustics*, **10**, Part 2, 281–288.

Hyde, J.R. and von Szeliski, J. (1986) Acoustics and theater design: exploring new design requirements for large multi-purpose theaters. *Proceedings of the Vancouver Symposium on Acoustics and Theatre Planning for the Performing Arts, August 1986*, pp. 55–60.

Izenour, G.C. (1977) *Theater Design*, McGraw-Hill, New York.

Krokstad, A. (1988) Electroacoustic means of controlling auditorium acoustics. *Applied Acoustics*, **24**, 275–288.

Lord, P. and Templeton, D. (1986) *The architecture of sound – designing places of assembly*. Architectural Press, London.

Luykx, M., Metkemeijer, R. and Vercammen, M. (2007) Variable acoustics of Theatre "De Spiegel" in Zwolle (NL). *Proceedings of the International Symposium on room acoustics,* Seville, Sept. 2007.

Nepomuceno, J.A., Solé, I. and Dupré, N. (2002a) Architecture, acoustics and design coordination for the Sala São Paolo, Brazil. *Proceedings of the Institute of Acoustics*, **24**, Part 4.

Nepomuceno, J.A., Blair, C. and Doria, D. (2002b) Variable-volume/coupled volume response at

Sala São Paolo. *Proceedings of the Institute of Acoustics*, **24**, Part 4.

Orlowski, R. (1999) The design of variable acoustics at the new Milton Keynes Theatre. *Proceedings of the Institute of Acoustics*, **21**, Part 6, 237–244.

Parkin, P.H. and Morgan, K. (1965) 'Assisted resonance' in the Royal Festival Hall, London. *Journal of Sound and Vibration*, **2**, 74–85.

Peutz, V.M.A. and Bernfeld, B. (1980) Variable acoustics of the IRCAM concert hall in Paris. *Proceedings of 10th International Congress on Acoustics, Sydney*, Paper E-1.3.

Peutz, V. and Klomp, A.J.G. (1986) Variable acoustics, its feasibility. *Proceedings of the Vancouver Symposium on Acoustics and Theatre Planning for the Performing Arts*, August 1986, pp. 19–20.

Poletti, M.A. (1994) The performance of a new assisted reverberation system. *Acta Acustica*, **2**, 511–524.

Prinssen, W. and Kok, B. (1994) Technical innovations in the field of electronic modification of acoustic spaces. *Proceedings of the Institute of Acoustics*, **16**, Part 4.

Reichardt, W. (1985) Die akustische Projektierung der Semper-Oper in Dresden. *Acustica*, **58**, 253–267.

Steinke, G. (1985) New developments with the Delta Stereophony System. *77th Audio Engineering Society Convention, Hamburg*, Preprint No. 2187.

Talaske, R.H. and Bonner, R.E. (eds) (1986) *Theatres for drama performance: recent experiences in acoustical design*. American Institute of Physics, New York.

Talaske, R.H., Wetherill, E.A. and Cavanaugh, W.J. (eds) (1982) *Halls for music performance, two decades of experience: 1962–1982*. American Institute of Physics, New York.

Valentine, J. and Day, C. (1998) Acoustic design and performance of the Bruce Mason Theatre. *Proceedings of the 16th International Congress on Acoustics, Seattle*, **IV**, 2463–2464.

Vian, J.-P. (2004) Carmen: a new acoustic enhancement system, based on the virtual wall principle. *Proceedings of the International Symposium on room acoustics: design and science 2004, Hyogo, Japan*.

de Vries, D. and Berkhout, A.J. (1988) Background and principles of the Delft acoustical control system (ACS). *Proceedings of the Institute of Acoustics*, **10**, Part 2, 369–382.

11 Multi-purpose halls in Britain

Conscious design for multi-purpose use is more recent in Britain than in North America and, among the examples considered here, inspired acoustic design is a fairly recent phenomenon. Eight British multi-purpose spaces will be discussed in detail; the earliest was completed in 1968. Whether these halls all deserve the label 'multi-purpose' is a matter of personal taste. Their inclusion here is based on the fact that during their design each had more than one intended use. In most cases the different uses required different acoustic conditions. Not all these halls cater for the widest change in acoustic characteristics as required for unassisted speech and orchestral music.

These eight British examples can be subdivided into five groups. Wembley Conference Centre and York University Central Hall are both semicircular in plan; both cater for music, but for speech use a public address system is nearly always used. The Butterworth Hall at the University of Warwick Arts Centre is most easily described as a poor man's concert hall, but no less successful for that. The Derby Assembly Rooms and the Reading Hexagon were each constructed with both music and drama in mind, but no flytower was built for the drama condition. Eden Court Theatre, Inverness, and the Theatre Royal, Plymouth, were designed principally as theatres with full flying facilities; stage enclosures are used to deal with the orchestral condition. Finally, in the Derngate Centre, Northampton, air castors enable towers to be moved around to modify the auditorium to four formats, including drama and concert. The basic details of these halls are given in Table 11.1, for each

hall scaled plans and sections and two photos are included. Measurements in these auditoria were conducted in the years 1981–6.

All halls have been tested objectively for music use; six have also been surveyed subjectively in the same manner as the concert halls discussed in Chapter 5. Objective measurements of speech conditions have been made in four of the auditoria. The objective speech intelligibility results for all these four auditoria are given in Figure 11.20 and speech level results in Figure 11.21.

11.1 Wembley Conference Centre

Wembley Conference Centre is owned and run by the same company as Wembley Stadium and associated buildings. These earlier buildings were originally erected for the British Empire Exhibition of 1924. The addition of facilities catering for conferences offered an extension complementary to their sporting venues. The conference business has now become a lucrative industry and the Wembley Conference Centre of 1977, designed by R. Seifert and Partners, was the first of the purpose-built centres in Britain. The main auditorium, or Grand Hall as it is called, holds an audience of 2500 with a large stage; it is employed for an extremely diverse range of uses. As well as conferences, the hall has proved popular for trade product launches, sporting events and popular music of all types. Its use for classical concerts is peripheral. In fact the architects commented in the prelude to Drury's appraisal (1977) that 'appropriate design methods which have

Table 11.1 Basic details of eight British multi-purpose halls. The quoted reverberation times are estimates of mid-frequency occupied values. The second element in the seat number of the University of Warwick Butterworth Hall is for choir seating.

Hall	Date	Seats	Auditorium volume (m³)	Reverberation time (s)	Acoustic consultant
Wembley Conference Centre	1976	2503	24,000	1.3	C.C. Buckle
Central Hall, University of York	1968	1064	7000	1.2	H.R. Humphreys and Bickerdike Allen Partners
Butterworth Hall, University of Warwick	1981	1152 + 177	12,100	1.8	Bickerdike Allen Partners
Assembly Rooms, Derby	1977	1478	8070*	1.0	H.R. Humphreys
Hexagon, Reading	1977				Sound Research Laboratories
Concert mode:		1454	8280*	1.1	
Drama mode:		960	5720*	0.9	
Eden Court Theatre, Inverness	1976				F.J. Fahy
Concert mode:		814	6200†	1.0	
Drama mode:		814	5050	0.9	
Theatre Royal, Plymouth	1982				Sound Research Laboratories
Concert mode:		1271	6490†	0.8	
Drama mode:		1271	5560	0.8	
Derngate Centre, Northampton	1983				Artec Consultants Inc.
Concert mode:		1400	13,500	1.7	
Drama mode:		1151	7180	1.1	

* auditorium volume below suspended ceiling only.

† includes approximate volume within stage shell.

been adopted ensure good acoustic conditions for orchestral performances'. While it is true that certain acoustic pitfalls were avoided in the design, this hall dramatically illustrates that room form is a significant determinant of sound quality.

In plan the hall is close to a 180° fan, so that with its single continuous seating bank, we have a form reminiscent of the classical Roman theatre (Figure 11.1). There is a small stage behind a proscenium opening, but this is not normally used for music performance when a screen closes off the opening and a thrust stage is used. The outer wall of the auditorium behind the seating is virtually circular, but the profiling of its surface together with the small vertical exposed height appear to be effective in suppressing echoes back to the stage. A balcony also runs round the perimeter. Though it contains little audience seating, the balcony will also contribute to breaking up sound reaching the concave curved wall. The suspended ceiling rises slightly towards the perimeter but over the thrust stage the ceiling is perforated; small panels in this area leave a high proportion of open area into the roof void.

The seating in the auditorium is to a particularly generous standard of 0.63 m²/seat. This is 34 per cent larger than the average for British concert halls and turns out to have serious acoustic implications. In simple terms it is equivalent to having an audience of 3350 at a normal seating density, which is a formidable capacity for orchestral music.

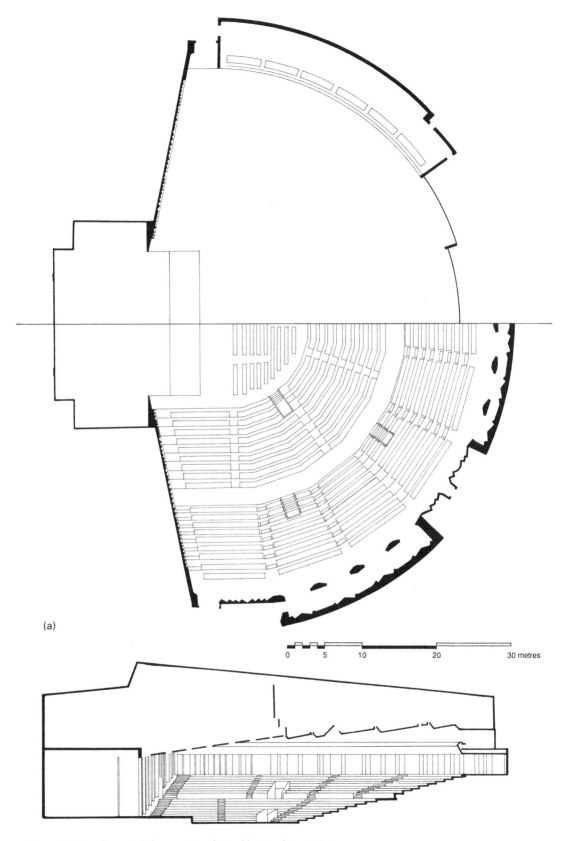

(a)

0 5 10   20   30 metres

Figure 11.1 (a) Plan and (b) long section of Wembley Conference Centre

Figure 11.2 Wembley Conference Centre

Figure 11.3 Wembley Conference Centre

Subjective characteristics

The response to a concert of 160 musicians (of the National Youth Orchestra) was not enthusiastic. This was certainly not helped by a remarkably noisy ventilation system. Musical sound was judged as dry, constricted, remote and weak, though the clarity was reasonable. Two seat positions were sampled and the position farther from the stage was unambiguously judged as inferior, with the major difference being the quieter, less intimate sound. The quality of the sound was also rather harsh with little sense of warmth. In spatial terms little was perceived from other than in front. This is a hall which does not flatter its performers.

Objective characteristics

With a reverberation time of only 1.3 seconds one anticipates criticism but not of such severity. The additional objective measures (Figure 11.4) indicate several peculiarities. The early decay time is close to 1 second, so one expects an even lower judgement for the sense of reverberation. In detail the early decay time decreases as one moves out away from the stage. The objective clarity is high; both this and the early decay time are caused by weak late sound at remote seats. A simple geometrical explanation can be provided for this late sound behaviour. Figure 11.5 compares measured late sound levels with expected values (section 3.10.5). The steep drop-off with distance can be explained if one considers the possible directions from which late reflected sound can come. The only major reflective surfaces are the front wall and the ceiling. In Figure 11.6 the ceiling

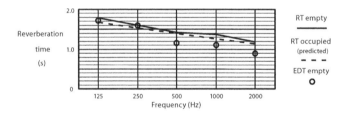

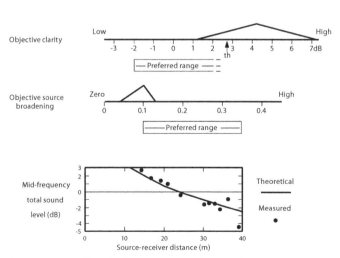

Figure 11.4 Wembley Conference Centre: objective characteristics

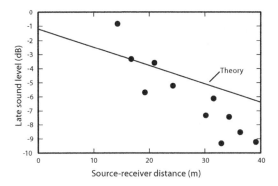

Figure 11.5 Measured late sound levels in Wembley Conference Centre compared with expected values according to revised theory

is imagined to mirror the hall, leaving the front wall and its image in the ceiling as the only surface from which reflected sound can arrive. Since the solid angle subtended by the wall decreases from a seat at A to one at B (Figure 11.6), the late sound level will drop.

The absence of any side walls leads to very low values for objective source broadening. Only for listeners near the front wall is a side reflection feasible. The major reflection comes from the ceiling and this helps to explain the harsh sound quality, which is often a characteristic of spaces with strong overhead reflections. When one looks at total sound behaviour, it does in fact match expected values quite closely. Unfortunately though, the expected levels fall below the criterion of 0 db beyond 24 m from the source. What has gone wrong? With an audience of 2500 and a volume of 24 000 m³ one expects music at least to be loud enough. And since the theory takes no account of form, the answer must lie in the reverberation time and the volume, or rather the total acoustic absorption which is related to them. If one calculates the expected reverberation time from the area occupied by audience (section 2.8.4), it turns out to be 1.6 seconds. That there is a difference relative to the measured reverberation time implies extra acoustic absorption in the space. If this could be removed by, for instance, sealing up the space above the stage, the 0 db criterion would move from 24 m to 32 m from the

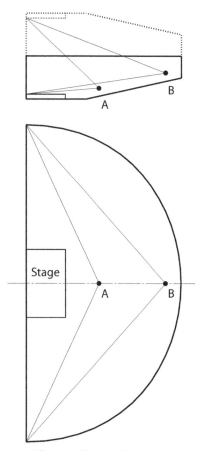

Figure 11.6 Effective solid angle for late sound in a semicircular plan hall

source. Further it is the area occupied by audience, rather than the number of seats, which determines the reverberation time. With a generous seating standard this area is high. With the same audience accommodated on standard seating, the reverberation time would rise to 2.1 seconds with the limiting distance now 56 m, well beyond the seating area. The acoustic penalties for luxurious seating and extra absorption are clearly high in a hall of this size. For multi-purpose use including conferences one must however add that a reverberation time of over 2 seconds would of course be unsuitably long, but excessive absorption is inappropriate in a large hall for unassisted speech or music.

Conclusions

The Grand Hall of the Wembley Conference Centre is a large space, indeed deceptively so. Its form conceals its large capacity. It was built as a conference space and it is tempting to imagine that a space which works so well for visual performance should also function for live music. It disappoints for two reasons: because of excessive acoustic absorption and because of room form. The absorption aspect has already been discussed. For a concert hall, acoustic energy should not be discarded lightly, whereas for conference purposes excessive absorption is of little concern when microphones are the norm rather than the exception. The matter of room form goes to confirm a prejudice among acousticians against the fan shape in plan. Fan-shaped concert halls tend to have apex angles in plan of around 60°; the shortcomings become particularly pronounced with a 180° fan. The fan shape does not furnish many reflections apart from a strong ceiling reflection. Of all the halls discussed here the Wembley Conference Centre perhaps demonstrates most clearly the differences in designing for alternative uses.

11.2   Central Hall, University of York

The Central Hall is a free-standing structure surrounded by an artificial lake on the campus of the University of York. Its architecture by Robert Matthew, Johnson-Marshall is very much of its time (1968), with the reinforced concrete structure clearly displayed on the exterior. The building has few frills with, for example, only a single-skin roof which leaves wind noise as a problem. Like the Wembley Conference Centre, the plan form is that of an amphitheatre, or 180° fan. But the capacity is much smaller, only 1064, and even from the rear of the thrust stage the furthest seat is at only 23 m. The hall was originally designed for lectures, examinations, films and ceremonial occasions. The internal volume was excessive for speech use so the ceiling was made partially absorbent by using exposed

woodwool slabs. This solution was also appropriate for the examination condition when, because bleacher seating at lower levels was removed, there was the risk of excessive reverberation.

The value of the hall for music performance gradually became apparent, hastened by developments in the cultural life of the city of York. But the unreverberant nature of the original space coupled with the paucity of early reflections meant that both audience and more particularly performers were dissatisfied with the acoustics for music. Two modifications to the hall were made to improve this situation. Between 1972 and 1974, the first commercial installation of the electronic assisted resonance system was made in this hall (section 10.5). Far fewer channels were used in York compared with the original Royal Festival Hall system (Parkin, 1974) and this led to problems in achieving an adequate increase in reverberation time at mid-frequencies, which is already a problem with assisted resonance. The system provided without excessive colouration a 40 per cent reverberation time increase at lower frequencies but offered no change to the 0.8 seconds reverberation time at 1 kHz and above.

For the first installation of a new system, the Central Hall was an unfortunate choice. It became apparent that in an amphitheatre-type space, many of the audience are too close to the stage to be aware of much reverberant sound in any case. The chosen location for the loudspeakers round the perimeter of the hall could not have helped in this regard. In brief, the response of the hall's users to the assisted resonance system was positive for audience round the edge of the auditorium, but the performers barely noticed an improvement.

A comprehensive series of remedial measures was undertaken in 1985 (Charles and Owston, 1985). It was felt that the reverberation time needed to be extended still further, so the exposed woodwool was treated to reduce its acoustic absorption. A treatment was required that avoided a total seal, which might have caused condensation problems. Three coats of a polymer-based paint were finally selected and applied to all unscreeded woodwool in the hall. This has raised the unoccupied mid-frequency

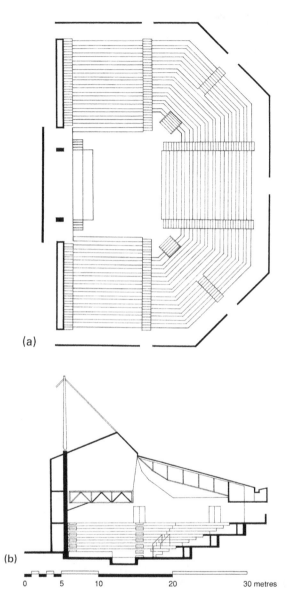

Figure 11.7 (a) Plan and (b) long section of Central Hall,
University of York

reverberation time to 1.3 seconds. Purpose-designed
acoustic reflectors have now been installed over
the stage, including an inverted trough to provide
cross-stage reflections and three-sided orthogonal

pyramids (Figure 11.10). Sound entering the pyramid
is reflected back in the direction it arrived, thus in
theory each pyramid will provide a reflection back to
each player of his own sound. It was found necessary

Figure 11.8 Central Hall, University of York

Figure 11.9 Central Hall, University of York

for pyramids used in the University of Warwick Butterworth Hall, which predates modifications in York, to make the base lengths of the pyramids 2 m to ensure full reflection at 1 kHz.

The assisted resonance system was also thoroughly overhauled in 1985 to extend the reverberation time at the bass frequencies where treatment of the woodwool was relatively ineffective (see

Figure 11.10 Stage reflector elements in the Central Hall, University of York

Figure 10.10). This second series of modifications was warmly greeted by performers and audience; they are comprehensively reported by Charles, Miller and Gwatkin (1987). The following discussion will restrict itself to the condition in 1985 with the assisted resonance system turned on.

Subjective characteristics

The Central Hall has been sampled in its present condition. It has a bright, clean sound character. Not only are listeners close to the orchestra but with the steep seating rake the situation is highly intimate. In this way several of the shortcomings of the amphitheatre form are compensated. Especially in richly scored works, there is a considerable excitement to be had in participating in the movement of the dominant musical line from performer to performer. The sound is loud, generally with good balance and rich string tone. In view of the short reverberation time, the sense of reverberation is inevitably lacking. Likewise, the plan form causes an absence of lateral reflected sound except at seats close to the front wall.

Objective characteristics

The objective nature of the hall is clear-cut and closely in line with subjective responses, Figure 11.11. The reverberation time rises significantly in the bass, which probably compensates somewhat for the very short mid-frequency value (see section 5.9). Objective clarity is high but reasonably in line with expectations on the basis of the reverberation time. Objective source broadening is low, but the total sound level is comfortably above criterion values. The measured total sound level is in fact slightly less than expectations. This is caused by the deficiency in late sound also found in the Wembley Conference Centre as well as a lack of early reflections. However, given the modest size of this auditorium no problems arise here of quiet sound.

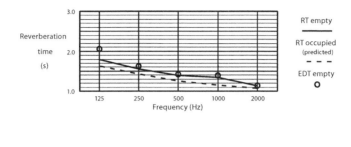

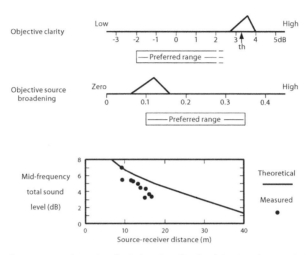

Figure 11.11   Central Hall, University of York: objective characteristics with assisted resonance switched on

Conclusions

The Central Hall was not built as a space for music and, although considerable improvements have been achieved, it is unfair to compare it with a purpose-built concert hall. The major fault is of too short a reverberation time, which is caused both by inadequate volume and by acoustic absorption in addition to the seating. Assisted resonance was able to extend the reverberation time at lower frequencies but cannot help above 1 kHz; the system has since been abandoned. The 180° fan-shape plan form is not ideal for music and only with additional reflector panels could one hope to provide some early lateral reflections. However for a hall at this scale with the steep seating rake used here, listening to music is rewarding and can be an exciting experience.

11.3 Butterworth Hall, University of Warwick Arts Centre

Of the various halls discussed here under the label of multi-purpose, the Butterworth Hall is the most surely dedicated to orchestral performance. The hall also has to cater for conferences, lectures, occasional films, pop music, examinations and spectator sports. The major element included to provide flexibility is bleacher seating for the Stalls. This allows for a 500 m² flat floor condition but imposes little by way of acoustic compromise. Adjacent to the hall is a theatre and conference room, which provide for major speech requirements. The client wanted this hall to become a major regional venue for orchestral concerts and had stipulated a reverberation time close to

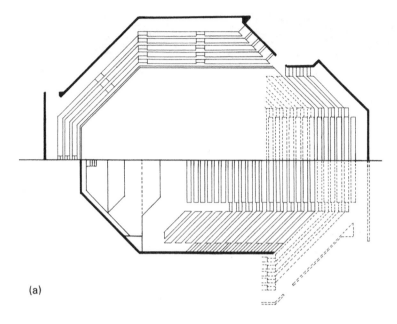

(a)

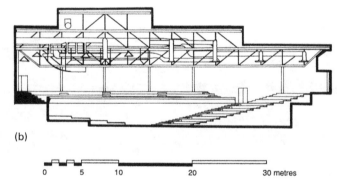

(b)

```
0        5        10                20               30 metres
```

Figure 11.12 (a) Plan and (b) long section of Butterworth Hall, University of
Warwick Arts Centre

2 seconds. For speech and other uses a sophisticated
distributed loudspeaker system is used. The major
constraint on design was that of cost and the hall
offers an interesting solution to the problem of good
acoustics for a medium-sized audience of 1330 at
modest cost. It was designed by the Renton Howard
Wood Levin Partnership, opening in 1981.

The plan form is hexagonal, indeed close to
a regular hexagon. This leaves a large maximum
width of over 36 m. By subdividing the audience, all
seats are kept at reasonable proximity to a reflecting
surface. A band of seating runs completely round
the perimeter, offering choir seating behind the
stage, while the raked stalls seating sits between
parallel walls which are the width of the stage apart.
The acoustic space includes much of the structural
support for the roof as well as extensive ventilation
trunking. No visual screen is provided to hide this

Figure 11.13 Butterworth Hall, University of Warwick

Figure 11.14 Butterworth Hall, University of Warwick

mêlée of ducts and girders. The roof level is stepped and roughly matches in reverse the depth of the floor below, so that the roof is highest over the stage.

Acoustic success relies on satisfying both audience and performers. The performers are more articulate and since they can air their views more easily, it is wise to satisfy them. Above the stage in this hall is an elaborate array of novel elements to provide support for the musicians, similar to those in the Central Hall, University of York (Figure 11.10). An annular plane reflector provides reflections back to the orchestra from all four sides, while an inverted trough provides cross-stage reflections. Three-sided orthogonal pyramids were also hung over the stage. A large curved surface provides good coupling between the choir and orchestra platform. This last item has been well received by choral groups and the overall performing conditions are praised by musicians. Whether for the orchestra a simpler array of horizontal panels covering the same percentage area would work as well is not known. The reflecting elements are at around 7.5 m above the stage.

The acoustic design of the hall is extensively described by Charles, Fleming and Miller (1985). The subjective assessment discussed here was independent of their work.

Subjective characteristics

The overall judgement of the acoustics of the Arts Centre Hall was 'Good', with a high degree of uniformity. This is a hall which responds naturally without excesses to mar the impression. The balance between clarity and reverberance was particularly appropriate, with the sense that the sound was arriving from many directions. In most seats and certainly in the main Stalls area the early lateral reflections are strong enough to produce a comfortable degree of spatial impression. The only consistent criticism concerned a slight lack of loudness towards the rear of the hall.

The major item at the concert we attended was a cello concerto performed by Rostropovich. The ability of this hall to hold a dramatic silence and support a virtuoso performance was amply demonstrated.

Objective characteristics

As shown in Figure 11.15, the measured reverberation times in this hall show a marked difference between unoccupied and occupied. The long unoccupied values occur due to the lightly upholstered seats, which when in the tip-up position have the absorbent seat pad obscured. The measured occupied reverberation time is 1.7 seconds at mid-frequencies, rising slightly in the bass. With severe cost constraints, selection of economical non-absorbent wall and ceiling elements was important. The blockwork walls were plastered and, for the ceiling, lightweight aerated concrete slabs were chosen in preference to pre-screeded woodwool.

Measured results of the early decay time and objective clarity are close to expectations based on the reverberation time. The highly acoustically scattering ceiling makes this no surprise. The measured total sound level provides some evidence for comments about absence of loudness towards the rear of the hall. It is probable that both the early reflected sound and later reverberant sound are influenced by the ceiling design. The highly scattering ceiling has been observed to influence the early sound (section 3.10.5) and it may be that the lower roof round the hall perimeter reduces the reverberant sound level as well. Whatever the reasons, the moderate size of the hall ensures that sound levels remain comfortably above criterion values. With the same design scaled up, this might no longer be the case.

Conclusions

The success of the Butterworth Hall can be ascribed to attention to basic concerns of acoustic design. The auditorium volume has been chosen as large enough, which together with selection of non-absorbent wall and ceiling surfaces has ensured a long enough reverberation time. The seating arrangement enables early reflections to reach all

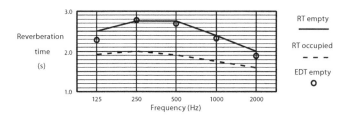

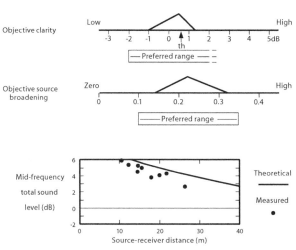

Figure 11.15 Butterworth Hall, University of Warwick: objective characteristics

seats in the hall. By including ventilation ducting within the acoustic volume, a highly diffuse sound field has been created which is generally a safe acoustic condition for music performance. Inclusion of ventilation ducting and structural elements was at the time the major untried element in the design. At the scale of this hall where no one is further than 30 m from the stage, there appear to be no significant disadvantages associated with this economical solution. Finally the designers have attended well to the needs of the performers.

11.4 Assembly Rooms, Derby

The Assembly Rooms and all remaining multi-purpose halls considered in this chapter are used for both speech and music. As has been discussed in the previous chapter, this imposes a severe conflict

of interest, particularly with regard to the choice of reverberation time. In the case of the Great Hall of the Derby Assembly Rooms, the reverberation time is appropriate to speech conditions leaving a dry acoustic for music. No variable acoustic elements have been included.

The design competition to replace the burnt-out eighteenth-century Assembly Rooms was won by Casson Conder and Partners. The brief was for two halls to accommodate virtually every event imaginable, from boxing to symphony concerts and theatre. Neville Conder commented that 'ours was the only design that solved all the problems' (*Architects' Journal*, 1977). A flat dance floor in the Stalls is converted for auditorium use by a bleacher seating system. The plan shape of the balcony is rectangular, arranged in an arena form with the stage at one end of the Stalls. There are no overhangs. An

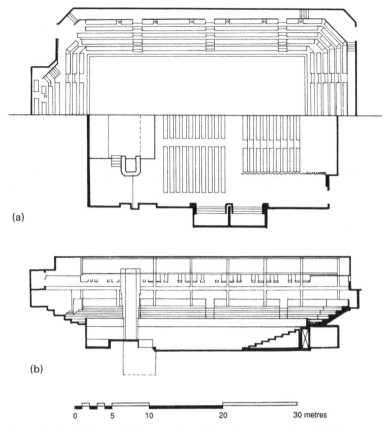

(a)

(b)

| 0 | 5 | 10 | 20 | 30 metres |

Figure 11.16 (a) Plan and (b) long section of Assembly Rooms Great Hall, Derby

elaborate plaster-on-lath suspended ceiling is perforated for lighting etc. to 27 per cent of its area. For theatre use, novel telescopic proscenium towers are lowered from above, but there is no flytower. The Great Hall considered here seats 1478 for concerts, up to 1800 maximum for other events, while the small hall (not considered here) seats around 500.

Subjective characteristics

The Great Hall has been sampled for orchestral music only. The responses were dominated by reactions to a dry acoustic that is clear but unreverberant. The sound was felt to be neither loud nor enveloping, with intimacy decreasing significantly

when one moved from the front towards the rear of the hall. The space only really succeeded in coming alive in *forte* passages.

Objective characteristics

The measured reverberation time is close to 1 second over the whole frequency range, a value normally associated with speech. Objective clarity for music and sound level are close to predictions based on the measured reverberation time and the room volume below the suspended ceiling (Figure 11.19). The suspended ceiling might have been a valuable acoustic element in the design by enhancing early reflections while allowing a larger space to

Figure 11.18 Assembly Rooms, Derby

Figure 11.17 Assembly Rooms, Derby

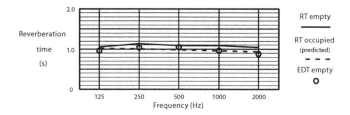

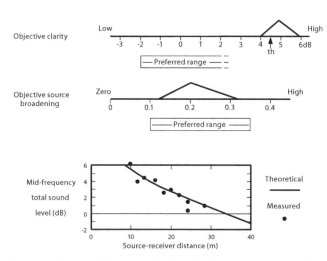

Figure 11.19 Assembly Rooms, Derby: objective characteristics in the concert format

contribute to the reverberation time. In reality, the space above the suspended ceiling contains many ducts with fibrous lagging which must absorb sound efficiently. As far as one can tell from limited measurements, the volume above the suspended ceiling makes no useful contribution acoustically and indeed the perforations in the suspended ceiling itself probably act as efficient acoustic absorbers. Sealing the perforations could raise the reverberation time by 0.2 seconds. Alternatively if the duct lagging were sealed with cement and the degree of perforation of the suspended ceiling were substantially increased, then the total room volume might contribute to give a longer reverberation time. However while raising the reverberation time would benefit music performance, it would compromise speech conditions.

Before leaving the music situation, two further points should be made. Firstly the sound level is only slightly above criterion at remote seats, although they are less than 30 m from the stage; the cause of this seems to be simply the short reverberation time. Secondly the objective source broadening is reasonable but this did not appear to be reflected in subjectively perceived spatial impression, again probably due to the reverberation time.

Turning to speech conditions, we find that the short reverberation time is not sufficient to guarantee good intelligibility. Measured results for the intelligibility measure, the early energy fraction (see section 7.3), are given in Figure 11.20 for all the multi-purpose halls used for speech. With the source oriented laterally (see Figure 7.17), there are several borderline positions in the Assembly Rooms. This

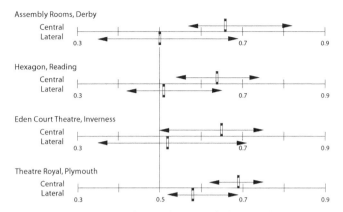

Figure 11.20 Measured values of the early energy fraction for a speech directional source in four multi-purpose halls. The range and mean of values is given for both central and lateral source orientations. The criterion is not less than 0.50 for satisfactory speech intelligibility

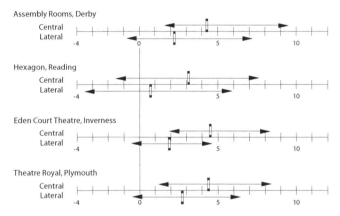

Figure 11.21 Measured values of the speech sound level (in dB) for a speech directional source in four multi-purpose halls. The range and mean of values is given for both central and lateral source orientations. The criterion is not less than 0 dB

however need not come as a surprise since the audience capacity here is large by theatre standards and seat distances from the stage extend well beyond the normal theatre limit of 20 m from the stage-front. Speech sound levels are plotted in Figure 11.21, but require no particular comment for this hall.

Conclusions

The arena plan form has many advantages as a performing arrangement, but it does not automatically function well acoustically. If in addition the hall is

to be used for both speech and music, the acoustic brief becomes particularly demanding. The reverberation time in the Great Hall of the Assembly Rooms is rightly biased towards requirements for speech, leaving particularly dry conditions for music performance. Measurements relating to speech use show that because of the auditorium form and large seat count, the intelligibility situation in some areas is borderline. As this hall has to cater for many non-acoustic events, it is debatable whether a more sophisticated acoustic approach should have been adopted. Current electronic enhancement systems now offer 100 per cent increases in reverberation time, which would seem ideal for handling the relatively rare unassisted musical events in a hall like this. Detailed design perhaps of reflectors might be able to improve conditions for unassisted speech.

11.5 The Hexagon, Reading

This hall and the Assembly Rooms, Derby, opened in the same year, 1977. They are each required to house an extremely diverse range of uses. Both have a proscenium facility but no flytower; both

have a suspended ceiling. The Hexagon, designed by Robert Matthew, Johnson-Marshall, holds a smaller maximum audience of 1500. Of the two halls, the Hexagon is clearly the more sophisticated, both architecturally and acoustically (Cowell, 1980). At the time of opening it used an electronic assisted resonance system to enhance reverberation for music use.

A hall which has to function as both a concert hall and a theatre is likely to have a seating design more appropriate to one than the other. For concerts there is only minor concern for angle of view

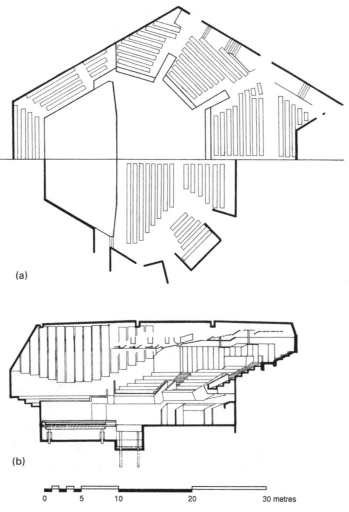

(a)

(b)

| 0 | 5 | 10 | 20 | 30 metres |

Figure 11.22 (a) Plan and (b) long section of the Hexagon, Reading

Figure 11.23 The Hexagon, Reading

Figure 11.24 The Hexagon, Reading

and an arena-type arrangement has advantages for other events such as boxing. But this leaves many seats unusable in a proscenium theatre configuration. In the appraisal of the hall by a theatre technologist in the *Architects' Journal* (Day *et al.*, 1979), the reader is left in no doubt that drama must have been viewed as a low priority here. Yet with 25 per cent of events requiring the proscenium stage, the demands of theatre surely deserve better accommodation, principally a more substantial, flexible and better-equipped stage.

The basic plan of the hall is, as its name implies, an elongated hexagon. The stalls area is substantially squatter than the balcony leaving commendably shallow overhangs. Seating blocks in the balcony are separated by slots, which help to 'justify' the absence for theatre performance of audience in the side blocks. The plan design contains a link with the Christchurch Town Hall, New Zealand (section 4.10), which was known to one of the senior architects on the design team. The use of inclined panels round the perimeter at high level may perhaps also have been inspired by the New Zealand hall. Potentially they could supply desirable lateral reflections. The adaptable components in the design include retractable seating in the Stalls and removable seats for the flat floor in front of the stage. The stage platform is on lifts, which also enables the whole Stalls floor to be set at a single level. Screens from above and from the side provide a 'picture frame' proscenium opening. Visible on the long section is the substantial suspended ceiling over the stage and auditorium; the suspended sections over the stage can be rotated from horizontal to vertical. The suspended ceiling has been designed to provide strong acoustic reflections, particularly to the balcony seats. Since the rear horizontal ceiling sections cannot perform this acoustic function, they have been left acoustically transparent.

Subjective characteristics

Listeners to a chamber orchestra concert in the Hexagon were significantly more enthusiastic about seats close to the stage in the Arena than seats in the rear Stalls and Balcony. Judgements ranged between 'Reasonable' and 'Good'. The sound was judged as clear but unreverberant, with average intimacy, spatial impression and loudness. To the author's ears the assisted resonance seemed inaudible in the Stalls but made a minor contribution in the Balcony; the electronic system still leaves an inadequate sense of reverberation.

Objective characteristics

The reverberation time in this hall is short and the measured change due to the assisted resonance is inadequate to convert between appropriate acoustic conditions for speech to good conditions for music (Figure 11.25). Objective clarity is high, in line with the reverberation time. Objective source broadening is also higher than average. But the total sound results show an interesting anomaly, in that all the Balcony seats have less than expected values. This deficiency proves to be principally associated with the early sound, suggesting that there are inadequate early reflections. The speech sound levels are also low in remote seats. It is however the dryness of the acoustics which dominated the subjective response to music.

In the theatre configuration the auditorium volume is reduced and the reverberation time falls to 1.0 second. The absence of early reflections is a minor deficiency with music owing to the small auditorium size. With speech, inadequate early reflections prove more serious. As usual there is no problem for actors facing into the auditorium, but for actors facing across stage the expected intelligibility is marginal. In section 7.6 the early reflection ratio concept was suggested, to indicate how many reflections arrive within 50 ms of the direct sound. In the Hexagon the mean value is only 1.6, which is particularly low and provides the major explanation for disappointing speech performance.

Conclusions

The examples of the Hexagon and the Derby Assembly Rooms indicate how it is possible to 'fall between

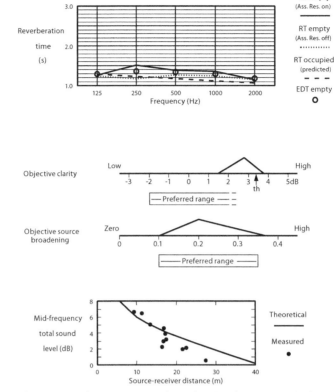

Figure 11.25 The Hexagon, Reading: objective characteristics in the concert format

two stools' when designing a multi-purpose space for music and speech. Modifying the reverberation time by passive acoustic means is unlikely to provide an adequate change, unless the technique of blocking off the flytower is used (section 10.4). If electronic solutions are acceptable, it appears better (and cheaper) to design the hall envelope for speech and extend the reverberation time for music. Current electronic reverberation enhancement systems can perform better than the one in the Hexagon, with reverberation time increases of 50 per cent or higher. Good acoustics for speech require a narrower auditorium design, with perhaps larger balcony overhangs; in other words a theatre design. But would it then be suitable for arena-type events such as boxing? The brief for these halls was certainly challenging.

11.6   Eden Court Theatre, Inverness

Inverness, the capital of the Scottish Highlands, is a modest-size town. The site of its major auditorium is on the bank of the River Ness adjacent to a Victorian Palace built by the wealthy Bishop Eden, hence its name. The Bishop's Palace is now used for offices, dressing rooms etc. for the theatre. The auditorium seats only 814 but it was specifically mentioned in the brief that it should retain its intimate atmosphere for smaller audiences. By selecting a horseshoe form derived from traditional eighteenth-century opera and playhouses, the architects Law & Dunbar-Nasmith have admirably fulfilled this requirement. The Stalls area can be comfortably used on its own with a capacity of 490. With

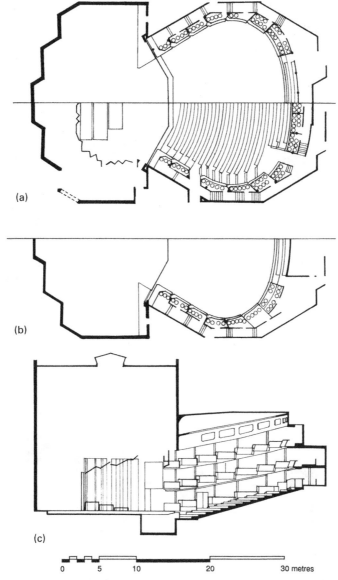

0 5 10 20 30 metres

Figure 11.26 (a), (b) Plans and (c) long section of Eden Court Theatre, Inverness

this plan form it was possible to include all seating within 22 m of the stage front and guarantee a good actor–audience relationship. The anticipated uses were for live drama, opera, ballet and orchestral music, together with amplified popular music and social events.

The auditorium has an open Stalls area surrounded by three tiers of boxes. There is no proscenium arch in the formal sense; the ends of the horseshoe tiers provide a frame. The structure is traditional with slim columns running through each box level, though the columns are placed behind

Figure 11.27 Eden Court Theatre with the orchestral shell, Inverness

Figure 11.28 Eden Court Theatre in the drama mode, Inverness

seating and do not obscure sightlines. The seating rakes are relatively steep, enhancing the sense of intimacy. Opposite the stage the balcony fronts step forwards slightly at higher levels to keep seating close to the stage. (The Barbican Theatre, London, also has this feature: see section 8.6.2.) There are no lighting slots in the main auditorium ceiling, which is certainly advantageous acoustically. The solution to multi-purpose use was basically to design a theatre-type space but to aim for a higher reverberation time than normal for theatre use of 1.3 seconds. For music performance a light-weight orchestral shell is erected in the stagehouse (Fahy, 1978). The front section of the stage can also be lowered to form a pit for opera etc.

Owing to its remote location, this auditorium has not been tested subjectively. Indeed reports on its characteristics are hard to come by, but the client appears to be well satisfied.

Objective characteristics

The measured reverberation time in the unoccupied concert condition at 1.1 seconds is slightly shorter than the design value; the theatre value is 1.0 second. The total auditorium volume is large for the seat capacity, but the question remains: how will sound behave in a space of this sort? Can sound which enters the box areas re-emerge with sufficient ease? In other words is the effective acoustic volume the total volume of the auditorium or just the volume between the balcony fronts? Since the latter is 72 per cent of the total, this might have a major influence on the reverberation time.

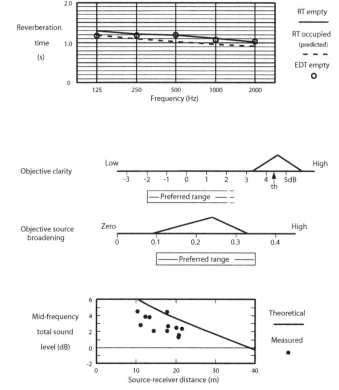

Figure 11.29 Eden Court Theatre, Inverness: objective characteristics in the concert format

Preliminary calculations suggest that the effective acoustic volume is only the space between the balcony fronts. In the case of orchestral performance, the large additional volume in the stagehouse as defined by the orchestral shell should increase the reverberation time. It appears though that the fit of the shell is loose, leaving many acoustically absorbent gaps. There are two consequences of this: a shorter reverberation time and reduced sound levels. The latter can be observed in the total sound level values in Figure 11.29, though it is unimportant in a small auditorium like this. Objective clarity is high, in line with the short reverberation time. Objective source broadening is above average, as one would expect in a narrow hall.

A potential problem with the horseshoe form is that of focusing by the balcony fronts. The acoustic consultant had wished to have the fronts inclined to direct sound down onto the Stalls, but this was rejected for architectural reasons. The balcony fronts have been given a backward tilt which increases progressively towards the back of the auditorium. The reflections from these surfaces therefore contribute to the reverberant part of the sound field. With surfaces as concave as these, complete elimination of echoes is problematic, and some evidence of focusing is found on the measured impulse responses. The subjective effects of this are probably minor, except for the case of theatre, when with an actor facing away across stage late reflections undermine intelligibility for box seats close to the stage. Conditions for speech are otherwise good.

Conclusions

The selection of a traditional horseshoe form is admirable for the creation of a compact, intimate auditorium, offering close contact between performers and audience. It is not without its acoustic implications, not least the risks of focusing by concave surfaces. Focused reflections have not been totally eliminated here, though by inclining the balcony fronts the problem has been minimized. In the concert format a simple orchestral shell has been used, but it does not manage to exploit the possibility of a significantly increased reverberation time for music use. In the case of Eden Court Theatre, such comments need to be seen in the context of what is only a modest-size auditorium. The very intimacy of the space should guarantee acoustic success.

11.7 Theatre Royal, Plymouth

The substantial war-time bombing of Plymouth had left it virtually without any large usable performance space. But by waiting so long to provide a replacement, the variety of demands which the new auditorium had to meet grew with the years, presenting a daunting task for the designers. The architects, the Peter Moro Partnership, had originally won a competition in the 1970s for a theatre and, in spite of many changes to the brief, they retained a clear view of their aims to produce a masterly solution for a wide variety of uses. The architect concludes his statement in Sugden *et al.* (1982) by saying that 'theatre design is based on hard facts and is rarely a matter of inventing new forms of theatre'. He presents a highly logical explanation for the major features of his design.

The brief specified a range of seating capacity from 750 to 1200, allowing at one extreme for small-scale repertory theatre, at the other for large-scale visiting productions. A pit was required for use for opera and musicals, but with no reduction in seat numbers while it is used. The auditorium in its large form was to be adaptable to use as a genuine concert hall. Finally the auditorium had to be appropriate for conferences.

The architects began from the concept of a theatre, with an electronic system being employed to modify acoustics for music. This route seems preferable from an acoustic point of view. The alternative of electronic assistance for speech in a space acoustically suited to music would be highly inimical for intimate drama. But to accommodate more than 1000 in a theatre necessitates several layers of audience with some degree of balcony overhang. Theatre design in recent years has tended to move away from the nineteenth-century picture frame

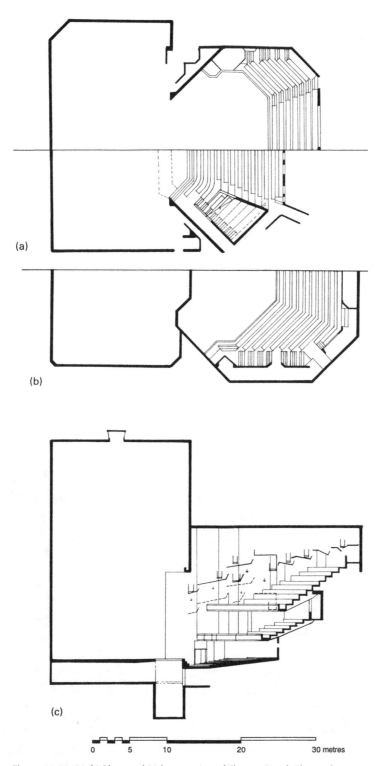

(a)

(b)

(c)

0   5   10   20   30 metres

Figure 11.30 (a),(b) Plans and (c) long section of Theatre Royal, Plymouth

towards more open stage forms. However, a proscenium arrangement was necessary for Plymouth in order for the theatre to be used for touring shows. The architects perceived their major concern as being to eliminate the division between stage and auditorium, to retain an intimate dramatic relationship in spite of the large proscenium opening and stage size. To this end, the top of the proscenium is defined by a suspended ceiling element rather than explicitly. There is a cranked forestage which provides some degree of encirclement by the audience. This projecting stage is reflected in the seating layout and contributes to the sense of self-awareness among the audience. The stage front steps down to the Stalls floor to provide a link rather than a barrier. The overall hexagonal plan form is the most natural given the form of the stage front. Continental-style seating has been used throughout, with a maximum capacity of 1271. As a multi-purpose theatre it is comparable to the type used extensively in North America.

The auditorium ceiling is not continuous, but is made of several loose-fitting suspended elements. The forward sections of this ceiling can be lowered to provide a visual barrier and block off the Upper Circle of seating. The proscenium opening is also adjustable in size. To create further homogeneity in the audience, seating on a ramp on the left of the auditorium only links the Dress Circle to the Stalls, a feature also valuable for conference use. For concert performance a series of screens are placed behind the orchestra and an inflatable plastic canopy can be suspended above. An assisted resonance system is used to extend the reverberation time for concert use, with in addition an electronic infill system under the balcony overhangs.

It is though as a theatre that much of the design strategy was directed. This use was also foremost in the mind of the acoustic consultant (Cowell, 1982a and 1982b), who used an acoustic scale model (tested by the author) to check speech intelligibility. The model exercise was particularly valuable for optimizing the orientation of the suspended ceiling elements. The ceiling reflection distribution for a typical acting position is illustrated in Figure 11.33. Unfortunately only limited subjective information about the Theatre Royal's performance is available.

Figure 11.31 Theatre Royal, Plymouth

Figure 11.32 Theatre Royal, Plymouth

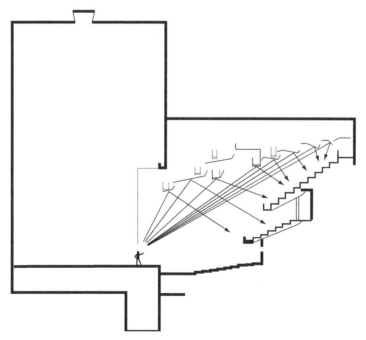

Figure 11.33 Ray diagram for reflections off the suspended ceiling in the Theatre Royal, Plymouth

Objective characteristics

Starting with speech conditions, the measured reverberation time in this mode was 0.8 seconds. A value below 1 second has been noted in section 7.5 as desirable for speech. The resulting values for the speech intelligibility measure are good for both source orientations (Figure 11.20). This is an impressive achievement when this auditorium is compared, for instance, with the Olivier Theatre (section 8.6.1) which has significantly fewer seats. The speech sound level was also found to be above criterion value with only one minor exception. It should be added that there was a substantial set on stage when these measurements were made. This probably had the effect of decreasing the intelligibility measure slightly but increasing the received

sound level, but the favourable conditions for speech are likely to prevail for more open staging.

The measured reverberation time (RT) for music performance in Figure 11.34 was only 0.9 seconds in the condition unoccupied with audience. A precise measurement of the effect of the assisted resonance system on the reverberation time was not made but it cannot be great. There is a correspondingly high degree of objective clarity. The predicted sound level on the basis of reverberation time and auditorium volume is also rather low, but measured sound levels are on average 2.5 db lower. The reason for this must lie in the orchestral shell arrangement. With so many gaps around the orchestral shell, much of the orchestra's energy will be lost in the basically absorbent flytower. This sloppy fit leaves several seat positions with resulting sound levels below

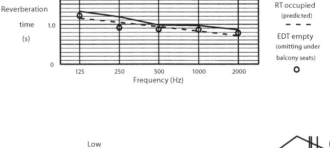

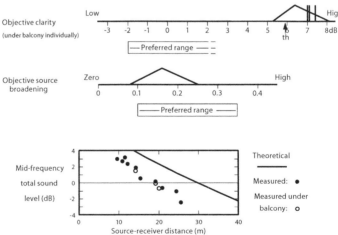

Figure 11.34 Theatre Royal, Plymouth: objective characteristics in the concert format

criterion. It also introduces additional absorption which lowers the reverberation time. Use of a tight fitting shell would give a valuable (estimated) 0.15 seconds increase in RT. The early lateral reflection situation is average, but because of the form of this theatre the better seats in this respect are towards the back of seating blocks, where there are some reverse-splay surfaces.

The objective measurements for the concert condition point to a particularly dry acoustic, and leave one considering that conditions for speech were achieved at the expense of music. The electronic enhancement which was intended to bridge the gap was certainly performing less impressively than expected. The reverberation time gain is mainly at low frequencies. This is characteristic of assisted resonance, whereas more recent systems could achieve greater reverberation time gains. On the question of whether the assisted resonance can be perceived, the answer is a personal one, but having had the experience of listening to live music and not knowing at the time whether the electronic system was on or off, it appeared to make the difference between acceptable and unacceptable listening conditions for music.

Conclusions

In the world of performing arts, there is an obvious division between professionals in theatre and music. The Plymouth Theatre Royal has been greeted with great enthusiasm, particularly among those associated with the theatre. It represents a thoroughly well-conceived solution to some difficult problems, with the additional virtue of two sizes of auditorium. The manner in which this variability has been achieved is simple enough for the facilities to be used properly. And on top of this the acoustics for speech are good.

The brief however specified the need for the auditorium to be adaptable to use as 'a genuine concert hall'. In constructional terms this requirement has been less lavishly endowed than the theatrical one. However, the major criticisms of the music acoustics are directed at the performance of the

variable elements employed. The electronic reverberation enhancement system is disappointing in its performance and alternative systems now exist which could provide a more significant increase in reverberation time. The orchestral shell could also be made more substantial with a tighter fit behind the proscenium opening. These are both changes which can be made in time should improvements be perceived as necessary. Perhaps with hindsight a little less emphasis might have been placed on perfecting speech intelligibility in, for instance, the design of the suspended ceiling, but this would have required considerable confidence in prediction of the acoustic performance.

The Theatre Royal of Plymouth seems to demonstrate a viable route to successful multi-purpose use for both speech and music. By designing a space as a theatre and using electronic and modest mechanical means to accommodate music performance, a highly workable and intimate auditorium has been created. The Lyric Theatre in the Hong Kong Academy of the Performing Arts (Lord and Templeton, 1986), designed by substantially the same consultants, is a clear descendant of the Theatre Royal, Plymouth.

11.8 Derngate Centre, Northampton

In the quest for multi-purpose auditorium form, many movable elements have been tried. Among the most elaborate of these have been movable ceiling elements. While these offer the chance to change room volume for acoustics, or to screen off seating areas, few halls have been built which provide the most valuable variation of all: to alter the plan form of an auditorium. The use of air castors has become common for scene shifting in theatres. The castor produces flotation by the hovercraft principle allowing massive elements to be manhandled easily. The Derngate Hall, by Renton Howard Wood Levin Partnership and Theatre Project Consultants, appears to be the first occasion on which this technology was used to change the internal layout of an auditorium (Reid 1984). It was clearly an exciting development.

Kit of parts

Derngate's kit of moving parts. The four formats shown below are created from differing assemblies of the 10 mobile towers and the range of seating wagons, coupled with varying heights of the elevated floor sections. All the boxes and wagons move on air castors. When not in use, seating wagons are stored underneath the fixed floors of the hall. Mobile box towers when not in use are stored in the scene dock.

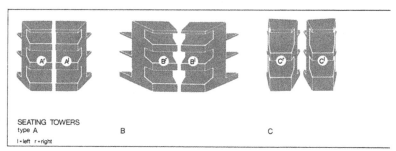

SEATING TOWERS
type A   B   C

l = left   r = right

Four formats

CONCERT

1 Stage area ceiling
2 Horizontal proscenium walls
3 Acoustic drapes
4 Catwalk
5 Upper gallery
6 Gallery
7 Technicians' gallery
8 Choir bleachers
9 Floor lifts
10 Rostra
11 Orchestra pit
12 Wagon storage

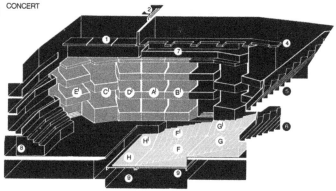

Section perspective showing one half of auditorium in 'concert' format (1400 seats). The layout for the other half is mirrored. All the box towers (shown in blue) and seating wagons are in use. One continuous rake of seats is created by

ARENA

positioning one group of seating wagons (shown in red) on the fixed floor at the back of the hall, and a further group on a lowered section of the elevated floor. The acoustic banners on the side wall behind the boxes are raised (for rock concerts

they would be lowered), so that the acoustic shell is formed by these walls and concrete roof above the catwalks. The one section of bleacher seating in the hall is pulled out behind the orchestra for use by a choir or members of the audience.

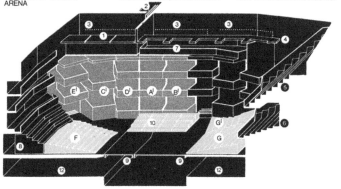

'Arena' format (1483 seats). Seating wagons have been rearranged, making two continuous rakes at each end of the hall. Tiered rostra with loose seats, not on air castors, are positioned on the remaining two sides. All the boxes

are in place. Unused seating wagons are stored below. Acoustic banners are likely to be lowered. The audience are splendidly focused onto the central performance area (see p49). The boxes increase this focus and by eliminating the bare walls

permit an intimacy that would otherwise be inconsistent with such a large seating capacity. Ceiling elements have been lowered to close off the flytower above the orchestra, and form a continuous ceiling level above the audience.

Figure 11.35 The four formats for the Derngate Centre, Northampton

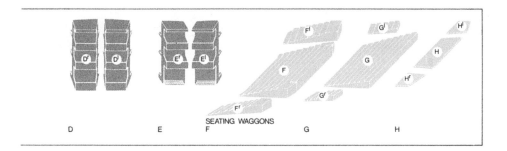

SEATING WAGGONS

D   E   F   G   H

LYRIC

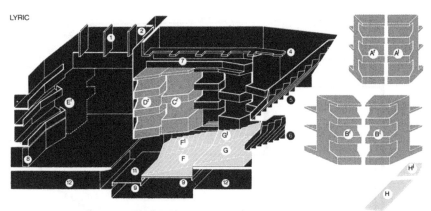

'Lyric' format (1151 seats without orchestra pit). Four boxes have been repositioned, the others have been stored away in the scene dock. The horizontal walls and vertical fire curtains which form the proscenium arch have been slid into position. The ceiling above the stage has been hoisted into the flytower. Bleacher seats are retracted. Most of the audience have clear sightlines and the shallowness of the boxes allows their occupants a reasonable view of the stage. Any loss of sight of the stage corners is more than compensated for by the intimacy and sense of inter-audience contact derived from 'hanging people on the walls'. Note the staggered effect of the rearranged boxes in contrast to the other formats.

FLAT FLOOR

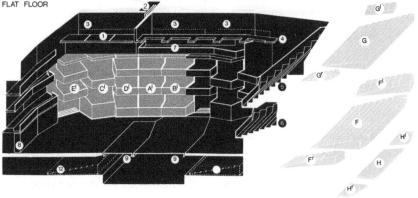

'Flat floor' format (648 m²), for civic receptions, banquets, dinner dances, exhibitions and trade shows. All the box towers are in place. The seating wagons are stored below. The ceiling level is continuous throughout the length of the hall.

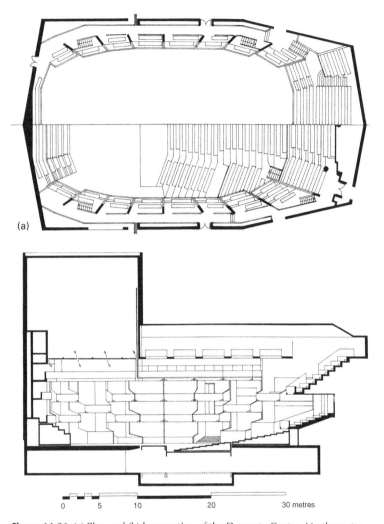

Figure 11.36 (a) Plan and (b) long section of the Derngate Centre, Northampton, in the concert format

The air-castor system allows blocks of seating to be moved as solid elements, avoiding the rather temporary feel of bleacher seating. To provide movable walls the designers have turned to the courtyard form, which is particularly suited to being divided vertically into towers. Each tower contains seating at two or three levels, like a set of vertically stacked theatre boxes. At the rear of each tower in the Derngate Centre is a platform which when linked up with neighbouring towers produce access aisles to the tower seating from behind. The mobile towers allow four formats to be created in this hall: concert, lyric, arena and flat floor (Figure 11.35). The fixed elements are the two gallery seating levels, adjacent side seating towers and shallow galleries behind the stage.

The transformation to the lyric format involves introduction of proscenium walls from the side, while the top is framed by the fire curtain. There are full flying facilities. In the concert format the

Figure 11.37 Derngate Centre, Northampton, in the concert format

Figure 11.38 Derngate Centre, Northampton, in the concert format

flytower is closed off with rotating sections, which are retracted into the flytower when not required. The lyric format is in fact little used for straight drama because the charming Royal Theatre of 1884 by Phipps forms part of the same arts complex. An open suspended ceiling in the main auditorium space conceals services and provides lighting positions while allowing for a large acoustic volume. The seat capacities for concert and drama are 1400 and 1151. There is an additional variable element in the form of acoustic banners which drop behind the towers. These can be used for performances which use sound amplification as well as for non-acoustic events. The tower design which allows these banners to be acoustically effective may however have acoustic penalties, as will be discussed below.

The survey of this auditorium has been limited to the concert mode, both for subjective and objective assessment. For the lyric mode it has only been possible to estimate the reverberation time, which has been calculated as 1.1 seconds unoccupied with audience. This bodes well for good speech intelligibility and is impressively different from the concert value. The reason for this is the substantial auditorium volume change between the two formats and the fact that the flytower is fully sealed off in the concert mode.

Subjective characteristics

The sound was sampled at two locations, at the rear of the Stalls and near the rear of the Upper Circle. The major observation about the subjective response in the Derngate Centre is that these two seats were judged differently. The Upper Circle was significantly preferred relative to the Stalls position.

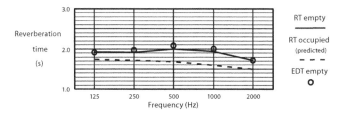

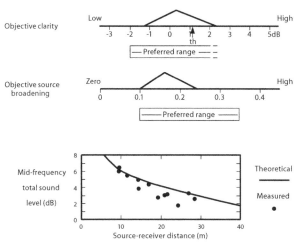

Figure 11.39 Derngate Centre, Northampton: objective characteristics in the concert format

Such a marked difference is unusual; for instance it occurred in none of the concert halls discussed in Chapter 5. The principal perceived deficiency at the Stalls position was lack of clarity; the degree of spatial impression was also smaller than in the Upper Circle. The sound in the Stalls has a slightly unreal character to it, as though looking through a gauze. As regards the remaining subjective scales, both positions were judged as somewhat unreverberant, as having average intimacy but with a loud sound. The mean assessment of the Upper Circle was of 'Good' acoustics, with good spatial impression and a loud sound. Individual instruments could be well differentiated there, probably helped by the steep seating rake and good sightlines.

Objective characteristics

Before discussing the detailed situation at the two seating areas used for the subjective tests, the mean objective behaviour should be reviewed. The mid-frequency reverberation times of 1.9 and 1.7 seconds in the concert mode unoccupied and occupied by audience is good by multi-purpose hall standards. The early decay time is equivalent to the reverberation time, so that no obvious objective explanation is available for the low perceived degree of reverberation. Objective clarity and source broadening are unexceptional, while some of the measured sound levels are a little below expectations.

The principal deviation from predictions is not immediately explicit in Figure 11.39, but it causes the mean objective clarity to be less than the theoretical. The early sound is the culprit here and hence there are some low sound levels overall. This observation of low early sound could explain the subjective problem in the rear Stalls; inadequate early reflections will produce poor clarity and frustrate the ability to relate subjectively to the performers. However the objective situation in the Upper Circle is not obviously superior to that in the rear Stalls. One is left wondering whether the objective differences between the two seats would widen when an audience is present, but this is pure speculation.

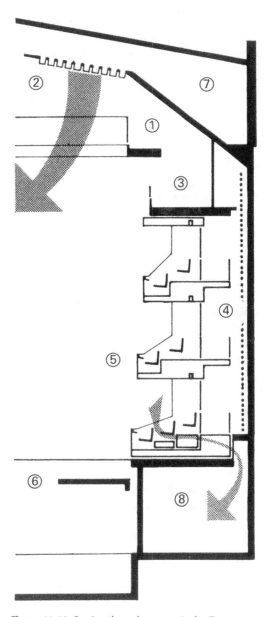

Figure 11.40 Section through a tower in the Derngate Centre, Northampton

The behaviour of sound when it enters a tower of the Derngate design may offer an explanation for a lack of early sound in the lower seating areas. As shown in a section through a tower (Figure 11.40), the surface immediately behind seating in the towers is of perforated metal, which is acoustically transparent. For sound to re-emerge from the tower, it has to be reflected from the masonry wall behind the tower. While this design provides a convenient location for the retractable acoustic banners, it does leave a highly obscured reflection path for sound entering the towers. The obscuring effect will be greater when audience is occupying the seats in the towers. This tower design is thus likely to inhibit reflections which normally come from the side walls. Unfortunately the objective evidence does not confirm this presumed behaviour adequately to explain the perceived subjective differences between the rear Stalls and the Upper Circle.

Conclusions

The use of towers which can be manoeuvred on air castors has proved itself to be a thoroughly viable technique for providing a high degree of variability in this auditorium. Many multi-purpose spaces can be criticized for being theatres which accommodate concerts uneasily or for being concert spaces which fail to work for drama. The air-castor system offers a greater degree of transformation than nearly all alternatives. In the Derngate Centre the variable auditorium has allowed the management to offer a highly diverse programme of entertainment and to run a most successful facility.

The discussion above suggests that the detailed design of the towers may be influencing their acoustic performance. One would expect that, with a solid surface immediately behind the seating in the towers, the acoustics would improve. Limitations on design forms introduced by the air-castor system do not at present look as though they should imply acoustic compromises. It has to be added that for music this hall performs better acoustically than most of the others discussed in this chapter.

11.9 Conclusions

Worldwide the successful acoustic design of multi-purpose spaces is a relatively recent development, initiated mainly in North America. The reasons for this are historical: in many European cities there are nineteenth-century theatres which are used for proscenium performances, while town halls have been used for music. With the original demand for performing centres arising later in North America, the inappropriateness of single-use facilities was apparent. The North American solution has however been little copied in Europe. Many North American halls do not attempt to work with unassisted speech, which leaves a lesser conflict in acoustic terms. The British examples are a heterogeneous group, but illustrate the diversity of demands and solutions in multi-purpose spaces.

In terms of the design aims of British multi-purpose halls, the major problem has been provision of variable reverberation time. The experience of these halls provides some valuable lessons from a variety of approaches. For the amphitheatre form with a semicircular plan, the situation for music performance proves acceptable for smaller auditoria. At larger scale the absence of nearby reflecting surfaces for many of the audience makes this an unsuitable form for concert use. The experience of the University of Warwick Butterworth Hall shows how an economical auditorium can achieve good acoustics. Here again the same form substantially enlarged is likely to encounter acoustic problems.

Five of the halls discussed are used for drama behind a proscenium opening. In the two cases in which the hall form approaches that of a concert hall, the acoustic performance for speech is less than ideal. Since in each case the reverberation time in the concert mode is close to 1 second, these designs by aiming for middle ground involve a compromise for all. With a seating capacity in each case of around 1000, it is necessary to consider a design based fundamentally on theatre principles, rather than on concert requirements. The Plymouth Theatre Royal uses this approach; it works well for drama but has been let down by the electronic assistance used for music performance. Design of a theatre implies attention to angles of view of the stage, limited auditorium width and, in the case of large seat numbers, close stacking of balconies in order to produce a total auditorium volume appropriate for speech acoustics. The major conflicts which arise when such a theatre form is to be used for orchestral music relate to the reverberation time being too short and the balcony overhangs too low.

That there is a way of providing a large reverberation time change between concert and drama conditions is confirmed by the example of the Derngate Centre, Northampton. It depends on a substantial auditorium volume change between the two formats. The drama stage needs to be acoustically absorbent but this absorption must be either removed or sealed off for the concert condition. For concerts the basic stage volume is required as part of the reverberant auditorium volume. Sealing off the flytower is unpopular with theatre designers as it takes up valuable flying space, but in many instances it is the only way to guarantee truly variable acoustics, suitable for all performers. The Derngate Centre also includes the fascinating innovation of movable elements on air castors to allow for a truly variable auditorium plan form.

Multi-purpose acoustic design is a particularly challenging area and in Britain much has been learnt over the last thirty years. With multi-purpose auditoria now becoming more common than single-use facilities, more evolution is bound to take place in the future. Electronic enhancement systems will also become more common as their sophistication grows. Electronics however offer no universal solution and should be designed no less carefully than more passive variable elements. Successful multi-purpose design involves confronting conflicts of interest but not simply conceding defeat by using compromise solutions.

References

General

Forsyth, M. (1987) *Auditoria – designing for the performing arts*, Mitchell Publishing, London.

Lord, P. and Templeton, D. (1986) *The architecture of sound – designing places of assembly*, Architectural Press, London.

References by section

Section 11.1

Drury, J. *et al.* (1977) Wembley Conference Centre. *Architects' Journal,* 23 March, 539–551.

Section 11.2

Charles, J.G. and Owston, R. (1985) Sound reflections at York's Central Hall. *Architects' Journal*, 18 September, 63–65.

Charles, J.G., Miller, J. and Gwatkin, H. (1987) Assisting the assisted resonance at the Central Hall, York, UK. *Applied Acoustics* **21**, 199–223.

Parkin, P.H. (1974) Assisted resonance at the Central Hall, York University. *Architects' Journal*, 31 July, 297–300.

Section 11.3

Architects' Journal (1981) Halls of change. 22 April, 738–739.

Charles, J.G., Fleming, D.B. and Miller, J. (1985) The hall of the University of Warwick. *Applied Acoustics* **18**, 195–234.

Section 11.4

Architects' Journal (1977) Civic pride. 7 December, 1104–1106.

Section 11.5

Cowell, J.R. (1980) The Hexagon, Reading. *Proceedings of the Institute of Acoustics, Spring Conference*, London 1980, pp. 45–49.

Day, B., Reid, F. *et al.* (1979) The Hexagon, Reading. *Architects' Journal*, 28 February, 415–430.

Section 11.6

Architectural Review (1976) Theatre, Inverness. October, 208–211 and 215.

Fahy, F. (1978) Eden Court Theatre, Inverness. *Proceedings of the Institute of Acoustics*, **4**, 14-6-1 to 14-6-4.

Section 11.7

Cowell, R. (1982a) Theatre Royal, Plymouth. *Proceedings of the Institute of Acoustics, Auditorium acoustics and Electro-acoustics meeting*, Edinburgh, September.

Cowell, R. (1982b) Electro-acoustic systems in the Harrogate Conference Centre and Theatre Royal, Plymouth. *Proceedings of the Institute of Acoustics, Auditorium acoustics and Electro-acoustics meeting*, Edinburgh, September.

Forsyth, M. (1987) *Auditoria – designing for the performing arts,* Mitchell Publishing, London, pp. 117–121.

Lord, P. and Templeton, D. (1986) *The architecture of sound – designing places of assembly,* Architectural Press, London, p. 172.

Sugden, D. *et al.* (1982) Building study: Theatre Royal, Plymouth. *Architects' Journal,* 13 October, 63–87.

Section 11.8

Davies, C. (1984) Interior design: Derngate Centre, Northampton. *Architectural Review*, April, 72–78.

Forsyth, M. (1987) *Auditoria – designing for the performing arts*, Mitchell Publishing, London, pp. 112–116.

Reid, F. (1984) Building study: municipal multi-form. *Architects' Journal*, 7 March, 49–64.

12 The art and science of acoustics

12.1 Precedence and scientific design

Before the twentieth century, auditorium acoustics owed very little to science. And it will always remain an art, involving such things as careful balancing of priorities and the incorporation of forms or surfaces beneficial to the acoustics but also acceptable architecturally. For a few nineteenth-century designers, the experience of building many auditoria inspired a feel for good acoustic design, but sadly they chose for their own reasons not to publicize their secrets. Valiant attempts at summaries of acoustic knowledge by interested architects such as Vitruvius (1960) in the first century BC and much more recently Smith (1861) failed to fulfil their promise as serious guides to design. Garnier (1880) for his design of the Paris Opéra concluded that acoustics was 'nothing but contradictory statements'. It was a fair assessment for the time.

For most designers acoustic design was a matter of heavy reliance on precedence. The achievements of this trial-and-error approach have been considerable. The Greek classical theatre, the Roman arena (as an organization of seating, not for its acoustics), the Roman theatre, the baroque theatre (with stacked balconies, generally with boxes), the traditional opera house, the English playhouse (with open galleries) and the so-called classical rectangular concert hall were all developed with almost no science of auditorium acoustics. But whereas another design based on trial-and-error, the Stradivarius violin, still retains some of its secrets and remains unsurpassed, there is now an advanced understanding of why the classical

designs perform well acoustically. If one wishes, one can reproduce classical halls and be virtually sure of replicating their acoustic character.

Sabine made the breakthrough with his reverberation time formula around 1900. He was able in the case of the Boston Symphony Hall to use precedence to advise him of the optimum room shape and proportions but used calculations to determine the optimum ceiling height. In less sensitive hands, some designers subsequently ignored the importance of room shape or form and designed auditoria which proved that a suitable reverberation time was a necessary but not a sufficient requirement for good acoustics. In particular the distinction between appropriate forms for theatre and concerts became blurred, with for instance some concert halls built in the period 1900 to 1950 based on theatre forms with many stacked balconies. The traditional forms of drama theatres and concert halls had developed quite independently, yet they each include features with acoustic significance for their particular type of performance. What makes the confusion understandable is that for smaller auditoria, design form is often unimportant. It is only near the capacity limits that acoustic design becomes critical. In crude terms these are above 800 seats for proscenium theatre, above 1000 seats for opera and above 1500 for concert halls; Table 10.1 lists maximum capacities.

As often happens, the application of science to a discipline previously treated as a mysterious art was greeted with great enthusiasm, only to be discredited in subsequent years when many acoustic problems remained unsolved. A great deal remained

Table 12.1 Some features which distinguish the acoustical design of concert halls, opera houses and drama theatres

	Concert hall	Opera house	Drama theatre
Reverberation time (seconds)	1.8–2.2	1.3–1.8	0.7–1.0
Scattering	Some	Yes, around stage	Not needed
Surfaces to provide early reflections	Yes	Yes, especially for singers	Yes, especially from above
Preference for early reflections from side	Yes	Some preference for orchestral sound	No preference
Balcony design, vertical angle of view*	$\theta > 40°$	$\theta > 30°$	$\theta > 25°$
Maximum distance of audience from stage	40 m	30 m	20 m

* as in Figures 3.23 and 9.11

to be discovered. But history has shown that Sabine's reverberation time is the foremost acoustic concern in rooms. Meanwhile the search for additional measures of acoustic quality has proved more laborious than expected. Reverberation time has several advantages which make it particularly convenient. In most auditorium spaces the reverberation time does not vary with position, which makes it a characteristic of the space as a whole. The measurement technique is also well standardized, subjective criteria are widely accepted and it can be predicted from architectural drawings. The newer measures generally vary with position in the auditorium and are more difficult to measure and predict.

In the first half of the twentieth century, auditorium design forms were haphazard, drawing inspiration from several directions but often proving disappointing. Major progress in research did not resume until around 1950 when subjective listening experiments were initiated mainly in Germany. The crucial realization was that our hearing system is highly complex. When confronted with a no less complex sound field, the ear interprets it in a highly selective way. Only by subjective tests can the important associated objective acoustic conditions be isolated. In the case of reverberation, the subjective effect, which may be called reverberance, is related to objective reverberation time, which is determined by room volume and the quantity of acoustic absorption. Until Sabine's systematic study, the link between the subjective effect of reverberance and details of architectural design had eluded many clever minds. In the same way, the new objective measures provide an invaluable link between subjective requirements and design implications. Present knowledge offers the possibility of auditorium design with far greater confidence than in the 1960s. We can also benefit from better documentation of existing halls, as found for instance in Beranek (2004).

The major concerns for different auditorium types are reviewed briefly below. As a summary, Table 12.1 lists the principal differences of acoustic significance between them.

12.2 Speech and theatre design

The requirement for speech in subjective terms is that it should be intelligible. This almost unidimensional requirement makes design resolution admirably suited to trial-and-error. The good acoustic performance of the many theatres built in the boom years before the First World War is an inspiration, which testifies to the extensive appreciation of acoustic factors by theatre architects of the time. It also gave a false sense of security to those responsible for theatre design of non-proscenium theatres in more recent years.

Intelligible speech depends on two considerations. The first of these is dominant in open-air environments: the speech sound must be loud enough relative to the background noise. This signal-to-noise problem is seldom critical in enclosed theatres where the disturbing sound is usually not noise but incoherent late reverberated speech sounds. In enclosed spaces, the proportion of early energy as a fraction of all received energy must be large enough for intelligible speech. This will occur close to the actor and in small auditoria. In large auditoria a reverberation time which is too long will provide too much late energy and intelligibility will be undermined. The usual recommendation for speech is a reverberation time of 1 second, but in critical situations shorter times down to 0.7 seconds are worthwhile. Adequate early energy is achieved by provision of strong early reflections arriving within 1/20th of a second after the direct sound. These are particularly important in theatres where members of the audience are regularly facing the back of an actor. Suitably designed suspended ceilings are valuable, though there is often a conflict between stage lighting and acoustic requirements.

There are several design features, especially in proscenium theatres, which are often taken for granted but which prove to be important for the acoustic design. A tight design in long section, with deep overhangs, both keeps the auditorium volume small, giving the desirable short reverberation time, and limits the distance to the back audience row, which is visually and acoustically valuable. Deep balcony overhangs prove not to be a great problem with speech, since the principal absence below an overhang is of late sound, which is itself detrimental to intelligibility.

Provision of adequate early reflections is the principal issue in acoustic design of theatres near the maximum seat capacity for that particular theatre type. One major problem is that the boundary between acceptable and unacceptable conditions is a narrow one so that accurate prediction can be essential. Recourse to modelling by computer or scale model is appropriate.

12.3 The concert hall

While the subjective situation with speech is relatively straightforward, it is now universally accepted that the listening experience with music has to be treated as multi-dimensional. What these dimensions are will long remain a topic of discussion. To the sophisticated listener, this compartmentalization of response into five dimensions or so will

seem very restrictive. But it does represent a considerable advance on the uni-dimensional approach of reverberation time which has caused so many disappointments. The practised designer will rely on his experience to design for the other nuances of subjective experience to music listening. Many comments of this nature have been made in the detailed analysis of individual halls.

The survey work discussed in this book was based on a consensus view of recent research results, with five principal subjective dimensions: clarity, reverberance, spatial impression, intimacy and loudness. Tonal balance is also important. Clarity refers to the ability to hear musical detail, reverberance to the sense of reverberation, spatial impression to the degree to which listeners sense they are in a three-dimensional space (see section 3.2), intimacy refers to the extent to which the listener identifies with the performance, while loudness needs no elaboration. Subjective tests show however that listeners do not place the same significance on the different subjective qualities. Some listeners, for instance, prefer good intimacy, while others seek full reverberance. For the best concert acoustics *it is therefore necessary to optimize the major different subjective attributes,* as far as is possible. It also follows that one should not rely on a single pair of ears for a definitive judgement of music acoustics! The relationship between these various subjective characteristics and objective measures is not entirely clear cut. The principal relevant objective measures in addition to reverberation time are: early decay time, the early-to-late index, the early lateral energy fraction and the total sound level. The probable relations between subjective and objective quantities are found in Table 5.2 (section 5.17).

The argument in terms of many subjective qualities and objective measures is logical, but it makes the business of concert hall design look formidable, a real juggling act. The reality is less forbidding but experience counts for much in concert hall design. Hence the stress in earlier chapters on discussion of individual examples analysed in detail. If one starts with the small hall, then reverberation time is certainly the major concern. Times of 1.8 – 2.2 seconds are optimum for symphony concerts, with shorter values being acceptable for halls with volumes below 10 000 m³. In most small halls the room surfaces often automatically provide suitable conditions to satisfy the other subjective concerns. The reasons for this are that in small halls the density of reflections is high, even within the important early period (of 8/100ths of a second after the direct sound), and the sound level is also loud. Adequate bass sound will always be a concern, which depends on a sufficiently massive auditorium shell.

Good acoustic design for large concert halls becomes progressively more difficult as the seating numbers increase. The criteria can be stated fairly simply: that the auditorium volume be chosen to provide a suitable reverberation time, that there are no excessive balcony overhangs and that there are adequate early reflections with a sufficient proportion arriving from the side. This raises the basic conflict in large auditorium design, that the large volume required for the long reverberation time tends to leave auditorium boundary surfaces so remote from the audience that they can no longer provide suitable early reflections. The early reflection requirement is further complicated by the need to accommodate the different directions in which individual instruments radiate most sound.

While the above refers to conditions for listeners, the other requirement of acoustic design is to provide support for the performers. Suitable surfaces are called for around the stage to reflect sound back so that musicians can both hear the result of their own efforts and monitor the playing of their colleagues. The design implications of satisfying the performers and listeners are for the most part different. For stage acoustics, the conflicts of interest are often with visual or lighting concerns.

It is tempting to say that concert hall design is going through a neo-classical phase at the moment. There is certainly an element of circularity over the centuries (Barron, 2006). The rectangular shoebox form used extensively in the nineteenth century is now the most popular today, after many excursions in other directions during the twentieth century. Current halls are more accurately described as parallel-sided and are also preferred by many performers. The terraced hall is currently the other preferred form. The (vineyard) terraced hall offers what might be called a more open sound for listeners, when compared with the parallel-sided hall. It will be very interesting to follow future developments of concert hall design into the twenty-first century.

12.4   The opera house

Of all auditorium types, opera house design is the most constrained. With a proscenium opening determining the limits of seating for visual reasons and the need for a pit separating the stage from the stalls seating, the options are highly limited. The two needs, to project the singer's voice with reasonable intelligibility and for orchestral sound to be supported by the acoustics of the auditorium, appear irreconcilable. Though more easily stated than realized in practice, the solution lies in providing a reverberant auditorium for the orchestral sound but designing the form to enhance the singer's sound as far as possible, principally with early reflections. With good acoustics, the presence of the singer's formant around 3000 Hz enables singers to compete and still be heard above an opera orchestra.

For reverberation time, values intermediate between speech and music of 1.3 to 1.8 seconds are appropriate, with the optimum being somewhat a matter of taste. In terms of form, nearly all opera house designs can be seen as intermediate between two extremes: the traditional baroque theatre and the raked fan-shape plan, first used at Bayreuth by Wagner. The baroque theatre with stacked boxes can have excellent acoustics for some, but for those sitting behind the front row in the boxes, the experience is much muted. Open galleries are more democratic than boxes, both socially and acoustically. Nineteenth-century designs often relied on deep overhangs, which have their disadvantages for seeing and hearing alike. The challenge in modern opera house designs is to avoid such deep overhangs and yet maintain acoustic balance in favour of the singer throughout the auditorium. It has been surprising in the early years of this millennium to find several new opera houses being built; most have been acoustically superior to their predecessors.

12.5 The multi-purpose hall

To summarize acoustics for multi-purpose halls is like attempting to hit a randomly moving target. The mix of uses and priorities for individual halls are seldom alike. It is notoriously difficult to achieve significant acoustic variability, so the temptation to compromise is considerable. While this is acceptable in halls seating audiences of a few hundred, it becomes progressively less acceptable the larger the scale. A compromise is not inevitable.

Though single-use or nominally single-use concert halls, theatres and opera houses have been built and developed over centuries, the purpose-built multi-purpose hall is a recent phenomenon. This has arisen from the need to optimize the usage of expensive auditoria and from the ability to supply more flexible facilities. Flexible acoustics have trailed behind the more obvious flexible staging and lighting, but they are now reaching the stage where worthwhile combinations are being successfully developed.

For physically variable acoustics, the most common elements are variable volume and absorption. Variable absorption is easy to include in small quantities, difficult in large quantities; in the condition with substantial absorption, there are risks of quiet sound and that vital reflections have been killed. Variable volume is often more appropriate but confronts severe architectural limitations. Both variable elements are likely to be included more often in the future. Manipulating early reflections also holds promise for variable acoustics. Sophisticated electronic systems ostensibly offer variability at the turn of a switch, with much less impact on the visual character of the auditorium. Systems which extend the reverberation time are particularly valuable for multi-purpose halls. Purists may criticize the sound quality, but it has improved with time.

Providing good acoustics for a single use is already a considerable challenge. As a measure of that challenge, Table 12.1 lists different requirements for different programmes. Some of these requirements dictate limiting values, as illustrated by the following. For a hall to be used for both concerts and drama, for example, one would prefer an open balcony for music, with a vertical angle of view below the balcony of not less than 40°, whereas a maximum audience distance of 20 m from the stage is applicable for drama. Neither a deep balcony overhang nor a long auditorium are acceptable for such a multi-purpose space. Several uses clearly demand much ingenuity. The state of the art of acoustics and auditorium design has reached the point where significant advances can be made in satisfying demanding briefs. The design of multi-purpose halls will probably be the most dynamic area of progress in the coming years.

12.6 The future

Acoustics occupies a curious place in the reputation of an auditorium, especially with concert and recital halls. The situation can become particularly critical at the opening of a new performing arts centre. On the first night the acoustics are the first criterion by which success or failure is judged. The critics will have already witnessed the marble in the foyers, the colour scheme in the auditorium and the comfort of the seats. In any case these issues are all considered matters of taste. On the opening night the invisible acoustic behaviour is finally revealed and judged (the possibility that acoustics is also a question of personal preference is generally ignored, but this view has barely reached the public forum). An adverse reaction to the acoustics on this occasion can blight the reputation of the hall out of all proportion to what was actually heard. If however all goes well, audiences often adapt to the acoustic character of a hall and acoustics takes its rightful place among the many concerns which influence attendances, such as the popularity of the performers, the programmes of the concerts, enthusiasm for the experience of a night out in this venue and promotions by the management.

Several decades ago auditoria were often judged as monuments to the affluence and mirrors to the culture of the societies that built them. Pride in temples to the arts made operational aspects of the buildings secondary in importance. In many places, this is now changing with audiences demanding higher standards. In the case of the acoustics, universal access to recorded and broadcast music with high-quality reproduction has made listeners more sophisticated and more critical of faults. Auditoria must now retain their place alongside the realism of modern electronics.

For all performing arts, live performance in front of an audience is critically necessary for their well-being and even survival. Acoustically, the spontaneous experience of a live production is different from reproduced sound. Auditoria should not and cannot emulate the acoustics which can be manufactured from many microphones. Excellence in auditorium acoustics is different but potentially even more exciting than the recorded variety. It is only by aiming for excellence in auditorium design that performing arts can expect to withstand the threats to their life blood.

To the layman who 'knows what he likes and likes what he knows', the arguments found in earlier chapters with questions of multi-dimensional this and energy fractions of that must appear irrelevant to the pleasures of hearing a full symphony orchestra or a professional acting group. Concepts like acoustic intimacy and envelopment may seem quaint but unrelated to experience. And in any case were not the world's greatest concert halls built before there was a science of auditorium acoustics?

Our unsophisticated layman may be correct in concluding that his enjoyment is scarcely enhanced by an ability to analyse acoustic sensations into compartments. But he would be wrong to imagine that he is insensitive to the concerns of this discussion. A sense of intimacy, for instance, is basic to any performance, without which that essential rapport between performer and audience is absent. If a concert hall lacks acoustic intimacy, the customer is fully justified in expressing his preference for his compact disc player.

The taunt that the science of acoustics has brought little to design is no longer valid. There are now modern concert halls with acoustics at least as good as the classical rectangular halls of 100 years ago. The fame of the old halls is perpetuated by tradition, tinged with some nostalgia. In any case reproducing concert hall designs of the nineteenth century and their acoustic characteristics is entirely possible, yet only rarely is this done. Most modern halls are required to meet demands in audience numbers and facilities which are far in advance of those even 50 years ago. Acoustic design by precedent has long ceased to be a necessity.

Mainly because of the complexities of the problems which had to be addressed, the science of acoustics has taken a while to attain some maturity. Of course more remains to be discovered, but the present conceptual framework looks a sufficiently robust base on which to build further advances. Meanwhile the design of a good auditorium remains no less an art, which demands diverse talents. In the 2400-year history of auditoria, there has been much conservative copying of precedents. Good acoustics were often a matter of luck, as Garnier also remarked. With a scientific basis to acoustic design, the fundamental requirements of auditoria can be reassessed and innovation can be applied to a design problem which, by its nature, will always be bounded by considerable constraints.

References

Barron, M. (2006) The development of concert hall design – a 111 year experience. *Proceedings of the Institute of Acoustics* **28**, Part 1, 1–14.

Beranek, L.L. (2004) *Concert and opera houses: Music, acoustics and architecture*, 2nd edn, Springer, New York.

Garnier, C. (1880) *Le nouvel Opéra de Paris*, Paris.

Smith, T.R. (1861) *Acoustics of public buildings*, Crosby Lockwood, London.

Vitruvius (translated by Morgan, M.H.) (1960) *The ten books on architecture*, Dover Publications, New York.

Appendix A

Sound reflection and reverberation calculation

A.1 Reflection from finite plane and curved surfaces

A.1.1 Reflection from finite surfaces

Diffraction behaviour is rightly considered complicated, but in the case of reflection from a suspended reflector the mathematics can be rendered quite simple. The finite reflector provides a valuable example where qualitative behaviour can be understood. This section considers just the reflection from the centre of a rectangular surface along the geometrical reflection path. Results have been taken from the more comprehensive analysis by Rindel (1986).

From Figure A.1, we consider a reflecting surface of dimensions $A \times B$, with angles of incidence and reflection of θ. The problem can be considered in two parts by taking first one and then the other dimension of the reflector in turn. We will first look

at reflection from a very long strip of width B (indicated by dotted lines in Figure A.1). We are aiming to calculate the level change ΔL in the reflection due to the finite size of the reflector, as compared with reflection from an infinite surface along the same reflection path.

The projected dimension of the reflector, as seen from the source or receiver, is $B\cos\theta$. The value of ΔL with a strip reflector depends on the wavelength, or frequency, of the incident sound. There is a limiting frequency, f_1, given by:

$$f_1 = \frac{c}{(1/s + 1/r) \cdot B^2 \cos^2\theta}$$

c, here, is the speed of sound and s and r are distances of the reflection point from the source and receiver. Above this limiting frequency the finite nature of the reflector can be more or less ignored, while below it the reflection amplitude decreases by 3 dB/octave.

Above the limiting frequency: $\Delta L = 0$ dB

Below the limiting frequency: $\Delta L = 10 \log f/f_1$

Introduction of the second dimension of the reflector introduces a second limiting frequency:

$$f_2 = \frac{c}{(1/s + 1/r) \cdot A^2}$$

The net effect for reflection from a finite reflector is shown in Figure A.2. At low frequencies diffraction effects attenuate by 6 dB/octave for decreasing frequency. The above equations can be used to determine the appropriate size for reflectors in auditoria.

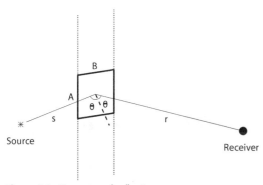

Figure A.1 Geometry of reflection

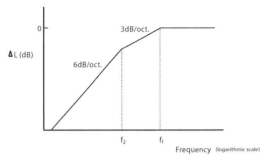

Figure A.2 Level change due to diffraction for reflection from a finite surface

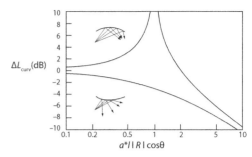

Figure A.3 Level change due to curvature of a reflector (from Rindel, 1985)

A.1.2 Reflection from curved surfaces

There are two dangers associated with the use of large plane reflectors, particularly for concert auditoria: false localization and tone colouration (see section 3.2). The risk of these potential faults is diminished when the reflection level is reduced. This can be achieved either by introducing scattering on the reflecting surface or by making the surface convex. Rindel (1985) has also provided a method to calculate the effect of curvature on reflection level. To calculate the net level of a reflection involves adding three effects: that due to inverse square law propagation, diffraction (above) and curvature. The curvature effect, ΔL_{curv} is independent of frequency.

In the case of reflection from a cylindrical surface of radius R, the following formulae are used when the source and receiver are in a plane perpendicular to the axis of the cylinder:

$$a^* = \frac{2sr}{s+r}, \quad \Delta L_{curv} = -10\log\left|1+\frac{a^*}{R\cos\theta}\right|$$

where s, r and θ are as in Figure A.1. a^* is a sort of mean source and receiver distance. The formula is for the case of a convex surface; for a concave surface R is made negative. Figure A.3 provides a graphical solution; the focusing effect with concave surfaces is clearly shown. For double curved surfaces there are two terms ΔL_{curv} which must be added.

A.2 The Quadratic Residue Diffuser

These diffusers rely on the phenomenon of interference, which can occur with all wave motions. When two sound waves are superimposed, their acoustic pressures are added. If the waves are equal but of opposite sign, a positive pressure will be added to a negative and the effect is known as destructive interference. Superimposition of entirely equal waves is called constructive interference. The difference between these two cases depends on the relative phase of the two waves. In simple terms, phase is the point on the cycle of a sine wave at the time of interest and relative **phase** is then the difference between the phases of each wave.

The Quadratic Residue Diffuser involves creating a series of slots which produce mini-reflections at different phases to each other (Figure A.4). In crude terms, the sound wave arrives at the top of a slot from some direction and then travels down the slot parallel to its sides. It is then reflected at the bottom

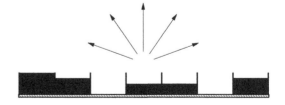

Figure A.4 Profile through a quadratic residue diffuser, which provides highly scattered reflections over a particular frequency range

of the slot and travels up the slot again. The phase of the wavelet leaving the slot depends upon the depth of the slot. Interference between wavelets determines the directional distribution of reflected sound from the diffuser.

Quadratic Residue number sequences are based on prime numbers. The code is derived as follows for the case of prime number 7:

Integer sequence:	0	1	2	3	4	5	6
Squared sequence:	0	1	4	9	16	25	36
Residue sequence:	0	1	4	2	2	4	1

The residue sequence is the remainder after division by 7 (the chosen prime number) of the squared sequence. If one continues the sequence, one finds it just repeats. The slot depths in Figure A.4 are given by multiplying the residue sequence by a design dimension. It can be shown mathematically that diffusers based on these sequences generate highly scattered reflections.

The process of designing a Quadratic Residue Diffuser (QRD) revolves around the choice of prime number, the maximum slot depth and the slot width. These influence the frequency range and the degree of scattering achieved. If the maximum slot depth is *d*, then in rough terms the diffuser will scatter sound down to a frequency corresponding to a wavelength of about 3*d*. The frequency range for diffusion is mainly determined by the aspect ratio: the ratio of maximum slot depth to slot width. The higher the aspect ratio the larger the frequency range. The degree of scattering is related to the prime number chosen.

For a diffuser which is highly diffusing over a large frequency range, it is necessary to use a high prime number with a high aspect ratio for the slots. Unfortunately it has been found that deep narrow slots tend to provide acoustic absorption, so that there are limits to how far QRD schemes can be exploited to achieve the ultimate sound diffuser. The interested reader is referred to Schroeder (1979, 1986), d'Antonio and Konnert (1984) and Cox and d'Antonio (2004).

A.3 Calculation of reverberation time

For the case of auditoria, the Sabine equation is almost universally used:

$$\text{Reverberation time (seconds)} = \frac{0.16V}{A},$$

where $A = S_1 \cdot \alpha_1 + S_2 \cdot \alpha_2 + \ldots$

V is the internal volume of the space in m³, while A is known as the total acoustic absorption with units of m² and is derived from summing for all room surfaces the product of their area (S) and absorption coefficient (α). The calculation is performed in octaves over the range of at least between 125 and 2000 Hz. There are strong reasons for tabulating the calculation as described below.

To perform the calculation it is necessary to calculate the volume of the auditorium. In the case of proscenium theatres there is the potential of a coupled space; if the flytower contains little absorbent material, double slope decays can be measured in the auditorium. In practice drapes and stage sets often fill the flytower space so that little sound energy entering the stage space returns to the auditorium. In other words, seen from the auditorium, the proscenium opening effectively presents a highly absorbent surface. If this is appropriate, the volume of the auditorium alone is calculated and the absorption coefficient for seated audience is applied to the proscenium opening.

The area of all internal surfaces needs to be calculated and appropriate figures for absorption coefficients, such as those in Table 2.4, are assigned to each surface. Absorption by audience is calculated on an area basis like other materials but an edge strip is added to obtain the effective absorbing area (section 2.8.4). The calculation proceeds by first establishing the total acoustic absorption in each octave and then calculating the reverberation time from the Sabine equation.

The sample reverberation time calculation in Table A.1 is for a rectangular recital hall of dimensions 20 × 10 × 10 m high. The audience occupies a rectangular area of floor 13 × 8 m while the lower 2 m of all wall surfaces is covered with timber panelling.

TABLE A.1 Sample reverberation time calculation for an occupied auditorium.

| Material | Area (m²) | \
Octave centre frequency (Hz) | | | | | | | | | |
|---|---|---|---|---|---|---|---|---|---|---|---|
| | | 125 | | 250 | | 500 | | 1000 | | 2000 | |
| | S | α | Sα | α | Sα | α | Sα | α | Sα | α | Sα |
| Audience (medium upholstered) | 126 | 0.62 | 78 | 0.72 | 91 | 0.80 | 101 | 0.83 | 105 | 0.84 | 106 |
| Orchestra | 24 | 0.62 | 15 | 0.72 | 17 | 0.80 | 19 | 0.83 | 20 | 0.84 | 20 |
| Thin wood | 120 | 0.42 | 50 | 0.21 | 25 | 0.10 | 12 | 0.08 | 10 | 0.06 | 7 |
| Plaster or thick wood | 730 | 0.10 | 73 | 0.06 | 44 | 0.05 | 37 | 0.05 | 37 | 0.05 | 37 |
| Air absorption ('α' = 4m m⁻¹) | Vol. | .000 | 0 | .001 | 2 | .003 | 6 | .004 | 8 | .009 | 18 |
| Totals: | 1000 | | 216 | | 179 | | 175 | | 179 | | 188 |
| V=2000 m³ | | | | | | | | | | | |
| Reverberation time (s) | | | 1.48 | | 1.79 | | 1.83 | | 1.79 | | 1.70 |

A small orchestra occupies an area of 5 × 3 m on stage. All remaining surfaces are either plaster on a substantial backing or thick wood. Including the 0.5 m edge strip, the 'acoustic seating area' is thus 14 × 9 m = 126 m² and the orchestra absorbing area is 6 × 4 m = 24 m².

The calculated reverberation time (RT) against frequency is normally plotted as a series of points joined by straight lines, as in Figure 5.6. One can comment on the result of the calculation in Table A.1 as follows. The mid-frequency reverberation time is suitable for a recital or chamber music hall but the drop at low frequencies is undesirable. Removal of the timber panelling would give a flatter RT response with frequency, which is more suitable for music.

References

Cox, T.J. and d'Antonio, P. (2004) *Acoustic absorbers and diffusers: theory, design and application*, Spon Press, London and New York.

d'Antonio, P. and Konnert, J. (1984) The reflection phase grating diffusor: design theory and application. *Journal of the Audio Engineering Society,* **32**, 228–238.

Rindel, J.H. (1985) Attenuation of sound reflections from curved surfaces. *Proceedings of 24th Conference on Acoustics,* The High Tatras, Czechoslovakia, October.

Rindel, J.H. (1986) Attenuation of sound reflections due to diffraction. *Proceedings of the Nordic Acoustical Meeting,* Aalborg, Denmark, August.

Schroeder, M.R. (1979) Binaural dissimilarity and optimum ceilings for concert halls: more lateral diffusion. *Journal of the Acoustical Society of America,* **65**, 958–963.

Schroeder, M.R. (1986) *Number theory in science and communication, with applications in cryptography, physics, digital information, computing and self-similarity,* 2nd edn, Springer, Berlin.

Appendix B
Objective measures for music auditoria

B.1 The impulse response and the new objective measures

Beyond reverberation time, the new objective measures of greatest interest have been listed in section 3.3 as follows:

- Early decay time (seconds)
- Early-to-late sound index or objective clarity (dB)
- Early lateral energy fraction or objective source broadening
- Total sound level (dB) (also known as 'strength')

Each quantity is measured at several seat positions in a series of octave frequency bands, though results are often presented as averages over a range of octave bands. All these quantities are described together with suitable measurement procedures in the standard ISO 3382. Attention also needs to be paid to several additional issues in order to acheive reliable measured values (Barron, 2005).

Before listing the mathematical definitions of these measures, it is valuable to mention the central position of the impulse response (Barron, 1984). The concept of the impulse response was introduced in section 2.4: it is the pressure response recorded at the receiver position of interest when a very short duration pulse is produced at the relevant source position. In ideal mathematical terms the pulse is very intense but very short, though practical pulses generated by, for instance, blank pistol shots or electrical pulses fed to loudspeakers can be adequate approximations. (More sophisticated signals to give better signal-to-noise ratios are now also available.) The impulse response is particularly attractive because it offers a complete description of transmission between two points, as long as we are not concerned with direction. All the measures mentioned above, with the exception of the early lateral energy fraction, ignore direction; they are omni-directional measures and can be calculated from the impulse response.

There is a further unifying aspect about the five measures, they are all energy measures. Acoustic energy is related to pressure squared (p^2), so the first operation on the pressure impulse response is to square it (Figure 2.14 shows typical results). The pressure impulse response involves positive and negative pressures, while the squared response is only positive.

The methods used to derive the **early-to-late sound index** and **total sound level** from the impulse response are given in section 3.3. If $p(t)$ is the impulse response at the receiver position with $t = 0$ the start of the direct sound, then the relevant equations are:

Early-to-late sound index,

$$C_{80} = 10\log\left(\frac{\int_0^{0.08} p^2(t)dt}{\int_{0.08}^{\infty} p^2(t)dt}\right) \ \text{dB}$$

Total sound level,

$$G = 10\log\left(\frac{\int_0^{\infty} p^2(t)dt}{\int_0^{\infty} p_A^2(t)dt}\right) \ \text{dB}$$

where $p_A(t)$ is the response at 10 m from the identical source in an anechoic environment. In other words, the total sound is measured relative to the direct sound level at 10 m. Some writers have normalized relative to zero sound power level for the source, but this gives uncomfortable numerical values. For a measurement relative to direct sound at 10 m, the measured values in concert halls are usually in the region 0 to +10 dB. The measurement can also be made with continuous noise signals, in which case the normalization level is measured with the same apparatus in an anechoic chamber.

The early-to-late index is derived from a proposal by Thiele in 1953 of a new measure *Deutlichkeit* (distinctness). This used the time interval of 50 ms which is now often employed for speech. For music an 80 ms period is now normally used, as proposed by Reichardt *et al.* (1975). Some writers have been concerned at the use of a sudden temporal cut-off, which is unrealistic for the working of the ears. Cremer (see Cremer and Müller, 1982, p. 434) made the ingenious suggestion of a quantity, known as **centre time**, which has units of time and thus avoids the sharp cut-off:

$$\text{Centre time} = \frac{\int_0^\infty t p^2(t)dt}{\int_0^\infty p^2(t)dt} \quad \text{seconds}$$

The centre time is the centre of gravity along the time axis of the squared impulse response. In practice, the centre time is found to be very highly correlated with early decay time (typical correlation coefficient = 0.975), so that it has lost favour for concert acoustics.

The **early lateral energy fraction** was derived from subjective tests with a simulation system (Barron and Marshall, 1981) as a linear measure of source broadening (then called spatial impression). It is measured by the ratio between the energy received by a figure-of-eight microphone with its null pointing at the source and the energy received by an omni-directional microphone at the same position.

Early lateral energy fraction,

$$\text{LF} = \frac{\int_{0.005}^{0.08} p^2(t)\cos^2\theta \, dt}{\int_0^{0.08} p^2(t)dt}$$

Ideally in the expression above, the upper term would contain $\cos\theta$, but the commonly available figure-of-eight microphone has a $\cos^2\theta$ relationship; θ is the angle between incident sound and the axis of maximum sensitivity of the microphone (equivalent to the axis through the ears of a listener). An alternative measure for source broadening is the maximum of the normalized cross-correlation function for a pair of spaced microphones. It can be shown (Barron, 1983) that the cross-correlation coefficient is sensitive to the choice of test signal, but that with an appropriate choice there should be high correspondence with the early lateral energy fraction.

The perceived degree of source broadening is not only determined by the spatial distribution of early sound (LF) but also by sound level, which varies of course during a piece of music but is also a function of the total sound level, *G*. Few suggestions exist for the relative importance of LF and *G*, this author's is as follows (Marshall and Barron, 2001):

Degree of source broadening

$$= \text{LF} + (\text{early level})/60$$

For the other spatial component, listener envelopment, Bradley and Soulodre (1995) have proposed the **late lateral level** as a suitable measure. In a similar way to the early lateral energy fraction, this measure is based on the ratio between energy measured with a figure-of-eight microphone with its null pointing at the source and the energy measured with an omni-directional microphone, though in this case the denominator effectively measures the source level. In this way the measure combines both the spatial and level components.

Late lateral level, LG

$$LG_{80}^{\infty} = 10\log\left(\frac{\int_{0.08}^{\infty} p^2(t)\cos^2\theta \, dt}{\int_0^{\infty} p_A^2(t)dt}\right) \quad \text{dB}$$

As a measure of conditions for performers mentioned in section 3.8.6, **objective support** (also called ST_{early} by Gade, 1989) is a further parameter based on acoustic energy. It is measured on stage with the microphone 1 m from the source, itself at a height of 1 m above the floor.

Objective support,

$$ST_{early} = 10\log\left(\frac{\int_{0.02}^{0.1} p^2(t)dt}{\int_0^{0.01} p^2(t)dt}\right) \; dB$$

The time interval for the denominator encompasses the direct sound alone and for the numerator the early reflections arriving from the stage enclosure and beyond. Measurements at three positions on stage are made and averaged. Values are taken at the four octaves between 250 and 2000 Hz and also averaged to give a single value for the particular stage.

The **early decay time** (EDT) is measured from what is known as the 'integrated impulse response'. Traditionally, reverberation time has often been measured with a noise source which is turned off; the sound level at the microphone is plotted against time on a level recorder. The reverberation time is measured from the slope of the decay. Even for a fixed source and receiver, the decay will look different in detail for each decay, because the precise signal before switch-off is different on each occasion. An average of several noise responses is therefore taken. Schroeder showed in 1965 that the average of many noise responses is identical to the integrated impulse function, as long as the integration is made in reverse time. The relationship between the noise responses, $n(t)$ (< > implies an average of very many), and the response to an impulse, $p(t)$, with identical source and receiver positions, is:

$$<n^2(t)> = \int_t^\infty p^2(\tau)d\tau$$

The integrated impulse function (see Figure 3.9 for example) has no superimposed random oscillations and can be used for measuring the decay rate of the first 10 db after the direct sound for the early decay time. The EDT is expressed identically to the reverberation time and in many spaces values are very similar (Barron, 1995).

Two further manipulations of the impulse response deserve mention. Firstly the true impulse response can be transformed mathematically (or computed) into a frequency response by a process known as Fourier transformation. In other words, the impulse response contains the information about the relative strengths of various frequencies, as regards the effect of the hall on the source signal.

Secondly we would expect the impulse response to be fundamental to the way the hall affects the musical signal produced by the performer. The 'multiplication' process involved is called **convolution**, so that given a sample of music the situation for the listener is derived by convolving the pressure signature associated with the music with the impulse response. It is thus no surprise that physicists of sound lay such store by the impulse response.

B.2 Prediction of occupied reverberation time

Measurement of reverberation time (RT) in fully occupied halls is always problematic, so one finds some auditoria where the measurement has never been made. A method to predict occupied reverberation times has had to be used in the case of a few auditoria surveyed here. For completeness the method used is briefly described in this section; it refers to halls surveyed before 1990 and was based on audience absorption figures published by Beranek in 1969. Beranek's figures for absorption by both occupied and unoccupied seating however did not appear entirely trustworthy, since the difference between occupied and unoccupied absorption coefficients seemed to be too large. Analysis of data to hand from British auditoria led to revised figures for unoccupied absorption, so that the following coefficients were used: occupied coefficient (after Beranek) of 0.48 at 125/250 Hz and 0.89 at mid-frequencies (500–2000 Hz), unoccupied coefficients of 0.41 at 125/250 Hz and 0.82 at mid-frequencies. For prediction of occupied RT these coefficients were used with the Sabine equation, but it is also significant that in the unoccupied state, measurements were made without an orchestra present, whereas an absorbing orchestra is present for occupied concerts.

As reproduced here in Table 2.4, Beranek with Hidaka revised his audience absorption figures in 1998 (which supported the concern mentioned above about his earlier figures) and with Hidaka and Nishihara he made proposals in 2001 for calculating occupied reverberation times.

B.3 Revised theory and correction of energy measures for reverberation time change

A revised theory of sound level in rooms was first alluded to in a theoretical investigation of geometrical behaviour in a rectangular space (Barron, 1973). Whereas in traditional theory the **reflected** sound level is considered to be constant throughout the space, it is apparent in actual halls that the reflected sound level decreases as one moves away from the source. Traditional theory divides the total sound into two components: the direct sound and reflected sound. The direct sound behaves according to the inverse square law, while the reflected sound is a function of the total acoustic absorption and independent of position. If for traditional theory the total level, L, is stated in terms of the direct sound level at 10 m, L_o, and the total acoustic absorption, A, is expressed using Sabine's equation ($A = 0.16V/T$), then:

$$L - L_o = 10.\log(100/r^2 + 31\,200T/V) \qquad (B.1)$$

where V is the volume, T is the reverberation time and r is the distance from the source.

When total reflected sound derived from measurements is plotted against source–receiver distance for individual halls, reflected sound turns out not to be constant with position but to decrease linearly with distance from the source (Barron and Lee, 1988). Revised theory was found to well represent average measured behaviour in auditoria; it can be used not only for total sound but also for the early-to-late sound index. Tests in scale models of diffuse spaces (not auditoria) show that revised theory is relevant in these reference acoustic spaces as well (Chiles and Barron, 2004).

The theory can be most easily explained on the basis of idealized integrated impulse curves in a room for three receiver positions, where a, b, and c are close, medium and far from the source (Figure B.1). In this figure the time $t = 0$ corresponds to the time at which an impulsive signal is emitted from the source. It is assumed that, at an instant in late time during a decay, the sound level is the same throughout the room. The decay is assumed to be linear, according to the reverberation time. The three decay curves are thus superimposed. Starting from late time at position c and moving forward in time, sound cannot arrive earlier than t_c, the arrival time of the direct sound. At this distant receiver position, the relative energy of the direct sound is so small that its presence is often not visible in the early decay trace. At position a, since $t_a < t_c$, the decay starts earlier and the integrated reflected energy is correspondingly larger; the direct sound is now also evident.

While the slope of the diagonal line in Figure B.1 corresponds to the reverberation time, its vertical position needs to be established. It was found that average measured behaviour placed the line at a position such that its intercept with the axis at $t = 0$ is the value for the reflected sound in the traditional formula (B.1). In other words, the traditional formula

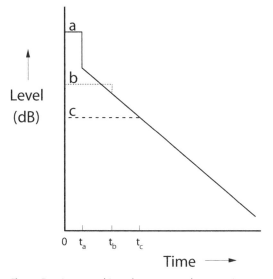

Figure B.1 Integrated impulse curves at three receiver distances from a source in a room according to revised theory. Here, t = 0 is the time when the sound was emitted from the source

is found to predict the reflected level correctly at the source, but beyond the source measured levels are less.

The simple geometric image model of sound in a rectangular space (Barron, 1973) leads to the following formula for energy arriving between time t and infinity (note, $t = 0$ is the time the sound leaves the source):

$$i_t = (31\,200T/V)\,e^{-13.82t/T} \tag{B.2}$$

This result can however be generalized since the analysis also predicts a linear decay according to the Eyring reverberation time equation The relationship (B.2) thus also applies to any linear decay as predicted by the Eyring equation; the result is not restricted by the severe assumptions associated with the simple geometric image model of sound in a rectangular space.

The relationship (B.2) leads to the following relationships for the early reflected sound energy within 80 ms of the direct sound, e_r, and the late sound later than 80 ms after the direct sound, l. The direct sound energy, d, is the same as in the traditional relationship (B.1):

$$d = 100/r^2$$

$$e_r = (31\,200/V)\,e^{-0.04r/T}.(1 - e^{-1.11/T})$$

$$l = (31\,200T/V)\,e^{-0.04r/T}.e^{-1.11/T}$$

The total sound level, $L - L_o$, and early-to-late index, C_{80}, are then

$$L - L_o = 10.\log{(d + e_r + l)}$$

$$C_{80} = 10.\log{[(d + e_r)/l]}$$

The values according to revised theory for a typical size hall are given in Figure 3.29. Theoretical values included for total sound and mean objective clarity in Figures 3.34, 5.6 etc. are all according to the formulae above.

Average behaviour for 17 British halls is found to correspond closely to revised theory (Barron and Lee, 1988), as shown in Figure B.2 for the total sound level. Good agreement is also found for total sound level for three classical halls (Bradley, 1991). The particular value of revised theory is that it provides a realistic point of comparison. That is not to say that deviations from theory necessarily imply poor subjective conditions but the theory enables one to comment, for instance, on whether the early reflection energy is typical for the size of hall.

Objective results in music spaces have been presented in Figures 3.34, 5.6 etc. Objective clarity and total sound level are measured in the unoccupied hall, but as mentioned in section 3.10.6 results for these two measures have been corrected for the change in reverberation time from unoccupied to occupied conditions. The correction procedure relies on revised theory.

If the superscript 'th' is used for the theoretical value according to revised theory above, while the subscript 'o' is for occupied and 'u' is for unoccupied conditions, then the correction formula for objective clarity is

$$C80_o = C80_u + C80_o^{th} - C80_u^{th}$$

From equations above, the theoretical reflected sound energy is

$$r^{th} = (31\,200T/V)\,e^{-0.04r/T}$$

This expression when converted to decibels becomes

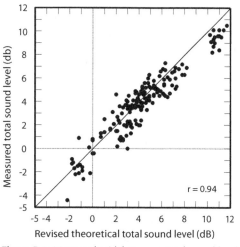

Figure B.2 Measured mid-frequency total sound levels in 17 British auditoria at seats not overhung by balconies plotted against predictions according to revised theory. The diagonal line represents perfect agreement

Reflected level $= 10 \log(31\,200T/V) - 0.174r/T$ dB

To correct the total sound level, it is assumed that the direct sound level corresponds to theory. The correction procedure is to convert the total sound level to energy and subtract the theoretical direct sound energy to give the reflected energy. This is then multiplied by (r_o^{th}/r_u^{th}) to correct for the reverberation time change. To give the corrected total sound level, the theoretical direct sound energy is added to this modified reflected energy and the sum is converted into level by taking logarithms.

The principle for correction of the early decay time for reverberation time change is mentioned elsewhere, though it happens not to have been used for Figures 3.34 etc. If D is the EDT, then

$$D_o = (T_o/T_u)\,D_u$$

References

Barron, M. (1973) Growth and decay of sound intensity in rooms according to some formulae of geometric acoustics theory. *Journal of Sound and Vibration, 27*, 183–196.

Barron, M. (1983) Objective measures of spatial impression in concert halls. *11th International Congress on Acoustics, Paris*, Paper 7.2.22.

Barron, M. (1984) Impulse testing techniques for auditoria. *Applied Acoustics, 17*, 165–181.

Barron, M. (1995) Interpretation of early decay times in concert auditoria. *Acustica, 81*, 320–331.

Barron, M. (2005) Using the standard on objective measures for concert auditoria, ISO3382, to give reliable results. *Acoustical Science and Technology, 26*, 162–69.

Barron, M. and Lee, L.-J. (1988) Energy relations in concert auditoriums, I. *Journal of the Acoustical Society of America, 84*, 618–628.

Barron, M. and Marshall, A.H. (1981) Spatial impression due to early lateral reflections in concert halls: the derivation of a physical measure. *Journal of Sound and Vibration, 77*, 211–232.

Beranek, L.L. (1969) Audience and chair absorption in large halls, II. *Journal of the Acoustical Society of America, 45*, 13–19.

Beranek, L.L. and Hidaka, T. (1998) Sound absorption in concert halls by seats, occupied and unoccupied, and by the hall's interior surfaces. *Journal of the Acoustical Society of America, 104*, 3169–3177.

Bradley, J.S. (1991) A comparison of three classical concert halls. *Journal of the Acoustical Society of America, 89*, 1176–1191.

Bradley, J.S. and Soulodre, G.A. (1995) The influence of late arriving energy on spatial impression. *Journal of the Acoustical Society of America, 97*, 2263–2271.

Chiles, S. and Barron, M. (2004) Sound level distribution and scatter in proportionate spaces. *Journal of the Acoustical Society of America, 116*, 1585–1595.

Cremer, L. and Müller, H.A. (translated by T.J. Schultz) (1982) *Principles and applications of room acoustics*, Vol. 1, Applied Science, London.

Gade, A.C. (1989) Investigations on musicians' room acoustic conditions in concert halls, II: Field experiments and synthesis of results. *Acustica, 69*, 249–262.

Hidaka, T., Nishihara, N. and Beranek, L.L. (2001) Relation of acoustical parameters with and without audiences in concert halls and a simple method for simulating the occupied state. *Journal of the Acoustical Society of America, 109*, 1028–1042.

ISO3382:1997 Measurement of the reverberation time of rooms with reference to other acoustic parameters.

Marshall, A.H. and Barron, M. (2001) Spatial responsiveness in concert halls and the origins of spatial impression. *Applied Acoustics, 62*, 91–108.

Reichardt, W., Abdel Alim, O. and Schmidt, W. (1975) Definition und Messgrundlage eines objektiven Masses zur Ermittlung der Grenze zwischen brauchbarer und unbrauchbarer Durchsichtigkeit bei Musikdarbietung. *Acustica, 32*, 126–137.

Schroeder, M.R. (1965) New method of measuring reverberation time. *Journal of the Acoustical Society of America, 37*, 409–412.

Thiele, R. (1953) Richtungsverteilung und Zeitfolge der Schallrückwürfe in Räumen. *Acustica, 3*, 291–302.

Appendix C

Further objective results in concert halls

C.1 Objective results in two New Zealand concert halls

Objective measurements, using identical procedures to those outlined in section 3.10, were undertaken in the New Zealand halls at Christchurch and Wellington (section 4.10) in 1983 and Segerstrom

Hall in California in 1986 (section 10.6). The New Zealand measurements are presented in Figures C.1 and C.2. Since the two halls are generically so closely related, they share many objective characteristics.

Objective conditions in the Christchurch Town Hall are discussed in some detail in section 4.10. While the reverberation times are long, the

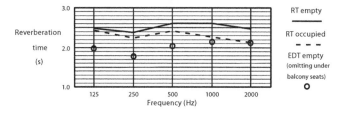

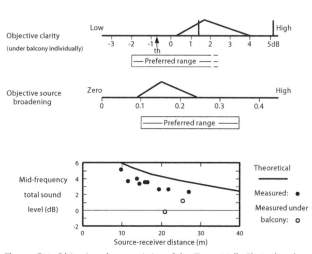

Figure C.1 Objective characteristics of the Town Hall, Christchurch, New Zealand

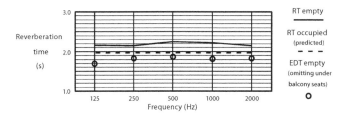

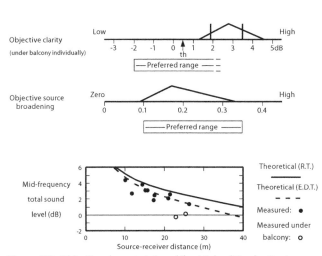

Figure C.2 Objective characteristics of the Michael Fowler Centre, Wellington, New Zealand

subjectively more significant early decay time (EDT) is much shorter (Barron, 1998). If the measured ratio of EDT to reverberation time of 0.82 in the unoccupied condition is applied to the occupied reverberation time, the predicted mean EDT at 500/1000 Hz with audience is 1.92 seconds. Objective clarity is high but acceptable, objective source broadening is reasonable and the total sound level at all positions is above criterion value. However, agreements between measured and theoretical objective clarity as well as total sound level are poor. With such a large difference between reverberation time and EDT, revised theory of sound level becomes inaccurate.

In the Michael Fowler Centre, Wellington, reverberation time values are more conventional, but with the EDT 83 per cent of the reverberation time, the predicted EDT in occupied conditions becomes

1.75 seconds. Other measured results are similar in the two halls, as is the disagreement between energy measurements and predictions according to revised theory. Revised theory uses reverberation time as a basic parameter. For both objective clarity and total sound level, agreement between measurement and theory is improved with a shorter reverberation time used in the theory. Included in the total sound level graph in Figure C.2 (dashed line) are revised theory predictions using the EDT in place of reverberation time. Agreement with measurement is now good. For objective clarity the theoretical mean using the EDT is within 0.5 db of the measured mean: again, good agreement.

The Christchurch Town Hall and Michael Fowler Centre are interesting both for their subjective characteristics and objective behaviour. The objective effect of large surfaces which reflect sound down onto audience seating is unmistakable. Most marked is the difference between the EDT and reverberation time and for an adequate sense of reverberance a long reverberation time appears necessary. In these directed reflection sequence halls, subjective and objective attributes are related to one another in ways which recent research has led us to expect.

C.2 Objective results in Segerstrom Hall, California

Segerstrom Hall, discussed in section 10.6, is radical in its asymmetry and in the way audience seating is subdivided into four areas. Whereas the New Zealand halls have free-standing acoustic reflectors, the envelope in Segerstrom Hall is profiled

to enhance early lateral reflections. The audience subdivision allows reverse splay plans to be used.

Reverberation time values in Figure C.3 are unexceptional and perfectly respectable. The ratio of mid-frequency early decay time to reverberation time is 0.98 in this hall. Objective clarity is quite high, mainly due to some high early sound levels. Objective source broadening is also high. Measured total sound levels are close to values according to revised theory. Given the unconventional design, this total sound level result indicates an impressive uniformity in acoustic performance. One can also observe in Figure C.3 that in overhung seats the objective clarity is typical and the total sound level is not attenuated compared with exposed seats. This confirms the wisdom of high overhangs and reverse-splay plans.

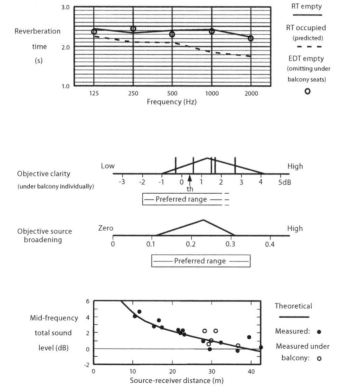

Figure C.3 Objective characteristics of Segerstrom Hall, Orange County Performing Arts Center, California

Not only are the audience numbers large by concert hall standards in Segerstrom Hall, but the design was further constrained by the requirement for good sightlines to a proscenium stage for musicals etc. Achieving objective characteristics such as these throughout the audience seating area is impressive.

Reference

Barron, M. (1998) Early decay times in the Christchurch and Wellington concert halls, New Zealand. *Journal of the Acoustical Society of America,* **103**, 2229–2231.

Appendix D

Objective measures for speech auditoria

D.1 Objective measures of speech intelligibility

For reasons given towards the end of section 7.3, much of the detailed discussion of theatre acoustics in Chapters 7 and 8 revolved around a single objective measure: the early energy fraction. Traditionally, acoustic measures of speech intelligibility have concentrated on only one of two concerns: either the signal-to-noise ratio or the impulse response. A valuable contender in the signal-to-noise group is the **articulation index**. Originally developed at Bell Telephone Laboratories, it has proved to be a powerful tool for predicting speech privacy in buildings (Cavanaugh *et al.*, 1962) and in open-plan offices (Pirn, 1971). It is not appropriate though for calculating speech intelligibility in rooms, where reflections from the internal surfaces need to be taken into account. In a room, the signal or speech level will be a function of the source–receiver distance, the speaker orientation and the room reflections, in addition of course to individual speaker differences.

Turning to the impulse response measures, the **early energy fraction** is based conceptually on a subdivision between useful and detrimental energy. In the case of the early energy fraction, the division occurs abruptly at 50 ms after the direct sound. In mathematical terms, using a notation similar to that in Appendix B.1:

Early energy fraction,

$$D_{50} = \frac{\int_0^{0.05} p^2(t)dt}{\int_0^{\infty} p^2(t)dt}$$

This can obviously be criticized since the ear does not operate with such a sharp temporal cut-off. However, the temporal cut-off proves to be much less of a problem than one might expect when applied to real room impulse responses. Recently Bradley *et al.* (1999) demonstrated that C_{50} is a linear measure for speech intelligibility; C_{50} is defined identically to the early-to-late sound index, C_{80}, mentioned in Appendix B.1, but with the time interval for early sound as 50 rather than 80 ms. There is a unique relationship between D_{50} and C_{50}; the limit of acceptable intelligibility for C_{50} is 0 dB.

Based on intelligibility experiments in simulated sound fields, Lochner and Burger (1961) introduced a weighting factor, $a(t)$, for the early energy. Useful energy was considered to arrive within 95 ms of the direct sound and its intelligibility contribution is given by multiplying the energy by the weighting factor depending on its delay (Figure D.1). Detrimental energy arrives after 95 ms. The **Lochner and Burger ratio** is expressed in decibels derived from the ratio of useful to detrimental sound:

$$\text{Lochner and Burger ratio} = 10\log\frac{\int_0^{0.095} \alpha(t)p^2(t)dt}{\int_{0.095}^{\infty} p^2(t)dt}$$

Alternatively Cremer has proposed a neat way of avoiding the need for a predefined temporal cut-off by making time the dimension of the measure. The **centre time** is discussed in Appendix B.1. Both the Lochner and Burger measure and the centre time are well correlated with speech intelligibility, and are slightly superior to the early energy fraction in

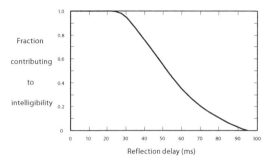

Figure D.1 Fraction of energy contributing to intelligibility as a function of its delay relative to the direct sound (as proposed by Lochner and Burger, 1961)

this respect. But this statement needs to be placed in context. For the 264 measurement situations in the 12 theatres discussed in the Chapter 8, these speech intelligibility measures are all correlated to each other with coefficients greater than 0.92. This makes selection of one or another somewhat arbitrary.

Ideally a speech intelligibility measure combines both the signal-to-noise and impulse response aspects. Two proposals deserve mention. Bradley (1986) has investigated **useful-to-detrimental ratios** including the effect of background noise. His equation for combining the early energy fraction with the signal-to-noise ratio proves to be particularly simple:

Useful/detrimental index,

$$U_{50} = 10\log\left[\frac{D}{1-D+n/s}\right]$$

D, here, is the early energy fraction defined above and *n/s* is the noise-to-signal ratio in energy terms (whereas signal-to-noise ratio is normally quoted in decibels, i.e. 10 log (*s/n*)). The subscript 50 of *U* refers to the 50 ms time limit for the early energy fraction. Bradley found the best correlations between measured speech intelligibility scores when he used an 80 ms temporal limit for useful sound, i.e. U_{80}.

The final measure discussed here, known as the **speech transmission index** (STI), is based on totally different concepts. It has been strongly promoted for predicting speech intelligibility and can be measured on site. The STI accommodates

both the signal-to-noise and impulse response aspects which affect intelligibility. The idea behind this measure is that for good speech intelligibility the envelope of the signal should be preserved (Houtgast *et al.*, 1980). To measure the envelope distortion, a noise signal is modulated in a sinusoidal fashion. The modulation frequencies used are between 0.4 and 20 Hz, which are the envelope frequencies relevant for speech sounds (the typical number of speech sounds has been stated in section 7.3 as 15 per second, i.e. a modulation frequency of 15 Hz). The noise signal is 100 per cent modulated so that for a 10 Hz modulation frequency there is an instant of silence every 0.1 seconds (Figure D.2). The received signal in a room will contain some sound in the quietest periods, owing to both the smearing effect of the room impulse response and any background noise. The modulation depth of the received signal is measured at a series of modulation frequencies, to give the modulation transfer function (MTF). The MTFs at the different signal frequencies are then used to calculate the STI. Clearly if the level of the sound source used for the measurement matches that of a human speaker, the measured STI takes account of both the impulse response and noise aspects of speech intelligibility.

MTF techniques have been widely used for some years in optics to measure such things as lens performance. The modulation transfer technique is persuasive for speech use though the fact that speech is still reasonably intelligible with no amplitude information is worrying. The STI is well correlated with speech intelligibility but not more so than other measures. Its gross disadvantage is that it is complicated and cannot be simply predicted for auditorium spaces. It can be predicted from the impulse response, but so can any measure. It can also be predicted on the basis of simple models for sound behaviour in rooms (Houtgast *et al.*, 1980). This prediction technique may be valuable for spaces like classrooms but it does not work for theatres. For instance, the minimum speech intelligibility predicted for a room with a reverberation time of 1 second is on the boundary between 'Fair'

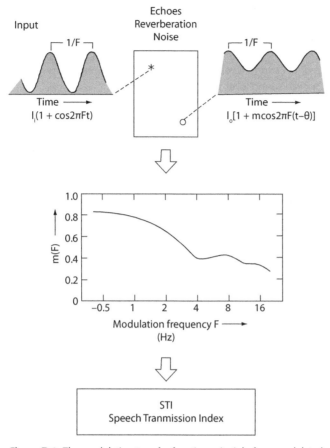

Figure D.2 The modulation transfer function principle for a modulated noise signal (Houtgast *et al.*, 1980)

and 'Good' using this STI prediction technique. Experience in some theatres shows definitely worse performance.

This discussion of various measures has illustrated the range of approaches taken to measure speech intelligibility. The useful-to-detrimental energy approach has a long history and gives results as accurate as any alternative. The speech transmission index is now widely used but it is not more accurate than alternatives. The early energy fraction is easily understood in terms of individual reflections.

D.2 The criterion speech level for theatres

A criterion for the required 'amplification' by the theatre on the spoken voice can be derived from data from three different sources. What has been called the **speech sound level** in Chapters 7 and beyond is the measure of this amplification. It is the sound level at an individual seat for an actor placed at a particular position and orientation on stage, **relative** to the average direct sound level at 10 m. We need an estimate of the minimum acceptable value for the speech sound level. All the following sound levels are based on a mean of values at the octave frequencies 500, 1000 and 2000 Hz.

From the mean of results in Figure 7.1, the mean speech sound power level of an actor is 70 dB. This corresponds to a **mean** direct sound level at 10 m of 39 db SPL. If S is the speech sound level, then the actual sound level at a seat is $(39 + S)$ db SPL.

A typical acceptable background noise level in theatres is NR25 (Figure 2.28). The mean value of NR25 over the relevant octaves is 25 db SPL. From considerations of speech in noise, for instance as discussed in papers on articulation index and modulation transfer function (STI), an acceptable signal-to-noise situation in a theatre is 12 dB. This is admittedly rather high by the standards of some speech transmission situations, but in theatres one may well have to contend with unfamiliar language as well as the need for the stage whisper to be heard.

Thus for intelligible speech transmission, we derive the approximate criterion:

$$39 + S > 25 + 12$$

or

$$S > -2 \text{ dB}$$

Several other considerations can be introduced into this estimation, particularly the presence of audience noise. As in Figure 2.28, audience noise is likely to exceed the background noise from ventilation, raising the criterion value for S. From experience of the theatres discussed in Chapter 8, a criterion of $S > 0$ db appears appropriate. It is also convenient that it is identical to the equivalent criterion for large concert halls.

D.3   Revised theory of sound level for use with speech

Revised theory for concert halls is described in Appendix B.3, and the same notation will be used here. It enables predictions to be made of average behaviour in concert halls, on the basis of the hall volume, the reverberation time and the source–receiver distance. A similar technique can be applied to theatres, though its validity is more tenuous.

There are three additional complicating features for theatres: that the human speaker is directional, that conditions are not sufficiently diffuse for simple prediction of the early sound and that in proscenium theatres energy is lost through the proscenium opening.

The directivity of a source is normally expressed relative to the value averaged over all directions. For the mean of frequencies 500, 1000 and 2000 Hz, the directivity index for a human speaker straight ahead is 3 db or a factor of 2. The following values apply at different directions:

Angle of listener to speaker:	0°	45°	90°	135°	180°
Directivity factor (δ):	2.0	1.67	0.87	0.44	0.70

The directivity factor obviously applies to the direct sound. In theatres, the early sound is not reliably predicted by the method used for concert halls. The chosen technique for dealing with the early reflections is to invoke the early reflection ratio, n, which is discussed at length in sections 7.4 and 7.6. Values for different types of theatre are given in Table 7.2.

The above gives the early energy, including the direct sound, as

$$\text{Early sound energy} = e \,(= d + e_r) = 100\, \delta\, n/r^2$$

With unobstructed direct sound, this expression is precise but begs the question of the true value of n.

For the late sound the basic formula as used for concert halls is employed (but a constant is changed since the time limit is now 50 ms). A further parameter is introduced for the case of a proscenium stage. It is assumed that no energy re-emerges after entering the flytower, as shown in Figure 7.11. If the source were omni-directional, an auditorium angle fraction, depending simply on the solid angle presented at the source by the auditorium (as opposed to the proscenium opening) would be appropriate. However, with a directional source the calculation becomes more complex; the relevant quantity can be called the auditorium energy fraction, f. Clearly in the case of a source on the line of the proscenium pointing **across** stage the fraction is 0.5. For the case of a source again on the line of the proscenium facing the auditorium, the fraction is 0.72.

Late sound energy, $I = (31\,200.f.T/V)\,e^{-0.04r/T}.\,e^{-0.69/T}$

The above can be used to calculate the early energy fraction and total speech level, and provides a simple model for sound level behaviour in theatres.

References

Bradley, J.S. (1986) Predictors of speech intelligibility in rooms. *Journal of the Acoustical Society of America*, **80**, 837–845.

Bradley, J.S., Reich, R. and Norcross, S.G. (1999) A just noticeable difference in C_{50} for speech. *Applied Acoustics*, **58**, 99–108.

Cavanaugh, W.J., Farrell, W.R., Hirtle, P.W. and Watters, B.G. (1962) Speech privacy in buildings. *Journal of the Acoustical Society of America*, **34**, 475–492.

Houtgast, T., Steeneken, H.J.M. and Plomp, R. (1980) Predicting speech intelligibility in rooms from the modulation transfer function, I: General room acoustics. *Acustica*, **46**, 60–72.

Lochner, J.P.A. and Burger, J.F. (1961) The intelligibility of speech under reverberant conditions. *Acustica*, **11**, 195–200.

Pirn, R. (1971) Acoustical variables in open planning. *Journal of the Acoustical Society of America*, **49**, 1339–1345.

Name index

Subject index

Printed and bound by CPI Group (UK) Ltd, Croydon, CR0 4YY

18/10/2024

01776204-0014